Christy Hoffman

Primate
Psychology

Primate
Psychology

Edited by Dario Maestripieri

Harvard University Press

Cambridge, Massachusetts

London, England

2003

Library of Congress Cataloging-in-Publication Data

Primate psychology / edited by Dario Maestripieri.

  p.   cm.

Includes bibliographical references and index.

ISBN 0-674-01152-X

1. Psychology, Comparative.   I. Maestripieri, Dario.

BF671.P75 2003

156—dc21   2003047853

Contents

Preface

I am a biologist by training. When I first became involved in behavioral research, I did not think that human behavior was intrinsically more interesting than the behavior of any other primate species, or that primate behavior was intrinsically more interesting than the behavior of any other animal species. My interests and views have not changed over the years. Why then a book titled *Primate Psychology*?

My interest in behavior can be defined in very simple terms. When I observe other behaving organisms, I want to understand what they are doing and why, in the broadest possible sense. During fifteen years of research, I have learned a lot about primate behavior. I feel, however, that when I observe a social group of primates, I understand perhaps no more than 50 percent of what they are doing and why. I suspect that there are few or no primate researchers who can confidently say that they understand 100 percent of what they observe. There are still too many behavior patterns whose causes and function we do not understand, too many signals whose meaning we do not understand, and too much environmental and interindividual variation whose origin and adaptive function we do not understand.

A full understanding of behavior can be accomplished only with an interdisciplinary approach. We cannot afford to ignore the contribution that other behavioral disciplines can make to understanding behavior simply because they have a different historical tradition or use unfamiliar concepts and terminology. Humans are the primates we know best, if nothing else, because we observe their behavior every single day of our lives and have a direct experience of how their minds work. Those of us who are interested in understanding the behavior and the mind of nonhuman primates cannot afford to ignore all the knowledge and understanding of human behavior and cognition that have been accumulated over hundreds of years of research. Behavioral primatology is a young discipline compared to psychology but has accumulated a great deal of information and knowledge in the few decades of its life. If psychologists hope to achieve a comprehensive understanding of human behavior and cognition, they cannot overlook what primatologists know about the behavior and cognition of other primate species and, in particular, of those closely related to us.

The goals of this book are to attempt an integration of behavioral research with human and nonhuman primates and encourage communication between students of primate and human behavior. This book is particularly directed to the young behavioral scientists who have not yet made up their mind about the species, the area of research, or the discipline on which they want to concentrate their research efforts. The message that the contributors hope to convey is that there is no good reason to limit themselves to only one of each. Comparative and interdisciplinary research can be a risky business, but the rewards are potentially great. The boundaries between different scientific disciplines, just like those between many other aspects of our lives, exist to be crossed.

My research and this book would not have been possible without the support obtained from institutions, funding agencies, colleagues, family, and friends. Among the institutions that provided training, hospitality, and research support, I am particularly grateful to the Yerkes National Primate Research Center and the University of Chicago. My research has been supported over the years by the Italian National Research Council, the Leakey Foundation, the Biomedical Resource Foundation, the Harry Frank Guggenheim Foundation, the McCormick-Tribune Foundation, and the National Institute of Mental Health (grants R03-MH56328, R01-MH57249, R01-MH62577, and K02-MH63097). I am very grateful to all of these funding agencies for their generous support. The friends, colleagues, research assistants, and students who have encouraged and supported my research in recent years are simply too numerous to mention. Several colleagues have read portions of this book and provided useful suggestions and criticisms: Irwin Bernstein, Horacio Fabrega, Howard Nusbaum, Russell Tuttle, Everett Waters, and Amanda Woodward. Thanks also to Frans de Waal, Steve Ross, and Mike Seres for allowing us to use their photos as illustrations for the book. Finally, I thank Elizabeth Knoll and Kirsten Giebutowski at the Harvard University Press for their support throughout all the stages of this project.

D. M.

Primate
Psychology

1

The Past, Present, and Future
of Primate Psychology

Dario Maestripieri

Psychology is the discipline that studies mind and behavior, and these days, this is taken to mean the mind and behavior of human beings. The study of the mind and behavior of other animal species is often referred to as comparative psychology, ethology, or simply animal behavior. Although psychology differs from other behavioral disciplines in some of its theories and experimental procedures, certain basic research questions about behavior (How do genetic and environmental factors interact in the development of behavior? What are the endogenous factors and external stimuli controlling and regulating behavior? How does behavior contribute to the survival and reproductive success of the individual?) are fundamentally the same regardless of the species being studied.

Arguments in favor of comparing human behavior to the behavior of nonhuman animals can be based on homology or analogy. Arguments of homology state that some aspects of the behavior of human beings and other animals may be similar by virtue of common descent. The probability that two species inherited a pattern of behavior from their common ancestor is higher the closer the phylogenetic relationship between the species. Thus, human behavior is more likely to be homologous to the behavior of other primates, in particular chimpanzees and the other great apes, than to the behavior of non-primate animals. For example, facial expressions of emotions in humans are more likely to be homologous to the expressions of other primates, in particular Old World monkeys and apes, than to those of other mammals (Darwin, 1872).

Arguments of analogy state that some aspects of the behavior of human beings and other animals may be similar by virtue of convergent adaptation to similar environments. Similarities in environments can also occur between species that are distantly related; therefore, in theory, to investigate the adaptive aspects of human behavior from a comparative perspective, honeybees can be as good a model as chimpanzees. In reality, however, there are many constraints on the type of adaptations to the environment that organisms can evolve through natural selection. Therefore, similarities in genetic, anatomical, physiological, and cognitive constraints increase the probability that organisms will evolve similar adaptations to the environment. Thus, human behavior is generally more likely to be analogous to the behavior of other primates than to the behavior of non-primate species. For example, maternal behavior in humans and other primates probably depends on experience acquired during development to a greater extent than in mammals with shorter life spans such as rats. Therefore, early interest in infants is likely to be an adaptation that human females share with other female primates but not necessarily with female rats (Maestripieri, 1999a; Maestripieri & Pelka, 2002).

The order Primates comprises over three hundred species and a great deal of variability in social and mating systems (Figure 1.1). This variability has led some researchers to believe that the choice of any particular primate species for the purposes of comparison with human behavior is arbitrary and unwarranted (e.g., Hinde, 1982; Lehrman, 1974). It is certainly true that any extrapolations from primate behavior to human behavior must be made with caution, and that all primate species are equally interesting and should be studied in their own right. It is also true, however, that it may be more meaningful to compare humans to some nonhuman primate species than to others. *Homo sapiens* is a species with a well-defined set of life-history characteristics including a long life span; slow rates of growth and maturation; the production of single offspring separated by long interbirth intervals; high levels of parental investment; moderate sexual dimorphism in body size; a terrestrial lifestyle; social units consisting of large multi-male, multi-female groups with kin-

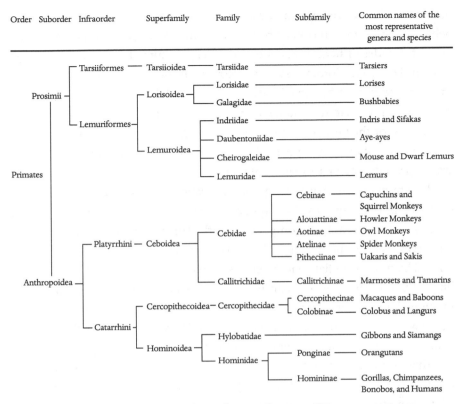

| Order | Suborder | Infraorder | Superfamily | Family | Subfamily | Common names of the most representative genera and species |

Figure 1.1 Primate taxonomy. (Redrawn from Dolhinow and Fuentes, 1999)

centered subgroups; a variable mating system that includes monogamy, polygyny, and promiscuity; and an omnivorous diet in which meat is an important component. Species that are phylogenetically closest to humans and share with humans most if not all of these characteristics have the highest probability of being behaviorally similar to humans, according to arguments of both homology and analogy. Thus, the fact that cercopithecine monkeys such as macaques and baboons, or great apes such as chimpanzees and bonobos, have been the preferred choice by researchers studying human behavior from a comparative perspective is not entirely arbitrary or unwarranted.

One argument against the usefulness of any kind of comparative behavioral research with primates (or any other animal taxon) rests on the claim that human behavior is entirely determined by the environment, and that the environment in which modern humans live is drastically different from that of both our primate ancestors and the extant primates. The claim that human behavior is entirely determined by the environment is simply untenable in the light of decades of behavior genetics research conducted with twin and adoption studies (e.g., Plomin, 1994). Moreover, an increasing number of genes associated with specific behavioral dimensions and disorders have been identified with molecular techniques (e.g., Plomin & Rutter, 1998). As for the uniqueness of the human environment, although it is undeniable that modern humans have altered their physical and ecological environment in a way that was unprecedented on this planet, it is also true that the social environment of modern humans still shares many characteristics with the social environment of other primate species. Like other primates, humans live in large social groups that cooperate and compete with other groups for access to and use of resources. Within these groups, humans form smaller family groups characterized by intense affiliative interactions between parents and offspring or siblings similar to those of other primates. Our social interactions with unrelated individuals of the same or the other gender are influenced by power and sex in a similar way as in other primate species (e.g., Alexander, 1979; de Waal, 1982). Thus, given the similarities between the social environment of human beings and that of other primate species, it is meaningful and useful to investigate the extent to which similarities between primate and human behavior represent adaptations to similar environments.

Every animal species has unique attributes that set it aside from other species. Human language and culture are, by definition, uniquely human. Language and culture are certainly responsible for a great deal of variability in the way modern humans think and behave. Language and culture, however, do not operate in a vacuum; they are produced and used in a social environment and for social purposes that are quite similar to those of other primates. To a psychologist or an anthropologist who is familiar only with the human species, the differences in human behavior associated with geographic and cultural variation may appear overwhelming and so great as to escape any attempts at generalization. Variability among groups or populations of the same species, however, is not a uniquely human characteristic. Such variability occurs in any animal species on this planet and has not prevented biologists from making meaningful generalizations about species-typical morphological or behavioral traits. Moreover, despite the intraspecific variability, it is clear that from a behavioral standpoint, humans as a species are much closer to other primates than to other non-primate animals. I suspect that a scientist from Mars who attempted to classify all the animal forms on our planet on the basis of behavioral traits alone would have no difficulty placing human beings and most of the other primates in the same taxonomic unit. Such a scientist might also be able to reconstruct correctly the evolutionary relationships within the Primate order down to the genus level, or perhaps even the species.

As with human language and culture, the human mind is also uniquely human. It is the product of an evolutionary history that added new structural changes and functional properties to the human brain (e.g., Donald, 1991). For the same reasons as those discussed for behavior, however, there are likely to be both homologies and analogies in the way humans and other primates think. Thus, some mental processes are likely to be similar in primates and humans because they are the product of regions of the brain that extant primates and humans inherited from their common ancestors. Likewise, some mental processes are likely to be similar in primates and humans because they represent adaptive responses to problems posed by the social environment, which is shared by humans and other extant primate species.

Studying the mind and behavior of nonhuman primates can potentially enhance our understanding of the proximate regulation and adaptive value of many human mental and behavioral processes. The converse is also true: the human species can provide a good model with which to study the mind and behavior of our closest relatives. There is therefore no good conceptual reason why the mind and behavior of humans and other primates should be the subject of study of different scientific disciplines. Rather, there are a number of reasons why the range of action of psychology should be expanded to include other animal species, beginning with those that are most similar to us. The human mind and human behavior are unique, just like the mind and behavior of any other animal species. Yet the evolutionary processes that gave rise to these unique products are fundamentally similar across all species living on this planet. Therefore, the creation of a discipline that studies the mind and behavior of all primates could be an important first step toward the creation of an integrative discipline whose goal is the study and understanding of mental and behavioral processes in all animal forms on this planet.

A Brief History of Primate Psychology

Psychologists played an important role in the birth and development of behavioral primatology in the twentieth century. In the first decades of primate behavior research, exchange of information between psychologists and primatologists was frequent and mutually beneficial. Such exchange became gradually less frequent, so that today many primate researchers are probably unfamiliar with recent theoretical and empirical developments in psychological science, while new advances in primate research are probably ignored by a large number of psychologists. In part, this is due to the objective difficulty of keeping up with progress in two rapidly expanding disciplines and their numerous subspecialties. For example, the existence of many specialized journals and books within both psychology and primatology and the extent to which these disciplines have grown and diversified over the past several decades have made it very difficult for researchers to keep up with the research progress outside their own field of specialization. Exchange of information between primatologists and psychologists has also been made difficult by the fact that a comprehensive and systematic integration of research on primate and human behavior has rarely been attempted. Several developments in behavioral primatology and psychology, however, have evidenced the need for an integration of these two disciplines, including primatologists' growing interest in complex social and cognitive processes and psychologists' growing interest in biology and evolution. Thus, there is reason to be optimistic about reestablishing those bridges between psychology and primatology that once allowed frequent crossing between these two disciplines. To understand the current state of affairs and appreciate the prospects for the future, it may be useful to provide a brief history of primate behavior research and its relationship with psychological science.

The implications of primate behavior research for understanding human behavior were first explicitly recognized by Darwin, who in his book *The Expression of the Emotions in Man and Animals* (1872) drew several parallels between the facial expressions of nonhuman primates and those of human beings. Given the paucity of data available, Darwin engaged in some comparative research himself, making several trips to the zoo to observe the facial expressions of the primates there and then comparing them with those of his own son William Erasmus (Desmond & Moore, 1991). Darwin's book certainly influenced the attempts to study primate behavior made in the first half of the twentieth century, even though such studies did not necessarily use evolutionary theory as the main explanatory framework.

One of the first psychologists to become systematically interested in primate behavior was Wolfgang Köhler. As a Gestalt psychologist, Köhler was interested in cognitive processes other than learning and was curious to see if apes could use "insight" to solve novel cognitive tasks. In a German research station for the study of great apes established at Tenerife in the Canary Islands, Köhler conducted over a period of four to five years (1913–1917) many elegant experiments with chimpanzees. A number of these experiments involved manipulation of the environment to obtain food rewards and the use of previously familiar objects in novel and instrumental ways (Köhler, 1925). Köhler's research questions and procedures were very innovative, and some of his findings are still frequently cited in contemporary research. He can certainly be considered one of the founders of modern research on primate cognition.

In the United States, the systematic study of primate behavior was begun mostly as a result of the efforts of Robert Yerkes, a Harvard-trained psychologist who established a primate laboratory in Orange Park, Florida (the laboratory was later moved to Atlanta, where it eventually became the Yerkes Regional Primate Research Center), with the goal of making primates available to many different kinds of scientific inquiry, most notably psychological research (Yerkes, 1925). Yerkes felt that research on the behavior and cognitive abilities of primates, and in particular the great apes, would help address some problems in psychology that had been very difficult to solve. His contributions to primate behavior research were many and ranged from studies of spatial cognition and problem solving to research on social and maternal behavior (e.g., Yerkes, 1933; 1934; Yerkes & Tomilin, 1935). Yerkes's awareness of the importance of primate research and his commitment to it were clearly expressed in the concluding sentence of an article he published in *Science* in 1916: "It does not seem extravagant to claim that the securing of adequate provisions for the systematic and long continued study of the primates is by far the most important task for our generation of biologists and the one which we shall, therefore, be most shamed by neglecting (cited by Bourne, 1971, p. 78).

In addition to the pioneer efforts by Köhler and Yerkes, other early attempts to study primate behavior and cognition were made in Russia, France, Cuba, and other parts of the world. The 1920s and 1930s also saw the beginning of attempts to teach language to chimpanzees. The first of such attempts was made in 1930 by Kellogg and Kellogg, who raised a young chimpanzee named Gua along with their son Donald (Kellogg & Kellogg, 1933). The Kelloggs' experiment turned out to be a failure, but it was followed by many others conducted by other researchers with the same and different species, and using similar or different strategies (see Chapter 14 for a complete history of this research).

Along with the growing recognition that primate behavior could be useful for understanding human behavior, the years before World War II were characterized by increasing interest in studying primates in their natural habitat and understanding the basic principles regulating their social organization (e.g., Bingham, 1932; Carpenter, 1934; Nissen, 1931; Zuckerman, 1932). During the war, research on primate behavior was interrupted for almost a decade, but in the early 1950s, and especially in the 1960s, there was renewed interest in studies of primate behavior all around the world. In the early 1950s, Japanese scientists began an extensive program of food provisioning and behavioral observation of many troops of Japanese macaques throughout their islands (Itani, 1975). This intensive and long-term study of macaque behavior led to the discovery of kinship systems and cultural traditions in macaque societies (e.g., Kawai, 1958, 1965). Primate behavior research in Japan, however, was originally conducted within the tradition of anthropology (e.g., Imanishi, 1965), and it was only later that such research established a strong connection with psychological science (e.g., Matsuzawa, 2001). The 1950s also witnessed the resumption of research with rhesus macaques on the island of Cayo Santiago, in Puerto Rico, where Carpenter had established a colony of these monkeys prior to the war (Rawlins & Kessler, 1986). The availability of genealogical information on the animals and the long-term observations of their behavior contributed, along with the work of Japanese primatologists, to the identification of the matrilineal structure of macaque society and the mechanisms underlying the acquisition of dominance (e.g, Altmann, 1962; Sade, 1965).

As more information on primate social behavior became available, the Harvard anthropologist Sherwood Washburn developed the conviction that extant primate species, and in particular savanna baboons, could provide important information on human origins and social evolution. He and his graduate students pioneered field studies of primate behavior in Africa and

Asia (DeVore, 1963; Jay, 1963), placing particular emphasis on aggressive and maternal behavior. These two topics dominated much of primate behavior research through the 1960s and 1970s. Interest in anthropology and human origins also motivated the paleontologist Louis Leakey to begin long-term studies of chimpanzees, gorillas, and orangutans, with Jane Goodall, Dian Fossey, and Biruté Galdikas, respectively.

Psychologists' interest in primate behavior rose dramatically with the resumption of research in captivity after World War II. Harry Harlow's research in Wisconsin played a pivotal role in this process. Although Harlow initially set out to study learning processes in rats, he was encouraged to work with monkeys by the fact that the rodent laboratory in his department was no longer available (Suomi, 1982a). After making important contributions to the study of primate learning (e.g., Harlow, 1949), Harlow concentrated his efforts on elucidating the nature of infant attachment and social development in rhesus monkeys. Harlow's well-known experiments with surrogate mothers demonstrated that the mother's ability to provide contact comfort is a more important determinant of infant attachment than her ability to provide milk, thus striking a fatal blow to secondary-drive theories of attachment (Harlow, 1958; see Chapter 5). Because Harlow's work touched on many areas of research that were very important to psychologists at that time (e.g., learning and motivation, attachment, normal and abnormal social development, the social origin of affective disorders), and because Harlow's academic career took place within the field of psychology, during the years in which most of his work was conducted and published, primate behavior research was very well known among psychologists. Harlow also trained a large number of behavioral scientists who pursued successful academic and research careers in the psychological and biomedical community.

Although Harlow was very effective in promoting the importance of primate behavior research in the scientific community and the general public (Suomi, 1982a), the person who made the most systematic effort to integrate ethology and primatology conceptually with other behavioral disciplines was probably Robert Hinde. Hinde was a British biologist with a research background in ornithology. His interest in primate research was sparked by John Bowlby, who encouraged him to set up a colony of rhesus monkeys in Cambridge and investigate mother-infant attachment processes (see Chapter 5). In addition to training and supervising a whole generation of primate fieldworkers, Hinde had a great impact on primate behavior research with his own work on mother-infant relationships in rhesus macaques (Hinde & Spencer-Booth, 1968). From the study of social influences on the mother-infant relationship, the scope of Hinde's research was gradually broadened and elaborated into a conceptual framework for the study of social processes, which distinguished three main levels of complexity: interactions, relationships, and social structure (Hinde, 1976). He made important conceptual contributions to the science of social relationships (Hinde, 1979, 1995), began his own line of research on social development, and for decades was one of the most articulate proponents of the conceptual integration between biological and psychological approaches to the study of behavior (Hinde, 1974b, 1987, 1990).

Thanks to the efforts of talented and charismatic scientists such as Harlow and Hinde and the success of early field studies of primate behavior begun in the 1960s, primatology reached a peak in popularity in the 1960s and early 1970s. In the mid- to late 1960s, in particular, behavioral research in the United States thrived at the newly established Regional Primate Research Centers, and most research proposals to study primate behavior were readily funded by federal and private agencies. In the same period a large number of articles and books on primate behavior were published, and primate behavior research was well represented in all branches of scientific psychology, including developmental, social, cognitive, and clinical. The heyday of primatology, however, did not last long. In the early 1970s, that is, only ten years after the establishment of the Regional Primate Research Centers, there were already significant cuts to research funding (Bourne, 1971).

The realization that evolutionary theory could be effectively applied to the study of social behavior (e.g., Hamilton, 1964; Trivers, 1971, 1972; Williams, 1966) gave a great boost to primate research in the field. Although anthropology and psychology had dominated primate behavior research up to the 1970s, ecology and evolutionary biology acquired a leading role in most subsequent research. The fact that behavioral ecologists were mostly interested in questions of ultimate causation whereas psychologists were mostly interested in questions of proximate causation was one of several factors that contributed to the growing separation between primate behavior research and psychological science that occurred in the 1980s and early 1990s. Another important factor was the rapid progress of biological disciplines such as genetics, molecular biology, and neuroscience and the growing popularity of scientific reductionism. In particular, the success of neuroscience led to the optimistic view that many important questions about behavior would eventually be answered by studies of brain anatomy and function, thus rendering behavioral research less necessary. One corollary of this view was the belief that comparative research with primates might not be as useful as research with other species, given the difficulty of conducting molecular work with primates. In addition to external factors such as changing trends in the scientific community, behavioral primatology probably suffered from internal problems as well. In particular, the lack of a strong theoretical focus or methodological rigor in some of the studies conducted in previous decades may have contributed to primatology's failure to project a strong image of itself as a "hard" science.

In the late 1980s and 1990s, primate behavior research went through a period of crisis. In the field, the natural habitats occupied by primate species were severely endangered by human action, thus leading to a dramatic shrinking of many primate populations and the need to place the highest priority on issues of conservation rather than research. In the United States and in Europe, primate behavior research was jeopardized by lack of support from research institutions and lack of academic job opportunities for primatologists. For a behavioral scientist entering the academic job market in the 1990s, the word "primate" on his or her CV was viewed as a liability. While research in psychology continued to thrive, references to primate behavior studies dropped dramatically in scientific publications in virtually all the branches of psychology.

Although it may be argued that primate behavior research is still very much in crisis, some favorable conditions have emerged for renewed cross-fertilization between primatology and psychology. The cognitive revolution that occurred in psychology in the middle of the century was followed, a few decades later, by a similar cognitive revolution in the field of animal behavior, and primate behavior in particular. Thanks to the efforts of pioneers such as Donald Griffin (1981, 1984, 1992) and Gordon Gallup (1970, 1985), once it became scientifically acceptable to ask whether animals have a sense of self and understand other individuals as having a mental life of their own, the field of primate cognition boomed. It became apparent that many of the questions traditionally addressed by cognitive psychologists could also be addressed, with similar or new experimental procedures, in monkeys and apes as well.

Thus, the border between nonhuman primate and human cognition research has been increasingly crossed by individuals on both sides of the fence, and often with remarkable results. Today, cognition is the branch of scientific psychology in which primate behavior research is best known and represented. Conversely, the recent theoretical and empirical advances in human cognition research are probably the area of psychology that most primate researchers are familiar with.

In addition to the cognitive revolution, there are two other major avenues through which communication between primate and human researchers could be reestablished. One is the growing interest in research on mind-body interactions and the success of new disciplines combining various branches of psychology and biology (e.g., developmental psychobiology, psychoneuroendocrinology, psychoneuroimmunology, psychophysiology, and cognitive and social neuroscience; Bunk, 2001; Cacioppo et al., 2000). The other is the increasing interest in the evolutionary origins and adaptive characteristics of the human mind and human behavior.

A growing number of psychologists are recognizing that human thinking and behavior are not entirely determined by the environment but are the product of complex bidirectional influences between biology and environment. Genes and their products can have direct influences on phenotypic traits such as temperament or some types of affective disorders. Conversely, the environment can influence gene expression and brain plasticity throughout the life span. Comparative research with nonhuman species is an indispensable tool for investigating the role of gene-environment interactions in the genesis and control of behavior, and more broadly the interaction between biological, psychological, and social processes. Although gene manipulations may be easier to conduct in rodents than in primates, research with primate models of human cognition and behavior is more ecologically valid than rodent research, owing to the similarities in the behavior and environment of human and nonhuman primates. Furthermore, recent advances in imaging techniques also allow the investigation of neural substrates of behavior without traditionally invasive procedures such as brain lesions. Thus, some of the limitations in the use of primates for biobehavioral research can be overcome with the development of new technologies that allow the assessment of biological variables without jeopardizing the well-being of the animals.

Along with the recognition that human thinking and behavior are not entirely determined by our social and cultural environment has also come the realization that the human mind and behavior have been, for much of our evolutionary history, under strong selective pressure to develop adaptive responses to problems associated with survival and reproduction. Some of these adaptations can be seen in our every day life. Unfortunately, researchers who study evolutionary aspects of human behavior from both a behavioral ecology (e.g., Smith, Borgerhoff Mulder, & Hill, 2000) and an evolutionary psychology (e.g., Daly & Wilson, 1999) perspective often underestimate the usefulness of the comparative approach and the value of primate research. Some evolutionary psychologists, in particular, dismiss the contributions made by phylogenetic analyses of animal behavior or by research with animal models of human behavior (Daly & Wilson, 1997), while others maintain that phylogenetic analyses may be useful for the analysis of neurobiological or physiological traits but not for behavior or cognition (Tooby & Cosmides, 1989). In reality, the importance of comparative and phylogenetic analyses of behavior has been recognized by a growing number of animal behaviorists (e.g., Martins, 1996). In particular, comparative and phylogenetic analyses of behavior and cognition have recently acquired a pivotal role in primate research (e.g., Di Fiore & Rendall, 1994; Dunbar, 1993; Nunn & Barton, 2001; Povinelli, 1993; Rendall & Di Fiore, 1995).

Although some of the psychological and cognitive adaptations that are the focus of evolutionary psychological research (e.g., language) probably arose after the pongid-hominid split, it is likely that many human psychological and behavioral adaptations, particularly in the domain of social and reproductive activities, have a much longer evolutionary history that can be traced back to the common ancestors of modern human and nonhuman primates. It is also likely that many such adaptations are still manifested even in the environment of modern industrialized societies. In fact, despite all the technological advances of industrialized societies, the social environment of modern humans and the problems it poses to them are less novel than people might think. The behavioral solutions that modern humans find to problems associated with cooperation, competition for resources, or access to mates are remarkably similar to those used by our closest relatives, cercopithecine monkeys and apes (e.g., Alexander, 1979; de Waal, 1982; Wilson, 1978). Thus, nonhuman primates can provide an invaluable resource for developing and testing hypotheses concerning the adaptiveness of human behavior and cognition and an indispensable tool for modeling their evolutionary history as well.

The contributors to this book review and discuss recent research on primate behavior and cognition in relation to similar areas of human research. Obviously, a truly comprehensive review of all major areas of primate behavior and cognition would be a difficult task to accomplish with a single volume; therefore, the chapters have been assembled according to specific selective criteria. Chapter 2, by J. Dee Higley, provides a discussion of aggressive behavior and its neurobiological substrates in primates and humans. In Chapter 3, Peter Judge synthesizes and integrates research on conflict resolution, emphasizing both similarities and differences across species and contexts. Chapter 4, by Kim Wallen and colleagues, focuses on primate sexuality and addresses sex differences in behavioral development as well as the social and neuroendocrine aspects of sexual behavior and sexual motivation in adolescence and adulthood. In Chapter 5, Dario Maestripieri discusses the conceptual premises of attachment theory and the parallels between attachment processes in primates and humans. In Chapter 6, Lynn Fairbanks takes an evolutionary approach to the study of parenting and emphasizes the adaptive value of individual differences in parenting style. In Chapter 7, James Roney and Dario Maestripieri contrast biological and socialization perspectives on behavioral development, including the effects of parenting styles on development and the determinants of behavioral sex differences. This chapter also compares different approaches to research on altruistic interactions and the relation between affiliation and social structure. Chapter 8, by Jesse Bering and Daniel Povinelli, addresses cognitive development from a comparative and evolutionary perspective and discusses the evidence in favor of and against the hypothesis that chimpanzees can form concepts about abstract things such as mental states or cause-effect relationships. Chapter 9, by Josep Call and Michael Tomasello, provides a different perspective on primate cognition, with particular emphasis on social processes and theory-of-mind abilities. In Chapter 10, Samuel Gosling and colleagues provide one of the first systematic attempts at discussing personality research from a comparative perspective, while Filippo Aureli and Andrew Whiten tackle the subject of emotions in Chapter 11. In Chapter 12, Lisa Parr and Dario Maestripieri review the research on face recognition and facial expressions, and then address some conceptual issues concerning the study of nonvocal communication in primates and humans. In Chapter 13, Michael Owren and his colleagues advance the hypothesis that much vocal communication in primates serves nonlinguistic purposes and draw specific parallels between primate calls and human nonlinguistic vocalizations. In Chapter 14, Duane Rumbaugh and colleagues tackle the question of defining language and review decades of research on language acquisition in the great apes. Chapter 15, by William Hopkins and colleagues, reviews the evidence for neuroanatomical and behavioral asymmetries in primates and discusses it in relation to the evolution of language and other forms of human communication and behavior, including handedness. In the last chapter, Alfonso Troisi addresses the issue of defining normal and abnormal behavior in primates and humans and identifying diagnostic criteria for psychopathology that may have validity across species.

The emphasis of this volume is clearly on social processes. Therefore, aspects of behavior and cognition that are not strictly social (e.g., foraging, spatial cognition) are not represented. With this book, we hope to extend and complement rather than replicate and reiterate the information provided by other recent books and essays on primate behavior, comparative psychology, and evolutionary biology. Therefore, information on the natural history, ecology, and socio-reproductive adaptations of different primate species is taken for granted or only briefly discussed. Instead, the focus is on behavior and social interactions. Because one of the main goals of this book is to encourage communication between primatologists and psychologists, we mostly rely on theories of behavior derived from psychology rather than ecology or biological anthropology. Likewise, the book also emphasizes data and concepts derived from primate behavior research that appear to be more easily extrapolated to humans rather than data and concepts that have limited cross-species applicability.

The chapters focus on some well-known selected genera and species of Old World monkeys and apes rather than attempting to represent all major taxonomic units within the Primate order. This is not because the contributors believe that macaques, baboons, or chimpanzees represent the "typical primate" (Strier, 1994). Rather, these genera and species are singled out for attention because of their genetic, social, and behavioral similarities with humans (who can hardly be considered typical primates). Moreover, previous research with specific genera and species of Old World monkeys and apes has already attempted integration with human research more than research with other primates. Therefore, studies of chimpanzees are more readily accessible to psychologists than studies of lemurs because the former are more likely to address questions of interest to psychological research than the latter. Although in recent years there has been a rapid growth of research with prosimians and New World monkeys, the majority of researchers studying these primates do not use psychological theories as a framework for their investigations, do not discuss their findings in relation to the findings of psychological research, and do not publish their findings in journals that are read by psychologists. It is our hope that this book will encourage researchers studying all primate species to relate their work to human behavior. As more information about lesser-known species becomes available, psychologists will also increasingly recognize the importance of these species for comparative research.

As with primate research, the psychological research discussed in this book is presented selectively. Obviously, the contributors to this volume do not believe that middle-class people living in Western societies represent the "typical human." It is simply a fact that most published research in psychology is biased toward middle-class people in Western societies. We hope that this preliminary effort to integrate information from a few primate species, on focal topics, with an even more selected body of work in psychology will encourage future attempts at broader integration. Integration of different disciplines and areas of research is difficult or impossible without effective communication among researchers. It is our wish that this book will encourage interest in interdisciplinary efforts and exchange of information, not only between primatologists and psychologists but among all researchers, students, and laypeople who are interested in understanding mind and behavior.

2

Aggression

J. Dee Higley

The study of aggression has been an important area of behavioral research with animals and humans for decades, if not centuries. Despite the fact that students of human and animal aggression often publish in the same scientific journals (e.g., *Aggressive Behavior, Aggression and Violent Behavior*) and attend conferences of the same professional societies (e.g., the International Society for Research on Aggression), there are significant gaps in communication between researchers working with humans and those working with animals, including nonhuman primates. Although lack of communication between human and animal researchers is not restricted to research on aggression, this problem is arguably more serious in this area than in the study of most other behavioral phenomena (Blanchard et al., 1999). In part, this problem may reflect the fact that some aspects of aggression are species-specific and closely associated with the social, communicative, and cognitive characteristics of a species (e.g., verbal aggression is arguably a unique human trait). As with other behavioral phenomena such as tool-using and cooperative hunting, however, studies have shown that forms of aggression that were believed to be unique to humans also occur in animals that are phylogenetically close to us such as the great apes. For example, wild chimpanzees engage in boundary patrols and conflicts with other groups that share some characteristics with human warfare (e.g., Manson & Wrangham, 1991; Watts & Mitani, 2001). Aside from species-specific differences in the expression of aggressive behavior, many of the general principles and theories that have been advanced to explain human aggression could also be applicable to animals and vice versa. This should be especially true for nonhuman primates such as Old World monkeys and apes, which are phylogenetically close to humans and share with us some social, communicative, and cognitive characteristics.

There are many different theories of aggression, and the best known of them can be grouped under the labels of "social learning theories" and "biological theories." Social learning theories emphasize the role of environmental variables and experience in the development and expression of aggressive behavior. In this view, aggression may be learned by children or adults through observation and modeling (e.g., when children observe the use of aggression within their family, their subculture, or the mass media), or through positive reinforcement (e.g., when children achieve a successful outcome such as obtaining a favorite toy with instrumental aggression) (Bandura, 1973). Biological theories address both the mechanisms underlying the expression of aggressive behavior (e.g., interactions between genetic and environmental variables, or the role of specific hormones and neurotransmitters) and its adaptive value (i.e., the contribution aggression makes to the survival and reproductive success of the individual; e.g., Archer, 1988). Although some early proponents of biological theories of aggression maintained that aggression is innate and inherently good for the individual or the species (e.g., Ardrey, 1966; Lorenz, 1966), biologists today recognize the importance of gene-environment interactions in the development of behavior and the potential occurrence of maladaptive processes even in behavioral phenomena that have been under strong selection pressure for their adaptive function such as mating, parenting, and aggression (e.g., Archer, 1988).

Aggression has been studied in nonhuman primates with observational and experimental methods, both in the field and in the laboratory. Observational studies have emphasized the functional and adaptive aspects of aggression, and in particular its role in mediating competition for access to resources or social status. By contrast, experimental studies of aggression have focused on the mechanisms regulating aggression, and in particular the role of particular hormones or neurotransmitters. The role of environmental variables and learning processes in the development and expression of primate aggression has received less systematic investigation both in the field and in the laboratory. Nevertheless, the primate data are consistent with the notion that environment and experience play an important role in the causation and development of aggressive behavior. Investigations of the neuroendocrine substrates of aggressive behavior have proceeded almost in parallel in human and nonhuman primates. Research in this area has been guided by similar hypotheses, used similar procedures, and obtained similar findings. Thus, primate and human studies of the neuroendocrine substrates of aggression can be directly compared and their findings integrated into a coherent explanatory framework.

The main goal of this chapter is to review and integrate current knowledge of the proximate regulation of primate and human aggression, with particular emphasis on neuroendocrine mechanisms. Since most primate research in this area has been conducted with Old World monkeys such as macaques and vervet monkeys, the discussion begins with a section providing some basic information on the characteristics of aggressive behavior in these primates. Some of the characteristics of aggressive behavior in Old World monkeys are also shared by some prosimians, New World monkeys, and apes. Unfortunately, we know very little about the proximate regulation of aggression in these primates, and therefore behavioral research with these species is not systematically reviewed here.

Aggression in Old World Primates

General Characteristics and Functions of Aggression

From the inception of studying nonhuman primate behavior, aggression has figured prominently as a major research topic (Altmann, 1962; Carpenter, 1942, 1974; Harlow, 1969; Harlow & Harlow, 1965; Koford, 1965). Bernstein and Ehardt (1985c) review data from a wide variety of species and report that aggression, on average, accounts for 2–5 percent of a monkey's daily activity budget. It is surprising that even among species such as rhesus macaques, which are viewed as particularly quarrelsome and truculent, Marriott (1988) reports that across a wide variety of study sites, aggression represents 2 percent or less of the average monkey's overall time budget. Nevertheless, most researchers investigating social behavior in nonhuman primates agree that aggression is a basic and integral part of societal functioning, and when appropriately expressed, it plays a number of important functions. Social assertiveness and defense of territory or safeguarding self or kin from physical danger are of critical importance in maintaining social status, ensuring access to resources, and preserving physical well-being. Mothers also use aggression to protect their threatened infants, and kin often use coalitions to defend threatened members of their matrilines. Studies have shown that aggression is integral to maintaining and enforcing societal norms and social rules, and it plays a major role in the maintenance of social cohesion and order (Bernstein, 1981; Bernstein & Ehardt, 1985c, 1986a, b).

The most frequent context in which aggression is seen is in defense of status. Unlike in rodents, among nonhuman primates the best fighters do not necessarily become the highest-ranking animals (Chapais, 1983, 1986; Higley, Linnoila, & Suomi, 1994; Higley & Suomi, 1989, 1996; Kaufmann, 1967; Raleigh & McGuire, 1994). Nevertheless, aggression is critical for defending and maintaining status, but such aggression seldom results in trauma or serious wounds (Bernstein, 1981; Higley, King, et al., 1996; Higley, Suomi, & Linnoila, 1996a). Aggression also plays an important role in the socialization process of developing primates, as mothers and other adults use it to punish and modify inappropriate behaviors (Higley & Linnoila, 1997b; Higley & Suomi, 1996; Itoigawa, 1993).

Of course, aggression is also at times dangerous and violent. Among most Old World monkeys, predators play only a minimal role in producing injuries and mortality; instead, the greatest danger to an individual monkey is aggression from other monkeys (Crockett & Pope, 1988; Higley, Mehlman, et al., 1992; Higley, Mehlman, Higley, et al., 1996; Lindburg, 1971; Simonds, 1965). While the average amount of time spent in aggression across individuals is rather low, some individuals exhibit excessive rates of aggression and violence, much higher than the norm, and use it in an unrestrained fashion that results in wounds, trauma, and even death (Chapais, 1986; Higley & Bennett, 1999; Higley, Linnoila, & Suomi, 1994; Southwick et al., 1965).

While severe aggression is more frequent between strangers, among some particularly aggressive individuals, and within groups of some species, it occurs among virtually all primate species studied. Both within troops in which individuals are familiar with one another and among more peaceable species such as bonnet macaques, at times aggressive encounters escalate out of control, producing injuries and trauma (Berard, 1989; Colvin, 1986; Crockett & Pope, 1988; Higley, Mehlman, et al., 1992; Higley, Mehlman, Higley, et al., 1996; Lindburg, 1971; Simonds, 1965; see also Steenbeek et al., 2000, for langurs). In both captive and free-ranging animals, old scars and wounds, at times even severe wounds, are frequently observed, especially among adult males (Rawlins & Kessler, 1986; Southwick et al., 1965; Steenbeek et al., 2000). Females also receive wounds, especially during the breeding season. For example, Chapais (1986) reports that among rhesus macaques, during the breeding season, fifteen of nineteen females were severely wounded, with the average number of wounds and slashes being 1.6 (range 1–6). Indeed, one of the dependent measures we have used to assess individual differences in aggressiveness among adolescent male rhesus monkeys involves counting the recent wounds and scars on the monkey's head, torso, and limbs (Higley, Mehlman, et al., 1992). Moreover, among food-provisioned free-ranging rhesus monkeys, aggression is the leading cause of premature death after infancy, at least among males (Higley, Mehlman, Higley, et al., 1996). It is of interest that among macaque species in which violence is less frequent and severe than in rhesus, for example, among Barbary macaques, the life expectancy is substantially longer (Paul & Kuester, 1988).

Social Dominance and Aggression

While aggression that results in prolonged noisy chases, trauma, and severe wounds is the most noticeable type of aggression, this type is rare, with biting, for example, accounting for less than 3 percent of all aggression within a rhesus macaque troop (Bernstein & Ehardt, 1985b). This is most likely because of the restraints that society places on how aggression can be expressed. Perhaps the most important modulating influence on aggression is social dominance. Although many aggressive encounters occur during competition over social dominance status, once dominance status has been established between two individuals, violent physical aggression is much less likely, and most challenges are settled using threats and avoidance (Bernstein, 1970, 1981; Bernstein & Ehardt, 1985b; Bernstein et al., 1974a; de Waal, 1986a; Kaufmann, 1967; Sade, 1967). Moreover, when dominance hierarchies become unstable, rates of severe, violent aggression may show marked increases (in baboons, for example:

Sapolsky, 1983). Nevertheless, knowing the social dominance rank of an individual monkey tells a researcher much about this individual's aggressive tendencies. For instance, although high-ranking rhesus monkeys are not necessarily the most aggressive individuals in a troop, they are more likely to initiate and win aggressive encounters than others (Chapais, 1983, 1986; Higley, Hasert, et al., 1994; Higley, King, et al., 1996; Higley, Linnoila, & Suomi, 1994; Kaufmann, 1967; Sade, 1967).

Chapais and others have shown that forming coalitions and maintaining social support are crucial to acquiring and maintaining a high social dominance rank (Chapais, 1986, 1988; Higley & Suomi, 1996; Raleigh & McGuire, 1986; Raleigh et al., 1991; Varley & Symmes, 1966). Nevertheless, at times males do use aggression and violence to bully and coerce their way to the top (e.g., in longtail macaques: van Noordwijk & van Schaik, 2001; and in gelada baboons: Dunbar, 1984). Such a strategy is risky, probably producing high levels of stress, and at times results in violence and dangerous contact aggression. In our longitudinal study of aggression, for example, free-ranging rhesus males who used violent aggression to achieve high social dominance status were more likely to be wounded or killed (Westergaard et al., submitted).

Life History and Social Influences on Aggression

Life history variables and the social environment in which primates live affect the probability that aggression will occur. Sooner or later, most if not all male macaques migrate from their natal groups to new social groups, typically during adolescence or early adulthood (e.g., see Higley & Suomi, 1996, for a review). Migration from the natal troop is a dangerous period of life for young male macaques, with adolescent males involved in many aggressive encounters (Bernstein & Ehardt, 1985c; Hausfater, 1972; Sugiyama, 1976). During this period there is a high rate of mortality from aggression such that between 30 and 50 percent of emigrating males are killed or disappear and are presumed to have died (e.g., Dittus, 1979; Drickamer, 1974; Higley, Mehlman, Higley, et al., 1996; Meikle & Vessey, 1988). At times, high rates of aggression precede the migration of a young male, leading researchers to conclude that young males are forced out of their natal troops by means of aggression (Kaufmann, 1967).

In natural environments, aggressive encounters with other monkeys are surprisingly frequent when dominance ranks are unstable or an alpha male is replaced (Simonds, 1965). The social setting in which aggression is most likely to occur is probably when two troops meet, or, in captivity, during the introduction of unfamiliar monkeys into an existing group, or the formation of a new group (Altmann, 1962; Bernstein et al., 1974b; Carpenter, 1974; Cheney, 1981; Poirier, 1974; Southwick et al., 1965, 1974). When unfamiliar males attempt to enter a new troop, the resident females may use aggression to control which males are allowed to migrate into their troop. This behavior, however, may be limited to species in which males and females are nearly the same size (Packer & Pusey, 1979). Highly aggressive males may use violence to force their integration into a troop. This strategy, however, is more successful in smaller troops where bullying can be used but it is unlikely to be successful in a larger troop where the bullies are outnumbered (van Noordwijk & van Schaik, 1985, 2001). Macaque males who use aggression to facilitate their introduction into a troop show signs that they are under more stress (Westergaard et al., submitted), and they have a shorter tenure in the troop than nonaggressive males who use coalitions to integrate themselves into a troop (van Noordwijk & van Schaik, 1985). In a related but opposite fashion, the presence of an aggressive adult male who assumes a "control role" within the group may also inhibit aggression among females (Coelho & Bramblett, 1981; Erwin, 1977). The temperament of the male is probably an important variable affecting his social role within the group because not all alpha males exhibit such "policing" behavior.

A number of studies show that aggression rates are higher under conditions of crowding (Anderson et al., 1977; Boyce et al., 1998; Erwin, 1977; Judge & de Waal, 1997; Westergaard et al., 1999; but see Aureli & de Waal, 1997, in chimpanzees, and de Waal, 1989c), although this is more likely for milder forms of aggression than for high-intensity aggression (Anderson et al., 1977; Judge & de Waal, 1993). In captivity, aggression, particularly violent aggression, is reduced when provision is made for a hiding place or when a barrier is available to divide the area so that the victim of aggression can flee and hide (Erwin, 1977; Westergaard et al., 1999). Victims of aggression during crowding are more likely to be highly anxious or inhibited (Boyce et al., 1998). Low-ranking individuals, however, are likely to reduce their general activity levels or increase their grooming behavior to minimize the risk of aggression during crowding (in longtail macaques: Aureli et al., 1995).

Sex, Age, and Species Differences in Aggression

While it is often believed that primate males are more aggressive than females, among many species of Old World primates, rates of aggression are surprisingly similar between the sexes. Nevertheless, some sex differences have been demonstrated. During group formation in captive rhesus macaques, fights between males often occur and aggression is more frequent for males than females. Males, however, rapidly form a dominance hierarchy, and once they form a hierarchy, females show higher rates of aggression (Bernstein et al., 1974a). Among Old World monkeys, males are more likely to exhibit aggression that results in observable injuries or trauma because of their larger body size and more developed canines (Hausfater, 1972; Lindburg, 1971; Paul & Kuester, 1988; Smuts, 1987). Rates of aggression vary, however, according to the age and sex class of individuals involved in the interaction. For example, among rhesus macaques, adult males seldom interact with one another, but when they do, the average rate of aggression, including displacements, threats and dominance-related mounting, is very high, particularly during the breeding season (Kaufmann, 1967). The presence of females in estrus and the resulting competition for mating affect the rate of male aggression. In patas monkeys, for example, males live together peaceably in all-male bands, but when males are in heterosexual groups, relations are more tense and males are more likely to act aggressively toward one another (Rowell & Chism, 1986).

Age also affects the probability and severity of aggression. While adults rarely attack young infants (rhesus macaques: Bernstein & Ehardt, 1985c; Harlow, 1969; vervet monkeys: Horrocks & Hunte, 1983a, b), older infants and juveniles frequently receive aggression and are the age class most likely to submit to aggression (rhesus macaques: Bernstein & Ehardt, 1985c; langurs: Jay, 1965). The presence of infants also reduces rates of aggression between familiar females (Erwin & Flett, 1974). Among macaques, baboons, and mangabeys, young infants are sometimes used as agonistic buffers to reduce aggression between males (Maestripieri, 1998c). Older infants and juveniles, however, may receive aggression because their rough-and-tumble play may place them in the path of less tolerant adults (e.g., in langurs: Jay, 1965). During intergroup encounters, when infants are involved in aggression, males are more likely to be involved than females.

In rhesus macaques, rates of aggression increase for the juveniles of both sexes, but more so for females than for males. Female rates of aggression do not increase further with age, but males do increase their rates of aggression so that another reversal occurs by adolescence, with adolescent males both giving and receiving the highest rates of aggression of any age class in both intra- and inter-troop aggression (Bernstein & Ehardt, 1985c). Adolescent males are also the age class most likely to receive mobbing and prolonged agonistic sequences (Bernstein & Ehardt, 1985c). Adolescent males are more likely to respond to threats than are adult males (Bernstein & Ehardt, 1985c; Bernstein et al., 1974a), but females are more likely to retaliate against aggression than males (Bernstein & Ehardt, 1985c; Bernstein et al., 1974a). By the time they become adults, rhesus macaque males' participation in agonistic episodes is less frequent, and when it does occur, it is typically silent, brief, and rarely involves biting.

Among Old World monkeys, there are species differences in temperamental traits that are associated with aggressiveness, with some species showing relatively pugnacious temperaments and other species being more docile. Even in closely related species, and when strains are compared within the same species, aggressive temperaments vary. In one study of three closely related macaque species, rhesus macaques were shown to be the most aggressive, longtail macaques the most fearful, and bonnet macaques relatively placid and friendly in their temperament (Clarke & Mason, 1988). In another comparison of three macaque species, rhesus macaques were found to engage in more severe aggression (bites) than Tonkean and longtail macaques (Thierry, 1985). Interestingly, while the Tonkean macaques virtually never bit one another, they engaged in the highest rates of milder aggression, exhibiting the highest incidence of slaps and grabs. Compared to the other two macaque species, longtail macaques were skittish, engaging in the highest rates of submissive behaviors. Similarly, we found that when pigtail and rhesus female macaques were compared, rhesus were more aggressive and pigtails were more gregarious and friendly (Westergaard et al., 1999). Within rhesus macaques, two strains have been identified: Indian and Chinese. The Chinese rhesus strain is more aggressive and emotionally labile compared to the more common Indian-derived strain (Champoux et al., 1994, 1997). Parallel to these differences in rhesus strains, when comparing two baboon subspecies, anubis and hamadryas, Kaplan and colleagues (1999) found that hamadryas baboon males engage in more friendly relations than their closely related but more hostile neighbors, anubis baboons, a difference that is correlated with differences in socioecological and neurobiological characteristics between species.

Developing the Skills to Use Aggression Appropriately

The development of primate aggression is relatively similar across very different environments, suggesting that the specific patterns of aggressive behavior need not be learned. For example, when monkeys reared in social isolation are exposed to pictures of an adult male displaying an open mouth threat, they show more interest in and a more vigorous response to the threat than to other nonaggressive depictions (Sackett, 1973; see also Chapter 12). Primates, however, must learn the skills for expressing aggression in the proper settings. Choices of timing, intensity, and targets are among the critical aspects of aggression that must be learned (Harlow, 1969; Higley, Linnoila, & Suomi, 1994; Higley & Suomi, 1996). Some animals fail to learn such skills and thus fail to recognize social cues. For example, Sapolsky (1992) describes a category of males who failed to acquire the social skills to discriminate between the threat of a male who slept nearby and that of a male who waited his turn to move up the hierarchy. Socially living primates must learn not only to recognize social cues of aggression but also to restrain and control their own impulses whenever necessary. In fact, achieving high dominance status within a troop may depend on inhibiting aggression as much as on expressing it.

The contextual use of aggression is learned during a prolonged period of development in which young primates have many opportunities to observe the behavior of older individuals within their group and practice their skills with peers. Infants first experience aggression early in life during mother-infant interactions. Aggressive punishment by the adult caregiver is probably of critical developmental importance because it is typically used to restrain and inhibit inappropriate infant behaviors such as biting and unrestrained acts toward other members of the group. Monkeys raised without their mothers and other conspecifics show inappropriate aggression, making poor social companions, and as a result, when placed into social settings, they are not preferred as social partners (Sackett, 1968, 1975). They are generally shunned by other monkeys (Capitanio, 1986; Sackett, 1968, 1975), and end up low in social dominance rank (Bastian et al., 2003).

By punishing unrestrained and inappropriate behaviors in the infant, mothers may provide the input that allows the infant to develop the cortical structures necessary for impulse control. In fact, early mother-infant interactions are associated with a major period of development for the frontal and prefrontal cortex. This area of the brain is of critical importance for the proper expression of impulse control. For example, monkeys with orbitofrontal damage exhibit inappropriate aggression and impaired social competence, and as a result are ostracized by other group members (Raleigh & Steklis, 1981; Raleigh et al., 1979). Social input during this critical period of development dictates which synapses in the brain will be maintained and which will be "pruned." As parentally deprived monkeys develop, they are more likely than parentally reared individuals to exhibit impulse control deficits, resulting in impulsive violence and self-aggression (Higley & Linnoila, 1997a; Higley, Suomi, & Linnoila, 1996a, b; Kraemer & Clarke, 1990). Moreover, monkeys deprived of parents during infancy are more likely to show deficits in their CNS (central nervous system) inhibitory systems (Higley & Linnoila, 1997b) and functional deficits in the frontal and prefrontal cortex region (Doudet et al., 1995; Heinz et al., 1998). For example, Lopez and colleagues (2001) found many parallels between the frontal cortical regions of peer-reared macaques and humans who commit suicide. Studies of humans and animals show that injuries in these areas of the brain often result in impulse control deficits and impulsive aggression (Davidson et al., 2000).

It is very likely that in human and nonhuman primates, parental input is of critical importance for developing CNS inhibitory cortical controls which function to restrain aggression (e.g., Doudet et al., 1995). For example, as the inhibitory influences of the frontal cortex are maturing and making connections, primate mothers punish inappropriate aggression in their infants, probably inducing cortical growth in the inhibitory control centers. Without such input, monkeys develop into hyperaggressive adults who act violently toward infants and other individuals (Suomi, 1982b). Equally important is the role of the mother in supporting and defending her infant during aggressive encounters with other monkeys. When yearlings are threatened, mothers are sought for comfort as well as support. Infants observe interactions between their mothers and other individuals, and such observations, along with maternal intervention in conflicts between infants and other individuals, are crucial for the inheritance of maternal dominance rank in Old World monkeys (Chapais, 1988; Sade, 1967). When infants learn from their mothers to whom they must defer and whom they can dominate, they acquire knowledge that is critical in determining when and how aggression can and cannot be expressed. Infants also learn from their mothers the skills to form coalitions as they interact with them in these coalitions. Such skills will become crucial when young monkeys can no longer count on their mothers' support, for example, when mothers die or after the young emigrate from the natal group (Sade, 1967).

During interactions with peers, developing primates build on what they have learned from their mothers, practicing and honing aggressive skills and learning their relative status in society. This occurs chiefly during play, and particularly in the context of aggressive play. During aggressive play, developing primates learn to use aggression, expressing it with appropriate intensity, inhibiting it when needed, and further learning toward whom they can and cannot express aggression (Harlow, 1969; Higley, Linnoila, & Suomi, 1994). In one study, we found that high rates of aggressive play in rhesus juveniles and infants were predictive of low rates of aggression and violence in adulthood; conversely, low rates of aggressive play were predictive of violence in adolescents, largely because aggressive play escalated into violence during an aggressive episode (Barr et al., submitted). In the absence of peer practice during formative stages, a juvenile monkey may attack a full-grown male twice its size or persist in aggression beyond what is necessary (Higley, Linnoila, & Suomi, 1994; Sackett, 1968; Suomi, 1982b). To act competently and to acquire a high social dominance rank, a monkey must develop the skills to recruit support from other individuals in times of social challenge (Altmann, 1980; Bernstein & Ehardt, 1985b, c; Chapais, 1986, 1988).

Excessive behavioral withdrawal from social challenges and inappropriate aggression hinder the capacity to acquire social competence. Developmental studies of humans indicate that traits characterizing behavioral withdrawal and aggression are present early in life, with behavioral withdrawal showing interindividual stability in infancy, and aggression showing interindividual stability by late childhood (Eron, 1987; Huesmann et al., 1984; Kagan et al., 1989). These studies also show that violent behavior and aggression are among the most stable personality traits, showing long-term interindividual stability (Lenzenweger, 1999a; Olweus, 1979). Aggression stabilizes early in life, perhaps as early as infancy (Keenan & Shaw, 1994), and early childhood aggression is predictive of late childhood, adolescent, and adult aggression (Cummings et al., 1989; Huesmann et al., 1984; Keenan & Shaw, 1994; Moskowitz et al., 1985; Olweus, 1980). Indeed, in an exhaustive review, Olweus (1979) concluded that aggression was one of the most stable personality traits, showing interindividual stability as strong as that observed for IQ. This may be true among nonhuman primates as well. In a longitudinal study of male aggression among free-ranging rhesus macaques, we recently found that impulsive aggression early in life was predictive of aggression in adults ten years later (Westergaard et al., submitted).

Biological Substrates of Aggression

Serotonin and Aggression

The investigation of the biological substrates of aggression has been a major focus of research in biological psychology and psychiatry for several decades. In recent years, much effort has concentrated on the role of the serotonergic system in the regulation of aggressive behavior. A number of studies with both human and nonhuman animals have shown that impaired CNS serotonin functioning, whether naturally occurring or pharmacologically induced, is related to violence and impulsive or unrestrained aggression (see Soubrié, 1986, for a comprehensive review).

The principal breakdown product of serotonin is the metabolite 5-hydroxy-indoleacetic acid (5-HIAA). This inactive metabolite is often measured in the cerebrospinal fluid (CSF) as a measure of serotonin activity, with low concentrations of CSF 5-HIAA thought to represent an underactive or impaired central serotonin system. Many studies of animals and humans have reported negative correlations between CSF 5-HIAA and various forms of impulsive aggression and violence. For example, studies of dogs show that aggressive animals that severely bite their owners and attack spontaneously, without warning, are more likely to have lower CSF 5-HIAA (e.g., Reisner et al., 1996). Among rhesus macaques, low CSF 5-HIAA concentrations are correlated with high rates of wounding, unprovoked and unrestrained violence, and violent deaths (Higley, King, et al., 1996; Higley, Mehlman, et al., 1992; Higley, Mehlman, Higley, et al., 1996; Higley, Mehlman, Poland, et al., 1996; Higley, Suomi, & Linnoila, 1996a, b; Mehlman et al., 1994, 1997). Finally, it is relatively well established that men with low CSF 5-HIAA concentrations exhibit increased unplanned aggression and impulsive violence (Brown et al., 1979, 1982; Lidberg et al., 1985; Limson et al., 1991; Linnoila et al., 1983; Roy et al., 1985; Virkkunen, Kallio, et al., 1994; Virkkunen, Rawlings, et al., 1994).

In addition to correlations, there is also experimental evidence showing that when central serotonin is enhanced, rates of aggression decrease, and when serotonin is diminished, aggression increases. For example, research with rodents shows that pharmacological agents that increase serotonin activity decrease aggression, while agents that decrease serotonin activity or block its action increase aggression (Miczek & Donat, 1990; Nikulina et al., 1992; Olivier & Mos, 1990; Olivier et al., 1990). Among vervet monkeys, individuals that consume experimental diets high in the serotonin precursor tryptophan (which increases serotonin in the CNS) exhibit decreased aggression, whereas individuals placed on diets low in tryptophan exhibit increased aggression, with the effects being stronger for males than for females (Chamberlain et al., 1987; Raleigh et al., 1985, 1986, 1991). Similar changes in primate aggression have been demonstrated with other serotonin treatments, including decreases in aggression with short-term administration of serotonin reuptake inhibitors (which augment serotonin levels in the synapse) (Chamberlain et al., 1987; Higley et al., 1998; Raleigh et al., 1980, 1985, 1986, 1991), and increases in aggression after administration of the serotonin synthesis inhibitor p-chlorophenylalanine, which acts to diminish serotonin production (Raleigh et al., 1980; Raleigh & McGuire, 1986). Likewise, long-term (but not short-term) treatment with fenfluramine, which decreases serotonin turnover in the brain, increases aggression in vervet monkeys (Raleigh et al., 1983; 1986).

Some studies suggest that differences in aggressive temperament between different strains or species are paralleled by differences in CNS serotonin functioning. Studies of CNS biochemistry in animals that are selectively bred for domestication and increasingly docile temperaments show that, as a program of selective breeding produces animals with less and less aggressive temperament, there is a concomitant increase in CNS concentrations of serotonin and its breakdown product, 5-HIAA (Namboodiri et al., 1985; Popova, Kulikov, et al., 1991; Popova, Voitenko, et al., 1991). With increasingly placid temperaments, there is also an increase in the enzyme responsible for converting the amino acid tryptophan into serotonin (tryptophan hydroxylase), with a parallel decrease in the enzyme that breaks down serotonin, monoamine oxidase type A (Popova et al., 1997). In nonhuman primates, Westergaard and colleagues (1999) found that CSF 5-HIAA concentrations were lower in an aggressive macaque species, the rhesus macaque, than in a less aggressive and more gregarious species, the pigtail macaque. Furthermore, when Champoux and colleagues (1997) compared different genetic strains of rhesus macaques, they found that the more aggressive Chinese rhesus macaques possessed lower CSF 5-HIAA concentrations than their more peaceable Indian-derived rhesus cousins. Finally, Kaplan and colleagues (1999) reported that anubis baboons, which are characterized by relatively high levels of inter-male aggression, had lower CSF 5-HIAA concentrations than the less aggressive anubis-hamadryas baboon hybrids.

A number of studies have assessed the serotonin system using a method originally called the prolactin challenge. This procedure involves administering a serotonin-enhancing drug, such as fenfluramine hydrochloride or another reuptake inhibitor, and then measuring levels of prolactin in the blood. High levels of prolactin are interpreted as evidence of an active serotonin system, and low levels of prolactin are seen as evidence of an impaired central serotonin system. Numerous studies using this method in humans have found evidence for impaired central serotonin functioning in aggressive men and violent patients (e.g., Newman ct al., 1998). Similarly, adult male longtail macaques who showed a lower prolactin response to the fenfluramine challenge were more likely to exhibit aggressive responses to pictures of humans making threatening gestures than were individuals with a higher prolactin response (Kyes et al., 1995).

Although the relationship between serotonin and aggression has not been systematically studied in women, nonhuman primate studies suggest that this relationship may extend to females as well as males. For example, adolescent and adult rhesus macaque females that exhibit high rates of impulsive aggression and unrestrained violence show lower CSF 5-HIAA than less aggressive females (Higley, King, et al., 1996; Higley, Suomi, & Linnoila, 1996a, b). Furthermore, in female longtail macaques, low prolactin in response to fenfluramine is associated with high rates of aggression (Botchin et al., 1993; Shively et al., 1995).

Serotonin and Impulsive Temperament

Soubrié (1986) suggested that aggression is higher in individuals with impaired CNS serotonin functioning because their impulse control capabilities are defective. In support of this hypothesis, evidence shows that, in rhesus macaques, low CSF 5-HIAA is not correlated with overall levels of aggression; instead, it is only spontaneous, impulsive aggression, aggression that tends to escalate out of control, that shows a negative correlation with CSF 5-HIAA concentrations (Higley, King, et al., 1996; Higley, Mehlman, et al., 1992; Higley, Mehlman, Higley, et al., 1996; Higley, Suomi, & Linnoila, 1996a, b; Mehlman et al., 1994). Moreover, in humans, individuals with impaired CNS serotonin functioning show a cluster of closely related behaviors and traits centering on impaired impulse control (Apter et al., 1990).

In his biosocial theory of personality, Cloninger (1988) reviews evidence suggesting that personality disorders related to dysfunctional social relationships, aggression, and diminished social bonding are neurobiologically based on diminished serotonin functioning and possibly reduced dopamine functioning. For example, studies of both human children (Kruesi et al., 1990) and nonhuman primates (Higley & Linnoila, 1997a; Higley & Suomi, 1996; Higley, Suomi, & Linnoila, 1996a) show that individuals high in social deviancy or who exhibit less competent social behavior have relatively low CSF 5-HIAA. Additionally, men with low CSF 5-HIAA concentrations exhibit evidence of impaired impulse control such as increased unplanned fire setting but not arson for hire (Virkkunen et al., 1987), and increased violent criminal recidivism (Virkkunen et al., 1989). In both rodents and nonhuman primates, serotonin-enhancing pharmacological treatments decrease alcohol consumption (Gill & Amit, 1989; Higley, Hasert, et al., 1992; McBride et al., 1989). Conversely, pharmacologically reducing CNS serotonin activity in rodents increases the frequency of performing a response despite the threat of punishment (Gleeson et al., 1989; Miczek et al., 1989; Soubrié, 1986).

Similar to humans, nonhuman primates with low CSF 5-HIAA concentrations are more likely to exhibit behaviors suggestive of impaired impulse control such as spontaneous, unprovoked long leaps at dangerous heights and repeated jumping into baited traps in which they are captured (Fairbanks et al., 1999; Higley, Mehlman, Poland, et al., 1996; Mehlman et al., 1994). In the laboratory, when a novel black box is placed in a room, rhesus macaques with low CSF 5-HIAA concentrations take less time to approach the box and to touch it than do monkeys with high CSF 5-HIAA concentrations (Bennett et al., 1998). Rhesus macaques with low CSF 5-HIAA concentrations are also more likely to overconsume alcohol once drinking begins (Higley, Suomi, & Linnoila, 1996b). In one study, Fairbanks and colleagues (2001) used a standardized test of impulse control to measure the relationship between impulsivity and CNS serotonin functioning. This test, known as the intruder challenge paradigm, measures the latency to approach an unfamiliar adult male. This study reported that vervet monkeys with low CSF 5-HIAA concentrations approached the potentially dangerous male more quickly than monkeys with high 5-HIAA. In this same study, vervet monkeys whose central serotonin was pharmacologically increased with the selective serotonin reuptake inhibitor fluoxetine were slower to approach the unfamiliar adult male than monkeys receiving a placebo.

To the extent that excessive aggression results from an impulse control deficit, one would predict that rates of impulsivity would be positively correlated with rates of aggression and violence. Consistent with this prediction, in two studies of rhesus macaques, we found that rates of spontaneous, unprovoked, dangerous long leaps were positively correlated with rates of violent behavior (Higley, Mehlman, Higley, et al., 1996; Mehlman et al., 1994). Similarly, Fairbanks and colleagues (2001) found that vervet monkeys who quickly approached the potentially dangerous male were also more likely to act aggressively toward him. Thus, the hypothesis that the relationship between low serotonin and aggression is mediated by impaired impulse control is relatively well supported by the primate data.

Serotonin and Sociality

Like humans, individual primates vary in sociality, with some subjects being more gregarious than others. Sociality is often considered a stable personality trait in Old World monkeys, perhaps equivalent to extroversion in humans (Chamove et al., 1972; McGuire et al., 1994; Stevenson-Hinde, Stillwell-Barnes, & Zunz, 1980a; see Chapter 10). Among rhesus macaques, very aggressive males are often avoided by other group members and seldom affiliate with adult females (Chapais, 1986; Mehlman et al., 1995, 1997). Because monkeys with low CSF 5-HIAA concentrations are more aggressive and exhibit impulse control deficits that are likely to interfere with social tolerance and cooperation, it seems reasonable to hypothesize that males with low CSF 5-HIAA concentrations would also show impairments in sociality.

Several studies of nonhuman primates have demonstrated that sociality is correlated with CNS serotonin function. For example, in a study of free-ranging adolescent rhesus macaque males, aggressive individuals with low CSF 5-HIAA concentrations exhibited reduced levels of four measures of sociality: time spent grooming other monkeys, time spent in close proximity to other group members, time spent in general affiliative social behaviors, and mean number of companions within a 5 meter radius (Mehlman et al., 1995). In the laboratory, low rates of positive social interactions among rhesus juveniles were also correlated with low CSF 5-HIAA concentrations in both sexes (Higley, King, et al., 1996; Higley, Linnoila, & Suomi, 1994).

The relationship between serotonin and sociality is also supported by experimental evidence For example, across repeated studies of captive vervet monkeys, Raleigh and colleagues found that enhancing serotonin functioning by administering the serotonin precursor tryptophan, the reuptake inhibitor fluoxetine, or the serotonin agonist quipazine decreased aggression while it increased positive social behaviors such as approaching and grooming other monkeys (Raleigh et al., 1980, 1983, 1985). When investigators reduced serotonin functioning by administering the tryptophan hydroxylase enzyme inhibitor PCPA to the monkeys, the inhibitor produced opposite effects. Thus, the monkeys withdrew from other individuals and avoided affiliative interactions (Raleigh et al., 1980, 1985; Raleigh & McGuire, 1990).

Serotonin, Social Competence, and Dominance

Studies of both human (Brown et al., 1982) and nonhuman primates (McGuire et al., 1994; Raleigh et al., 1989) have shown that individuals high in social deviancy, or who are rated as low in competent social behaviors (Kruesi et al., 1990), have relatively low CSF 5-HIAA concentrations. In many primate species, social dominance is acquired and maintained through the formation of affiliative bonds and coalitions with other troop members (Packer & Pusey, 1979; Raleigh & Steklis, 1981; Smuts, 1987; Walters & Seyfarth, 1987). Affiliation and coalition formation require competent social skills, and, given the relationship between serotonin and sociality, it is reasonable to predict that monkeys low in CSF 5-HIAA concentrations would be more likely to be low in social dominance. Indeed, several studies of rhesus macaques and vervet monkeys have reported that naturally occurring low CNS serotonin functioning (Higley, King, et al., 1996; Higley, Linnoila, & Suomi, 1994; Higley, Suomi, & Linnoila, 1996b; Raleigh et al., 1983; Raleigh & McGuire, 1994; Westergaard et al., 1999), or pharmacologically reducing CNS serotonin functioning (Raleigh et al., 1983, 1986), is linked to low social dominance ranking. Some studies of longtail macaques, however, have not found a positive correlation between high rank and impaired CNS serotonin (Botchin et al., 1993; Shively et al., 1995).

Other Behavioral Correlates of Low Serotonin

Aggressive monkeys with low CSF 5-HIAA concentrations also exhibit high levels of stereotypy, perhaps reflecting their irritable natures (Erickson et al., 2001), and dysregulations in daily activity and circadian cycles. For example, juvenile rhesus monkeys with low CSF 5-HIAA concentrations take longer to fall asleep at night and exhibit higher motor activity during the day. In the field, the aggressive macaques with low CSF 5-HIAA concentrations wake up more often and spend more time in motor activity during the night than do their high CSF 5-HIAA conspecifics. They also nap more often during the daytime (Mehlman et al., 2000; Zajicek et al., 1997).

Cholesterol and Aggression

In the late 1980s and early 1990s, as investigators studied cardiovascular disease, an interesting phenomenon emerged. As the data were published from cholesterol-lowering trials, it was clear that high cholesterol was a contributing factor to cardiovascular disease. When rates of death were examined, however, the data showed that overall, men with high cholesterol were not more likely to die than men with low cholesterol. Further analysis of causes of death, however, showed that the underlying reasons for death were quite different in men with high as opposed to low cholesterol. Men with high cholesterol were indeed more likely to die of heart disease, whereas men with low cholesterol were more likely to die of suicide, violence, cancer, and accidents, with suicides accounting for the majority of premature deaths (Muldoon et al., 1993). This finding led to a series of studies designed to assess possible mechanisms producing premature death in men with low cholesterol. A study of longtail macaques showed that individuals that were fed a diet low in cholesterol were significantly more likely to engage in high rates of aggression than were individuals fed a diet high in cholesterol (Kaplan et al., 1991).

A subsequent study showed that monkeys fed the low cholesterol diet were also more likely to have low CSF 5-HIAA concentrations (Kaplan et al., 1994), suggesting that the relation between low cholesterol and aggression may be mediated by the serotonin system. While further studies of primates are needed, the relationship between low cholesterol, anger, hostility, and aggression has been confirmed by a number of human studies (Buydens-Branchey et al., 2000; Golomb et al., 2000; Hillbrand et al., 2000; Richards et al., 2000). Some of these studies have also replicated Kaplan et al.'s (1994) finding of impaired CNS serotonin in monkeys with low cholesterol (Buydens-Branchey et al., 2000; Hibbeln et al., 2000; Steegmans et al., 1996).

Testosterone

A great deal of research investigating the neuroendocrine substrates of violent aggression has focused on the male hormone testosterone. This is largely because in many animal species, aggressive behavior can be drastically reduced or eliminated with surgical or chemical castration and can be reinstated or enhanced with testosterone treatment (Archer, 1991). More recently, however, it has become clear that testosterone is probably not the major determinant of violent aggression. Instead, testosterone probably underlies a drive for competitiveness or an overall aggressive motivation rather than violent behavior itself (Archer, 1991; Buchanan et al., 1992; Christiansen & Knussmann, 1987; Olweus, 1984, 1986). Studies showing a positive correlation between aggression and testosterone often fail to replicate (Archer, 1991; Olweus, 1986), but when significant correlations between testosterone and aggression are found, they most often occur when both are measured in competitive situations, during challenges to social status (Mazur, 1983; Scaramella & Brown, 1978), or in response to provocation or threat (Olweus, 1986; Olweus et al., 1980, 1988). Although a positive correlation between aggression and testosterone has been reported among violent criminals (Archer, 1991; Olweus, 1986), this correlation may be mediated by personality characteristics such as impaired impulse control (see below). Indeed, most individuals with high testosterone are not violent but express their aggressiveness in competitive situations and using methods that are socially acceptable. Moreover, high testosterone is also correlated with other behavioral traits such as toughness (Dabbs et al., 1987), social dominance (Booth et al., 1989; Christiansen & Knussmann,

1987; Ehrenkranz et al., 1974; Lindman et al., 1987), social assertiveness (Lindman et al., 1987), and competitiveness and physical vigor (Booth et al., 1989; Mattsson et al., 1980). Therefore, testosterone appears to be more directly related to motivation for competition and social status than to aggressive and violent behavior per se.

Among seasonal breeding nonhuman primates such as rhesus and Barbary macaques or mandrills, both aggression and testosterone peak during the breeding season (Bernstein et al., 1977; Gordon et al. 1976, 1978; Kaufmann, 1967; Kuester & Paul, 1992; Paul, 1989; Rose, Bernstein, et al., 1978; Wickings & Dixson, 1992). This parallel increase in testosterone and aggression, however, is more likely to reflect increased sexual motivation and competition over females than a direct effect of testosterone on aggression. As in studies of humans, a number of primate studies have found no correlation between aggression and testosterone across individuals (Eaton & Resko, 1974; Gordon et al., 1976; Rose et al., 1972, 1975). Administration of testosterone had no effect on either aggression or acquisition of social dominance rank in talapoin monkeys living in an all-male group (Keverne et al., 1983). In another study of rhesus macaques, adult males received weekly injections of chorionic gonadotropin, which stimulates testosterone production. While there were significant increases in testosterone, there was essentially no effect on aggression (Gordon et al., 1979). By contrast, when longtail macaques were injected with testosterone propionate, rates of aggression increased, but the rate of aggression varied across individual subjects, with higher rates in dominant than in subordinate monkeys (Rejeski et al., 1988). A follow-up study with the same subjects showed that testosterone was associated with an increase in aggression among dominant individuals but reduced aggression among the subordinates (Rejeski et al., 1990). These findings suggest that the relationship between testosterone and aggression is complex and mediated by social dominance rank.

Perhaps the most consistent finding concerning testosterone and aggression is that during competition, winning augments testosterone and losing reduces testosterone. For example, Steklis and colleagues (1985) failed to find a relationship between overall aggression and testosterone, or social dominance and testosterone, among captive vervet monkeys. In that same study, however, on days when males fought, testosterone was higher in the winner of the conflict than in the loser (Steklis et al., 1985). In a study of group formation in captive talapoin monkeys, levels of preexisting testosterone failed to predict which monkeys became dominant; the monkeys who became dominant, however, showed the highest levels of testosterone when the dominance hierarchy became established (Eberhart et al., 1985). In other studies of talapoin monkeys, males who successfully challenged other males and increased in rank showed an increase in testosterone levels (Yodyingyuad et al., 1982), whereas males who were attacked and retreated exhibited a drop in testosterone (Martensz et al., 1987). Similar findings were obtained with rhesus macaques. When rhesus males were placed into a social group, the male who became dominant showed progressive increases in testosterone, while the males who became subordinate showed decreases in testosterone levels (Rose et al., 1975). In other studies of rhesus macaques, defeats in conflicts with other males were accompanied by a fall in testosterone, especially if such defeats resulted in loss in social status (Bernstein et al., 1979; Rose et al., 1972).

Testosterone and aggression are most likely to show a relationship during periods of social instability and competition for social status. For example, Rose and colleagues (1971) found that during the first nine months following group formation in captive rhesus macaques, aggression was correlated with testosterone. Similarly, a study of free-ranging baboons reported that, during a period of social instability, the most dominant males initiated the most fights and had the highest testosterone levels (Sapolsky, 1983). Long-term studies of stable social groups, however, often failed to find a significant correlation between testosterone and aggression, or testosterone and dominance rank (e.g., Eaton & Resko, 1974; Gordon et al., 1976). For example, in a long-term study of hormones and behavior in male mandrills, Wickings and Dixson (1992) found that the adult males who were peripheralized from the social group had lower rank and lower plasma testosterone than the males who were well integrated into the group. Among the males within the group, the male with the highest rank initially had the highest testosterone, but over time, the relationship between dominance and testosterone disappeared.

Sapolsky (1992) suggested that testosterone may be related to aggression during the process of dominance acquisition but that this relationship may disappear among animals that have lived together for more than one year and whose dominance relationships are well established. This hypothesis is supported by a study of longtail macaques in which adult males were paired and followed for eight months (Clarke et al., 1986). About half of the pairs quickly settled their conflicts and established clear and stable dominance relationships. The other half either continued to fight or had reversals in their ranks. In the half with an established, stable dominance hierarchy, there was no relationship between social dominance and testosterone, but in the half that were still acting aggressively, and in which dominance rank was still being aggressively contested, higher-ranking animals had higher testosterone.

The Relation between Testosterone, Serotonin, and Aggression

Primate research suggests that testosterone and serotonin may affect different forms of aggression through different mechanisms. An early study of rhesus macaques by Rose, Gordon, and Bernstein (1978) reported that although both testosterone and mild, noncontact aggression peaked during the breeding season, violent aggression showed no seasonal variability. Moreover, testosterone was positively correlated with day-to-day mild aggression but not with violent aggression. More recently, we reported that, on the one hand, among free-ranging rhesus macaques, CSF free testosterone was positively correlated with overall aggressiveness and with competition for status and overall aggressiveness, but not with measures of impulsivity such as spontaneous long leaps or rates of capture (Higley, Mehlman, Poland, et al., 1996). On the other hand, CSF 5-HIAA was negatively correlated with impulsive behaviors and severe, violent aggression, but not with competitive overall rates of aggression. High rates of impulsive behavior were positively correlated with severe, unrestrained aggression but not overall rates of aggression. Finally, we reported that individuals with both low CSF 5-HIAA and high CSF testosterone had the highest rates of aggression, except for the most severe forms of aggression, which were unrelated to testosterone.

These findings and others obtained in both animals and humans (e.g., Archer, 1991; Olweus et al., 1980, 1988; Soubrié, 1986) suggest that testosterone is associated with mild or competitive aggression, while low serotonin is associated with impulsive, severe aggression. More generally, it is reasonable to hypothesize that testosterone enhances the motivation for competition and aggression, whereas serotonin exerts an inhibitory effect on the expression of aggressive behavior. Put simply, testosterone may provide the push to act competitively, and serotonin may provide the brakes that determine the threshold, timing, and intensity of aggression. Thus, individuals with low testosterone would be unlikely to engage in aggression, regardless of their serotonin levels, but if they had impaired central serotonin functioning, they might act impulsively in other behaviors. By contrast, individuals with above-average testosterone but normal serotonin may express aggression in a variety of settings but would not generally express violence or unrestrained aggression. They would also be expected to exhibit the assertive behaviors that characterize socially dominant males, such as threats, displacements, or mounting of other male monkeys. Individuals with lower-than-average serotonin would be expected to exhibit impaired impulse control, resulting in a low threshold to display aggression. Subjects with low serotonin would also have difficulty stopping the aggression before it escalated into violence. Individuals with both high testosterone and low serotonin would be expected to have high motivation for aggression, a low threshold for the initiation of aggression, and impaired ability to control and end aggression once the behavior has been expressed. Therefore, such individuals would be expected to show high rates of aggression and violent behavior.

The hypothesized relationship between testosterone, serotonin, and aggression can be further tested in human and nonhuman primates as well as in other animals. Most research on the neurobiological substrates of primate aggression has concentrated on a few species, and among humans, almost exclusively on males. Therefore, it would be important to assess whether the findings of this research can be extended to other species and be applicable to both sexes. Future research on the biological substrates of aggression should also pay more attention to the experiential and environmental factors that contribute to the expression of aggressive behavior and its developmental changes across the life span. Given the close phylogenetic relationship between human and nonhuman primates and the many potential similarities in the causation, development, and function of aggression in primate and human societies, future efforts in this area of research should be aimed at developing theories of aggression that incorporate both biological and environmental variables and that have broad validity and applicability both within and across species.

3

Conflict Resolution

Peter G. Judge

Conflicts can be as subtle as two children disagreeing over the color of crayon to use in a drawing or two monkeys jockeying for the same sunny spot in a tree. Alternatively, conflicts may be as involved and intense as human global wars or the lethal "warfare" seen between males of different chimpanzee communities (Boehm, 1992). Broadly defined, a conflict occurs whenever individuals interact with incompatible objectives. Since most primates are highly social and interact often, conflicts are an inevitable aspect of their daily life. Therefore, long-term relationships within primate societies are characterized by a continuous cycle of conflict, resolution, and aftermath until the next inevitable conflict.

Strategies for resolving conflicts can range from blunt force in which one party prevails over another to complex negotiations in which both parties benefit mutually. Conflict and resolution are not necessarily conspicuous events, however. Conflict can be defined as any incompatibility over objectives. Resolution reduces or eliminates the incompatibility. Conflicts are not necessarily aggressive, and most conflict resolution involves simple acquiescence. The aftermath of conflict resolution is also of interest because the type of conflict and the means of its resolution can influence future interactions between individuals and the quality of their relationship. For example, the aftermath of conflict can include elaborate forms of appeasement and reconciliation.

Although human social interactions appear to be more complex than those of other primates, human and nonhuman primates often have conflicts of interest over similar objectives. Before attempting comparisons between human and nonhuman primates, we must examine how different perspectives attempt to explain the occurrence of conflict resolution. These perspectives range from ultimate explanations of conflict resolution to more proximal explanations of behavior. For example, an evolutionary perspective emphasizes the adaptive value and the evolution of conflict resolution. A cognitive-developmental perspective views changes in conflict resolution behavior over time in relation to learning within an individual's lifetime. A social process perspective examines each step in the sequence of the social interactions involved in conflict and conflict resolution and attempts to describe and explain the patterns of behavior that constitute such interactions.

Perspectives on Conflict Resolution

An evolutionary perspective emphasizes the functional significance of behavior and the contribution it makes to an individual's reproductive success. Any behavior has both fitness costs and benefits to an individual, and natural selection acts to maximize the difference between the benefits and the costs. If two individuals can resolve a conflict without excessive costs or injury, they both benefit. For example, it is not beneficial to persist in an aggressive conflict if one will continue to lose. Likewise, the winner of a conflict should not risk injury by pressing an advantage past the point where an adversary has withdrawn. Mechanisms that allow animals to resolve conflicts to their mutual benefit would be favored through natural selection (Silk, 1998).

Most primates live in stable social groups and are thought to derive benefits from group living such as protection from predators and increased competitive ability over neighboring groups. By defraying the costs of conflict and using successful resolution strategies, primates offset the costs of competition within the group. The increased dependence on others and the need for co-existence in social groups promote resolution that is not costly to the individual. In the many species in which close kin live in the same group, the benefits of effective conflict resolution are magnified. If animals reduce the cost of conflict with kin through effective resolution strategies, those strategies are selected for through kin selection. Even among non-kin, nurturing a valuable relationship with a partner through reciprocity and exchange increases an animal's competitive ability within its group (Nöe et al., 1991). Such cooperative long-term relationships require competent conflict resolution skills.

A cognitive-developmental perspective emphasizes adaptive social functioning and the developmental changes that occur as individuals learn to adjust to conflict-producing situations (Cords & Killen, 1998). Strategies develop as individuals learn the complexities of interpersonal interactions and acquire the competence to deal with them. For example, Laursen (1993) found that children's conflict resolution strategies changed as a function of their age. Young children (two to ten years) used coercive strategies designed to produce one-sided outcomes, whereas older children used more negotiation to resolve conflicts. Some developmental changes in conflict resolution strategies are associated with the acquisition of linguistic skills and the development of abstract moral reasoning (Hakvoort & Oppenheimer, 1993; Killen & de Waal, 2000; Osterman et al., 1997; Potegal & Davidson, 1997). The development of conflict resolution skills is also influenced by third parties such as parents, peers, siblings, and teachers (Grammer, 1992; Ross et al., 1990; Siddiqui & Ross, 1999).

The development of conflict resolution strategies in nonhuman primates has been poorly studied, but primates appear to acquire conflict resolution skills at young ages (Cords, 1988; Cords & Aureli, 1993; Judge et al., 1997; Schino et al., 1998; Weaver & de Waal, 2000). Although developmental changes in conflict resolution strategies over time have not been described in detail, young monkeys have been shown to change their conflict resolution behavior based on previous interactions with others (Cords & Thurnheer, 1993). Young monkeys are also influenced by observing third parties. For example, young rhesus macaques housed with stumptail macaques appeared to adopt the stumptail macaques' characteristic tendency to resolve conflicts through friendly behavior, a strategy uncharacteristic of rhesus macaques (de Waal & Johanowicz, 1993). Such changes due to social exposure indicate that primates probably do learn to resolve conflicts from third parties or their own experience. This aspect of nonhuman primate development is relatively unstudied, however.

Differences in cognitive skills can account for some developmental changes in conflict resolution strategies as well as for differences between human and nonhuman primates. For example, the ability to take another individual's perspective and consider that individual's needs and desires may be expected to play an important role in the development and use of effective conflict resolution strategies. Young children develop the ability to take another individual's perspective after three years of age (Povinelli & deBlois, 1992), and this change is likely to affect their conflict resolution skills and strategies. There is some debate over whether this ability occurs at all in nonhuman primates, or if it occurs in apes but not in monkeys (e.g., Hare et al., 2001). Thus, conflict resolution in human and nonhuman primates may or may not be associated with similar cognitive processes.

A third approach to studying conflict resolution is a social process perspective. This descriptive approach attempts to identify the structure of social conflict and its constitutive elements such as the identity of the individuals involved, the reason for their conflict of interest, their strategies of conflict resolution, the outcomes of the elected strategy, and the aftermath of resolution. The social process perspective emphasizes the issue of conflict management or the steps that can be taken to prevent the initial conflict of interest from occurring. Conflict and its resolution can be depicted as a process model (Figure 3.1). This model is a simplified version of a model formulated to describe human conflict (e.g., Fry, 2000). The simplification eliminates complex human third-party interactions, such as legal arbitration, that cannot occur in nonhuman primates. Using this perspective, conflict resolution can be understood in terms of the factors that contribute to each component of the model and how they interact to produce particular outcomes.

Similar models have been applied to particular stages of a conflict in nonhuman primates. For example, de Waal (1996a) proposed a "Relational Model" that describes the aftermath of conflicts resolved through aggressive coercion. In the aftermath of aggression, conflicts may be "reconciled" through affiliative responses between former adversaries. The model predicts the likelihood of reconciliation based on the characteristics of the individuals involved in the conflict, such as the quality of their relationship, and the immediate costs and benefits of the conflict to each individual. A similar model proposed by Aureli, Cords, and van Schaik (2002) also uses the value of a relationship and the costs and benefits of renewed aggression to predict the likelihood of reconciliation following aggression in nonhuman primates. Thus, process models of conflict resolution attempt to describe relationships between the elements of a conflict and understand the proximal variables that influence the resolution process.

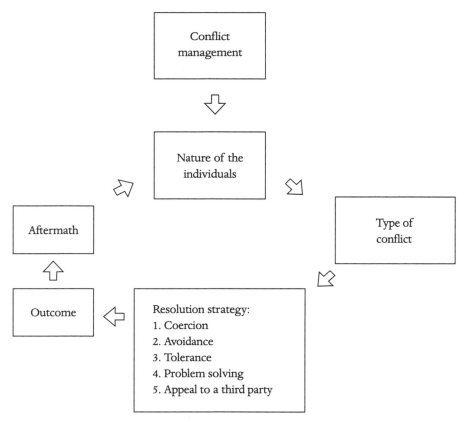

Figure 3.1 A "process model" depicting the factors involved in dyadic conflict resolution and the circular nature of conflicts.

In the comparative discussion of conflict resolution in human and nonhuman primates, I emphasize the social process perspective because most research in this area has not yet extended much beyond a descriptive phase. Whenever possible, however, I also discuss evolutionary and cognitive aspects of conflict resolution. For example, when primate studies show that there are marked interspecific differences in the rates of post-conflict friendly reunions (Aureli & de Waal, 2000b), it is important to recognize that such differences probably reflect adaptations to the social and ecological environment that evolved through natural selection. Furthermore, when describing conflict resolution behavior that appears similar in human and nonhuman primates, one must consider that the cognitive processes underlying the interaction may differ considerably.

A Process Model: The Structure of Conflict

In this section I attempt to apply primate data and comparable human data to a simple process model (Figure 3.1). But first, each step must be explained in some detail. "Nature of the individuals" refers to specific qualities that may affect interactions with others. The characteristics of individuals involved in the conflict will dictate the conflict and its resolution. These include variable qualities inherent in the individual (i.e., personality) that may influence the individual's reaction to a conflict situation. As shown in Chapter 10, personality plays an important role in the social interactions of nonhuman primates. Species-typical differences in the social environment will also influence the pattern of conflict resolution between individuals. For example, species characterized by rigid, strictly enforced dominance hierarchies resolve conflicts differently than do species with more relaxed dominance systems (Aureli et al., 1997). The past history of interactions leading up to a particular conflict can also influence resolution strategies. Some relationships are more highly valued than others, and they are influenced differently by conflict. The nature of the individuals involved in the conflict expands to include third parties as well. A third party may be a partisan of one of the individuals involved in the conflict and thus influence outcomes. Finally, the context of a conflict can be assessed in terms of the item being contested and the social and physical environment in which the conflict occurs. Contested items can be material (e.g., food, water, space, or social partners) or intangible (e.g., social status, statements of fact, ideological differences).

I have classified conflict resolution strategies into five general types based on discussions by Fry (2000), Rubin and colleagues (1994), and Schellenberg (1996). "Coercion" occurs when one party attempts to resolve the conflict by imposing its wishes on another. A clear winner and loser are identified, and one party acquires the object of conflict and the other yields the object. This type of interaction is a "zero-sum" outcome because the winner receives benefits (+), the loser incurs an equal cost (−), and the sum is zero. Aggression is typically associated with the coercive resolution strategy, but coercion is relatively rare. In most cases one party simply complies with the other. Coercion as a conflict resolution strategy has also been referred to as a "unilateral" solution in which only one party benefits, as opposed to a "bilateral" solution in which both parties benefit (Fry, 2000).

An "avoidance" strategy occurs when a conflict of interest arises between two individuals but both withdraw from the conflict. There is no attempt at resolution by either party, and conflicts are "resolved" by effectively denying that a conflict occurred. "Tolerance" is a similar strategy in that the conflict remains unresolved, but in this case attempts are made by both parties to resolve the conflict. Tolerance occurs when both parties agree that the conflict cannot be resolved and "agree to disagree."

The "problem solving" strategy involves active participation by both sides to come to an agreement over the conflict. Such bilateral solutions (Fry, 2000) are positive-sum outcomes because each party benefits, at least to some degree. Examples of problem solving include compromise and negotiation in which both parties agree to settle for a portion of the desired objective or exchange one commodity for another.

A fifth strategy of conflict resolution involves intercession by a third party. A third party can coerce one or both of the conflicting individuals, promote avoidance through distraction or separation, aid in achieving tolerance, and join in problem solving. The influence of third parties may vary as a function of whether they are partisan or not and the degree of their intervention. The involvement of authorities to varying degrees (mediation, arbitration, adjudication) gives human third-party interaction several levels of complexity that are not observed in nonhuman primates.

The outcome of a conflict is defined here as the distribution of the contested item. Simply put, the item can be distributed to one party (coercion), both parties (problem solving), or neither party (avoidance and tolerance). The trichotomy is easily understood when the contested item is an object such as food, but the distinction may also be applied to abstract entities such as status or ideas. For example, in an argument over ideas, one party may convince the other party, both may accept the other's point of view, or each may remain unconvinced by the other. In some studies, an "outcome" is defined not by the distribution of the contested item but by the effect of the conflict on the interaction status of the two contesting parties. Outcomes would therefore be classified according to whether the participants remain interacting or apart (e.g., Sackin & Thelen, 1984). In the model described here, this aspect of the conflict would be considered part of the aftermath.

The aftermath includes the effects of the conflict and its resolution on the behavior of and the relationship between the conflicting parties. The aftermath of a conflict can influence the nature of the dyadic relationship and thus the next ensuing conflict in a circular process. Conflicts in which aggression is used as a resolution strategy provide interesting aftermaths because aggressive coercion effectively produces an additional conflict in the social relationship between the opposing parties. For example, a conflict over an item of food that is resolved through aggressive coercion can produce a new conflict that threatens the social relationship.

Conflict management can be considered differently from conflict resolution. Whereas conflict resolution occurs at the time of a conflict, conflict management involves behavior between individuals before a conflict occurs that may reduce or preclude conflict. Although it is relatively unstudied in nonhuman primates, monkeys and apes do appear to use conflict management mechanisms. Even though this chapter mostly focuses on conflict resolution, I offer a brief treatment of conflict management in nonhuman primates.

The model outlined above provides a basic framework with which to compare conflict resolution in human and nonhuman primates. The model views conflict and conflict resolution as the outcome of characteristics of individuals, those of their social interactions, and those of the social environment in which they live (Schellenberg, 1996). Given the variety of primate social systems, social structure is expected to play an important role in conflict resolution. For example, in gorilla societies, females tend to avoid conflicts and depend on the silverback male in their group to mediate conflicts (Watts, 1994). Resolution strategies in this species would be expected to differ from those in macaque societies, in which there is a complex network of female-female relationships mostly independent of male influences.

Considerations for Applying the Model to Both Humans and Other Primates

The process model of conflict resolution assumes that the nature of the relationship between individuals has the potential to influence resolution strategies and that the outcome and aftermath of a conflict influence later interactions. Primate societies are characterized by long-term associations between individuals who recognize one another and who form social relationships in which earlier interactions are remembered and influence later interactions. These have been described as the minimum attributes necessary for reconciling aggression (de Waal & van Roosmalen, 1979), but they can also be considered minimal criteria for application of the model.

There are at least four areas in which the scope of human-nonhuman comparison is strictly limited, however. First is the arena of interaction. Human primates interact on a vast scale compared to nonhuman primates. With increasing levels of complexity, humans interact at an interpersonal level, a community level, an institutional level, and an international level (Schellenberg, 1996). At the community or intergroup level and beyond, the sphere of human interaction becomes immensely different from that of non-human primates. Using written and oral agreements, humans have developed a system of social institutions for the expression, management, and resolution of conflict between individuals and groups that has no parallel in nonhuman primates. Much of the complexity involves resolution by third parties in disputes established by evolving laws (Yarn, 2000). Nonhuman primates exhibit a rich array of triadic interactions for resolving conflicts, but they do not compare to the multilayered aspects of human third-party interaction (see Fry, 2000). Nonhuman primate interactions are restricted to mostly interpersonal relationships within current groups and intergroup interactions with immediate neighbors (Cheney, 1987).

As mentioned above, cognitive differences may limit the types of resolution strategies that nonhuman primates can employ. The "problem-solving" strategy in which positive-sum outcomes are attained through negotiation and compromise may necessitate a "theory of mind" in which one individual can take the point of view of another and consider the other party's interests. Current evidence suggests that only great apes may possess this ability, and only in a rudimentary form (e.g., Hare et al., 2001; see Chapter 10). Primates may also be limited in their ability to remember interactions with many individuals. For example, studies of group fission suggest that groups are likely to split apart when individuals can no longer effectively track relationships with all members of their group (Henzi, Lycett, & Piper, 1997).

Human language as a medium of communication produces substantial differences in the potential for conflict resolution in human and nonhuman primates. Most human conflicts, even those of young children, are resolved through verbal exchange and rarely become overtly aggressive (Eisenberg & Garvey, 1981; Shantz, 1987). That many human conflicts arise over simple statements of fact highlights how pervasive language is in the interpersonal conflicts of our species. Use of language produces different strategic levels of negotiation and compromise that are not available to other primates (Eisenberg & Garvey, 1981). Language allows negotiation without the more physical gestures that nonhuman primates must use, and these gestures may promote escalation to aggression. Humans also use verbal apologies in the aftermath of conflicts to regulate relationships (Ohbuchi et al., 1989). There is evidence that nonhuman primates use vocalizations to manage or reconcile conflicts (Cheney & Seyfarth, 1997; Cheney et al., 1995), but the role of signals in conflict resolution has not been studied systematically.

Finally, cross-cultural differences in conflict resolution occur among human populations that may not have parallels in nonhuman primates (e.g., Barnland & Yoshioko, 1990; Butovskaya et al., 2000; Fry & Bjorkqvist, 1997; Fry & Fry, 1997). Nonhuman primates are thought to exhibit cultural differences in behavior, particularly with respect to tool use (e.g., Wrangham et al., 1994), but cultural differences in conflict resolution behavior across populations have not been documented. Different groups of the same species do exhibit differences in conflict resolution behavior (Castles et al., 1996; de Waal, 1996a), but these differences can be attributed to demographic factors rather than culture.

The above differences between human and nonhuman primates restrict the comparison of conflict resolution process to the level of interpersonal interaction. Additionally, given the differences in communicative and cognitive processes, the most valid avenue for comparison may be an ethological approach that emphasizes objective observation of behavior. Unfortunately, this approach excludes much of the research on human conflict resolution because many studies involve interviews and assessment scales describing the motivations, emotions, or thought processes that occur in association with conflict situations. Many studies also arrange experimental situations in which subjects are evaluated during staged verbal disagreements (Nelson & Aboud, 1985), a technique that relies on language. An ethological approach also by necessity includes mostly studies of human children because most objective investigations of interpersonal conflict in humans have been conducted on children (see reviews by Cords & Killen, 1998; Shantz, 1987; Verbeek et al., 2000).

The Process of Human and Nonhuman
Primate Conflict Resolution

Ideally, to understand the evolution and use of conflict resolution strategies in the Primate order, one should quantify how often each of the five strategies (coercion, avoidance, tolerance, problem solving, and third-party involvement) was used in a variety of conflict types across many primate species and classify the events that occur in the aftermath. The information necessary to accomplish this task, however, is largely unavailable. For example, the relative use of the five strategies has rarely been documented in nonhuman primates. An avoidance strategy, by definition, is difficult to measure since no observable interaction may occur. Furthermore, the context of a conflict is not always recorded and is often unknown in nonhuman primates. Although conflict resolution has been examined in many species, few studies attempt to examine each stage of the process. Studies of the aftermath of aggressive conflicts provide the most data, undoubtedly because aggression is a conspicuous behavioral marker from which to track subsequent events. The study of affiliative behavior in the aftermath of aggressive conflicts (i.e., reconciliation) is perhaps the closest approximation to a standardized comparative approach, since many studies have been conducted in this area using the same methodology.

In the next section, I review and discuss some of the human and nonhuman primate data on conflict resolution in relation to each stage of the process model described above. Any similarities between primate and human behavior are not necessarily interpreted as homologies, although this possibility is addressed in the concluding remarks. I start by describing the types of resolution strategies in more detail to form a basis from which to examine the other stages of the model.

Conflict Resolution Strategies

Conflict resolution strategies are classified according to whether one party receives all the benefits of the contest (coercive, unilateral, or zero-sum outcomes), neither party receives benefits in the contest (avoidance and tolerance), or both parties receive at least some benefit during the contest (problem solving, bilateral, or positive-sum outcomes). In the case of problem solving, the benefits need not be distributed equally. Lopsided outcomes are positive-sum if one party still benefits in some small way. Classifying conflicts into one of these three categories is sometimes difficult because the benefits gained in a conflict cannot always be measured simply as commodities lost or gained.

Coercion

A coercive strategy resulting in a zero-sum outcome (i.e., winner-loser) is a common type of conflict resolution in nonhuman primates, probably because most primate societies are organized around an established dominance hierarchy. Once an individual learns to cede to another during conflicts of interest, the pattern of submission occurs consistently in all subsequent conflicts. Typically, one animal simply allows the other animal to obtain the item without any overt aggression by the dominant animal. Even if aggression is used to resolve the conflict, it often involves mild threats that do not escalate into conspicuous full-scale fighting. It is important to emphasize that coercive resolution does not require aggression; all that is necessary is that one of the two parties involved in the conflict of interest gain all of the contested item. In humans, affectionate cajoling can lead to coercive zero-sum outcomes.

Rhesus macaques are arguably one of the most belligerent species in the Primate order, yet they illustrate the principle that coercion is not necessarily aggressive. In approximately 58 percent of agonistic interactions in captive groups, one animal simply submits or avoids a dominant animal without threats or aggression. Of the remaining 42 percent of agonistic interactions, 31 percent are noncontact aggression that include behavior such as threats. Only 11 percent of agonistic interactions include physical contact such as bites (Bernstein & Ehardt, 1985b). If one considers that all these agonistic interactions account for only about 2 percent of a rhesus macaque's daily activity budget (Bernstein et al., 1993b), then overtly aggressive coercion is a fairly infrequent event. Nevertheless, aggressive coercion by rhesus macaques is a fairly successful strategy: 93 percent of aggressive interactions produce a clear winner, with only about 7 percent of aggression being contested or ignored (Bernstein & Ehardt, 1985b).

The pattern of conflict resolution observed in rhesus macaques is not universal, however. Among female gorillas, aggressive coercion is often used as a strategy to resolve conflicts, but the strategy is rarely successful (Watts, 1994): 82 percent of aggression is followed by counteraggression or is ignored, while only 18 percent of aggressive conflicts produced a clear winner (Watts, 1994). In some primate species coercion is almost nonexistent. Strier (1992) observed only thirty-one agonistic interactions in over 1,200 hours within a group of woolly spider monkeys, and these were mostly displacements (71 percent). Strier concluded that this species actively manages conflicts by spreading out while feeding to avoid competition. In addition, males maintain such close relationships that they do not contest mates.

Humans, of course, frequently resolve conflicts with zero-sum outcomes in which one individual dominates another. Children under two years old form dominance relationships with peers and begin to show consistent dominance outcomes with particular partners (Hay & Ross, 1982). By the time they are in preschool, children may form linear dominance hierarchies in which coercive strategies are usually successful (Strayer & Strayer, 1976). Preschoolers in other studies were found to use a yielding submissive strategy in from 66 percent (Sackin & Thelen, 1984) to over 78 percent (Vespo & Caplan, 1993) of their conflicts.

The rate at which children use aggression to resolve conflicts appears to be fairly low, but surprisingly, general rates are rarely reported, and methods used to categorize aggressive resolutions are extremely diverse. Hay and Ross (1982) reported that forcible contact was used in 16.5 percent of struggles over objects in toddlers. Object struggles, however, lend themselves to aggressive resolutions. The authors did not report aggression rates for conflicts in other contexts. Killen and Turiel (1991) report physical harm as the source of conflict in 3 percent of structured peer group interactions and 29 percent of free play interactions in children, but the definitions used make it unclear whether these interactions were actually the source of the conflict or coercive resolutions to conflicts. In a study of children's verbal exchanges, Eisenberg and Garvey (1981) report that only 1 percent of conflicts entail physical force. When agonism is used as part of a coercive strategy, however, rates of contact aggression can be rather high. In preschoolers, 53 percent of agonistic conflicts involved physical attack, while 47 percent included threats or noncontact aggression (Strayer & Strayer, 1976). Of these cases, 21 percent involved counterattack.

Avoidance and Tolerance

As children age, they shift from overt aggression to less physical means of coercion (Strayer, 1992). Children also shift from coercive strategies to using avoidance and tolerance in resolving conflicts. In avoidance and tolerance, a conflict of interest arises between two individuals but there is no resolution. In avoidance there is no attempt at resolution, while in tolerance there is an unsuccessful attempt at resolution. The two strategies are grouped together here because they are usually not distinguished in the literature. Young siblings (two to six years) end 74 percent of their conflicts with "no resolution" and have coercive submissive outcomes in 20 percent of their conflicts (Siddiqui & Ross, 1999). Conflicts between children and their parents follow the same general pattern, indicating that "no resolution" is also a common resolution strategy for adults. For example, most conflicts between mothers and preschoolers typically end without resolution (Eisenberg, 1992). Adolescent conflicts with parents also follow this general pattern (Montemayor & Hanson, 1985). Nonhuman primates, by contrast, tend not to avoid resolution or tolerate non-resolution once a conflict arises. As noted above for rhesus monkeys, "ties" in aggressive disputes are rather rare, and most species that establish dominance hierarchies strictly enforce the status quo during conflicts of interest.

Problem Solving

Problem-solving strategies result in positive-sum outcomes in which both parties gain some portion of the contested item. The strategy usually involves some form of negotiation or compromise. Children begin using problem-solving strategies before they are three years old, but this strategy accounts for less than 3 percent of conflict resolutions (Siddiqui & Ross, 1999). Studies of adolescents indicate that use of the problem-solving strategy increases with age and may be utilized up to 45 percent of the time to resolve conflicts with close peers (Laursen, 1993).

Nonhuman primates diverge markedly from humans in the use of problem-solving strategies. It is difficult to provide examples of positive-sum outcomes beyond anecdotal reports, and there is no accumulated body of evidence to suggest that this strategy is used by nonhuman primates. Perhaps one example is the sharing of food by chimpanzees in response to begging (Goodall, 1986). Two individuals with incompatible interests interact, and the conflict is resolved with both parties obtaining a portion of the desired item. One reason why problem solving may not be widespread in nonhuman primates is that an individual must be able to consider the needs of another to bring about a positive-sum outcome. There is debate whether nonhuman primates are able to take another's perspective, and if they can, it may be in only a rudimentary form (e.g., Hare et al., 2000, 2001; see also Chapter 9). Some degree of empathy with others would also promote problem-solving resolutions, but whether nonhuman primates are capable of empathy is also being debated (de Waal & Aureli, 1996; see also Chapter 11).

"Negotiations" have been reported in nonhuman primates (Colmenares, 1991; de Waal, 1996a), but these seem to represent a gray area that combines differing degrees of tolerance with zero-sum coercive outcomes. For example, de Waal (1996a) describes a dominance dispute between chimpanzees in which a dominant animal occasionally ceded to the objectives of the other. De Waal suggests that "each party gives in to some of the demands or objectives of the other so as to make room for a relationship that in the long run benefits both" (p. 169). Thus, conflict resolution is assessed not on the basis of single win-loss outcomes but on an accumulation of concessions that may produce a cooperative relationship (Nöe et al., 1991).

Involvement of a Third Party

Conflict resolution involving third parties occurs when other individuals join the conflict to implement or augment any of the four resolution strategies mentioned above. At an interpersonal level, involvement can be partisan or impartial, aggressive or peaceful, selfish or altruistic. A third party can choose to join the conflict or be enlisted by one of the conflicting parties. Third-party interactions are so complex that patterns representing most combinations of these four dimensions probably occur in both human and nonhuman primates. Rather than attempt to examine all of these possibilities, I briefly review the most common types of third-party interventions observed in human and nonhuman primates and then discuss a few less common patterns.

Third-party interventions in conflicts by nonhuman primates are likely to involve agonistic coercion and are not likely to involve problem solving. A common pattern of intervention in nonhuman primate disputes is to aid a higher-ranking aggressor against a lower-ranking target. Aid may be provided to reinforce the dominance status of the aggressor or to maintain and reinforce the third party's own status. A second common pattern is to attempt to defend a victim against an aggressor. These patterns of aggressive intervention have been described in detail in nonhuman primates (e.g., Chapais, 1995; Ehardt & Bernstein, 1992; Harcourt & de Waal, 1992; Widdig et al., 2000), and similar intervention patterns also occur spontaneously among preschool children (see Grammer, 1992, and Strayer & Noel, 1986, for detailed analyses).

Since most nonhuman primate interventions are partisan affairs that are motivated by an actor's own interest or that of its kin, impartial agonistic interventions are interesting cases. Typically, in such situations, a high-ranking animal will disrupt an ongoing aggressive conflict and terminate a fight (Ren et al., 1991; Watts, 1997). Such interactions have been called male policing since they usually involve dominant males (Watts et al., 2000). For example, a dominant male gorilla may intervene in fights between the females in his group. A proposed function of these interventions is to prevent conflicts from escalating to the point where a female might leave the group, thus diminishing male mating opportunities (Watts et al., 2000). Parent interventions into sibling conflicts have been shown to serve similar functions (Perlman & Ross, 1997). Parents tended to intervene in fights that were prone to escalation, and parent involvement promoted more positive forms of behavior between the children.

Nonhuman primates also perform peaceful interventions in which a third party intervenes without using aggression (Petit & Thierry, 2000). For example, gorillas interpose themselves between two antagonists (Sicotte, 1995), and individuals in a variety of species direct affiliative and appeasement behavior toward one of the antagonists during ongoing aggression (de Waal, 1988; Petit & Thierry, 1994; Petit et al., 1997; Ren et al., 1991). Interestingly, Petit and Thierry (2000) found that peaceful interventions were more effective than aggressive interventions in terminating fights. They suggested that nonaggressive intervention was a means for an animal to protect a victim from harm without simultaneously jeopardizing the social relationship with the animal performing the aggression. De Waal (1982) reports a case of peaceful intervention in chimpanzees in which the adult females in a captive group persistently succeeded in bringing two feuding males into closer proximity until the males resolved their conflict. He suggested that the females were motivated by a desire to keep the peace within their group. This interaction is thus a case of true mediation as defined in humans in which a neutral third party attempts to bring conflicting parties together to force an agreement (Fry, 2000).

Another tactic for resolving ongoing aggression that involves third parties is redirection of aggression. Redirection occurs when the recipient of aggression responds by attacking a third party. Redirection can resolve conflicts because it sometimes succeeds in diverting the attention of the aggressor away from the original recipient and toward the new target (Aureli & van Schaik, 1991a; de Waal & van Hooff, 1981). In a special case of redirection, macaques were shown to redirect aggression against the kin of their attackers (Aureli et al., 1992; Judge, 1982). This kin-based revenge may inhibit the original aggressor from initiating future attacks. Redirection also occurs in human children, accounting for 23 percent of triadic conflicts in preschoolers (Butovskaya & Kozintsev, 1999; Straycr & Noel, 1986).

Nature of the Individuals

The characteristics of an individual or the quality of a relationship between two individuals can influence the process of conflict resolution.

Individual Characteristics

Differences in personality traits such as hostility, ease to anger, and social competence can influence the course and outcome of a conflict. Children less than two years old already exhibit personality traits that influence the initiation and termination of their conflicts (Hay & Ross, 1982). Young monkeys also exhibit traits that influence their tendency to engage in social interactions (Suomi, 1997). Sex differences are evident in the conflict resolution strategies of children (Hartup et al., 1993; Osterman et al., 1997; Verbeek & de Waal, 2001) and monkeys (de Waal, 1984), and social status has been shown to influence children's conflict resolution styles as well (Bryant, 1992; Putallaz & Sheppard, 1990).

Nonhuman primates exhibit a varied potential for interpersonal interaction because group size and stability differ across species. In species such as tamarins and marmosets groups may be rather small, containing just a mated pair and their offspring (Goldizen, 1987), while in baboons and macaques groups might contain up to one hundred potential partners (Melnick & Pearl, 1987). Groups may also remain very stable over time or undergo regular change in membership. In "female-bonded" species, females and their off-spring typically spend their entire lives in the group into which they were born (Wrangham, 1980). Within these groups, most interpersonal interactions occur among close maternal kin and a subset of preferred non-kin individuals (Silk et al., 1999). Thus, female interpersonal relationships involve a relatively fixed number of companions. At the other extreme are fission-fusion societies in which membership of males and females changes regularly (e.g., McFarland Symington, 1990; Pope, 1998) and animals routinely form new relationships with strangers. In a way, the species differences in nonhuman primates mimic the diversity of interpersonal interactions within human societies, in which interactions can range from small, relatively closed groups to extensive networks of close kin, friends, acquaintances, and strangers.

Much variation in primate social systems may be the result of selective pressures involving food distribution, predator defense, and male mating strategies (Hill & Lee, 1998; Sterck et al., 1997; van Schaik & van Hooff, 1983; Wrangham, 1980). The level of food competition within and between groups may also influence the nature of social relationships among individuals (van Schaik & van Hooff, 1983). For example, species that evolved under highly competitive feeding conditions might develop rather despotic patterns of social interaction characterized by aggressiveness and rigid dominance hierarchies. Under such highly competitive conditions, animals might also have evolved nepotistic behavior that promotes the survival of kin. Species evolving in less competitive ecological environments may have developed more egalitarian, less aggressive, and more tolerant styles of interpersonal relationships that do not rely on kin support to achieve a competitive advantage (van Schaik & van Hooff, 1983).

Thus, some primate species are inherently more antagonistic and more likely than others to use aggressive strategies of conflict resolution. Rhesus and stumptail macaques are an excellent example of this phenomenon. The two species are closely related and sympatric, yet rhesus macaques have evolved a despotic and intolerant style of social interaction that includes strict enforcement of an established dominance hierarchy, while stumptail macaques have evolved a more egalitarian and tolerant mode of social existence (de Waal, 1989b). Accordingly, rhesus macaques are more likely than stumptail macaques to use coercive aggressive solutions during conflicts of interest and less likely to reconcile aggression (de Waal & Ren, 1988). Parallel differences are also found between closely related species of lemurs (Pereira & Kappeler, 2000).

Relationship Quality

Many studies have shown that the quality of a dyadic relationship is likely to influence conflict and its resolution in both human and nonhuman primates. The quality of a relationship can be measured in many ways, but in human studies investigators typically contrast relationships between friends versus non-friends. Acknowledging that a friendship relationship develops over a long period and is itself a product of a series of resolved or unresolved conflicts, we can look for differences in conflict resolution strategies between established friends and non-friends (see review by Hartup, 1992). Hartup concluded that in open "free play" situations, friends engaged in conflicts as often as non-friends, but that young nursery school–aged friends tended to resolve conflicts differently than non-friends. Specifically, friends resolved conflicts by discontinuing the conflict sooner and had more positive-sum outcomes versus zero-sum winner-loser outcomes than non-friends (Hartup et al., 1988). Friends were also more likely than non-friends to continue interacting following conflicts (see also Fonzi et al., 1997). In other studies of both children and adolescents, compared to non-friends, friends used more negotiation and compromise as their conflict resolution strategies (Adams & Laursen, 2001; Laursen, 1993).

Conflict resolution in nonhuman primates is also influenced by quality of relationship (see reviews by Cords & Aureli, 2000; van Schaik & Aureli, 2000). A high-quality relationship in primates typically includes some combination of high levels of friendly affiliative behavior, relatively low levels of aggression, and a high frequency of mutual aid in fights. Since kin in primate groups typically conform to this general pattern, it is often convenient to classify kin as having high-quality relationships in order to test the influence of this variable. Some non-kin, however, also have high-quality relationships.

Third-party interventions provide excellent examples of the influence of relationship quality on nonhuman primate conflict resolution. Kin are likely to come to the aid of their relatives by either joining them in attacking others or defending them from attackers (e.g., Bernstein & Ehardt, 1985a; Kaplan, 1977b; Massey, 1977; Prud'homme & Chapais, 1993). The strong influence of high-quality kin relationships is particularly evident when an animal defends its relative against a higher-ranking aggressor. In despotic species such as rhesus macaques, attacks contrary to the established dominance hierarchy are met with fierce reprisals, yet defending a relative against a high-ranking aggressor was found to be the most common type of triadic intervention in this species (Bernstein & Ehardt, 1985a).

Context of the Conflict

In nonhuman primates, most conflicts appear to arise over material resources such as food, water, space, or mates. Other sources of conflict include dominance status and weaning. It has also been argued that conflicts occur owing to infractions of societal rules or moral codes (de Waal, 1996b). The context of aggression in primate societies, however, is often "unknown," since many conflicts appear to erupt without apparent cause. Many of these conflicts may result from violations of established patterns of social interaction that human observers may not understand.

Within the framework of the process model, it is useful to know the source of a conflict because different items of contention can lead to different resolution strategies even between the same two individuals. In some species, fights over food are not as often reconciled as other fights (Aureli, 1992; Castles & Whiten, 1998a; Matsumura, 1996; Verbeek & de Waal, 1997), probably because food competition is a routine part of daily life that does not seriously threaten social relationships (Aureli, 1992). Serious challenges to the status quo, however, may be dealt with quite harshly. Apparently, a young adult male chimpanzee was killed by the other males in his group because he flouted mating privileges reserved for older, higher-ranking males (Fawcett & Muhumuza, 2000).

The types of children's conflicts are similar to those of nonhuman primates, with most involving the possession and use of objects (Blurton Jones, 1967; Cords & Killen, 1998; Killen & Turiel, 1991). For example, Hay and Ross (1982) found that 88 percent of toddler disputes were over objects. It is important to note, however, that at this young age some conflicts (12 percent) were already being used to establish interpersonal relationships. The nature of human disputes changes quickly as children age and become more socially motivated. The aggressive conflicts of middle school children rarely involved object disputes (2.5 percent) but tended to involve teasing, rules of games, and dominance (Boulton, 1993). As with nonhuman primates, the most common context for observed aggressive conflicts is "unknown" (Boulton, 1993). The type of conflict can also be influenced by environmental context. For example, children's conflicts occurring in structured peer group settings can be different from those occurring in free-play settings (Hartup, 1992; Killen & Turiel, 1991).

Outcome and Aftermath

The transition between the outcome of a conflict and its aftermath can be considered as the period before and after the contested item is distributed. Such a transition may not be easily identified, but it is nevertheless important from a conceptual standpoint. One reason why the transition may be difficult to identify is that conflicts are often settled subtly, quickly, and without discord. Even outcomes resulting from coercive strategies often do not interrupt ongoing social interaction. Nevertheless, the aftermath of a conflict can include a disruption of the social relationship of the individuals involved and result in emotional distress, especially if the conflict was resolved through aggressive coercion. When such disturbances occur, individuals may take action to restore their relationship with their former opponent. The term "reconciliation" is typically used to describe these post-conflict events. A large body of work has been devoted to understanding this process in nonhuman primates, and more recently the same methods have also been applied to humans (Verbeek et al., 2000). In the brief overview of reconciliation presented below, it is important to note that aggressive conflict not only jeopardizes a social relationship but also can produce emotional distress. Therefore, in addition to restoring a damaged relationship, reconciliation with an opponent can also alleviate this distress.

Reconciliation can be defined as an exchange of affiliative behavior between opponents shortly after an aggressive conflict. Assessment of whether reconciliation has occurred should include a comparison of the level of affiliation between two partners after aggression and the level of affiliation between the same two partners outside of an aggressive context. Using this technique, reconciliation has been documented in almost every nonhuman primate species in which it has been investigated (Aureli & de Waal, 2000a). Reconciliation may serve many functions. Silk (1997) suggests that the proximate function of reconciliation is to signal that a fight is over. Systematic studies have shown that post-conflict affiliation can reduce the likelihood of further aggression (Aureli & van Schaik, 1991a; Castles & Whiten, 1998a; Cords, 1992; Watts, 1995), reduce the anxiety levels of participants (Aureli & van Schaik, 1991b; Castles & Whiten, 1998b; Das et al., 1998), and restore relationships between former opponents (Cords, 1992). Presumably, reconciliation serves similar functions in humans. The study of reconciliation in humans, however, is a fairly recent endeavor, and few studies compare post-conflict rates of affiliation to those during baseline periods. Nonetheless, one suggested function of reconciliation in humans is to maintain the integrity of social relationships (Hartup et al., 1988).

The rate at which different species of primates reconcile conflicts is quite variable. In some species there is a friendly reunion after almost half of all aggressive conflicts (e.g., golden monkey: Ren et al., 1991; stumptail macaque: de Waal & Ren, 1988; bonobos: de Waal, 1987), while in other species no reconciliation is reported (red-bellied tamarin: Schaffner, 1991; Schaffner & Caine, 2000). On average, the species in which reconciliation has been studied ($N = 26$) reconcile 30 percent of their conflicts (range = 0–55 percent; Aureli & de Waal, 2000a). Human children exchange affiliative behavior following conflicts (Dunn & Herrera, 1997; Killen & Turiel, 1991: Ljungberg et al., 1999; Sackin & Thelen, 1984; Siddiqui & Ross, 1999; Verbeek & de Waal, 2001; Vespo & Caplan, 1993). Recent studies indicate that children reconcile approximately 50 percent of their conflicts, putting humans at the upper end of the primate range (Butovskaya et al., 2000; Butovskaya & Kozintsev, 1999). These studies also suggest that children reconcile more conflicts as they get older.

The quality of a relationship between two individuals influences the likelihood that they will reconcile a conflict. Numerous studies of nonhuman primates have shown that kin are more likely to reconcile than non-kin (e.g., Aureli et al., 1989; Aureli et al., 1997; Castles et al., 1996; Castles & Whiten, 1998a; Judge, 1991). An explanation is that primates are most likely to reconcile relationships with partners with whom they have valuable relationships and share strong social bonds (Cords & Aureli, 2000). As mentioned, kin are typically an animal's most frequent social companions, and they are valuable partners because they provide aid in fights. For example, male chimpanzees form strong social bonds with other males and rely on one another for defense against other groups. Female chimpanzees have loosely defined relationships and typically transfer from group to group. As expected, male chimpanzees are more likely to reconcile aggression than females (Goodall, 1986). Perhaps some of the best evidence for the valuable relationship hypothesis is a study of long-tailed macaques in which the value of a relationship between pairs of individuals was manipulated experimentally (Cords & Thurnheer, 1993). Animals whose relationship was made more valuable through cooperation in a food acquisition task were more likely to reconcile aggression. Since there are only a few systematic studies of human dyadic reconciliation, the effect of relationship quality on the process is largely unknown. Unlike in nonhuman primates, who are more likely to reconcile fights with high-quality partners, in children no consistent relationship has been found between the tendency to reconcile and friendship (Butovskaya et al., 2000; Butovskaya & Kozintsev, 1999).

Reconciliation following aggression may also involve third parties. In pig-tail macaques, the kin of a victim are likely to affiliate with the victim's attacker after an aggressive interaction (Judge, 1991). These interactions differ from the peaceful interventions described above in that they occur after the initial aggression is over. Since the victim's kin takes the initiative on behalf of the victim, this pattern of response could be interpreted as reconciliation "for" the victim by the victim's kin. In addition, the aggressor is likely to affiliate with the kin of its victim. Since kin are likely to join fights on behalf of a relative, relationships between entire groups of kin are at risk following aggression between two unrelated individuals. Triadic affiliation after conflicts may serve to reconcile entire groups of kin. Similar triadic post-conflict affiliative interactions have been observed in long-tailed macaques (Das et al., 1997). Triadic post-conflict affiliative interactions also occur in human children (Verbeek et al., 2000) and represent a promising avenue for understanding human conflict resolution, but there are no systematic investigations of third-party reconciliation in children.

Conflict Management

Conflict management is considered here as behavior undertaken to reduce the likelihood that a conflict of interest will develop in a dyadic relationship. Conflict management behavior influences the relationship between individuals before a conflict occurs, therefore affecting the "nature of the individuals" during the chain of events that constitute a conflict (Figure 3.1). I have focused only on nonhuman primates in this section partly because the literature on human conflict management is too vast to cover in this chapter, but also because so little systematic investigation has been conducted into the mechanisms of conflict management at the dyadic level in either human or nonhuman primates. Successful management is difficult to describe and investigate experimentally because it is generally difficult to assess the nonoccurrence of an event (i.e., a conflict). Ideally, a test should be conducted in which the likelihood of a conflict between two individuals is assessed after conflict management behavior and then compared to conflict rates between the same two individuals during periods when no conflict management behavior is exchanged. A reduced rate of conflict following management behavior would indicate a successful conflict management strategy.

Aureli and van Schaik (1991b) have used this approach to demonstrate that reconciliation of aggressive conflicts is a form of conflict management. Long-tailed macaques were less likely to resume fighting when aggression was followed by reconciliation than when no reconciliation occurred. Cords (1992) also found that two long-tailed macaques would tolerate each other's presence after a reconciled conflict more than two animals that did not reconcile aggression. Although such systematic methods are a potentially powerful tool for investigating conflict management in primates, they are rarely used.

Another method used to study conflict management is to examine the behavior of animals under conditions known to increase the risk of conflict and assess whether they modify their behavior in high-risk situations in ways that suggest attempts to manage conflict. Grooming behavior in nonhuman primates has been shown to have a calming influence on the recipient (Aureli et al., 1999; Boccia et al., 1989), and primates tend to groom more in tense situations (de Waal, 1984; Schino et al., 1988). Appeasement gestures and submissive displays are also used to signal subordinate status and preclude overt aggression (de Waal & Luttrell, 1985; Nöe et al., 1980). Under conditions of potential conflict animals may increase grooming, appeasement, and submissive displays to reduce the chances of conflicts. This coping model (de Waal, 1989c) has been tested in nonhuman primates by examining behavior under two conditions of increased risk: anticipated food competition and crowding.

Captive primates fed on a predictable schedule appear to anticipate the competition associated with the presentation of food and increase affiliative responses among one another during the time prior to feeding. Koyama and Dunbar (1996) found that prior to feeding, chimpanzees spent more time grooming and affiliating with their preferred social partners, and that such affiliative interactions were followed by co-feeding with the same individuals. Similar results have been found in stumptail macaques, in which individuals were observed to increase grooming of the alpha male in their group prior to scheduled feedings (Mayagoitia et al., 1993). In both studies, the affiliative increases were interpreted as attempts to reduce the likelihood of conflicts during subsequent feeding. A similar positive relationship between pre-feeding affiliation and later co-feeding has been found in another chimpanzee group (de Waal, 1992) and in bonobos (de Waal, 1987).

Crowding is also a situation with high potential for conflict, and nonhuman primates may use conflict management mechanisms under these conditions. For example, captive rhesus macaques living at high density for many years exhibited higher levels of affiliative behavior and submissive displays than those living at lower density (Judge & de Waal, 1997; Novak et al., 1992). These differences in behavior have been interpreted as a proactive strategy to ease social tensions and reduce the likelihood of conflict under high-risk conditions. Like the feeding studies mentioned above, the increases in conflict management behavior were directed to the individuals (e.g., adult males and animals of other kin groups) that had the highest potential for escalated aggression (Judge & de Waal, 1997). In addition, the increases in submissive displays occurred outside of aggressive contexts, an indication that animals may have been attempting to control aggression (Judge & de Waal, 1997).

Results of crowding studies are not as consistent or clear-cut as the above description implies, however. Aggression often increases during crowding despite apparent efforts to manage conflict. In addition, many variables appear to influence behavioral changes observed during crowding, such as the species studied, the duration of crowding, and the procedures used to produce crowded conditions (see review by Judge, 2000). For example, primates respond differently to long-term crowding than to short-term crowding and also appear to use different conflict management strategies depending on the duration of crowding. During long-term crowding involving years or months, primates may use the more proactive conflict reduction strategy described above. During short-term crowding involving hours or days, however, primates may inhibit their behavior (Aureli & de Waal, 1997) or avoid any interactions with others that may lead to conflict (Aureli et al., 1995; Judge & de Waal, 1993). Thus, evidence supports the hypothesis that nonhuman primates take steps to manage conflicts before they occur. A tendency to withdraw from social interaction during short-term crowding has also been reported in studies of conflict management on humans (see review in Judge, 2000).

Comparative research on conflict resolution in human and nonhuman primates is still in an information-gathering stage of understanding. If an ethological analysis is used and applied to a simple process model of conflict resolution (Figure 3.1), then many nonhuman primate conflict resolution responses can also be found in humans (Table 3.1). The similarities may be superficial, but they may also reflect shared phylogeny and/or similar adaptations to group life. Nonhuman primates exhibit numerous mechanisms for resolving conflicts within their groups that range from seeming indifference (tamarins) to despotic domination (rhesus) to prolonged and complex status disputes that depend on the support of others (chimpanzees). Humans use this full range of strategies.

To pursue the model, future research should continue documenting the behavior that occurs at each stage of the conflict resolution process. More basic knowledge of the component parts would allow investigation of how these parts interact to produce the circular process. A true comparative approach to conflict resolution would undertake the formidable task of standardizing data collection and investigating each aspect of the model for many species. This would include quantifying the rate and types of conflict that occur in each species, quantifying the relative rates with which each type of resolution strategy is used, and operationally defining the different strategies. For example, precise definitions of problem solving in terms of gains and losses may help identify when and if this strategy occurs in nonhuman primates. Careful analysis of negotiation-like tolerance behavior may reveal positive-sum outcomes. It will also be necessary to delineate more clearly the

Table 3.1 Aspects of conflict resolution exhibited by human and nonhuman primates

Phenomenon	Nonhuman primates	Humans
Resolution strategies		
Coercion	Yes	Yes
Relationship quality	Yes	Yes
Avoidance	Yes	Yes
Tolerance	Yes	Yes
Problem solving	No—unstudied	Yes
Relationship quality	Yes	Yes
Third-party involvement	Yes	Yes
Aggressive aid	Yes	Yes—unstudied
Redirection	Yes	Yes
Affiliative intervention	Yes	Yes—unstudied
Relationship quality	Yes	Yes—unstudied
Aftermath		
Reconciliation	Yes	Yes
Relationship quality	Yes	Yes—unstudied
Triadic aftermath	Yes	Unstudied
Relationship quality	Yes	Unstudied
Conflict management	Yes	Not reviewed

Note: A "Yes" indicates that the phenomenon has been observed and has been subjected to at least two rigorous investigations among those cited in the chapter. "Yes—unstudied" indicates that the phenomenon has been described only anecdotally or has just one citation documenting its occurrence. "Unstudied" indicates that I have not found investigations of this topic in researching this review. "No" indicates that available evidence indicates a lack of this behavior. "Relationship quality" under a main heading addresses whether aspects of the phenomenon are influenced by the quality of relationships among the individuals involved. Designations are descriptive and fairly subjective interpretations by the author.

beginning, end, and aftermath of a conflict in order to make comparisons, a distinction that is blurred in many studies.

Comparisons between nonhuman primates and humans are limited by the lack of ethological studies of human adults. Studies of children may underestimate our capacity to resolve conflicts effectively and misrepresent the typical strategies used by humans. A difficulty with studies of adults is that adult humans may have developed such subtle and effective resolution strategies based on language that conflicts and outcomes cannot be easily documented. Adults also do not tend to aggregate in settings like schoolyards for convenient behavioral investigation. In addition, human conflicts are mostly nonaggressive, whereas nonhuman primate studies typically focus on aggressive events.

This is unfortunate since aggression is also a relatively rare event in nonhuman primates. Most day-to-day resolution and management of conflicts probably involve subtle communication and gestures that human observers have not yet learned to decipher, although new revelations occur regularly (e.g., Cheney & Seyfarth, 1997).

Evaluation of a process model of conflict resolution is dependent on research advances in other aspects of primate and human behavior examined in this volume. For example, research on social cognition may reveal that what appear to be sophisticated resolutions based on compromise or negotiation may in fact not require complex cognitive skills such as a theory of mind. Successful conflict resolution also implies emotional homeostasis, so models of emotion that assess physiological changes must be linked to the behavior described in the process model. Finally, investigation of developmental changes in resolution strategies in nonhuman primates would help us understand the extent to which conflict resolution skills are acquired through experience and learning.

At a dyadic interpersonal level, humans are a very conciliatory species compared to most nonhuman primates. At a young age humans already exhibit rates of reconciliation that are as high as in the most conciliatory nonhuman primates, and humans appear to become more accommodating as they age. Human resolutions also involve more positive-sum outcomes in which both parties benefit. Our use of language to express intentions and to negotiate, and our advanced cognitive abilities that allow us to take another's perspective and calculate the relative value of negotiated outcomes may limit meaningful comparisons between human and nonhuman primates. Yet because there are so many similarities in patterns of behavior between human and nonhuman primates, much of the work currently being conducted on human conflict resolution has been inspired by work conducted on monkeys and apes. Thus, regardless of the specific similarities in the cognitive or emotional mechanisms underlying conflict resolution in human and nonhuman primates, exchange of information and collaboration between primate and human researchers can stimulate progress in this area of research, both conceptually and empirically.

4

Sexuality

Kim Wallen, Julia L. Zehr, Rebecca A. Herman,

and Franklynn C. Graves

Primate sexuality reflects developmental history, current social context, and hormonal state. The development and expression of sexuality in nonhuman primates parallel much of what we know about human sexuality, with the added advantage that experimental manipulations are possible in nonhuman primates that are ethically impossible in humans. This ability to do experimental work has resulted in a more detailed understanding of the relationship between biological factors and social context in primate sexuality than is possible in humans. This chapter presents a life-span approach to understanding primate sexuality. We first describe the occurrence of juvenile behavioral sex differences and their relationship to adult sexuality. We then consider the adolescent period and the relationship between the onset of adult endocrine function and the emergence of adult sexual behavior. The final section explores the relationship between gonadal function and adult sexuality.

Although it is tempting to describe behavior as resulting from either socialization processes (nurture) or biological processes (nature), this artificial dichotomy ignores the basic truth that all biological processes are inherently the result of environmental interaction and that all behavior is inherently biological. It is not possible to separate one influence from the other, as all behaviors emerge from the interaction between biological predispositions and environmental conditions (Fausto-Sterling, 2000; Wallen, 1996). That is not to say that it is not important to identify specific biological processes or environmental factors affecting behavioral development and expression. Such elucidation is crucial to developing a full understanding of behavior. It must be kept in mind, however, that these processes do not operate in an independent fashion but always interact. Thus, this chapter focuses on the interaction between early social environment and developmental hormonal influences in shaping the development and expression of sexual behavior. By integrating a developmental perspective into the behavioral neuroendocrinology of sexual behavior, we explore the tight linkage between social context and biological factors in primate sexuality. Rhesus monkeys are the only primate species besides humans about which there is sufficient information to address these issues and are therefore the principal focus of this chapter.

Juvenile Behavioral Sex Differences

Behavior in a variety of primate species is sexually differentiated both during development and in adulthood. Chapter 7 presents a broad overview of developmental sex differences in affiliation and play and their possible genesis. This section focuses on the relationship between juvenile behavioral sex differences and adolescent and adult sexual behavior.

Relation of Female Passivity to the Development of Sexuality

Harry Harlow (1965), a pioneer in studies of the effects of early experience on social development, reported that infant rhesus monkey females were submissive to males, showing passivity, withdrawal, and rigidity, and that males threatened females and directed pelvic thrusts toward both males and females more than did females. Harlow regarded these behavioral sex differences as reflecting the learning process that led to normal adult heterosexual behavior. An adult male rhesus monkey mates by clasping a female's legs with his feet and her hips with his hands while the female stands still and supports the male's weight (together these postural adaptations are referred to as a "foot-clasp mount"). Harlow suggested that the passivity and rigidity of infant females help infant males "discover" correct sexual positioning through trial and error in mounting and thrusting on various parts of the female body. These results, based on observations of singly caged infants placed together in small groups for twenty minutes a day, without adults or older siblings, offered the first coherent description of the relationship between juvenile sex differences and adult behavior. No mechanism was proposed, however, to explain how an active male interacting with a passive and rigid female partner would eventually develop the complex cooperative behavior of the foot-clasp mount (Wallen, 1996).

Unfortunately for Harlow's theory, subsequent studies of the monkeys he studied as infants found that they failed to engage in typical rhesus monkey adult sexual behavior at sexual maturity. Thus, whatever function the infantile sex differences reported by Harlow serve, they do not appear to facilitate the development of adult heterosexuality. Twelve males reared in Harlow's peer groups failed to show foot-clasp mounts when paired with feral females, and seventeen of eighteen peer group-reared females paired with feral males similarly did not show the normal female pattern of presenting and supporting a male's weight during a mount (Harlow, 1965). It subsequently became apparent that the infantile sex differences that Harlow reported were strongly influenced by the rearing conditions that the developing monkeys experienced (Wallen, 1996). For example, rigidity, seen as a hallmark of female development by Harlow, was never observed in young monkeys housed in large indoor-outdoor enclosures with their mothers and other monkeys of both sexes and different ages (Lovejoy & Wallen, 1988). Similarly, under social group rearing conditions, withdrawal and threatening behavior were not sexually differentiated. Studies that carefully manipulated the extent of peer exposure in early development demonstrated that the heightened male aggressiveness and female passivity described in Harlow's studies resulted from very limited opportunities for peer interaction within the Harlow rearing paradigm (Goy & Wallen, 1979; Wallen et al., 1981). In contrast to Harlow's contention that the foot-clasp mount was not in the juvenile male rhesus monkey's behavioral repertoire (Harlow & Lauersdorf, 1974), males reared with extensive opportunities for peer interaction readily displayed the foot-clasp mount and copulated normally as adults (Figure 4.1).

Harlow's discovery of these sex differences in more limited environments is partly explained by the role these behaviors play in juvenile dominance interactions and not in their relation to adult sexuality. Harlow reared monkeys

Figure 4.1 Juvenile rhesus males displaying a double foot-clasp mount. The form of the mount is like that shown by adult males and is displayed more frequently by males than females. At this age, mount partners are as likely to be males as females. This behavior is very sensitive to social rearing conditions and is rarely displayed when rhesus monkeys are reared under socially restricted conditions. (Photo: K. Wallen)

with mothers, or surrogate cloth mothers, in their home cages and typically allowed them access to peers for thirty minutes or less per day with no older monkeys present. Young monkeys placed together for brief periods of interaction, without the "safe haven" of their mothers, engage in exaggerated social competition that does not resolve into stable social relations in the brief time available for social interaction. Instead of amicable social relations, agonistic interactions prevail, accompanied by displays of submission, including rigidity, withdrawal, and passivity. Any tendency for males to develop physically faster than females is translated into higher social rank in this "Lord of the Flies" context. In contrast, when infants develop in social groups of monkeys that more closely mimic their natural social environment, dominance relations exist but reflect maternal rank and are hard to identify from infant social interactions.

While sex differences in rigidity and withdrawal are dependent for their expression on a very specific rearing environment, it is not the case that all juvenile behavioral sex differences are similarly socially labile. Sex differences in the hindquarters present (presenting), mounting, and various forms of affiliation and play have been reported under a variety of social conditions. Thus, an understanding of the relationship between juvenile sex differences and adult sexual behavior necessitates consideration of the effects of social and hormonal conditions on the expression of these behaviors.

Social Context and Sex Differences in Presenting Behavior

Presenting behavior, in which one individual offers or "presents" its genitalia to a partner, is typically part of the adult sexual behavior repertoire but is also used by both sexes in dominance interactions (Wallen, 1996; Figures 4.2, 4.3). Adult females permit normal mating behavior by displaying presents, which allow the male to achieve penile intromissions during mounting. In restricted social environments, presenting deviates from the adult form, and its expression can be exaggerated, with its occurrence varying with social rank and the sex composition of the group (Wallen, 1996). Thus, presents by juveniles appear not to be strictly sexual but are part of the dominance and submission repertoire.

Figure 4.2 An adult female displays a genital present to an adult male. This pattern of behavior is displayed by both juvenile males and females and is related to social subordination. Adult females primarily display this behavior, and it is used principally as a sexual solicitation, although it is also used as a submissive signal by both adult males and females. (Photo: K. Wallen)

Figure 4.3 Adult display of a double foot-clasp mount in a large social group of rhesus monkeys. Mounts are typically displayed in mating contexts in response to sexual solicitations from females. They are also used in social contexts, as illustrated here, where the highest-ranking female (the mount recipient) has solicited a new male to mount when she is not fertile. The male became the highest-ranking of four males following this mount series initiated by the highest-ranking female. (Photo: K. Wallen)

The form of the present itself is affected by rearing conditions. Harlow (1965) described infant females, reared without mothers and with limited peer access, who displayed abnormal presents, characterized by raised buttocks and accompanied by whole body rigidity. In juvenile monkeys (some mother-reared and some cloth surrogate–reared) with limited peer experience, both sexes showed normal presents, but females presented more frequently than did males. In contrast, in heterosexual age-graded social groups, only normal presents were observed, and male and female infant and yearling rhesus presented at equal but low frequencies (Brown & Dixson, 2000; Lovejoy & Wallen, 1988; Wallen et al., 1995). The characteristics of presenting do not appear to stem completely from access to peers but probably reflect the level of agonism that infants experience. For example, rhesus infants who had limited peer access but continuous access to mothers, and could thus retreat to the safety of their mothers when in conflict with peers, did not show abnormal present postures or significant sex differences in presenting (Wallen et al., 1981).

The sex differences in presenting reported in monkeys without mothers and limited peer access probably reflect a sex difference in social rank in these small artificial groups. Under such rearing conditions, males typically outrank females, and low-ranking infants and juveniles present at high frequencies whereas high-ranking individuals rarely present (Goldfoot, 1977; Goldfoot et al., 1984). Thus, the higher frequency of presenting by females may simply reflect their lower social rank. Male juveniles reared in isosexual groups (i.e., groups in which all individuals are of the same sex) presented significantly more frequently than did heterosexually reared males (Goldfoot & Wallen, 1978; Goldfoot et al., 1984). This difference was accounted for by a higher frequency of presenting by low-ranking males. Thus, male presenting is strongly influenced by social rank, and its expression is a submissive behavior. Isosexually reared females, however, did not present less than heterosexually reared females. This raises the possibility that juvenile females may use presenting in a different manner from males and have a predisposition to present more than do males. Under social conditions that increase agonism and produce lower female social rank, presenting by females may be exaggerated while male presenting is suppressed. Although this issue cannot be completely resolved, it is apparent that presenting is quite sensitive to the social conditions encountered during development.

Social Context and Mounting Behavior with Emphasis on the Foot-Clasp Mount

Mounting is another juvenile behavior that is used by adults in both sexual and dominance contexts. With mounting, however, unlike presenting, rhesus males have been found to mount more often than females in every social environment that has been studied (Wallen, 1996). In addition, a similar sex difference has been observed in other species of macaques (stumptail macaques: Nieuwenhuijsen, Slob, & van der Werff ten Bosch, 1988; pigtail macaques: Zehr, 1998; Japanese macaques: Eaton et al., 1990). Although the social environment has little effect on the frequency of male mounting, the environment markedly affects the form of the mounts themselves. Restricting peer access early in life eliminates the display of foot-clasp mounts by juvenile males (Goy & Wallen, 1979; Harlow, 1965). Instead, these young males direct their mounts to various parts of the partner's body, including the side or head, and they may or may not accompany such "abortive mounts" with pelvic thrusting. The occurrence of foot-clasp mounting is so rare under these restricted rearing conditions that Harlow stated that it was not part of the juvenile male behavioral repertoire (Harlow & Lauersdorf, 1974). Subsequently, it was demonstrated that the amount of time spent with peers during the first year of life predicted whether or not juvenile males would display foot-clasp mounts (Goy & Wallen, 1979; Wallen et al., 1981). Males reared in complex social groups routinely display foot-clasp mounts and rarely display abortive mounts (Brown & Dixson, 2000; Lovejoy & Wallen, 1988; Wallen et al., 1995). The occurrence of the foot-clasp mount is so sensitive to early social conditions that it serves as a reliable indicator of adequate early socialization (Goldfoot, 1977). It is also a predictor of adult sexual behavior as young male rhesus who do not foot-clasp mount peers will not show normal levels of appropriate mating behavior as adults (Goy & Wallen, 1979; Wallen et al., 1977).

Access to peers has been demonstrated to be a key influence on whether or not young males display foot-clasp mounts. Male infants reared with their mothers are more likely to show foot-clasp mounts and do so at much greater frequencies if they have continuous access to peers than if access is limited to thirty minutes a day (Wallen et al., 1981). The failure of young males reared in socially restricted environments to display foot-clasp mounts is not explained by a lack of the necessary motor skills. For example, males raised with mothers and limited peer access do not foot-clasp peers but do foot-clasp mount their mothers (Wallen et al., 1981). Similarly, males reared with surrogate mothers and limited access to peers display fear or aggression to females and do not foot-clasp mount peers but will foot-clasp mount full-size models of adult female monkeys (Deutsch & Larsson, 1974). Motivation to mount is also not lacking in males reared in socially restricted environments, as total mounting (abortive mounts and nonfoot-clasp mounts included) is the same as that displayed by males reared with continuous opportunities to interact with peers (Wallen et al., 1981). Instead, the failure of males with limited peer interaction to display foot-clasp mounting most likely reflects the unique place that the foot-clasp mount occupies in the behavioral repertoire of juvenile monkeys. Unlike other behaviors, such as threatening, presenting, or even grooming, the foot-clasp mount requires active cooperation and coordination between the two animals involved in the mount. The mounter must execute the foot clasp, and the recipient must allow the mounter in close contact, support his weight, and remain upright and still for at least a few seconds. Infants unaccustomed to peers probably do not have the necessary social skills to develop the trust required for cooperative mounting. It is likely that foot-clasp mounting arises from the nature of social relations developed through other patterns of social interaction. The expression of foot-clasp mounts then very likely contributes to further developing and maintaining cooperative and trusting social relations between peers (Deutsch & Larsson, 1974; Wallen et al., 1977, 1981).

Although males consistently mount more frequently than do females across social environments, the magnitude of the sex difference varies with the early rearing environment. Isosexually peer-reared males foot-clasp mount less frequently than heterosexually peer-reared males, and isosexually peer-reared females foot-clasp mount more than their heterosexually peer-reared counterparts (Goldfoot et al., 1984). Isosexually peer-reared females, however, still mount less than isosexually peer-reared males. The direction of these effects again suggests a role for rank. Regardless of sex, higher-ranking individuals mount more than do lower-ranking individuals of the same sex. Thus, some of the reduction in mounting by isosexual males and some of the increase in isosexual females could result from males and females occupying social ranks that are atypical for their sex in restricted laboratory environments. Yet social rank is not the only possible explanation for the effect of altering the sex composition of the social groups. Juvenile males typically mount male and female partners equally (Wallen, 2001). Thus, for an isosexually grouped male, a large proportion of his potential mounting partners are absent from his social environment. This view, of course, presupposes that males distribute their mounts across social partners relatively equally and do not increase their mounting of males, for example, when females are not present. As in the case of threatening behavior, it appears that the sex of the partner interacts with the predispositions of the males to determine the final frequency of mounting.

Other Sex Differences in Juvenile Behavior and Their Functional Significance

Threats are part of the agonistic behavioral repertoire of rhesus monkeys and have been reported to be sexually differentiated in young rhesus monkeys, with males threatening more often than females (Harlow, 1965). When juvenile rhesus monkeys are observed as part of large age-graded social groups, however, there is no evidence of a sex difference (Lovejoy & Wallen, 1988; Wallen et al., 1995), suggesting that male and female threatening patterns are affected by social context.

Sex differences in juvenile affiliative behavior including (1) rates of contact, proximity, grooming and play; (2) partner choice; and (3) relative occurrence of different types of play (e.g., solitary versus social, rough-and-tumble versus play mothering) have been observed in many species of monkeys and apes (see Chapter 7). Such differences have also been observed across different social environments and rearing conditions. For example, unlike presenting and mounting, isosexual rearing does not affect the sex difference in frequency of play. Isosexually reared females played slightly less than heterosexually reared females, while isosexually reared males played as much as heterosexually reared males (Goldfoot et al., 1984). Thus, rhesus monkey juvenile males have been reported to play more than females under all social conditions studied.

Many developmental sex differences in behavior can be attributed to the different life histories of males and females in primate societies. In female philopatric species (i.e., species in which males but not females leave the group), young males engage in behaviors that prepare them for emigration, alliance formation, and competition for females, while females prepare to become part of the already established matriarchal society and to care for infants. In male philopatric societies, some of the same sex differences in juvenile behavior can still be observed, while others are absent or reversed.

Many sex differences in juvenile behavior are functionally related to acquiring social competence rather than learning specific patterns of adult behavior. For example, instead of learning how to copulate per se, as was once thought, juvenile males likely learn to interact with potential partners and read social cues. Similarly, instead of learning how to hold and carry infants, juvenile females likely learn how to encourage or discourage their behavior, or how to protect infants from the environment. In fact, although isolation rearing clearly does not produce normal females, some of these females do show basic patterns of maternal care under appropriate conditions (Ruppenthal et al., 1976).

Juvenile primates clearly need not be aware of the long-term consequences of their behavior. Instead, genetic predispositions and other biological factors interact with social and environmental processes in determining the extent to which juvenile males and females engage in different activities and find them rewarding.

Hormonal Influences on Juvenile Behavioral Sex Differences

It is well established that juvenile patterns of social behavior can be affected by the prenatal hormonal environment in which males and females differentiate (Wallen, 1996). These environments differ naturally as the testis of the male rhesus monkey is differentiated by gestational day 40 and maintains testosterone at levels higher than in female fetuses throughout gestation (Resko et al., 1980). During the first few months of life, this sex difference continues with elevated testosterone secretion in males (Mann et al., 1984) before falling to baseline for the remainder of the juvenile period.

Prenatal Hormonal Manipulations in Females

Administration of androgens to females pregnant with female fetuses mimics the hormonal environment of male fetuses. Female rhesus monkeys exposed to high levels of prenatal androgens show increased frequency of mounting when reared in small groups with or without their mothers (Wallen, 1996). High doses of prenatal androgens that begin early and are continued through most of gestation masculinize both juvenile behavior and female genitalia, producing penises and scrotums that are hard to distinguish from those of genetic males (Goy, 1981). If however, shorter prenatal treatments are administered, beginning at either the second trimester (early) or third trimester of gestation (late), the effects of androgens on genital differentiation can be separated from the effects on behavior (Goy, Bercovitch, & McBrair, 1988). Early treatments of high doses of testosterone for a duration of twenty-five days produce complete genital masculinization. If androgen treatment is instead limited to later in gestation (beginning on day 115 of the approximately 168-day gestation), female genitalia are not masculinized. Treated females, regardless of timing of treatment, mount peers at elevated frequencies. Females treated late in gestation, however, unlike early-treated females, play at male-like frequencies, and only early-treated females show significantly less grooming of mothers than control females. The combination of these findings indicates both that behavioral masculinization can occur without accompanying physical masculinization and that the timing of sensitivity to androgen during development can differ by behavioral endpoint. In Japanese macaques, treatment of socially living pregnant females with high doses of androgen implants produced females with masculinization of genitalia and somewhat elevated mounting frequencies, but frequency of play and grooming was unaffected (Eaton et al., 1990). These treatments ended at day 100 of gestation and may have missed the period of sensitivity for masculinizing play behavior.

Because these studies of both Japanese and rhesus macaques either involved small social groups without infants or did not investigate interactions with infants by juvenile subjects, the effects of high doses of prenatal androgens on female play-mothering were not carefully investigated. A more recent study of lower doses of prenatal testosterone indicates that masculinization of interest in infants may also have limited critical periods (Herman et al., 2003). Juvenile females administered androgen late in gestation showed no decrease in interactions with infants, whereas females receiving androgen early in gestation embraced and kidnapped infants less than control females at three years of age. Another study of prenatally androgenized females investigated the interactions between a single infant and a post-pubertal subject in a cage together (Gibber & Goy, 1985). On those behaviors in which frequency of interactions differed between control males and females, androgen treatment of females had no effect. Nevertheless, because of the rarity of sex differences found and the artificial test situation, the validity of these results is uncertain.

Hormonal Manipulations in Males

Prenatal hormonal manipulations. The majority of studies investigating prenatal administration of androgen have not demonstrated an effect on sexually differentiated behaviors in males (Eaton et al., 1990; Goy, 1981). Recent data from our laboratory indicate, however, that prenatal treatment with androgen can actually hypermasculinize mounting and play behavior in male subjects in large social groups (Wallen et al., 2001). Interestingly, this study used lower doses of testosterone than in previous studies that found no behavioral effects in males. We have also investigated the effects of prenatally blocking androgen exposure with the androgen-receptor blocker flutamide. Prenatal treatment with flutamide failed to prevent masculinization of behavior in male subjects (Wallen et al., 2001), even though early treatment with flutamide altered male genitalia (Herman et al., 2000). Paradoxically, males receiving flutamide late in gestation demonstrate hypermasculinization of mounting behavior, mounting at frequencies equal to those of androgen-treated males (Wallen et al., 2001). These surprising results raise the possibility that the treatment did not effectively block androgen exposure late in gestation. One other study has attempted to eliminate prenatal androgen exposure in male rhesus. Phoenix (1974b) found that a rhesus male who was castrated at gestational day 100 displayed behavior clearly similar to that of a control male peer and different from that of a control female. One difference between the control male and the prenatally castrated male was noted, however. The castrated male showed no foot-clasp mounts at 7 months and very few at 17 months, whereas the control male frequently foot-clasp mounted peers at both ages. Although no conclusions should be drawn from this case study, it raises the possibility that testosterone during the last trimester of pregnancy is not crucial for masculinization of behavior but may alter the time course of development.

Neonatal hormonal manipulations. Unlike the prenatal hormonal environment, neonatal hormone exposure has generally been shown to have very little if any effect on masculinization of sexually differentiated behaviors in primates. Neither suppressing testicular function during the first four months of life with a gonadotropin-releasing hormone (GnRH) antagonist nor producing supraphysiological levels of androgen during the same period altered mounting or play behavior of yearling males in social groups (Wallen et al., 1995). Neonatal treatment of males with a GnRH agonist, which also suppresses testicular function, and females with testosterone implants for the first six months of life similarly did not alter mounting or play behavior (Nevison et al., 1997). Manipulations of neonatal testosterone can, however, have subtle effects on proximity interactions in rhesus monkeys. Control yearling females and neonatally testosterone-suppressed yearling males spent time with their mothers, followed other individuals, and initiated proximity with other individuals significantly more often than neonatally androgen-treated yearling males (Wallen et al., 1995).

Both hormonal exposure and social context influence behavioral sex differences. The ability of prenatal testosterone to affect the display of sexually differentiated juvenile behavior is clear, as is the importance of the rearing environment. These two influences undoubtedly interact to produce behavioral differentiation. For example, prenatal exposure to testosterone may act to alter neural systems, making rough-and-tumble play enjoyable and rewarding. If opportunities for such behavior arise, both young male and female rhesus monkeys may engage in a bout of play, but the experience will be enjoyable for males and frightening or unnecessarily exhausting for females. When experience with peers is limited and play relationships are not easily established, males' predisposition for play may be seen as agonism, and when females cannot escape play groups because of limited space or the absence of mothers, their attempts at avoidance may take the form of submissive behaviors. In this manner, biological predispositions interact with experience to produce observable sex differences.

Comparison of Nonhuman Primate Studies with Hormonal Influences on Human Development

Because nonhuman primate juvenile sex differences depend on the life histories of males and females, we should not expect sex differences in humans to parallel nonhuman sex differences exactly. Men and women certainly do not have the same life roles as male and female rhesus monkeys. Nevertheless, behavioral sex differences are consistently identified in human boys and girls (Chapter 7). Boys rough-and-tumble play more than do girls, girls show more interest in infants and doll play than their male peers, and both sexes show sexually differentiated toy preferences and tend to interact with peers of the same sex (Maccoby, 1998). Traditionally, these sex differences have been attributed to socially taught ideas of sex-appropriate behavior (Money & Ehrhardt, 1972). According to this social construction theory, gender identity is completely malleable at birth, and sex differences result from differential treatment which itself is based on perception of the child's sex from his or her genitalia. Although social experiences certainly play a role in determining human behavior, just as they do in nonhuman behavior, evidence indicates that biological predispositions do exist and children are not born gender neutral (Diamond & Sigmundson, 1997; Fausto-Sterling, 2000; Reiner, 1996). Instead, as in nonhuman primates, prenatal events predispose individuals to engage in specific behaviors. Sex differences in the prenatal hormonal environment lead to sex differences in behavioral predispositions, which are then augmented or suppressed by social experience.

Because controlled hormonal manipulations in children are unethical, human studies investigate behavior of individuals who through medical disorders or treatments are exposed to abnormal hormone levels during fetal life. These studies rely on individuals whose mothers were given certain synthetic progestogens in an attempt to prevent pregnancy complications, and on individuals with congenital adrenal hyperplasia (CAH). CAH is caused by an enzyme defect in the adrenal gland, which results in the accumulation of precursor hormones, including androgens (White et al., 1987). Both girls and boys with CAH have elevated androgen levels during prenatal life, but the condition is normally diagnosed shortly after birth, and hormone levels are then controlled. Genital virilization is common in girls with CAH but is also typically corrected with surgery at an early age.

Girls with CAH show a greater tendency as children to classify themselves and be classified as tomboys, to show less interest in weddings, marriage, doll play, and infant care, and to be less satisfied with their sex role as girls, as compared to matched controls (Ehrhardt et al., 1968). CAH girls also spend more time playing with boys' toys and less with girls' toys than non-CAH female relatives (Berenbaum & Hines, 1992), and tend to prefer boy playmates more than control females (Berenbaum & Snyder, 1995). Leveroni and Berenbaum (1998) gave parents of CAH children a questionnaire about the nurturing behavior of their children toward infants and found that CAH girls showed less interest in infants than control females and did not differ from control males. No effect was observed in CAH boys.

The behavioral effects of hormone treatments administered to pregnant women depend on the specific hormones used in treatment. Some synthetic progestogens related to androgens have been shown to masculinize some aspects of behavior (Collaer & Hines, 1995), while estrogens and other progestogens may actually produce some hypomasculinization of play behavior in both boys and girls (Meyer-Bahlburg et al., 1988). These latter compounds are known to show some anti-androgenic activity.

Although evidence in humans remains sparse, there is a surprising degree of agreement between more experimental studies of macaques and the "experiments of nature" in humans. While the extent to which behavioral sex differences are canalized by experience remains an active area of investigation, it is no longer questionable whether prenatal hormones produce behavioral predispositions that lead to juvenile behavioral sex differences (Fausto-Sterling, 2000; Maccoby, 1998). As is the case in nonhuman primates, these behavioral sex differences appear to be intimately linked to life history differences between males and females, although the exact nature of this relationship remains to be elucidated.

Pubertal Endocrine Activation and the Transition to Adult Sexuality

Historically juvenile behavioral sex differences have been either studied in isolation or related to adult social and sexual behavior. The developmental progression from infancy through adolescence and into adulthood has rarely been articulated. It is in adolescence, however, that juvenile patterns of behavior often disappear, adult patterns emerge, and adult sexuality begins. Thus adolescence represents an important and unique developmental phase for understanding the psychology of primate sexuality.

Reproductive physiology changes rapidly during adolescence, and these changes define this developmental period. In male and female primates, the hypothalamic-pituitary-gonadal (HPG) axis is largely quiescent during the juvenile period, and the onset of increased HPG axis activity marks the beginning of adolescence. The end of adolescence is reached when an individual has the functional reproductive physiology of an adult. The rapid changes in reproductive physiology at adolescence are accompanied by dramatic changes in the sociosexual behaviors that are the crux of this chapter.

The changes seen in adolescence depend on the specific behavior. For example, play behavior declines gradually during adolescence (Rose, Bernstein, et al, 1978; reviewed in Caine, 1986), whereas other behaviors such as female sexual solicitation and male ejaculation occur for the first time during adolescence (Hanby & Brown, 1974; Liang et al., 2000; Wolfe, 1978). Mounting, a behavior which is seen in male nonhuman primates throughout development, changes in terms of frequency (e.g., Hanby & Brown, 1974) and sex of partner (Wallen, 2000a, 2001) but not in form (Goldfoot, 1977) during adolescence. Thus, behavioral changes during adolescence may involve a change in the frequency with which a behavior occurs, a modification of a juvenile behavior, or the new occurrence of behaviors not in the repertoire of juvenile animals.

Adolescent behavior differs from adult behavior in ways that may reflect differences in the activity of the adolescent neuroendocrine system. For example, adolescent female rhesus monkeys exhibit longer periods of mating ⚥ within the ovarian cycle than do adult females (Wilson & Gordon, 1980). This behavioral difference led to the discovery that adolescent females have a longer period of preovulatory estradiol secretion than do sexually mature females (Wilson et al., 1982a). Similarly, the absence of sexual behavior in callitrichids of pubertal age led to the discovery that ovulation was often suppressed in the family context of these primates, even though the underlying neuroendocrine system would function normally if the young female was moved to another social context (Abbott, 1984; Saltzman et al., 1997; Ziegler et al., 1990).

Many studies show changes related to adolescence but do not contain enough detail to elucidate fundamental principles during the pubertal transition. Thus, this section focuses on those studies that demonstrate the social and neuroendocrine influences on sexual behavior particular to adolescence. First, we describe the effects of pubertal neuroendocrine changes on sexual behavior. Then we discuss how the social environment and previous experience affect the expression of adolescent sexual behavior. Finally, possible differences in the attractiveness of adolescents as sexual partners and the role of learning in first sexual experiences are considered.

Hormonal Influences on Adolescent Sexual Behavior

Females

Puberty in female primates is marked by changes in hypothalamic activity, gonadotropin secretion, and steroid hormone secretion (reviewed in Terasawa & Fernandez, 2001; Wilson, 1992). During the juvenile period, gonadotropin secretion in female primates is in part limited by the negative feedback produced by low levels of gonadally secreted estradiol (Plant, 1986; Pohl et al., 1995; Suter et al., 1999). At the onset of puberty, gonadotropin secretion increases as a result of maturational changes both in hypothalamic activity (Watanabe & Terasawa, 1989) and in reduced sensitivity to the negative feedback effects of estradiol (Chongthammakun & Terasawa, 1993; Rapisarda et al., 1983; Wilson et al., 1986). The increased activity of the HPG axis results in increased circulating levels of estradiol, endometrial development, and menarche, and culminates in first ovulation. In many species of primates, circulating estrogens cause an edema of the perineal skin, resulting in the onset of sexual swelling at puberty (Anderson & Bielert, 1994; Hisaw & Hisaw, 1961). Maturational changes in the HPG axis are not as dramatic in callitrichid species that show social suppression of reproduction in the natal group (e.g., common marmosets: Abbott, 1984; Evans & Hodges, 1984; cotton-top tamarins: Savage et al., 1988; red-bellied tamarins: Küderling et al., 1995). In cotton-top tamarins, urinary levels of conjugated estrogens and luteinizing hormone are higher during puberty than during the juvenile period but do not show cyclical changes similar to the adult ovarian cycle (Snowdon et al., 1993). The ovaries of both pubertal and nonreproductive post-pubertal saddle-back tamarins show clear developmental changes from the juvenile period but are unlike those of reproductive adult females (Mansdotter et al., 1992).

These changes in hormone secretion have a direct effect on sexual interest in adolescent females. In humans, administration of estrogen to hypogonadal adolescent females resulted in an increase in kissing behavior (Finkelstein et al., 1998). Additional studies have found evidence that changes in testosterone during puberty were related to the timing of first intercourse in adolescent females (Halpern et al., 1997). In rhesus monkeys, adult females living in naturalistic social groups show periods of intense sexual solicitation that last for several days and cease around ovulation (e.g., Carpenter, 1942; Cochran, 1979; Gordon, 1981; Wilson et al., 1982a, b; Wallen et al., 1984; Zehr et al., 2000). Adolescent female rhesus monkeys have a longer period of sexual solicitation, lasting an average of fifteen to sixteen days compared to an average of eight to ten days in adults (Carpenter, 1942; Wilson & Gordon, 1980; Wilson et al., 1984). The extended rhesus adolescent female mating period results from longer periods of increased estradiol secretion prior to ovulation (Wilson et al., 1982a). Adolescent females also present to adult males more frequently per hour of observation than do adult females when averaged over the estrus period (Perry & Manson, 1995; Rasmussen, 1983). This difference in sexual initiation frequency could also result from the extended adolescent period of preovulatory estradiol secretion, since presentation frequencies in group-living rhesus females are correlated with estradiol levels (Wallen et al., 1984; Zehr et al., 1998). After first ovulation, length of follicular estradiol secretion and behavioral estrus in adolescents are the same as in adult females (Wilson et al., 1984).

Males

At puberty, the HPG axis of males also awakens from its quiescent juvenile state (reviewed in Plant, 1994; Terasawa & Fernandez, 2001). Nocturnal pulses of luteinizing hormone and testosterone are the first indicator of pubertal hypothalamic activity (Suter et al., 1998; Wu et al., 1993). Pituitary gonadotropin secretion results in increased testosterone secretion and sperm production (reviewed in Bercovitch & Goy, 1990; Bernstein et al., 1991). These neuroendocrine changes are also associated with testicular descent (e.g., Alberts & Altmann, 1995; Nieuwenhuijsen et al., 1987) and increases in testicular volume (Lunn et al., 1994; Mann et al., 1989, 1998). Although female callitrichids living in their natal groups show a suppression of HPG activity during puberty, male callitrichids appear not to be physiologically suppressed (Dixson, 1986; Ginther et al., 2000, 2001).

Adolescent male primates of most species show increased levels of sexual behavior at puberty coincident with increases in testosterone secretion. In humans, testosterone increased sexual activity in hypogonadal adolescent boys (Finkelstein et al., 1998). In addition, testosterone levels correlate with sexual interest (Udry et al., 1985) and initiation and frequency of sexual activity (Halpern et al., 1998) in pubertal boys. In nonhuman primates, male mounting increases and is accompanied by intromissions and ejaculation at adolescence (e.g., Hanby & Brown, 1974; Nieuwenhuijsen, Bonke-Jansen, et al., 1988; Wallen, 2000a). Studies of nonhuman primates have not demonstrated that testosterone is directly responsible for pubertal increases in male sexual activity; the temporal association of changes in behavior and changes in physiology suggests, however, that the two are linked (Glick, 1980; Rose, Bernstein, et al., 1978).

Social Influences on Adolescent Sexual Behavior

Adolescence is a unique time in the social development of male and female primates. Human adolescents may be disconcerted by the physiological changes that take place during puberty, and the same may extend to nonhuman primates. In hierarchical primate societies, males and females often undergo rank changes around puberty. For example, during puberty, young male stumptail macaques become integrated into the adult male hierarchy, and females gradually move up in the hierarchy to a rank just below their mothers (Nieuwenhuijsen, Bonke-Jansen, et al., 1988). Adolescent primates either integrate themselves into the adult social world or emigrate from their natal social group.

Furthermore, adolescence is clearly a transition to adulthood in which behavior is like that of neither juveniles nor adults. For example, adolescents mate more often than juveniles but less often than adults (Hanby & Brown, 1974) and play more often than adults but less often than juveniles (Rostal & Eaton, 1983). It is not clear, however, whether adolescent male monkeys are considered by other monkeys to be juveniles or adults. Adult stumptail macaque males discriminate between adolescents of different maturational stages (Nieuwenhuijsen, Bonke-Jansen, et al., 1988). Prior to testicular descent, stumptail adolescent males can mate with females in their social group without interruption by adult males. Mating attempts by adolescent males after testicular descent, however, are quickly interrupted by adult males. As a result, adolescent males mate only surreptitiously after testicular descent (Nieuwenhuijsen, Bonke-Jansen, et al., 1988). Thus, testicular descent, a visible marker of puberty onset, seems to be used by other monkeys in the social group as a marker of changing reproductive status. Interestingly, adolescent females have an exaggerated sexual swelling compared to that of adult females in many primate species (Anderson & Bielert, 1994). Perhaps this visible marker of puberty onset is used by other monkeys to gauge adolescent female reproductive status. It remains to be discovered how adolescents are viewed in a variety of primate species.

Rank Effects

In hierarchical species of primates, one of the most important social changes that an individual experiences during development is the acquisition of adult rank (reviewed in Pereira, 1995). In macaque females, this process is completed around puberty (Chikazawa et al., 1979; Datta & Beauchamp, 1991; Nieuwenhuijsen, Bonke-Jansen, et al., 1988). Macaque females eventually attain a rank just under their mothers, but during puberty, females may rank below other adult females who are lower ranked than their mothers (Nieuwenhuijsen, Bonke-Jansen, et al., 1988). In macaque males, emigration from the natal group often results in a low-ranking, peripheral status (van Noordwijk & van Schaik, 1985). If they do not emigrate, male macaques become integrated into the adult hierarchy during puberty (Manson, 1993), and higher-ranked adolescent males mate with more partners than do low-ranked males (Colvin, 1985). During this process, male adolescent ranks remain linear with respect to other adolescents, and the adolescent male hierarchy becomes integrated into the existing adult male hierarchy (Nieuwen-huijsen, Bonke-Jansen et al., 1988).

Rank has been hypothesized to affect the rate of puberty onset and first reproduction; results have been mixed, however. In savanna baboons, the highest-ranking females have been reported to have an early onset of puberty as measured by first menses and by first sexual swelling (Altmann et al., 1988; Bercovitch & Strum, 1993). Age at first parturition may also be earlier in higher-ranking baboon females (Altmann et al., 1988; but see Bercovitch & Strum, 1993). Some studies of macaques have also shown earlier ages of first reproduction in dominant females (Bercovitch & Berard, 1993; Drickamer, 1974; Paul & Thommen, 1984; Sade et al., 1976). In Japanese and stumptail macaques, however, there was no effect of social rank on age of first ovulation or first parturition (Gouzoules et al., 1982; Nieuwenhuijsen et al., 1985). In males, the effects of rank on puberty onset have been more consistent. In baboons, rank predicts the age at testicular descent and enlargement (Alberts & Altmann, 1995). In pubertal rhesus monkeys, maternal rank is correlated with testicular volume and testosterone levels (Bercovitch, 1993; Dixson & Nevison, 1997; Glick, 1979; Mann et al., 1998).

Suppression in the Natal Group and the Effects of Conspecifics

There have been scattered reports of the direct effects of conspecifics on adolescent physiology and behavior, the most dramatic effect being that seen in callitrichids. In these species, adolescents live in natal groups and do not engage in sexual behavior during puberty. In wild populations of saddle-back tamarins, both females and males delay emigration and reproduction until well after they are likely to have completed adolescence (Goldizen & Terborgh, 1989). Thus, while adolescents are living in the natal social group, expression of sexual behavior is suppressed by the presence of related conspecifics. In males, reproductive behavior but not circulating testosterone is suppressed

(Ginther et al., 2001). In females, investigation of the hormonal correlates of callitrichid adolescence showed that urinary conjugates of estradiol and luteinizing hormone increase during puberty but ovulation is usually suppressed (Saltzman et al., 1997; Snowdon et al., 1993). When post-pubertal females are removed from their natal group, these females can ovulate and conceive within days of pairing with an unfamiliar male (Evans & Hodges, 1984; Ziegler et al., 1990). Thus, maturation of the HPG axis is occurring without the expression of sexual behavior and without the expression of cyclical neuroendocrine changes seen during the puberty of Old World primates and callitrichids not living in family groups.

Suppression of adolescent sexual behavior in the natal group may also occur in some Old World monkey species; this suppression of behavior does not, however, include suppression of reproductive physiology as in callitrichids. Although adolescent female macaques do not appear to be subject to behavioral suppression by conspecifics, a lack of mating opportunities for adolescent males is one of the factors hypothesized to result in male emigration. This hypothesis is supported by reports that sons of high-ranking females achieve high rank among the male dominance hierarchy and delay emigration from their natal group (Koford, 1963). In addition, the few sons of low-ranking rhesus females that had high copulation frequencies also delayed emigration, whereas those who copulated infrequently did not (Manson, 1993). Among the many macaque species, pubertal males may have different social constraints on sexual behavior in their natal group. While longtail macaque males do not mate with any females in their natal groups (van Noord-wijk, 1985), pubertal Barbary and rhesus males do not mate only with closely related females (Manson, 1993; Paul & Kuester, 1985). Mating by adolescent stumptail males in their natal group is actively prevented through interruption by higher-ranking individuals (Nieuwenhuijsen, Bonke-Jansen, et al., 1988). In contrast to this evidence of developmental suppression in the natal group, adolescent male mandrills that remain closely associated with their natal social group have higher testosterone levels and greater testicular development than adolescent males who live peripherally to their natal social group (Wickings & Dixson, 1992).

Pubertal maturation in male macaques may be hastened depending on the social composition of the group. Rose, Bernstein, and colleagues (1978) found that adolescent male testosterone levels at the onset of puberty directly relate to the number of adult males in the social group. Pubertal males in a social group with only two adult males had adult levels and an adult seasonal pattern of testosterone secretion; pubertal males in a social group with seven adult males had lower levels of testosterone secretion and a delay in the seasonal peak of testosterone, and males living in a social group with twelve adult males had incomplete suppression of testosterone secretion at the same age (Rose, Bernstein, et al., 1978). Similarly, the number of adult females, and thus mating opportunities, can affect male sexual maturation. For example, once male baboons achieve their adult dominance rank, they enter into their first sexual consortship faster when there are more cycling females in the group (Alberts & Altmann, 1995).

Does it matter whether adult ♂♂? are related to pubertal males?

Role of Learning and Previous Experience

One possibility is that adolescent sexuality differs from that of adults owing to a lack of experience or learning that takes place during adolescence. This question has been little investigated, however, and supporting evidence for this idea is sparse. Early studies of adolescent males suggested that males learn intromission behavior (Erwin & Mitchell, 1975; Michael & Wilson, 1973; Wolfe, 1978). These studies showed that with experience, males were more efficient at achieving intromissions (fewer mounts without intromissions) and that adolescent mount series including mounting, intromissions, and ejaculations became similar in pattern but not in frequency to those of adults (Hanby & Brown, 1974; Michael & Wilson, 1973; Wolfe, 1978). The mounting series of adolescent males included more mounts to ejaculation and had longer latencies to ejaculation (Michael & Wilson, 1973; Wolfe, 1978). Perhaps more interesting are the behavioral changes that males appear to undergo after experiencing intromissions and ejaculations. Prior to achieving intromissions, adolescent males mount male and female partners almost equally often. After achieving intromissions, adolescent males mount female partners almost exclusively (Wallen, 2000a, 2001), suggesting that males learn the rewarding stimulus properties of intromissions. Thus, the transition from a juvenile pattern of indiscriminate mounting of both sexes to one of exclusive mounting of females may rely on actually achieving full heterosexual intercourse for the first time.

The pubertal transition to adult sexuality depends not only on the maturation of the neuroendocrine system but also on the rearing history of the individual. Male rhesus monkeys that have limited peer experience during development undergo puberty at the same developmental time as socially experienced males, yet they do not make the transition to adult sexual behavior (Goy & Wallen, 1979). Instead, males reared with limited access to conspecifics may become hyperaggressive and socially incompatible with heterosexual interactions. It appears that social rearing is the critical factor and not the sex of the rearing companions. When tested as adults, males reared with extensive social experience in all-male groups were indistinguishable from males reared in mixed-sex groups (Bercovitch et al., 1988). Similarly, males reared only with other males showed no preference for males as sexual partners as adults (Erwin & Maple, 1976; Slob & Schenck, 1986). Thus rearing experience, as described earlier, is critical for the performance of the adult mounting pattern but is apparently not critical for the adult male preferences for female mounting partners (Wallen & Parsons, 1998b).

Attractiveness of Adolescents as Sexual Partners

Adolescent primate females go through a period of adolescent sterility. Following menarche, females may have several menstrual cycles that are anovulatory (Hartman, 1931; Resko et al., 1982; Smith & Rubenstein, 1940). In addition, some menstrual cycles may be ovulatory but have insufficient progesterone for a full luteal phase (Foster, 1977). Thus, there is a period of time when adolescent females are undergoing cyclic estradiol changes but may not be fertile, possibly reducing their attractiveness to adult males as sexual partners (Anderson & Bielert, 1994).

Several lines of evidence suggest that adult males are less attracted to adolescent females than to adult females. Male chacma baboons who have visual access to freely behaving females respond more to the visual stimuli of adult females than to adolescent females (Bielert et al., 1986). This suggests that something in the behavior or the physical characteristics of the adolescent was less attractive to the adult male baboons. Smaller size of adolescent sexual swellings is one possible factor for the difference in male responsiveness; a second possible factor, the length of adolescent perineal tumescence, did not differ from that of adults (Bielert et al., 1986). Macaque males may also find adolescents less attractive than adults. Male rhesus monkeys ignored the presentations of adolescent females more frequently than they did the presentations of adult females (Perry & Manson, 1995). In this study, however, the adolescent females also presented more frequently than adult females (Perry & Manson, 1995), and the authors did not evaluate whether the proportion of adolescent presentations that were ignored differed. In some species, the sexual swelling of adolescents is exaggerated in size over that seen in adulthood. Some researchers have hypothesized that the exaggerated sexual swellings seen in adolescence are an evolutionary response to decreased adolescent attractiveness (Anderson & Bielert, 1994).

Adolescent Sexuality: A Complex Transition to Adulthood

Unlike juvenile development, adolescent development reflects a maturation of both neuroendocrine and behavioral systems. Without either system the transition to adult sexuality is unlikely to occur. This area remains one of the most incompletely investigated in primate psychology. This is partly because it almost demands longitudinal studies, which are time-consuming, expensive, and fraught with difficulty. Critical events occur rapidly and at different times in different individuals. The patterns seen in adolescence and adulthood reflect, to some extent, the early rearing history of individuals, but only in terms of whether or not "normal" adult sexuality is achieved. The relationship between early social experience and individual differences in adult sexuality remains unknown. Such differences are small enough, however, that generalizations about adult sexual behavior can be discerned as described in the section that follows.

Adult Sexuality

Nonhuman Primate Females

In primates the ability to engage in sexual behavior is not under strict hormonal control. This is in stark contrast to rodents and other mammals in which hormones regulate the capacity to mate. For example, the female guinea pig's vaginal opening is normally closed by a membrane that disappears only when she is fertile (Young, 1937). In other species, such as rats and mice, sexually receptive females must display the lordosis response, a concave arching of the back with a rotation of the pelvis, in order for the male to achieve intromission. This behavior is under strict hormonal control and occurs only when the female is fertile (Pfaff, 1999). The female additionally shows enhanced sexual motivation concurrently with her ability to mate. In most mammalian females, estrogen and progesterone regulate both this capacity to mate and interest in mating. In contrast, primate females are capable of engaging in sexual activity at any time, but the motivation to do so changes with the underlying hormonal environment (Wallen, 1990, 2001). This distinction between the ability to engage in sexual behavior and the motivation to do so is also represented in male primates, in which testicular hormones are not required for sexual behavior but enhance sexual motivation (Wallen, 2001). Humans share with their primate cousins this capacity to engage in sex at any time, making it difficult to reach agreement on the relationship between hormones and adult sexual behavior. The complexity of the relationship is best illustrated by studies of female primates.

Substantial evidence suggests that primate female sexual behavior varies with the ovarian cycle, peaking at midcycle (rhesus macaques: Czaja & Bielert, 1975; Gordon, 1981; Wallen et al., 1984; Wilson et al., 1982b; capuchins: Linn et al., 1995; gorillas: Nadler, 1982; common marmosets: Dixson, 1987). This peak in behavior occurs near ovulation when estrogen and androgen levels are highest and during the presumed period of maximal fertility. The situation is less clear in humans in that the clearest change in sexuality in relation to the ovarian cycle is in sexual desire and not sexual intercourse (Wallen, 2001). This reflects the fact that, in both human and nonhuman primates, the extent to which sexual behavior varies with the ovarian cycle is strongly influenced by social context. For example, in studies of heterosexual pairs of rhesus monkeys, ovariectomy reduces but does not completely eliminate sexual behavior (Johnson & Phoenix, 1976; Wallen & Goy, 1977). In contrast, when ovariectomized rhesus monkey females are tested in stable social groups with multiple males and females, mating behavior ceases completely if there is no hormonal replacement therapy (Pope et al., 1987; Zehr et al., 1998). Similarly, the degree to which mating behavior is coupled to the female's ovarian cycle varies according to social conditions, with much weaker coupling in pair tests and much stronger coupling in environments with multiple females (Wallen, 1990). Similarly, social context strongly affects the occurrence of sexual behavior in humans, making the identification of consistent hormonal influences difficult. The large body of research done on this issue suggests, however, that hormones modulate important psychological mechanisms that in turn regulate sexual behavior. The critical issues have arisen from studies of rhesus monkeys, which have been studied in the widest range of environments with the most detailed behavioral and physiological work.

Early field studies of rhesus monkeys found that sexual behavior occurred for limited periods of time and near the midcycle portion of the female's ovarian cycle (Carpenter, 1942; Conaway & Koford, 1964). This view was not widely accepted because studies of laboratory pair tests, in which great control was possible, found neither clear cyclical changes in female sexual behavior nor limited periods of mating (Goy, 1979; Michael & Zumpe, 1970). This discrepancy was resolved when captive groups of monkeys were studied under conditions that approximated the social complexity of the wild while allowing accurate measurement of behavior and hormonal changes (Gordon, 1981, Wallen et al., 1984; Wilson et al., 1982b). The difference between the two environments stemmed in part from the much smaller areas used for laboratory pair tests (Wallen, 1982) and in part from the number of females present in the testing situation (Wallen & Winston, 1984). Because of the small areas used in laboratory pair tests, it appeared that males controlled sexual interactions, leading some investigators to conclude that "the male's ejaculatory frequency can be a more consistent and sensitive index of the female's sexual status than the overt behavior of the female herself, at least as far as can be discerned by human observers" (Michael & Bonsall, 1979, p. 296). When rhesus monkeys were studied in large areas with multiple females, however, it became evident that females initiated the majority of sexual interactions, and the primary influence of the female's hormones was to modulate her initiation of sexual activity (Wallen et al., 1984). Under these complex social conditions, it also became clear that increases in female behavior were positively correlated with estradiol levels and negatively correlated with progesterone. The degree of correlation between estradiol levels and sexual initiation, however, is modulated by rank within the social group, where the sexual behavior of the highest-ranking female shows the least hormonal modulation (Wallen, 1990). Thus the extent to which hormones modulate sexual behavior is affected not only by social context but also by social relationships within the context. Why would this be so?

In the rigid hierarchy of the rhesus macaque social groups, sexual initiation entails social risk (Wallen, 2000b). By closely associating with males, which is uncommon outside a mating context (Wallen & Tannenbaum, 1997), the female places herself at risk both physically and socially. During periods of increased mating behavior, as seen near ovulation, females experience increased threat and aggression from other females, particularly those that are higher ranking. Under these conditions, females who are not particularly attracted to a male or particularly interested in sex simply avoid sex and the negative social consequences it can engender. Thus, under complex social situations, sexual motivation critically modulates sexual behavior, and to the extent that gonadal hormones regulate sexual motivation, they also regulate sexual behavior. It is therefore not surprising to find that the sexual behavior of low-ranking females, who are at the greatest social risk, is much more tightly coupled to hormonal state than is the behavior of high-ranking females, who are at liberty to engage in sexual behavior at any time with whoever interests them (Wallen, 1990).

While peak estrogen levels coincide with increased sexual behavior, increasing levels of progesterone appear to inhibit sexual behavior. Baum and colleagues (1977), using data from pair tests, reported that progesterone treatment decreased the sexual behavior that a female receives, perhaps by making her less attractive to the male. Studies of females within a larger social group found that sexual behavior abruptly stopped when endogenous progesterone levels increased (Gordon, 1981; Wallen et al., 1984; Wilson et al., 1982b). In addition, female initiation is inversely correlated with endogenous progesterone levels in both intact females (Wallen et al., 1984) and ovariectomized females treated with estradiol (Zehr et al., 1998).

Adrenal Androgens and the Sexuality of Women

The work already summarized provides evidence that ovarian hormones, particularly estrogen and progesterone, modulate female sexual desire and thus female sexual behavior. In contrast to other mammalian females, there is evidence among primates, both human and nonhuman, implicating androgens, particularly androgens of adrenal-cortical origin, in the modulation of female sexual behavior (Baum et al., 1977; Sherwin & Gelfand, 1987; Waxenberg et al., 1959). While adrenal androgens could not produce the cyclical fluctuations in female sexual initiation and desire typical of human and nonhuman primate females, the notion that ovarian androgens may be critical for female sexuality is widely held and, if true, would set primate females apart from other mammalian females in which estrogens and progestins regulate sexual behavior.

The idea that androgens affect female sex drive comes from studies dating from the 1930s and 1940s on the effects of synthetic androgens on female sexuality. Treatment with large amounts of testosterone, frequently accompanied by clitoral growth and deepening of the voice, increased libido in women (Foss, 1951; Geist, 1941; Loeser, 1940; Salmon & Geist, 1943). Such findings, when combined with a study claiming that ovariectomy produced no diminution of libido in 88 percent of the subjects (Filler & Drezner, 1944), led to the view that androgens and not the ovary were important to female sexuality. The source of endogenous androgens was unclear, as it was unknown at that time that the ovary secreted large amounts of androgen at midcycle. It was known, however, that the adrenal cortex secreted androgens, and this became the focus of research on hormonal influences on female sexuality.

The view that the adrenal cortex was the source of the androgens responsible for female libido derives from a study which reported that adrenalectomy plus ovariectomy almost completely eliminated sexual activity in terminally ill breast cancer patients, whereas ovariectomy by itself had little effect (Waxenberg et al., 1959). Although this study has had a marked impact on thinking about female sexuality in both human and nonhuman primates, there are several reasons to be skeptical about its conclusions. First and foremost is that the authors focused on the occurrence of sexual intercourse instead of interest in sex, which is most likely to be affected by hormones in women. Not surprisingly, they found little effect of ovariectomy on sexual intercourse, but the decrease in sexual interest following ovariectomy was as large as the decrease they found following ovariectomy and adrenalectomy (Waxenberg et al., 1959). Thus, this study actually provided the first evidence that ovariectomy reduced female sexual motivation but had minimal effects on the capacity to have sexual intercourse. Furthermore, the authors concluded that the adrenal cortex is critical to female sexual function, but their study lacked the critical experimental group of women receiving only adrenalectomy. This group is needed to demonstrate that the adrenal cortex is necessary for female sexuality when a functioning ovary is present. Thus, the design of this study precluded demonstrating a critical role for the adrenal hormones in female sexuality. Surprisingly, since 1959, there have been no other studies of the psychological effects of adrenalectomy, yet the notion that adrenal hormones are important remains commonly held in human sexuality.

Following this study in humans, the role of adrenal secretions in rhesus monkeys was investigated using pair tests. These studies suggested that adrenalectomy (or suppression of the adrenal gland with the synthetic gluco-corticoid dexamethasone) diminished or abolished female sexual receptivity (Everitt & Herbert, 1969, 1971; Everitt et al., 1972; Johnson & Phoenix, 1976). Testosterone and, in some cases, androstenedione reinstated female sexual behavior in animals without adrenal androgens. It is significant, however, that only aromatizable androgens (e.g., testosterone or androstenedione) rein-stated sexual behavior, while the non-aromatizable androgen 5-dihydrotestos-terone was ineffective (Wallen & Goy, 1977). Thus, female sexual behavior is reinstated only with large, nonphysiological amounts of androgens that can be converted to estrogen (Johnson & Phoenix, 1976). In the musk shrew, the only other mammalian female in which androgens have been suggested to be important for female sexuality, androgens are important for female sexual receptivity but only if they are aromatized (Rissman, 1991; Veney & Rissman, 2000). In contrast to rhesus monkeys, adrenalectomy has no detectable effect on female sexual behavior in pairs of marmosets (Dixson, 1987) or stumptail macaques (Baum et al., 1978; Goldfoot et al., 1978). Although species differ-ences cannot be ruled out, it is possible that the effects reported in rhesus monkeys reflect the use of pair tests and ovariectomized subjects.

Adrenal suppression, via chronic dexamethasone treatment, in group-liv-ing rhesus monkey females with intact functioning ovaries had no detectable effect on female sexual behavior (Lovejoy & Wallen, 1990). This treatment reduced androgen levels by more than 75 percent but, because ovarian func-tion remained, had no effect on estradiol levels. In contrast, ovarian sup-pression using a GnRH agonist, which produced ovariectomized levels of estradiol, almost completely eliminated female sexual initiation and stopped sexual activity in group-living females (Wallen et al., 1986). This dramatic decrease in female sexual behavior occurred even while the females experi-enced normal levels of adrenal androgens. Thus ovarian suppression, but not adrenal suppression, reduces or eliminates female sexual activity in group-liv-ing rhesus monkeys, and the effect is more related to estradiol than to andro-gen levels.

Finally, cyclic fluctuations in sexual desire are incompatible with the tonic sex steroid secretion produced by the adrenal. As evidence accumulates of cyclic changes in sexual interest, there is little reason to continue to hold the notion that adrenal function is critical to normal sexual functioning in women. Still, the question of whether ovarian androgens or estrogens are the primary modulators of female sexuality remains controversial because human studies have failed to find robust correlation between either group of steroids and female sexual behavior. This is partly because studies rarely measure both estrogens and androgens within the same population and partly because human studies do not allow the frequent hormonal and behavioral sampling necessary for the detection of correlations with the rapidly changing levels of estradiol.

Correlative Studies of Androgens and Estrogens and Female Sexual Behavior

In one study of rhesus monkeys, female sexual initiation correlated positively with estrogen and negatively with progesterone, while no pattern of female sexual behavior correlated significantly with testosterone (Wallen et al., 1984). In a different study, when female estradiol, progesterone, and testosterone levels were used to predict when copulation occurred in female rhesus monkeys living in a social group, a model using estradiol and progesterone accounted for more variance ($R^2 = 0.49$) than either estradiol, progesterone, or testosterone alone (Wilson et al., 1982b). The addition of testosterone to the model did not significantly increase the variance accounted for ($R^2 = 0.52$: Wilson et al., 1982b). Together these findings support the idea that ovarian function regulates female sexual motivation and that estrogens, not androgens, are the critical gonadal steroids. These findings cannot rule out an effect of androgens, as in all studies some estrogens and androgens were present. These studies clearly demonstrate, however, that androgens are not sufficient to increase female sexual motivation.

Few studies in humans have attempted to measure both androgens and estrogens in relation to female sexual behavior, and those that have done so have produced mixed results. Persky and colleagues (Persky, Charney, et al., 1978; Persky, Lief, et al., 1978) reported no significant relationships between changes in either estrogen or testosterone and sexual behavior in married couples. Yet the average peak estradiol values, which reflect preovulatory estradiol, correlated very strongly with the wife's average sexual initiation score ($r = 0.68$, $n = 11$, $p = 0.02$). In contrast, the women's testosterone levels were not significantly correlated with their sexual initiation scores ($r = 0.37$, $n = 11$, $p = 0.26$). This analysis, derived from data presented in the published paper, suggests that estradiol may have an effect on female sexual initiation that is masked in the daily variance of the cycle. Interestingly, in other studies that measured estradiol from daily urine samples, intercourse frequency was significantly higher on the day preceding peak urinary estrogen levels, the presumed day of peak blood estrogen levels (Hedricks et al., 1987, 1994). Other investigators, however, have not found a significant relationship between daily urinary estrogens and a measure of female sexual interest (Dennerstein et al., 1994).

A more recent study investigating this issue (van Goozen et al., 1997) measured estrogens and androgens three times per week and reported that average androgen levels across the cycle correlated significantly with frequency of sexual intercourse, masturbation, and average sexual interest. This study also reported, however, that ovulatory androgen levels were not significantly correlated with sexual interest even though female sexual initiation was highest at this time. Furthermore, these significant correlations pertained only to a subgroup of approximately one-half of the subjects who had premenstrual complaints, but not the other women who experienced no premenstrual problems. Average estradiol levels were not significantly correlated with any measure of sexual behavior, but the use of average cyclic values instead of peak values prevented detecting periovulatory changes in estrogens and behavior. Surprisingly, even though this study included alternate-day estradiol measurements, no attempt was made to correlate changing estradiol levels with behavioral measures, although inspection of the behavioral and hormonal figures presented in the article suggests a tighter fit between female sexual initiation and estradiol than between behavior and testosterone (van Goozen et al., 1997). Thus, after more than thirty years of research, there is still no human study that has correlated daily changes in ovarian hormones with daily ratings of sexual desire and looked at both estrogens and androgens. Existing evidence suggests weak but sometimes significant correlations with testosterone. These measures of testosterone, however, often reflect average levels across the cycle and not peak levels or cyclic variation. Available evidence also suggests that estrogens strongly influence sexual motivation in women, but unfortunately this possibility has been less investigated than the potential role of androgens in female sexual motivation. The same imbalance is evident in studies of hormonal replacement therapy (HRT).

Postmenopausal Hormone Replacement Therapy in Women

Perhaps the largest body of research on the effects of androgens on female sexuality comes from studies of HRT for postmenopausal women experiencing decreased sexual desire. It is now well established that ovariectomy markedly reduces female sexual desire (Dennerstein et al., 1977; Leiblum et al., 1983; Sherwin, 1985). The question of which hormones restore sexual desire remains controversial, however. In one of the first studies of surgically menopausal women, Dennerstein and colleagues (1977) suggested that estrogen replacement therapy did not affect overall sexual behavior, although there was a specific decrease in pain from intercourse, which might reflect increased vaginal lubrication. In a later study this same laboratory reported increased sexual desire and orgasm frequency in surgically menopausal women during estrogen therapy in comparison to either a placebo or a progestational control (Dennerstein et al., 1980). More recently, Sherwin and colleagues (Sherwin 1985, 1991; Sherwin & Gelfand, 1987; Sherwin et al., 1985) have investigated the effects of estrogen therapy or combined estrogen-androgen therapy in surgically menopausal women. A double blind, placebo-controlled study by this group reported that women receiving a combination of estrogen and androgen had higher levels of sexual desire, arousal, and fantasy than those receiving estrogen alone or placebo (Sherwin et al., 1985). The results, however, are not completely consistent with androgen regulation of female sexual desire, since the surgical control group, which had naturally low androgen levels and elevated estrogen levels, expressed levels of sexual desire comparable to those in the group receiving the estrogen and androgen HRT, which had substantially higher androgen levels but comparable estrogen levels. Furthermore, during the placebo month, the control group showed higher levels of desire and fantasy than any other group, despite having lower androgen levels than these other groups (Sherwin et al., 1985). Because subjects had recently undergone ovo-hysterectomy, lingering effects of surgical trauma may have affected these results.

A later study in long-term ovo-hysterectomized subjects controlled for this effect and appeared to show a marked enhancement of sexual desire in women receiving a combined estrogen-androgen therapy in comparison to either estrogen alone or no HRT (Sherwin & Gelfand, 1987). But the results of this study are also not consistent with androgen regulation of female sexual desire. All subjects had previously received either the estrogen-androgen HRT, estrogen HRT, or no HRT treatment. Prior to the start of data collection, all subjects stopped their specific hormonal therapy for eight weeks. At baseline, after eight weeks without HRT, all subjects reported very low levels of sexual desire and sexual fantasy (Sherwin & Gelfand, 1987), demonstrating that ovariectomy reduces female sexual motivation. When HRT therapy was reinstated, sexual desire and fantasy increased markedly in the combined estrogen-androgen HRT group, but not in the estrogen HRT and no HRT groups. While this study appeared strongly to support the importance of androgen in combination with estrogen for reinstating female sexual desire, it also provided clear evidence that androgen by itself was not capable of increasing female sexual desire. At baseline, all three subject groups reported low levels of sexual desire and had uniformly low basal estradiol levels. Yet the basal testosterone levels in the group that had previously received the estrogen-androgen HRT were significantly higher than in the other two groups of subjects and four to five times higher than peak midcycle levels in non-ovariectomized women (Sherwin & Gelfand, 1987). Thus, even though all three groups of women reported uniformly low levels of sexual desire, one group had circulating androgen levels well above peak physiological levels without any apparent effect on sexual desire. It was only when additional estrogen was given to these women that sexual desire increased, strongly suggesting that androgen alone does not increase female sexual desire.

Further support for little or no effect of testosterone on female sexual desire comes from a study of sixty-five surgically menopausal women experiencing low sexual desire and sexual satisfaction (Shifren et al., 2000). All subjects received conjugated estrogens and either a placebo or 150μg or 300μg of daily testosterone in a counterbalanced double-blind design. None of the treatments increased an index of sexual thoughts or desire to the level reported for nonmenopausal women, but all treatments, including the non-hormonal placebo, increased sexual thoughts or desire above the baseline. There was some evidence that the 300ug testosterone dose increased the composite score of sexual functioning above the placebo, but this treatment also produced serum levels of testosterone that were almost twice peak endogenous levels (Shifren et al., 2000). Furthermore, for the thirty-one women in the study under forty-seven years old, there was a significant effect of the placebo on their sexual functioning score but no additional effect of either testosterone treatment. Thus, although this study purports to show an effect of testosterone on female sexuality, it finds only a weak effect with nonphysiological doses and provides evidence of a larger effect of the placebo than of either testosterone treatment. Interestingly, subjects under all treatment conditions showed significantly elevated luteinizing hormone and follicle-stimulating hormone levels, suggesting that the chronic estrogen treatment all subjects received was not sufficient to produce negative feedback suppression of gonadotropin secretion. The possibility remains that this estrogen treatment was insufficient to influence female sexual desire.

Data from naturally, not surgically, menopausal women suggest that estrogen by itself increases sexual desire and arousal in postmenopausal women relative to a period without hormonal administration (Sherwin, 1991). Progesterone added to estrogen under these conditions did not alter libido, although it increased scores of negative psychological symptoms. Several other studies have also found a positive effect of estrogen administration on female sexuality in menopausal women (Dennerstein et al., 1980; Iatrakis et al., 1986; Sherwin, 1991). These data from naturally menopausal women are in contrast to studies suggesting that estrogen therapy alone (or in combination with progesterone) has little effect in improving sexual desire or behavior (Campbell & Whitehead, 1977; Coope, 1976; Furuhjelm et al., 1984; Sherwin, 1985; Utian, 1972). Other authors have reported that long-term androgen levels, but not estrogen levels, correlate with postmenopausal sexual interest (Bachmann & Leiblum, 1991; Bachman et al., 1985; McCoy & Davidson, 1985). Dow and colleagues (1983), however, could show no advantage of estrogen plus testosterone administration over estrogen alone, whereas Sarrel and colleagues (1998) found a significant increase in sexual desire using a combined estrogen-androgen HRT in women dissatisfied with estrogen HRT.

The reason for the discrepancies among these studies is unclear but may be related to the role that sex hormone binding globulin (SHBG) plays in regulating free hormone levels. It has been suggested that SHBG serves as a biological servomechanism that regulates the relative bioavailability of estrogens and androgens through differential binding and the differential effects of these two classes of hormones on SHBG production (Burke & Anderson, 1972). Sarrel and colleagues (1998) reported that SHBG levels increased during estrogen HRT but decreased during combined estrogen-androgen HRT. They also reported that free testosterone increased during the combined HRT, but they did not measure free estradiol, which also binds to SHBG and which also would increase with decreased SHBG levels. Work in our laboratory found that androgen administration to chronically estradiol-treated ovariectomized rhesus monkeys increased the unbound fraction of estradiol in the blood. This increase in unbound estradiol was related to increased sexual initiation in female rhesus monkeys, who had been sexually inactive on estradiol treatment alone (Wallen & Parsons, 1998a). This finding suggests that androgens may play a crucial role in regulating estrogen availability through dynamic interactions with serum-binding proteins and that estradiol is likely to be the hormone most responsible for modulating female sexual desire.

The focus on androgens in human studies has meant that little attention has been paid to the dynamics of free estradiol and its relationship to the presence of androgens. In this regard it is interesting that in intact cycling females the strongest correlations between hormones and behavior are found with estradiol, and in ovariectomized women the strongest effects on female sexual desire are obtained with combined estrogen and androgen treatments, not with either estrogen or androgen alone. Both estrogens and androgens vary in concert during the female ovarian cycle. It seems likely that their coordinated fluctuation is an adaptation that regulates the relative bioavailability of either or both to influence fertility and sexual behavior. The resolution of this issue requires investigations of the dynamics of both estrogens and androgens in relation to serum binding and changes in sexual desire. It does seem, however, that primate females are unlikely to be remarkably different from other mammalian females in relying solely on androgens for sexual motivation.

Nonhuman Primate Males

Primate males do not radically differ from female primates in the role that gonadal hormones play in regulating their sexual behavior. Like that of females, the male's ability to engage in sexual behavior is not under gonadal control. Years after castration, male rhesus monkeys can achieve vaginal intromission and show the ejaculatory reflex (Phoenix, 1977). Similarly, long-term castrated men continue to engage in sexual intercourse, even though they have extremely low levels of androgen (Heim, 1981). In rodents, by contrast, the ability to attain an erection, achieve intromission, and ejaculate is closely tied to testicular hormonal effects on penile morphology (Beach & Levinson, 1950; Phoenix et al., 1976). Again, as in female primates, hormones modulate the motivation to engage in sexual behavior, and in males this results from the actions of androgens. The main difference between male and female primates in the modulation of sexual behavior is that male gonadal hormones produce constant levels of sexual motivation whereas hormones produce cyclical motivational changes in females.

In male primates, castration decreases the frequency of sexual behaviors, but the male is still capable of the full suite of sexual behavior, including ejaculatory reflexes. In males that either are castrated or have testosterone suppressed to castration levels, social context affects the speed with which sexual behavior declines and how long it continues to be displayed. In pair tests males display full sexual behavior for weeks longer than they do in group-testing environments (Wallen, 2000b). Since mating behavior in a pair test poses little or no risk to the male, little motivation is required to initiate sexual behavior. Conversely, in group tests, male rhesus face competition from other males, and low-ranking males face much social risk by engaging in sexual behavior, making sexual motivation a critical factor for the occurrence of such behavior.

Suppression of testosterone or the elimination of testosterone by castration may not limit the ability to engage in sexual behavior, but it does result in a marked decrease in sexual activity. Castration reduces the overall frequency of sexual behavior in monkeys when tested with a single female but does not eliminate it even over a year later (Phoenix et al., 1973). The administration of androgens can restore the sexual behavior of males to pre-castration levels (Phoenix, 1973, 1974a). Again, this suggests that androgens enhance the motivation of males to engage in sexual activity. Just as was seen in females, the presence of a socially dominant individual may decrease the sexual activity of a subordinate male, even if the dominant individual is merely caged in the same room (Zumpe & Michael, 1990). In multi-male social groups, the degree to which the removal of testicular hormones affects the expression of sexual behavior may be mitigated by social rank. Following the administration of a GnRH antagonist, which effectively suppressed testicular function, the sexual behavior of all seven of the males in a multi-male, multi-female group decreased significantly after a single week of low androgen (Wallen, 2000b). When the rank and sexual experience of these males was taken into consideration, however, the modulatory role of testosterone was more apparent.

Among the sexually experienced males, the highest-ranked male showed little to no reduction in sexual activity, despite his low levels of testosterone. In contrast, the lowest-ranked male showed a tight coupling of his sexual behavior with his testosterone levels, engaging in sexual behavior only when his testosterone levels were high (Wallen, 2000b).

Human Males

In men, as in other nonhuman primates, the capacity for erection is not under hormonal control. Hypogonadal males (i.e., males who have castrate-like levels of testosterone as a result of illness or accident) display erections to sexual stimuli as readily as do control males with normal testosterone levels (Bancroft & Wu, 1983; Kwan et al., 1983). Although testicular androgens do not regulate the ability to engage in sexual behavior in human males, they do powerfully influence the motivation to engage in sex. For example, long-term castrated males engaged in sexual intercourse much less frequently than they did prior to castration, suggesting a decrease in sexual motivation (Heim, 1981). Interestingly, these males reported a substantially greater decline in masturbation than in sexual intercourse. Sexual intercourse reflects the sexual interests of both partners, and thus its occurrence may reflect responses to social pressures to engage in sex that do not reflect the motivation of both partners. In contrast, masturbation more strictly reflects the sexual desire of the individual. Further support for androgens modulating male sexual motivation comes from studies investigating GnRH antagonists as a possible male contraceptive. As was observed in the rhesus monkey, treatment with a GnRH antagonist suppresses testicular function, rapidly producing castrate levels of testosterone (Bagatell et al., 1994). While testosterone levels were suppressed, the males reported decreased sexual desire, intercourse frequency, and masturbation (Bagatell et al., 1994). Concomitant administration of testosterone to gonadally suppressed males prevented the decline in sexual desire and activity. The evidence currently available supports the idea that gonadal hormones in primates, both males and females, human and nonhuman, primarily affect sexual motivation and not the physical capacity to engage in sex.

Primate sexuality reflects individual life history, current social context, and neuroendocrine state. In addition, it is likely that there are species-specific differences that are the product of phylogenetic history or ecological adaptation.

At this time the weight of the evidence supports the importance of social factors in the development and expression of primate sexual behavior. Early experience shapes behavioral sex differences, which in turn facilitate male and female life histories. Restrictive early socialization can completely eliminate adolescent and adult mating behavior, with males more easily and more profoundly affected than females by restrictive early experience. Early experience, however, does not act on a completely malleable substrate. Prenatal hormones produce specific behavioral predispositions in males and females that are subsequently articulated by early social experience.

Adolescence, a period of some of the most rapid changes in physiology and behavior in a primate's life, integrates the behavioral systems expressed during development with neuroendocrine influences initiated at puberty. Adult sexuality results from the maturation of this behavioral and neuroendocrine integration, which produces a hormonally modulated sexual motivation system, making fertile mating more likely. This behavioral and neuroendocrine integration, however, does not prevent mating for nonreproductive social reasons. Primates appear to be uniquely adapted to engage in sex at any time, if the social conditions either require or permit such behavior.

The evolutionary reasons for such behavioral flexibility remain elusive, although explanations ranging from protection against infanticide to tension reduction have been offered as the selective pressures leading to the evolution of this adaptation. The fact that sexual motivation is under strong hormonal regulation in primates as it is in nonprimate mammals illustrates the crucial importance of psychological mechanisms to understanding the regulation of behavior. How and when the influence of hormones on the capacity to mate in primates disappeared remains poorly investigated. It is apparent that the changes that resulted in the dramatic evolution of complex cortical mechanisms were accompanied by increased psychological influence on sexual behavior. Whatever the ultimate explanation for this evolution, it is abundantly clear that a complete understanding of primate sexuality requires knowledge of the tight integration of early experience, current social context, and hormonal history.

❖

Preparation of this chapter was supported in part by grant MH 50268-K02 (KW) and NSF Fellowships to JLZ, RAH, and FCG.

5

Attachment

Dario Maestripieri

In a well-known trilogy of volumes published in the 1960s and 1970s, the British psychoanalyst and psychiatrist John Bowlby developed a new theory addressing the processes underlying the formation and breaking of socio-emotional bonds between infants and their caregivers as well as those underlying the role of early social experiences in the development of personality (Bowlby, 1969, 1973, 1980). Bowlby's ideas were stimulated by observations of children's responses to separation from their parents and were developed in reaction to the psychoanalysts' claim that infants become attached to their mother because she meets their basic physiological needs, notably food. His attachment theory was a monumental body of work, incorporating notions from psychoanalysis, cognitive psychology, psychiatry, control systems theory, ethology, and evolutionary biology. Attachment theory had a major impact on research on parenting and development and is still one of the major conceptual paradigms for research in psychology.

According to Bowlby, infants have a biological predisposition to become attached to a caregiver, which manifests itself in behaviors aimed at maintaining proximity or stimulating interaction (e.g., crying, smiling, following, or

108

clinging). Human infants are totally dependent on their caregivers for protection from the environment. Thus, Bowlby argued that the attachment system probably evolved by natural selection in what he referred to as the environment of evolutionary adaptedness as a set of psychological and behavioral adaptations that promoted infants' survival by enhancing their proximity to and interaction with a caregiver. The behavioral expressions of the attachment system change with age, but the function of the system and the way its components are organized tend to remain consistent.

really?

Bowlby argued that attachment and other motivational/behavioral systems (e.g., fear/wariness, exploration, affiliation, and sex) have different set goals and specific activating and terminating conditions. In the case of attachment, the set goal of the infant is not an object (the mother) but a state: the maintenance of contact with or proximity to the mother. The attachment system is activated when the infant is separated from the mother and is terminated when contact or proximity is achieved. The system thus works like a thermostat that measures the current temperature and, after comparing it with a preset standard, makes adjustments. Unlike in the thermostat analogy, however, the terminating conditions vary according to the intensity of the activation. When a system is intensely active, nothing but physical contact will serve to terminate it. When intensity is lower, the sight or voice of the caregiver will be sufficient to terminate it. The set goal itself varies in relation to both exogenous (e.g., distance from the mother, natural clues to danger) and endogenous factors (e.g., illness or fatigue, hunger).

really?

Although Bowlby emphasized proximity to the caregiver as the set goal of the attachment system, other attachment theorists subsequently emphasized the cognitive aspects of attachment. Ainsworth and colleagues (1978) argued that when a mother leaves her child, it is not the departure in itself that matters so much as the child's evaluation of the mother's departure based on previous experiences (i.e., the child's ability to interpret the meaning of the departure and predict the mother's subsequent behavior). Thus, there is an important cognitive component in the attachment system related to the child's appraisal of the caregiver's availability and responsiveness. Ainsworth et al. (1978) thus proposed the use of the adjective "secure" to describe a child's confidence that the caregiver will be accessible and available, while Sroufe and Waters (1977) proposed that "felt security" should be considered

secure attachment not always most adaptive

the set goal of the attachment system. In this view, any threats to felt security cause anxiety, and repeated experiences of threatened felt security lead to the formation of insecure attachment.

According to Bowlby, the attachment system interacts with other systems in a complex manner. The attachment system can be activated simultaneously with the fear/wariness system, and they both inhibit the exploration system. Thus, when the infant is frightened or anxious, she wants to be near the mother and does not explore the environment or play. When the infant feels secure, she will explore and play. To investigate how the different systems mutually interact and how the mother can serve as the infant's secure base for exploration, Ainsworth et al. (1978) developed an experimental procedure called the Strange Situation Test (SST). This procedure takes place in a labora-tory room with toys and involves three individuals: the child, the caregiver, and an unfamiliar individual (stranger). The test consists of seven episodes, each lasting three minutes or less. During these episodes, the child is in the room alone, alone with the caregiver, alone with the stranger, or with both the caregiver and the stranger in a predetermined sequence. The tests are video-taped, and both attachment behaviors and exploration/play are recorded. The child's behavior, and in particular her response to separation and reunion with the caregiver, are scored and later classified as Secure (B), Avoidant (A), or Ambivalent-Resistant (C). Different attachment subtypes are recognized for each category, and a fourth attachment category (D; see below) has recently been introduced. Most children tested with the SST are classified as secure. The occurrence and relative proportion of children classified into the insecure categories (A, C, or D) show some cross-cultural variability (van Ijzendoorn & Sagi, 1999). The SST can be used with children who exhibit clear attachment behaviors to their primary caregivers, who have achieved complete or almost complete locomotor abilities, and who are disturbed by a novel environment and the presence of a stranger. Thus, this test is commonly used with children between one and three years of age. The SST is not used before attachment is established (see below) or with children who are too old to be responsive to this particular experimental situation.

According to Bowlby, human infants are born with the predisposition to respond to particular stimuli, especially those provided by the caregiver. In his view, attachment develops in four stages. Stage 1 ("orientation and signals without discrimination of figure") occurs from birth to between eight and twelve weeks of age. In this stage there is no attachment to caregivers, although by four weeks of age infants already show a preference for looking at the human face and listening to human voices. Babies respond to stimuli in a manner that increases the likelihood of continued contact with other humans and behave the same way toward many people. The baby understands that certain interactions have predictable outcomes, but at this stage, contact and proximity between infant and caregiver are mostly maintained by the caregiver. Stage 2 ("orientation and signals directed toward one or more discriminated figures") begins at eight to twelve weeks and ends at six months of age. This is the stage for pre-attachment. In this stage there occurs a restriction of effective activating and terminating conditions, and the infant's behavior patterns are differentially directed to the caregiver. Infants gradually learn the perceptual attributes of the caregiver and manifest their innate bias toward approaching whatever is familiar. Learning by positive and negative reinforcement, mostly in response to the behavior of the caregiver and other individuals, also contributes to the attachment behaviors being primarily directed to the caregiver. At this stage infants often initiate interactions with other individuals rather than just being passive and responding to their behavior. Infants cannot yet understand, however, that the caregiver is someone with a separate existence from their own experience.

Stage 3 ("maintenance of proximity to a discriminated figure by locomotion and signals") begins at six to nine months and ends at two to three years of age. This is when attachment takes place. This stage is characterized by new attachment behaviors (e.g., following the caregiver) and changes in the organization of attachment behaviors under the infant's intentional control. The infant has an internal image of an end state (set goal) that she would like to achieve and is capable of differentiating means from ends. The internal image of the attachment figure is now independent from perception, and the infant can think about the caregiver even in her absence (i.e., the infant has reached the stage of object permanence; Piaget, 1954). At this stage, the infant also has separate mental images (i.e., working models) of the self and the caregiver, although she thinks about other individuals only in terms of their behavior and its short-term consequences. In other words, the infant does not yet understand that other individuals have long-term goals and that their behavior may be aimed at achieving these goals. Stage 3 is characterized by the development of separation anxiety and fear of strangers. Thus, as the attachment system is established, the exploration and the fear systems become active as well.

The last stage of attachment formation is Stage 4 ("formation of a goal-corrected partnership"). In this stage, the child uses complex goal-directed systems in the management of her relationship with the caregiver and the external world. The caregiver is now perceived as an individual with her own goals and plans to achieve them. Attachment takes place at the cognitive rather than at the behavioral and emotional level. From now on the child becomes dependent on the working models that she has constructed from the experienced interaction patterns with her principal attachment figure. These working models can be used by the child to feel secure even when the caregiver is not present, and they serve to regulate, interpret, and predict both the caregiver's and the self's attachment-related behavior, thoughts, and feelings. The developing working models of the self and of the caregiver are believed to be complementary. For example, a working model of the self as a valued and competent individual is constructed in the context of a working model of a caregiver as an emotionally available individual who is supportive of explorative activities. In contrast, a working model of the self as a devalued and incompetent individual is the counterpart of a working model of the caregiver as rejecting or ignoring attachment or interfering with exploration. Bowlby acknowledged that developmental change in working models is required for continued security. He also emphasized, however, that working models tend to be resistant to change because representations of prior interactions bias what individuals expect and, within limits, regulate the perception of future interactions with the caregiver. Bowlby thus placed great emphasis on the intergenerational transmission of working models of attachment through parent-child interactions. Consistent with mainstream psychoanalytical theories, he viewed personality as being greatly influenced by early attachment experiences via the working models. Attachment theory therefore is not only a framework for understanding early parent-child interactions but also a theory of personality development. Accordingly, through its effects on personality, attachment is believed to play an important role not just during infancy and childhood but throughout the entire life span of an individual.

Developments and Extensions of Human Attachment Research

Bowlby's account of attachment mostly focused on the normative processes underlying the formation and breaking of emotional bonds during infancy, childhood, and beyond. Although most of his hypotheses and assumptions have withstood the test of time, some aspects of his original formulation of attachment theory have been revised or updated (in some cases by Bowlby himself) on the basis of new research findings or theoretical developments. Since the development of the SST in the 1970s, research has increasingly focused on understanding the causes and developmental consequences of individual differences in attachment. This research has pointed to early parent-infant interactions, particularly parental responsiveness to the infant's needs and demands, as one of the main determinants of different attachment styles (Ainsworth et al., 1978). A large body of evidence has also identified the positive social, emotional, and cognitive features of development associated with secure attachment as well as the risks and vulnerabilities associated with insecure attachments. Following the shift in emphasis from the behavioral to the cognitive aspects of attachment that occurred in the 1980s and 1990s, research has increasingly focused on the cognitive representations of attachment relationships in late childhood and adolescence (Bretherton, 1985; Main et al., 1985). Following the introduction of the D attachment classification ("Disorganized/Disoriented"; Main & Solomon, 1990), attachment has also been increasingly investigated in clinical populations.

Parental attachment (i.e., the attachment system complementary to infant attachment) was long neglected by researchers but is now receiving increasing attention (Bell & Richard, 2000; George & Solomon, 1999). Attachment theory has also moved beyond the study of parent-child interactions and branched into social psychology. In 1994, Hazan and Shaver suggested that attachment could serve as a theoretical framework for the study of adult romantic relationships (see also Weiss, 1982). Since then adult attachment has been a rapidly growing area of research, where both the conceptual issues and the methodological techniques of investigation have increased exponentially. Finally, attachment has also attracted the attention and interest of evolution-oriented psychologists, who have reinterpreted the significance of individual differences in attachment and linked research on attachment and social development to life history theory and sexual selection theory (Belsky, 1999; Simpson, 1999).

Primate Research on Attachment

Ethological concepts and primate research played a pivotal role in Bowlby's formulation of attachment theory. Bowlby derived from ethology (e.g., Tinbergen, 1963) the notion that a comprehensive theory of attachment should account for this phenomenon at different levels of analysis: causation (the endogenous conditions and external stimuli that activate attachment), ontogeny (the development of the attachment system in infancy and throughout the life span), function (the adaptive value of the attachment system), and phylogeny (the evolutionary continuity in attachment across humans and the other primates). The concept of behavioral system used by Bowlby is also very close to that used by ethologists to refer to a set of behavioral responses serving a particular biological end (Hinde, 1982). In addition to the conceptual contributions of ethology, observations of mother-infant interactions in rhesus macaques conducted by Hinde and co-workers provided important insights into developmental changes in the attachment system, responses to separation, and secure base phenomena (e.g., Hinde & Spencer-Booth, 1967). Finally, Hinde introduced Bowlby to the work of Harlow, which experimentally demonstrated the primary role of the caregiver as a source of contact, comfort, and protection rather than nutrition (Harlow, 1959, 1974). Thus, research on primate behavior played a crucial role in outlining the normative aspects of attachment theory, and throughout the late 1960s and early 1970s, interactions between primate and human researchers were frequent and mutually beneficial.

Unfortunately, such interactions have since become rare. Kondo-Ikemura and Waters (1995) suggested that the focus on individual differences in human attachment in the 1970s was an important source of the divergence between primate and human research. Whereas many attachment studies since the 1970s have concentrated on understanding individual differences in attachment, primate researchers have, in many cases, continued to refer to Bowlby's original formulation of attachment theory, particularly his emphasis on normative aspects of attachment and separation. Specifically, most empirical efforts by primatologists have gone into investigating the short- and long-term responses of monkey infants to maternal separation. Such responses have been viewed as an important indicator of the strength of the attachment relationship or of the role of the mother in social development.

Attachment was initially conceptualized as a behavioral system and was therefore studied with behavioral measures in both human and nonhuman primates. Over the years, however, primate research has gradually shifted from behavioral to endocrine and neurobiological measures of attachment and separation, whereas the study of attachment in humans has become increasingly cognitive. Methodological changes in the study of primate attachment have been accompanied by attempts to reconceptualize attachment as a neurobiological system (Kraemer, 1992; Mason & Mendoza, 1998; Reite & Capitanio, 1985) and identify new or broader biological functions for this system (see below). Most empirical research on primate attachment, however, has been narrowly focused on the consequences of maternal separation and loss, and the reconceptualization of attachment has been strongly influenced by this work. The narrow focus of primate attachment research has had three negative consequences. First, primate researchers interested in attachment have largely ignored most of the theoretical and empirical developments taking place in human attachment research since the 1970s. Second, other primate researchers interested in the study of parenting and development have moved away from attachment theory as a framework for their research. Finally, human attachment researchers have also found it increasingly difficult to relate their own research to primate studies and therefore have lost interest in the comparative study of parenting and development. Consequently, among many psychologists, knowledge of primate research on parenting and development including attachment is limited to the work done in the 1960s by Hinde and Harlow and the elaboration of this work by their former students or collaborators.

Primate research on attachment since that time has not been a homogeneous body of work. In the primatological literature, the term "attachment" has often been used in a very broad sense, to refer to any behavioral interactions between mothers and infants or any pairs of individuals (e.g., Erwin et al., 1973; McKenna, 1979; Mitchell, 1968; Suomi, 1995). Furthermore, attachment between two individuals has often been inferred from their responses to separation, without any information on their interactions before or after the separation (e.g., Andrews & Rosenblum, 1993; Reite et al., 1989).

This approach is implicitly based on the notion that primate attachment can be studied without a comprehensive or sophisticated understanding of the mother-infant relationship and of how such a relationship is embedded in the social environment.

Although primate and human researchers on attachment have gone in different directions for a long time, it may not be too late to attempt to bridge the gap between them. Exchange of information and interaction between primatologists and psychologists could be mutually beneficial and lead to new theoretical and empirical advances, just as it did when attachment theory was first formulated. In order to reestablish the links with human attachment research, it is important to review our current knowledge of attachment processes in primates and trace a pathway through which primate research could get closer again to human research. It is to be hoped that this will encourage a complementary effort on the part of human attachment researchers to incorporate the comparative and biological perspectives provided by primate research into their work.

Infant Attachment in Primates: Normative Aspects

Like human babies, primate infants are born with basic perceptual and cognitive skills that predispose them to respond selectively to environmental stimuli and classify them into broad qualitative categories (Mason & Mendoza, 1998). Such perceptual and cognitive skills can be inferred from a number of early postnatal reflexes and behaviors, some of which become important components of the attachment system. Primate infants can cling, suck, and cry virtually from the moment they are born, and develop the ability to follow other individuals much earlier than human babies (e.g., Hinde, 1974b). The appearance of smiling in the second month of life is an important contribution to human infants' ability to elicit proximity and caregiving behavior. Monkey infants do not have a facial expression comparable to the human smile, but a smile-like expression is not uncommon in chimpanzees (Maestripieri & Call, 1996).

Early attachment behaviors can be initially directed to a variety of conspecifics other than the mother, as well as to other animals or inanimate objects with the proper stimulus characteristics (e.g., Harlow, 1959; Mason & Capitanio, 1988). The range of stimuli capable of eliciting attachment responses, however, is gradually restricted. The formation of a specific social bond with the primary caregiver (in most cases the biological mother) takes time to develop and probably requires the maturation of specific perceptual and cognitive skills. Specifically, the occurrence of preferential responsiveness to the mother requires the ability to discriminate her from other individuals. Since rhesus infants are capable of discrimination learning by ten to twenty days, Harlow and Harlow (1965) believed that infants can recognize and bond with their mother as early as fifteen to twenty days after birth. Early experimental studies in which macaque infants were tested in an experimental choice apparatus, however, suggested that infants first express a preference for their mothers at three months of age (Rosenblum & Alpert, 1974; see also Sackett et al., 1967). A more recent study of Japanese macaques confirmed that infants can visually recognize their mothers by the age of eight to twelve weeks (Nakamichi & Yoshida, 1986). Thus, visual discrimination of the mother seems to be contingent on the maturation of the infant's visual acuity (Boothe & Sackett, 1975). Vocal recognition of the mother, however, occurs earlier. Masataka (1985), for example, provided evidence that vocal discrimination of the mother can occur as early as three to four weeks of age.

Most information on the development of attachment in primates comes from studies of changes in mother-infant contact and proximity during the first months of infant life. Such research was pioneered by Hinde and collaborators in the 1960s (e.g., Hinde & Spencer-Booth, 1967; see Chapter 6). Hinde and co-workers documented the quantitative changes in time spent in contact and proximity, and in the mother's and infant's roles in maintaining contact and proximity (e.g., Hinde & Atkinson, 1970). These studies had a great impact on the field, such that most subsequent studies of primate mother-infant relationships focused almost exclusively on measures of contact and proximity. These studies have shown that the developmental changes in mother-infant interactions during the first few months of infant life are remarkably similar across different populations and environments, suggesting that there are modal developmental curves for mother-infant interactions that are characteristic of all Old World monkeys (Maestripieri, 2001c).

In the first few days after birth, macaque infants are in almost continuous contact with their mothers; early maternal attempts to break contact and encourage independent locomotion are met with distress calls (Maestripieri, 1994a, 1995a). Initially, mothers are responsible for both breaking and making contact with their infants, and contact is typically interrupted for only a few seconds. Infants first begin to crawl over their mother's body, and beginning from the second week of life, they venture on short forays around their mother. Infants gradually increase the frequency with which they leave their mother, but mothers are still mostly responsible for reestablishing contact. At about six weeks of age, the frequency with which infants break contact reaches a plateau, and about half of the departures are followed by infants' returning to their mothers on their own (Hinde & Atkinson, 1970). From this age on, infants' exploration of the surrounding environment takes the form of short radial trips from their mother, which suggests that infants begin to use their mother as a secure base to explore the environment. Departure from the mother is inhibited or reestablishment of contact is accelerated if the infant is afraid or in distress, suggesting that the attachment, exploration, and fear systems begin to be mutually dependent. Chimpanzee infants reared by human caretakers use their primary caregiver as a secure base to explore novel environments, and younger individuals are usually less explorative and more susceptible to distress than older ones (Miller et al., 1986). In macaques, infants become primarily responsible for both breaking and making contact with mothers in the second or third month of life (Hinde & Atkinson, 1970). From now on, mothers become progressively less motivated to be in contact with their infants, and behavioral conflict over time in contact and suckling becomes intense (Hinde & Spencer-Booth, 1967; Maestripieri, 2002; van de Rijt-Plooij & Plooij, 1987).

When young monkey infants begin to explore the social environment, they typically bump into other individuals of all ages and both sexes without showing obvious signs of fearfulness or anticipation of danger. In fact, they enjoy a period of temporary immunity from adult-like patterns of aggression, although their social environment is by no means safe (Maestripieri, 1993a). Suomi (1999) suggested that the reversal of proximity maintenance with the mother that occurs between two and three months of age is associated with the development of fear of strangers. A causal relationship between development of fear of strangers and changes in proximity relationships with the mother, however, remains to be demonstrated.

Observational studies alone do not provide adequate information on the onset of fear of strangers because, in group-living monkeys, infants become gradually acquainted with other family and group members through their mothers' networks of social relationships (Berman, 1982a) and are unlikely to be suddenly exposed to unfamiliar individuals. Experimental studies suggest that avoidance and fear of strangers may not emerge until several months of age. Cross-fostering studies conducted with macaques show that infants do not show resistance to adoption by an unrelated lactating female for at least one to two months (Maestripieri, 2001a). Harlow (1974) suggested that for surrogate-reared rhesus infants, initial fear responses to stimuli placed in the home cage appear at about twenty-two days of age, whereas according to Sackett (1975), no fear responses develop until eighty days of age. Other studies of laboratory-reared infants suggested that fearful responses first appear between three to four weeks of age (Rosenblum & Alpert, 1974). In chimpanzees, nursery-reared infants begin to show wariness of human strangers at six months of age, and wariness diminishes by two years of age (Miller et al., 1990).

Fear of strangers is a component of attachment that is expected to vary significantly among different primate species. First, we may expect some differences between group-living species in which infants are constantly exposed to other conspecifics and semi-solitary species such as orangutans, in which infants may not be exposed to individuals other than their mother for several years (Horr, 1977). Second, we may expect variability in infant fear of strangers also among group-living species in relation to the frequency and quality of infants' interactions with other individuals. For example, in species such as langurs, in which infants are handled and carried by other adult females from birth (Jay, 1963) and in which most interactions are benign, fear of strangers may occur later, if it occurs at all, or be less intense. Conversely, in species such as rhesus macaques and chimpanzees, in which other individuals pose a threat to infants and mothers are very restrictive, fear of strangers may occur earlier and be more intense. The development of fear of strangers in human infants is consistent with the notion that humans have lived in groups for most of their evolutionary history and that strangers may have been dangerous to young infants.

Although the assumption of primary initiative by the infant in maintaining contact with the mother occurs several weeks after infants begin their first forays, the mother probably serves as a "safe haven" right from the earliest manifestations of infant independence. Distress calls and other signs of behavioral agitation shown by infants who have lost visual contact with their mothers or have been hurt by other group members cease almost immediately upon reestablishment of contact with the mother (Maestripieri & Call, 1996). This suggests that contact serves an important reassuring and soothing function for the infant. In fact, primate infants spend a longer amount of time on their mother's nipple than they would need for nutritive purposes, suggesting that nonnutritive sucking serves an important social function (Brown & Pieper, 1973).

Some studies have reported that initiation of ventral contact with the mother promotes rapid decreases in the activity of the hypothalamic-pituitary-adrenal (HPA) axis and in sympathetic nervous system arousal, along with other physiological changes commonly associated with soothing (Suomi, 1999). Such studies, however, have mostly documented the infant's physiological responses to reunion with the mother after human-induced separation (e.g., Levine & Wiener, 1988). To my knowledge, there are no data on the infant's physiological responses to reestablishment of contact with the mother in group-living monkeys. Studies of adult monkeys, however, have shown that being groomed by another individual reduces heart rate and is associated with a release of endogenous opioids (Boccia et al., 1989; Keverne et al., 1989), suggesting that maternal behavior, in addition to infant contact and sucking, can contribute to soothing and reassuring the infant. Although young primates use their mother as a safe haven more than any other individual, other family and group members including older sisters, aunts, grandmothers, and even unrelated adult males often protect and comfort infants (Hrdy, 1976). There is no evidence that mothers are unique in their ability to reduce their infants' physiological arousal, and it is very likely that any individual's potential to serve as a safe haven to an infant is dependent on the quality of that individual's relationship with the infant rather than on their genetic relatedness.

In summary, there is good experimental and observational evidence that primate infants (1) gradually discriminate their mother, or more generally their primary caregiver, from other individuals; (2) selectively direct their attachment behaviors toward this individual; (3) show distress and anxiety when separated from this individual; (4) use this individual as a secure base for exploring the environment; and (5) use this individual as a safe haven when they are frightened or in distress. Taken together, these findings suggest that primate infants possess an attachment system whose normative aspects are very similar to those of human attachment. In particular, the attachment system appears to serve a similar biological and social function in primate and human infancy and seems to be regulated by similar exogenous and endogenous factors. There may also be some obvious differences, however. The extent to which primate infants develop a representational model of the caregiver in relation to the self and the external environment is not clear. Although primates have excellent skills in observing and predicting the behavior of other individuals, they are unlikely to possess the cognitive abilities necessary to adopt another individual's perspective and understand that other individuals may have thoughts, knowledge, and goals different from one's own (Tomasello & Call, 1997). Other differences between primate and human attachment likely involve the ontogeny of the attachment system and the timing with which the system and its components mature and/or become activated. The ontogeny of the attachment system, and to some extent also its proximate regulation and adaptive function, are likely to vary not only between humans and other primates but also among the over three hundred primate species in relation to the life history characteristics of the species and its specific ecological, social, and reproductive adaptations. Before primate and human attachment are discussed from a broad comparative and evolutionary perspective, the study of primate attachment through separation responses and surrogate mothers will be reviewed, and the conceptual extrapolations of this work will be discussed.

The Study of Normative Attachment Processes
through Responses to Separation

The early observations of the responses to separation exhibited by institutionalized children (Robertson, 1953; Spitz, 1946) and Bowlby's (1952) review and discussion of the deleterious consequences of maternal separation and deprivation prompted the study of infant responses to separation in monkeys. Early separation studies were conducted with rhesus macaques by Harlow (1959) and Hinde (Hinde et al., 1966), with pigtail macaques by Jensen and Tolman (1962), and with pigtail and bonnet macaques by Kaufman and Rosenblum (1967). Separation studies have subsequently been conducted with virtually every species of monkeys and apes available in research laboratories around the world.

Early studies focused on the infant's behavioral responses to separation.

Such studies documented the occurrence of an initial phase of "protest" followed by a phase of "despair," similar to those observed for institutionalized children (e.g., Kaufman & Rosenblum, 1967). Although agitation following separation was ubiquitous, not all species in every environment showed a despair response (Rosenblum & Kaufman, 1968). A third phase of detachment observed in children following the reunion with their caregivers has not been observed in primates. Early descriptive studies of separation in monkeys were followed by a large number of studies investigating the effects of variables such as the number and length of separations (e.g., one versus repeated separations; separations lasting a few minutes versus those lasting days or months), the separation procedure (e.g., removing the infant from a group versus removing the mother), or the environment of separation (e.g., the home cage versus a novel environment; captivity versus natural environment) (Levine & Wiener, 1988; Mineka & Suomi, 1978). Separation studies have gradually incorporated more physiological variables (e.g., heart rate and body temperature, hormones and neurotransmitters, immunological measures; Laudenslager & Boccia, 1996; Levine & Wiener, 1988) and more sophisticated techniques of measurement (e.g., implantable multichannel telemetry systems; Reite, 1985). Some developments of this research have included postmortem studies of brain anatomy in maternally separated and/or socially deprived monkeys (e.g., Ginsberg et al., 1993).

The rationale underlying the study of infant responses to separation was, at least initially, the notion that the occurrence of a response to separation would provide information about the existence and strength of an attachment bond. Consistent with this view, the behavioral agitation and high rates of infant vocalizations following separation were viewed as serving an attachment-related function: increasing the probability of locating the mother (Kaufman & Rosenblum, 1967; Levine et al., 1987). Many studies have shown, however, that the neuroendocrine, immune, and neuroanatomical effects of separation are very similar to those observed following other forms of physical or psychosocial stress (e.g., Laudenslager & Boccia, 1996; Levine & Wiener, 1988), leading researchers such as Levine to argue that infant responses to separation are best understood within the framework of theories of stress and coping rather than attachment theory (Levine & Wiener, 1988).

Response to separation from the caregiver is an important component of attachment theory, and the occurrence of a separation response may indeed provide evidence for the existence of an attachment relationship. Beyond showing that separation is stressful, however, the contribution made by separation studies to research on attachment is unclear. For example, Insel (1992) remarked that there is no evidence demonstrating that the neurobiological systems altered by separation play a role in the formation or maintenance of social bonds. He and others (e.g., Kagan, 1992) therefore warned against research that uses concepts and measures of attachment that rely too heavily on separation, or research that tends to view attachment simply as the opposite of separation.

Swartz (1982) pointed out that separation measures alone are not sufficient to characterize primate attachment and emphasized the need for more integrated approaches. Unfortunately her suggestions have been largely ignored. The study of separation in primates and humans made an important contribution to the original formulation of attachment theory. But since modern attachment theory places little emphasis on separation responses as a way to investigate attachment processes, it is unlikely that a coordination of efforts of primate and human researchers in the study of separation will produce exciting new information that will advance our understanding of attachment.

The Study of Normative Attachment Processes with Surrogate-Reared Monkeys

In primates, just as in other animals, early infant reflexes and behaviors such as rooting, sucking, clinging, crying, and following are normally directed toward the mother. Similar to what studies of imprinting had done with birds, early research on attachment in primates investigated which specific stimuli from the mother's body or behavior elicited filial responses in the offspring (e.g., Mason et al., 1974). Thus, many studies were conducted, particularly with rhesus macaques and squirrel monkeys, in which different infant responses could be directed to different objects or surrogate mothers.

Well-known experiments with cloth and wire surrogate mothers conducted in Harlow's laboratory assessed the relative contributions of clinging and nursing to the formation of attachment (Harlow, 1974). Mason's and Kaplan's subsequent work with other inanimate surrogate mothers or live animals further explored the role of visual, tactile, and kinetic characteristics of caregivers in attachment formation (Kaplan 1977a; Mason et al., 1974). This research showed that infants reared with live surrogates such as dogs or same-aged monkey infants (peers) exhibit behavioral abnormalities that are not as dramatic as those exhibited by infants raised with inanimate surrogates or in total isolation (e.g., Mason & Capitanio, 1988).

The early groundbreaking studies of rhesus infants reared with cloth and wire surrogate mothers stimulated thirty years of research with surrogate mothers, in which monkey infants were separated from their mothers at birth and exposed to a variety of rearing environments and experimental manipulations. According to Kraemer (1997), rearing monkeys in progressively richer social environments (i.e, total isolation, partial isolation, surrogate rearing, and mother rearing) allows us to assess "the effects of the addition of components of social stimuli that the infant would usually experience, without providing the full component of usual peri- and postnatal stimulation until the attachment object is the real mother . . . Hence, when we compare surrogate- and peer-reared monkeys to mother-reared monkeys, we are comparing the effects of exhibiting attachment behavior towards an unresponsive object, to the effects of being attached to a living being, to the effects of being attached to a living being and being mothered" (p. 404). In reality, isolation-reared, surrogate-reared, and peer-reared infants are not just individuals who lack various degrees of the mothering experience. Instead, these are individuals who experience a traumatic event after birth (i.e., separation from their own mother); spend most of their infancy in artificial environments such as incubators, nursery rooms, and playpens; interact with their human caregivers in different ways and for variable periods of time; are deprived of multiple sensory, motor, and social stimuli; and are exposed to novel stimuli in a manner that is virtually impossible to control for. In fact, the research comparing peer-reared and mother-reared monkeys has produced a large number of contradictory findings that are very difficult to interpret (Kraemer, 1997). Even when studies reported clear differences between peer-reared and mother-reared monkeys, the meaning of these differences and their implications for human attachment were rarely discussed.

It is paradoxical that in all the years of research on primate attachment with surrogate mothers, the interactions between infants and their biological mothers have, at best, been used for comparison purposes but have never themselves been studied. Moreover, in many cases, the mother-reared infants that were used for comparison were themselves reared in species-atypical environments (e.g., alone in a small cage with their mothers or in small social groups lacking the species-typical sex- and age-graded structure). One of the unfortunate consequences of this approach is that most of the information that is often used to describe normative patterns of attachment in primates (e.g., Suomi, 1995, 1999) was derived from atypical populations of individuals (e.g., individuals with aberrant early experience or individuals not living in a species-typical social environment). This is equivalent to attempting to understand normative patterns of human attachment solely with information obtained from clinical populations. Like research on separation, the study of surrogate-rearing is unlikely to be an area in which the coordinated efforts of primate and human researchers could lead to significant advances in our understanding of attachment processes. Primate researchers focusing on separation and surrogate-rearing, however, maintain that this research has led to a reconceptualization of attachment processes and a better understanding of the function of attachment (e.g., Kraemer, 1992). In their view, the main function of attachment is not to ensure protection of the infant from the environment or facilitate healthy social development but to promote psychobiological attunement (or synchrony) between two individuals.

Attachment as Psychobiological Attunement

According to the psychobiological attunement theory of attachment, caregivers play an important role in the development and regulation of their infants' physiological and behavioral systems. Much of the scientific evidence used to elaborate and refine this theory was provided by research with rats conducted by Hofer and collaborators (see Polan & Hofer, 1999, for a review), who showed that the mother's presence and behavior affect many behavioral and physiological functions in the offspring, including urination/defecation, thermoregulation, and autonomic and endocrine response to stress. Consistent with this theory are studies of behavioral synchrony and rhythmicity in mother-child interactions (e.g., Field, 1985; McKenna, 1990).

Maternal stimulation of offspring is characteristic of many mammalian species, and therefore it is plausible that primate mothers share their circadian and other biological rhythms with their offspring through contact and behavioral interaction similar to what mothers rats do (Kraemer, 1992; Reite et al., 1989; Suomi, 1999). Primate researchers have noted that maternal separation and surrogate rearing bring about, among other things, an impairment of the regulation of physiological and behavioral systems (Kraemer, 1992; Reite & Capitanio, 1985). These findings have been interpreted as evidence in support of the psychobiological attunement theory (Kraemer, 1992). Although Field (1985, 1996) recognized that an attachment model derived from research on mother-infant separation is limited because separation models do not specify what is ordinarily present in a relationship that is then missing during separation, the primate version of the psychobiological attunement theory is entirely based on separation research (Kraemer, 1992). Therefore, the conceptual and methodological weaknesses of research on maternal separation and surrogate rearing also extend to the psychobiological attunement theory.

The hypothesis that mammalian caregivers exert regulatory functions on their offspring's development makes intuitive sense and is supported by evidence from both animal and human research. The issue, however, is whether primate and human attachment processes can be reduced to mutual stimulation and synchronization of physiological rhythms or whether there is more to attachment than psychobiological attunement. Bowlby's (1969) initial suggestion that caregivers provide an important protective function for the offspring, along with the human research showing the complex effects of attachment on social and cognitive development, suggest that although psychobiological attunement may be an important component of early parent-infant interactions (and possibly also of other social and affective bonds), this is only one of many aspects of attachment.

Psychobiological attunement between mother and infant, and more generally the study of parental influences on biobehavioral development, is an area of research where collaboration between biologists and psychologists working with primates and humans could be profitable. The use of separation procedures and surrogate mothers, however, has already shown its limitations in this arena. Therefore, to understand fully the regulatory role played by mothers in their infants' bio-behavioral development, primate researchers need to start investigating the psychobiological aspects of mother-infant interactions rather than just focusing on infants without their mothers. To this end, they need to use sophisticated techniques to record and quantify mother-infant interactions at both the behavioral and physiological level. Such techniques could be exported to human research, along with their background of biological and comparative knowledge, and contribute to expanding the study and understanding of attachment processes in humans. In this view, research on psychobiological attunement would complement and extend the study of attachment processes rather than provide an alternative framework to attachment theory.

Infant Attachment in Primates: Individual Differences

Four decades of research with monkeys and apes has shown that there is a great deal of variability in mother-infant relationships within social groups and populations (Fairbanks, 1996a; see Chapter 6). There is now good evidence that individual differences in parenting styles are remarkably stable across different offspring and even across generations. Individual differences in infant behavior, however, have been poorly investigated, in part because such differences are less apparent and more difficult to quantify than those in maternal behavior. Individual differences in infants' attachment to their mothers have also been difficult to assess owing to the lack of standardized procedures and scoring methods comparable to those used in the SST or the Q-sort (see below for a few exceptions). Thus, primate researchers studying mother-infant relationships in group-living monkeys have generally refrained from interpreting their findings in terms of attachment theory. In contrast, primate researchers studying infant responses to separation and the development of surrogate-reared infants have consistently used attachment concepts and terminology, such that any interindividual variability in behavior has automatically been linked to attachment.

Studies of Separation and Surrogate Rearing

Studies of maternal separation have documented differences in both behavioral and physiological reactions to separation among individuals of similar age, rearing background, and separation paradigm (for reviews, see Laudenslager & Boccia, 1996; Suomi, 1995). For example, there appears to be great variability in the intensity of the "protest" response to separation and in the extent to which such response is followed by withdrawal and depression (Levine & Wiener, 1988). Individual differences tend to remain consistent across repeated separations and reunions and stable throughout major periods of development. Stable individual differences in behavioral and physiological responses to social separation have also been found among rhesus macaques reared with peers (Kraemer, 1997). Suomi (1995) has argued that these individual differences in response to separation are accounted for by both heritable and experiential factors, but there has been little empirical research in this area.

Based on studies of separation and surrogate mothers, Suomi (1999) has maintained that "there exist compelling parallels in rhesus monkey attachment relationships to each of the major human attachment types, and at least arguable similarities for most of the classical subtypes" (p. 188). Specifically, he claimed that a subgroup of "high-reactive" monkeys (constituting about 15–20 percent of both wild and captive populations studied to date, according to Suomi's estimates), who consistently respond to separation with behavioral and physiological signs of intense stress, tend to have attachment relationships with their mothers that are "C"-like (ambivalent). Another subgroup of "highly impulsive" monkeys (constituting approximately 5–10 percent of populations studied to date, according to Suomi) instead "typically develop difficult attachment relationships with their mothers" (p. 189), and "in Ainsworth's strange situation terminology, these infants tend to form "A"-like (avoidant) and "D"-like (disorganized) attachment relationships" (Suomi, 1999, p. 189). Finally, most individuals in monkey populations appear to exhibit secure attachment to their mothers, a subgroup of them being classified as being "unusually secure" (Suomi, 1999). According to Suomi (1995, 1999), early individual differences in infant attachment carry over into adolescence and adulthood, with insecurely attached individuals exhibiting difficulties in their social interactions with their peers and other group members. Furthermore, differences in attachment would also be transmitted across generations, the consistency in attachment being equally strong whether the young monkey's early attachment experience was with her biological mother or with another conspecific (Suomi, 1999).

Although individual differences in bio-behavioral reactivity to stress have been well established not just in primate research but in virtually all animal research on stress (e.g., Moberg, 1985), the notion that these differences reflect differences in attachment styles similar to those observed in humans is problematic given the lack of any data on the quality of these infants' relationships with their mothers or other caregivers. Moreover, given the lack of standardized procedures for the categorization of infant attachment in primates, the assignment of individuals to different attachment subgroups extrapolated from human attachment research appears to be arbitrary. Finally, although longitudinal and cross-generational consistencies in parenting styles have been well established by studies of mother-infant relationships in group-living monkeys (Fairbanks, 1996a), it is not at all clear that such differences are mediated by the attachment system.

In addition to identifying differences in attachment among individuals with similar rearing history, Suomi (1995, 1999) has also argued that rhesus infants reared with peers differ, as a group, in attachment style from mother-reared infants. Specifically, "because peers are not nearly as effective as typical monkey mothers in reducing fear in the face of novelty, or in providing a 'secure base' for exploration, the attachment relationships that these peer-reared infants developed were almost always 'anxious' in nature" (Suomi, 1999, p. 190; see also Suomi, 1995). In this view, there are explicit parallels between the attachment styles of peer-reared monkeys and those of high-reactive mother-reared monkeys, as well as implicit parallels between these two types of monkeys and children classified as insecurely attached in the SST.

Part of the problem with research focusing on maternal separation and surrogate rearing is that such research has focused on the effects of stressful manipulations in highly artificial laboratory conditions while neglecting the study of mother-infant relationships in natural or semi-natural conditions. Mildly stressful perturbations such as brief separations are also used to assess differences in human attachment. The assessment of responses to separation, however, would be largely insufficient to evidence different patterns of organization of the attachment system if the analysis of separation were not accompanied by the study of parent-infant interactions at reunion and in the home environment. Thus, the hypothesis that the individual differences in behavior and physiology in response to maternal separation or surrogate rearing have anything to do with attachment still remains to be empirically tested.

Manipulations of the Foraging Regime and Infant Attachment

Laboratory studies of pigtail and bonnet macaque mother-infant dyads conducted by Rosenblum, Andrews, and their colleagues have suggested that experimental manipulations of the mother's foraging regime produce infants with insecure attachment (e.g., Andrews & Rosenblum, 1991, 1993). In an early version of this paradigm, mother-infant pairs were exposed to different foraging treatments: Low Foraging Demand (LFD), High Foraging Demand (HFD), and Variable Foraging Demand (VFD) (Rosenblum & Paully, 1984). In the LFD condition, animals had ad libitum access to food, and such food could be retrieved without effort. In the HFD condition, animals had access to six times less food than the LFD animals and had to make some effort to retrieve it. Finally, in the VFD condition, animals were exposed to a two-week alternation of HFD and LFD. The treatment period began when infants ranged from four to seventeen weeks of age and lasted fourteen weeks. Shortly after the foraging treatments, mothers and infants were tested for one hour in a novel room for four consecutive days. Most of the subsequent research with this paradigm has compared dyads exposed to the LFD and VFD treatments.

Andrews and Rosenblum (1991) reported that LFD infants engaged in more exploratory behavior in a novel environment than VFD infants. The behavior of VFD infants was interpreted as indicative of insecure attachment to the mother and inability to use the mother as a secure base for exploration. Thus, in this and other studies, the VFD infants were explicitly compared to insecurely attached or behaviorally inhibited children (Coplan et al., 1996; Rosenblum et al., 1994). Owing, however, to the lack of standardized criteria to classify monkey infant attachment and the lack of a comprehensive investigation of mother-infant interactions before, during, and after the novel environment test, the view of VDF infants as being insecurely attached relies on the equation of low exploratory behavior with insecure attachment.

The exploratory behavior exhibited by the VFD infants was assumed to result from alterations in the mother's behavior. Specifically, Rosenblum and colleagues hypothesized that, following the foraging manipulation, VFD mothers became more anxious, erratic, and dismissive, and less responsive to their infants' signals (Andrews & Rosenblum, 1991, 1993; Coplan et al., 1996; Rosenblum et al., 1994). Such changes in maternal behavior presumably resulted in reduced infant perception of maternal availability and therefore in insecure attachment. These explanations are plausible but remain largely speculative given the lack of clear effects of the foraging treatment on mother-infant interactions and the lack of any information on maternal anxiety or maternal responsiveness to infant signals.

The findings by Rosenblum and colleagues are potentially interesting but were obtained with bonnet macaques living in small cohorts of three mother-

infant pairs and therefore in a social environment very different from that typical of macaque societies (i.e., large age-graded groups with matrilineal structure and a few unrelated males; Lindburg, 1991). The few attempts to replicate
the effects of foraging manipulations in different species have provided inconsistent results. For example, in squirrel monkeys, the HFD treatment affected
plasma cortisol levels of the mothers more than those of their infants (Champoux et al., 2001; Lyons et al., 1998), and the uncertainty and unpredictability
presumably associated with the VFD treatment did not result in elevated cortisol (Champoux et al., 2001). Therefore, the extent to which the findings
obtained with the VFD paradigm can be generalized and extended to humans
appears to be contingent on their replicability in different environments and
with other populations or species of primates.

The Study of Individual Differences in Primate Attachment with the SST or the Q-Sort Method

The only attempt to use an experimental procedure similar to the SST to
assess individual differences in primate attachment was made in an unpublished study by Bard (1991). In this study, sixteen hand-reared chimpanzees
between twelve and fifteen months of age were each tested with their favorite
human caregiver as "the mother" and an unfamiliar female as "the stranger" in
an unfamiliar room with toys. Bard (1991) classified twelve infants as secure
(five of them tentatively labeled as dependent, subtype B4), one avoidant, and
three resistant. She concluded that attachment processes in chimpanzee
infants parallel those in human infants. Because of the lack of detailed information on the procedures and findings of the study, the extent of these similarities remains unclear.

The only attempt to date to assess individual differences in primate attachment with the Q-sort method has been made by Kondo-Ikemura and Waters
(1995). These authors modified the human Q-set to adapt it to the behavior of
Japanese macaques and aimed to evaluate the hypothesis that patterns of
secure-base behaviors, postulated to index attachment security in human
research, are associated with concurrent patterns of supportive maternal
behavior in macaques. Twenty-four mother-infant pairs living in semi-natural
enclosures in Texas were studied and five characteristics of insecure infants
were identified: they adopt an awkward and uncomfortable posture when

held, are easily annoyed with their mother, expect that their mother will be unresponsive, become distressed when the mother moves away, and their transition from contact to exploration is executed awkwardly. The study reported no differences in security in relation to sex or age. A strong relation between maternal and infant behavior emerged, however. Specifically, three categories of maternal behavior were correlated with infant security: maternal supervision and active involvement in caretaking, sensitivity to infant signals and availability, and rank and social adjustment.

Kondo-Ikemura and Waters (1995) acknowledged that closely coordinated face-to-face interactions analogous to those typically defining maternal sensitivity in humans are not a distinctive feature of infant-mother interactions in macaques. They suggested that the mother's willingness and ability to organize her behavior around the infant and to serve as a secure base is probably the critical factor organizing and maintaining the infant's secure-base behavior. In particular, maternal protection emerged as one of the major dimensions of parenting influencing infant security. This is understandable given that macaque infants are at serious risk of injury from conspecifics (Maestripieri, 1993a). In contrast, in human research there is a relative lack of emphasis on variables such as maternal status and vigilance (Kondo-Ikemura & Waters, 1995).

Kondo-Ikemura and Waters (1995) warned that the fact that attachment theory can be mapped onto primate behavior does not guarantee that the security concept affords a particularly powerful perspective on behavior. They acknowledged, however, that this is an empirical question that needs to be addressed in future studies. Thus, the use of the Q-sort method is a promising avenue of research on primate attachment because this method allows one to survey a wider range of behaviors as well as to capture details and summarize the function of complex behavior patterns. These features can expand the reach of primate attachment research and help rebuild some of the bridges that once linked primate and human research.

Separate? → Caregiving Attachment

Although Bowlby (1969) briefly mentioned that the infant attachment system is complemented by a parental attachment system in the caregiver, the caregiver system occupied a minor role in the original formulation of attachment theory. In a later essay, Bowlby (1984; see also Bowlby, 1988) expanded his discussion of the caregiving attachment system, although his views of parenting were still strongly influenced by the basic psychoanalytic tenet that adults reproduce in their parental role their own early experiences with their caregivers. Thus, by overemphasizing the influence of early experiences on caregiving behavior and the inversion of roles between parent and child, Bowlby failed to appreciate fully that caregiving attachment may be an evolutionary adaptation completely independent from infant attachment.

This distinction has not yet been fully appreciated by most of the more recent research on the caregiving attachment system (e.g., George & Solomon, 1999; but see Simpson & Rholes, 2000). The potential independence of the caregiving and infant attachment systems, however, has been emphasized by some primate researchers. For example, Mason and Mendoza (1998) pointed out that there appears to be much greater variability in parental behavior than in infant behavior both within and among primate species.

As with infant attachment, there are no standardized measures of parental attachment in primates. Thus, the presence and activation of the parental attachment system is inferred from the selectivity with which primate parents (in most species, mothers) selectively nurture their own infants, their nurturing behavior itself with particular regard to warmth and protectiveness, the occurrence of separation anxiety when young infants break contact with the mother, and the exhibition of behavioral and physiological responses to infant separation, which suggests that this is a stressful event not only for the infant but also for the mother. Any one of these items in itself would probably be insufficient to assess maternal attachment. The co-occurrence of all these characteristics in particular individuals, however, suggests that such individuals possess a motivational-behavioral system whose primary goal is to maintain proximity to the offspring and protect it from environmental hazards.

Maternal recognition of offspring is a well-established phenomenon in most primate species in which it has been studied, although a few species such as titi monkeys do not appear to discriminate their own infants from same-aged strange infants in choice tests (Mason & Mendoza, 1998). Offspring recognition occurs through multiple sensory channels, and olfactory cues are probably recognized earlier than visual or auditory ones. Mothers are able to discriminate their offspring from other infants within a few hours or days after giving birth (see above), and offspring recognition is generally maintained throughout their lives, even if individuals are separated for months or years (Maestripieri, 2001a).

Nurturing behavior is not necessarily limited to one's own biological off-spring. Female primates in most species are attracted to infants in general and often engage in caretaking of unrelated infants (e.g., carrying or grooming). But all forms of caregiving behavior by lactating females, in particular suck-ling and protective behavior, are mostly directed to their own offspring. Both naturalistic observations of infant adoption and infant cross-fostering experi-ments in macaques suggest that there is a postpartum sensitive period for care-giving motivation such that (1) when mother and infant are separated during the sensitive period, the mother is likely to accept her own infant or an alien infant with similar characteristics if reunion occurs before the end of the period; (2) when mother and infant are separated during the sensitive period, the mother is likely to reject her own infant and any other infant if reunion occurs after the end of the period; (3) when mother and infant are separated after the sensitive period and later reunited, the mother is likely to accept her own infant but reject any other infant (Maestripieri, 2001a). This postpartum sensitive period for caregiving motivation is probably dependent on the neu-roendocrine changes associated with late pregnancy, parturition, and early lac-tation (perhaps involving hormones and peptides such as estrogen, oxytocin, prolactin, or endogenous opioids; Maestripieri, 1999a). It can be viewed as a sensitive period for mother-infant bonding in the sense that the new mother's heightened motivation for caregiving can facilitate the formation of a strong bond with the offspring (Maestripieri, 2001b). It may not be a sensitive period for mother-infant bonding, however, in the sense that small differences in amount of time spent in contact during this period have necessarily negative consequences for subsequent parenting or offspring development (i.e, the way mother-infant bonding has been interpreted in the psychological literature; see Klaus & Kennell, 1976).

Female attraction to infant stimuli in primates may be considered a behav-ioral predisposition similar to infant attraction to caregiver stimuli. Attraction to infants may be conducive to the formation of an emotional bond but does not guarantee it (Mason & Mendoza, 1998). For example, attachment to infants may also be present among monkey mothers who abuse their infants, and abuse often co-occurs with intense nurturing behaviors (Maestripieri & Carroll, 1998). Among lactating females, protective responses toward offspring are present from the moment offspring are born, even before mothers are able to discriminate them from other infants. Furthermore, in species in which off-spring remain in their natal group, mothers continue to nurture and support them throughout their life.

Most if not all macaque mothers display separation anxiety when contact with their infants is temporarily interrupted during the first months of infant life (Maestripieri, 1993a, b). In macaques, maternal separation anxiety is expressed with an increase in the frequency of maternal displacement activi-ties (notably self-scratching) and vigilance (Maestripieri 1993a, b; Figure 5.1). Such behavioral changes are probably accompanied by the activation of the autonomic nervous system and the HPA axis (see Chapter 11). There are both individual and species differences in the intensity of maternal anxiety (Maestripieri, 1994a). Maternal anxiety decreases sharply with infant age, as infants become more independent and less vulnerable to environmental haz-ards (Maestripieri, 1993a; Figure 5.2). Finally, when infants are forcibly sepa-rated from their mothers by human experimenters, mothers in many primate

Figure 5.1 Rhesus macaque mother scratching herself when her infant is out of contact and close to a potentially dangerous individual. (Photo: D. Maestripieri)

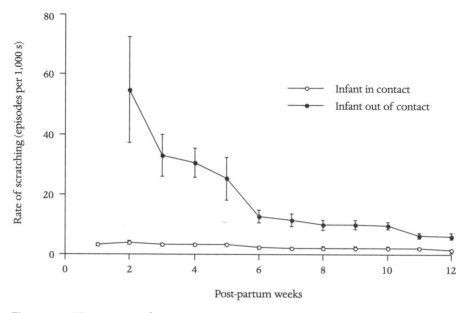

Figure 5.2 Time course of maternal scratching when the infant is in contact and when the infant is out of contact during the first twelve weeks of infant life in rhesus macaques. (Redrawn from Maestripieri, 1993a)

species show behavioral agitation and neuroendocrine responses similar to those of their infants (Champoux & Suomi, 1994; Mendoza et al., 1978). In titi monkeys, however, parents do not resist when an infant is removed, and they show no behavioral or physiological stress responses following separation (Mason & Mendoza, 1998). Differences in physiological maternal responses to separation in relation to offspring age have not been systematically investigated, but in group-living macaques, most mothers continue to show strong behavioral reactions to the capture and removal of their adult offspring throughout their life (Maestripieri, 2001a).

Based on these observations, the normative aspects of primate maternal attachment can be sketched as follows: primate females have strong predispositions to be attracted to and to nurture infants, which are the product of biological (e.g., genetic, neuroendocrine) and experiential factors (e.g., observation of caregiving and direct experience with infants during development). Such predispositions facilitate the formation of a strong social and emotional bond with their offspring after giving birth. Giving birth is a powerful (but not a necessary) activator of the caregiving attachment system, which can be viewed as a set of motivational and behavioral responses aimed at providing nurturance, support, and protection of the offspring. The basic functions of the caregiving system are different from, but often complementary to, those of the infant attachment system, and the causal factors underlying maternal and infant attachment are probably different as well, despite similarities in the organization of the two attachment systems (Mason & Mendoza, 1998). As the infant attachment system can be viewed as a developmental adaptation evolved to guarantee survival during infancy, the caregiving system can be viewed as a reproductive and parenting adaptation evolved to guarantee offspring survival during the postpartum period. Similar to the infant attachment system, the caregiving system can also be viewed as a reproductive and lifelong adaptation. In fact, this system can remain active throughout the female's lifetime, although the relative expression of its different components is likely to vary over time.

Among primates, the characteristics of the caregiver system are likely to vary in relation to other ecological, reproductive, and social adaptations. Mason and Mendoza (1998) have argued that in species in which maternal emotional attachments are the mode, animals live in large social groups, are probably subject to predation, and are likely to contain many immature individuals of similar age. In the monogamous titi monkeys, who lack some of these ecological and social characteristics, neither of the parents shows unequivocal evidence of forming a specific attachment to their infants (Mason and Mendoza, 1998). Finally, there is no firm evidence for any primates that adult males establish a bond with a specific infant that can be classifiable as attachment, whether or not these infants are their biological offspring.

Other Attachment Systems

One view promoted by early primate research was that there are multiple independent attachment systems such as infant-mother, mother-infant, father-infant, peer-peer, and heterosexual adult attachment (Harlow & Harlow, 1965). An alternative view is that a single neurobiological system subserves social attachment and affiliation in primates, with the specific form and direction of its actions being determined or modulated by ontogenetic or environmental events (e.g., Reite & Capitanio, 1985).

Research with surrogate-reared monkeys has often referred to the peer attachment system to describe the social bond occurring between two or more rhesus monkey infants who are permanently separated from their mothers and hand-raised together in a cage (e.g., Higley, Hopkins, et al., 1992). Such monkey infants engage in mutual clinging behavior and seek comfort in each other under conditions of stress. Because "peer-rearing" does not occur in the natural environment, the peer attachment system could be an artifact of the laboratory environment. Of course, free-ranging monkey infants develop social relationships with their peers superficially similar to those of human children. To argue, however, that these relationships may reflect the manifestation of a specific peer attachment system would be equivalent to arguing that any social relationship between any individuals has its own unique attachment system.

An early experiment by Patterson et al. (1975) suggested that rhesus infants show specificity for safe-haven individuals and do not use their peers as mother substitutes in a fear-arousing situation. In this study, rhesus infants were exposed to a fearful looming stimulus and tested in two conditions: when both the mother and a preferred peer were present, and when the peer alone was present. Without the mother, the infants spent similar amounts of time with the peer and in an empty cage. When the mother was present, however, the infant always approached the mother and never the peer for comfort. Thus, it may be argued that although in the absence of a biological parent other individuals may act as surrogate mothers, this does not constitute evidence that distinct attachment systems have evolved to allow infants to bond with different surrogates.

Cooperatively breeding primates such as marmosets and tamarins represent an interesting case because in these primates infants spend a lot of time being carried by their fathers, older siblings, or nonreproductive helpers. Infants have different relationships with different caregivers, and each relationship has its own developmental trajectory (Cleveland & Snowdon, 1984; Locke-Haydon & Chalmers, 1983). When infants are exposed to a fearful stimulus, they seek protection from their father, who is often the individual most involved in infant carrying (e.g., Kostan & Snowdon, 2002). If fathers are removed from the group, however, infants show little or no distress as long as their mothers remain available (Arruda et al., 1986). This suggests that in this social system in which parental care is shared by different group members, infants can form multiple attachments with their various caregivers. Independent attachment relationships with mother and father are also likely to occur in bi-parental primate species such as titi monkeys (Hoffman et al., 1995).

In addition to the issue of attachment with multiple caregivers, a question arises as to whether affiliative or sexual relationships between adults reflect the expression of an attachment system similar to that operating between parents and children. Mason and Mendoza (1998) have argued that evidence of adult attachments is overwhelming for monogamous primates such as titi monkeys but virtually nonexistent for polygynous or promiscuous primates such as squirrel monkeys, macaques, and chimpanzees. For example, paired adult titi monkeys follow each other closely, groom each other, and often sit together with their tails intertwined. They show specific preference for each other in preference tests, and are reluctant to interact with strangers. They exhibit agonistic displays toward strangers and increases in heart rate and plasma cortisol. The reaction is particularly prominent when strangers are of the same sex, and it is stronger in males than in females. These findings suggest, however, that this reaction may be best viewed as a form of mate guarding, not as a form of attachment. The reactions of titi monkeys to separation from their pairmates parallel those of infants and their mothers: they exhibit increased vocalizations, locomotion, and heart rate, and higher levels of plasma cortisol. In contrast, none of these responses can be observed among squirrel monkeys (Mason & Mendoza, 1998).

In nonmonogamous species of primates, strong emotional attachments between adults are unusual and have seldom been examined systematically. In fact, Kondo-Ikemura and Waters (1995) have pointed out that although macaques have been used as a model for attachment research since Harlow's early studies, the case for using these primates as a model of human social attachment has rarely been examined in detail. They also argued that, given that the central hypothesis in human attachment theory is that early attachments serve as prototypes for adult attachment relationships, one might question the relevance of focusing on a primate species that does not form adult pair bonds. Of course, this "prototype" hypothesis is far from being unequivocally supported by empirical evidence, in humans or in other primates.

The issue of whether affiliative or sexual relationships between primate adults reflect the operation of an adult attachment system that develops from infant attachment has not been systematically examined in nonhuman primates. This issue needs to be further explored in humans as well. Comparative research with primates can help to elucidate the nature and function of social and emotional bonds between adults and contribute to clarifying whether these bonds are best understood from the perspective of attachment theory or within some alternative theoretical framework.

Evolutionary Issues

In addition to clarifying issues of ontogeny, causation, and function, primate research on attachment can make the unique contribution of elucidating the evolutionary history of the human attachment system. Evolutionary psychologists look for the evolutionary origins and adaptive value of human psychological traits in the "goodness of fit" between the design of the trait and the function the trait is supposed to serve (e.g., Tooby & Cosmides, 1992). In doing so, however, they tend to overlook the contribution made by comparative research, most notably with primates, to the reconstruction of the evolutionary history of the human mind and human behavior (see Povinelli, 1993, for a different approach; see also Chapter 1).

A general issue to be addressed is whether primate attachment and human attachment serve similar functions but evolved independently from each other (i.e., they are analogous), or whether they also share a common evolutionary history (i.e., they are homologous). Bowlby appears to have believed that the basic features of attachment are essentially homologous in rhesus monkey infants and human babies (Suomi, 1999). He did not fully appreciate, however, that mother-infant relationships in rhesus monkeys may not necessarily be representative of the whole Primate order. Suomi (1995, 1999) has argued that in prosimians and New World monkeys, the existence of infant attachment is uncertain. He further speculated that attachment phenomena may represent a relatively recent evolutionary adaptation among primates (Suomi, 1995, 1999). Although Suomi's hypothesis may be correct, there has been very little research on attachment-related processes in prosimians, while some evidence of infant attachment processes in New World monkeys is certainly available.

Mason and Mendoza (1998) have argued that the probability of homology versus analogy between primate and human attachment is different for different attachment systems. On the one hand, they argued that "the striking commonalities among disparate primate species suggest that filial attachments are based on homologous processes, probably relating to the critical contribution of attachment to survival" (p. 773; this exact argument, however, could be used to support the analogy hypothesis). On the other hand, because of differences among primate species in the tendency to form maternal attachments, and in the specific patterns of parental behavior, Mason and Mendoza (1998) believe that the caregiving attachment systems are unlikely to be homologous, except among closely related species. Finally, according to Mason and Mendoza (1998), adult attachments in primates probably evolved in response to a different set of selective pressures than other forms of attachment. In their view, among monogamous primates, attachment should be viewed as an integral part of the species' territorial/monogamous/biparental social system, and pair-bonds among different primate species are not based on homologous processes.

Reite and Capitanio (1985) concluded that there is insufficient information to make a strong statement about the homology versus analogy of primate and human attachment. They noticed that social attachments are present in mammals in which limbic structures are first prominent phylogenetically, and suggested that attachment evolved in birds and mammals after their split from their reptilian ancestors. Thus, there could be homologies at the level of neural substrates of attachment even among relatively distant species. Reite and Capitanio (1985), however, also argued that, on the basis of marked dissimilarity in brain structure and function in birds and mammals, attachment may serve similar functions but be regulated by different mechanisms.

As soon as information on attachment processes becomes available for a wider range of species, it may be possible to assess potential homologies between the behavioral patterns of primate attachment and human attachment with analytical techniques used in studies of behavioral phylogeny (Atz, 1970). Some evolutionary psychologists would argue that the evolutionary analysis should focus on the psychological and motivational processes underlying attachment rather than on behavior itself. Natural selection, however, ultimately acts on behavior, and it is very likely that attachment-related behaviors were under strong selective pressure for their contribution to infant survival irrespective of whether or not they were subserved by similar psychological mechanisms. Thus it is possible that there may be behavioral homologies between human attachment and the attachment systems of other animal species, even if the latter lack some of the cognitive components of human attachment (e.g., the ability to take other individuals' perspectives and to mentally represent relationships). For example, the behavior of domestic dogs in the SST (Topal et al., 1998) suggests that the kinds of relationships many humans have established with their pets may have been facilitated by the existence of behavioral predispositions for attachment-like processes in the dog.

The Integration of Primate and Human Research on Attachment

The information reviewed in this chapter suggests that there is a great deal of potential overlap between primate and human research on attachment at both the theoretical and the empirical level. One promising area where comparative research on attachment would be particularly fruitful is the study of parent-offspring social relationships. The comparative study of parent-offspring relationships in human and nonhuman primates could produce interesting new information on normative aspects of infant and caregiving attachment as well as on individual differences in attachment styles. In terms of normative aspects, primate research can be useful to investigate further the ontogeny, causation, function, and evolution of the attachment system, the way Bowlby himself had envisaged. By investigating the development of the attachment system in primate species characterized by different rates of maturation, different caregiving systems, and different social and cognitive adaptations, primate research can help us understand the relation between the development of the attachment system and other aspects of social and cognitive development. By experimentally investigating the endogenous and exogenous conditions that activate and terminate the attachment systems in other primate species, primate research can expand our knowledge of the proximate regulation of attachment. The comparative and psychobiological study of attachment can also help elucidate the multiple functions of attachment and its significance both as a developmental adaptation (i.e., as a system evolved to solve problems specific to stages of development) and a life history adaptation (i.e., as a system evolved to favor the individual's adaptation to its environment across the life span). In this regard, primate research offers the opportunity to perform biological and environmental manipulations that would be difficult or impossible in humans. Primate research also provides the opportunity to investigate attachment processes longitudinally across an individual's life span and across generations in a relatively short period of time, allowing researchers to investigate consistencies in attachment with prospective longitudinal studies rather than with retrospective studies of attachment memories in adults. Finally, by studying and comparing attachment processes across a number of primate species in relation to their phylogenetic history, primate research can help us understand the extent to which the characteristics of the human attachment system are the result of evolutionary adaptations to unique aspects of human evolutionary history or the by-product of phylogenetic inertia (i.e., the notion that closely related species may maintain biological and behavioral similarities because of their common evolutionary history and despite their recent divergence and adaptation to different environments). Thus, integration of primate and human research could revitalize the study of the normative aspects of attachment in humans and stimulate progress in areas other than the cognitive-linguistic aspects of attachment.

In addition to enhancing our understanding of normative aspects of attachment, comparative research can produce further knowledge about the determinants and consequences of individual differences in attachment styles. The study of variability in mother-infant relationships in primates is already well advanced and sophisticated (see Chapter 6). To make this and other research accessible to human attachment theorists, primate researchers must develop standardized techniques to obtain information on attachment styles from the observation of mother-infant interactions in both naturalistic environments and experimental situations similar to the SST. Furthermore, studies of mother-infant interactions must increasingly incorporate biological variables and the experimental approaches used in the fields of behavioral endocrinology and neuroscience. Endocrine and neurobiological aspects of parent-offspring interactions also can and should be increasingly taken into consideration by human attachment researchers. Finally, research on attachment could benefit from testing evolutionary hypotheses concerning the functional significance of attachment derived from studies of primates and other animals. Clearly, the attachment system is not an evolutionary novelty that emerged with the human species. Studying attachment from a biological and comparative perspective can lead us to an understanding of this phenomenon above and beyond the specific form that this phenomenon takes in our own species or culture.

6

Parenting

Lynn A. Fairbanks

Mothers can be warm and responsive, devoting almost all of their time to feeding, carrying, and caring for their infants. They can also be neglectful and rejecting, pushing their infants away or temporarily abandoning them to seek new mates. Fathers can be nurturing or indifferent. Research on primate parenting has sought to understand the reasons for this kind of variation in behavior. Over the years, this work has increased our understanding and appreciation of the complex processes influencing primate maternal and paternal behavior, and has also contributed new ways of looking at problems in human parent-child relationships.

Discussions of human parenting typically concentrate on the influence that different parenting styles or practices have on offspring behavior and development (Collins et al., 2000). This chapter, in contrast, focuses on parenting as a dependent variable, describing factors that produce variation in the form and quality of primate parental behavior in natural and semi-natural social settings. Three basic approaches used in nonhuman primate parenting research are presented: ethological, comparative, and evolutionary biological. The ethological approach has provided detailed, quantitative, longitudinal descriptions of mother-offspring interactions and relationships. The comparative method focuses on understanding species differences in parenting behavior in the context of ecology and social organization. The evolutionary biological approach uses the theory of natural selection to predict and understand why parents behave the way they do (Clutton-Brock, 1991).

Studies of nonhuman primates have provided support for the idea that parenting behavior is influenced by complex trade-offs between costs and benefits for both parents and offspring. This chapter presents evidence of these trade-offs in mother-offspring interactions as they change over time, in the role of fathers in infant care, and in the effects of social and demographic factors such as mother's age, the presence of grandmothers, and family status on maternal care. Parental investment theory provides a coherent explanation for the balance between nurturance and conflict in each of these domains.

Ethological Descriptions of Changes in Mother-Offspring Relationships over Time

Mothers and Infants

In the 1960s and 1970s a number of researchers in the laboratory and the field set out to describe maternal behavior in primates. Field studies of baboons, langurs, and vervets documented typical patterns of mother-infant interactions in natural contexts (Altmann, 1980; Jay, 1963; Struhsaker, 1971). At the same time, Robert Hinde and his colleagues were contributing detailed analyses of changes in mother-infant interactions over time for rhesus monkeys living in captive social groups (Hinde, 1974a). These studies had the advantage of being able to observe behavior directly, longitudinally and quantitatively. They established a common set of ethological methods that have been used to describe mother-infant interactions ever since. Ethological studies have demonstrated that primate mother-infant interactions follow a similar course across a wide range of species and contexts. For example, in the first month of life, rhesus monkey mothers and infants spend most of the time in close ventral contact (Hinde & Spencer Booth, 1971). The infant clings to the mother's fur and has continuous access to her nipples. Mothers hold, groom, and inspect their infants during this time, and protect them from unwanted attention from other group members (Rowell et al., 1964). The infant matures rapidly and by the second month is climbing around on the mother's body and beginning to explore the surrounding area.

Mother-infant contact time declines predictably with infant age from the first to the sixth month of life, and the nature of mother-infant interactions changes too. Figure 6.1 illustrates these changes for 127 mother-infant pairs living in large outdoor social groups at the Vervet Monkey Research Colony (VMRC) (Fairbanks & McGuire, 1987). In this colony, the percentage of time that mothers and infants spend in close proximity to each other declines from more than 90 percent of the time in the first month of life to about 25 percent in the sixth month. At month 2, the infant begins to make short excursions away from the mother, using her as a secure base for further exploration (see Chapter 5). This is reflected in the infant's increase in both leaving (moving more than 1 meter away) and approaching the mother (Figure 6.1a). The mother is more likely to follow the infant and reestablish proximity than to leave it at this stage. By month 3 the dynamic changes, and the mother begins actively to promote infant independence. Her rate of moving away from her infant doubles from the second to the third month, while her rate of approaching stays the same or declines slightly (Figure 6.1b). At three months, the mother's tolerance for carrying and feeding the infant also wanes, and

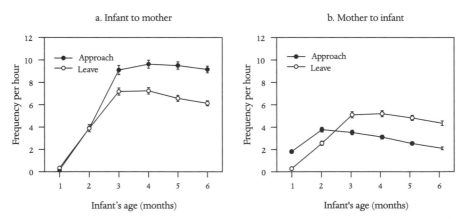

Figure 6.1 Mother and infant contributions to mother-infant proximity. Mean (± s.e.) frequency per hour that (a) vervet monkey infants approach and leave close proximity with their mothers and (b) vervet monkey mothers approach and leave close proximity to their infants, by month of life.

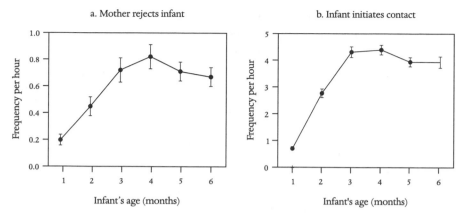

Figure 6.2 Mother-infant conflict over ventral contact. Mean (± s.e.) frequency per hour that (a) vervet monkey mothers reject infants and (b) vervet monkey infants attempt to initiate ventral contact with mothers, by month of life.

mothers are increasingly more likely to reject or prevent the infant's attempts to initiate ventral contact or gain access to the nipple (Figure 6.2a). The infant responds to these rejections by increasing its attempts to make ventral contact (Figure 6.2b). Sometimes the infant 'wins' these weaning conflicts and sometimes the mother prevails. Disputes over access to the nipple continue as the total amount of mother-infant contact time declines from three to six months of age. This results in an increase in the rate of rejection per hour in contact as the infant matures.

Similar developmental changes in mother-infant interactions have been widely reported across primate species. Vervets have also been observed in undisturbed groups in the field (Hauser & Fairbanks, 1988; Lee, 1984; Struhsaker, 1971); rhesus and Japanese macaques have been studied extensively in relatively small captive social groups and in free-ranging provisioned troops (e.g., Berman, 1990a; Maestripieri, 1993a, b; Schino et al., 1995); and baboons have been followed at a number of different field sites (Altmann, 1980; Lycett et al., 1998). In all of these species and settings, the same basic patterns are observed. Mothers maintain high rates of contact with their infants in the first few months after birth. As the infant grows, the mother progressively reduces her availability, and the infant compensates by playing an increasing role in maintaining the relationship. When the infant is approximately six months old, the mother enters her first postpartum estrus and begins mating. She sharply reduces contact time and actively rejects her infant's approaches (Berman et al., 1993; Worlein et al., 1988). The infants respond with evidence of distress and an increase in their attempts to gain nipple contact. High levels of maternal rejection and low levels of nipple contact during the first postpartum estrus have been shown to increase the mother's chance of conception (Berman et al., 1993; Johnson et al., 1993).

Interpretation of these changes in the mother-infant relationship can take two basic forms. The approach used in most psychological and psychobiological research attempts to identify intervening psychological or physiological processes that explain the observed developmental outcomes. Examples include linking hormonal changes to changes in maternal behavior, or identifying maturing characteristics of the child that trigger new maternal responses. These are mechanistic or proximate approaches that focus on how a response is produced. The reader is referred to several excellent reviews of research on nonhuman primate parenting that use this approach (Coe, 1990; Pryce, 1996). Our understanding of primate maternal behavior has been expanded by incorporation of a fundamentally different approach, derived from evolutionary biology, which focuses on why a behavioral system is designed the way it is. Evolutionary biology considers how a behavior functions to promote survival and reproductive success in accordance with our modern understanding of natural selection (Mayr, 1963).

Darwin's theory of natural selection consists of three basic postulates. First, there is variation among individuals in a population (e.g., in form, in physiological reactivity, in behavioral tendencies). Second, at least part of that variation is heritable (i.e., it is passed from parents to offspring). Third, because of these differences, some individuals have a greater chance of surviving and reproducing in the current environment than others. The result is that the certain individuals will leave more descendants in the next generation than others. The incorporation of information on molecular genetics into Darwin's theory allowed natural selection to become a quantifiable theory with internal logical consistency conducive to mathematical modeling (Mayr, 1963).

In an attempt to explain the existence of altruism, Hamilton (1964) recognized that genetic representation in future generations may be accomplished not only through direct descendants (children) but also through reproduction of individuals who share a portion of the same genes (e.g., siblings). He derived a simple equation, referred to as Hamilton's rule, which has widespread implications for social interactions and parental behavior. Hamilton predicted that altruism should evolve whenever the cost to the giver (C) is less than the benefit to the recipient (B), times the coefficient of relatedness (r), a term that defines the probability of both parties sharing the same genes by common descent $(C < B \times r)$. Both costs and benefits are measured in terms of the impact of the action on the individual's lifetime reproductive success.

Robert Trivers (1972) used Hamilton's rule to understand and predict changes in mother-infant relationships over time. He defined parental invest-ment as "any investment by the parent in an individual offspring that increases the offspring's chance of surviving (and hence reproductive success) at the cost of the parent's ability to invest in other offspring" (p. 139). In doing so, he turned the focus on parental behavior away from an almost exclusive emphasis on the effects on the infant toward an appreciation for the effects on the parent. He recognized that there are limits to the effort a parent can give, and that energy devoted to rearing one offspring takes away from what is available for rearing other offspring and for maintaining the parent's own health and welfare.

By modeling the benefits and costs of parental investment for mothers and offspring over time, Trivers discovered that there is an inherent conflict between the optimal amount of investment that mothers should give to each offspring and the optimal amount of investment that each offspring would like to receive (Trivers, 1974). This is because the mother is equally related to all of her offspring $(r = 0.5)$, but each offspring is more closely related to itself $(r = 1.0)$ than to its siblings $(r = 0.5)$. According to Hamilton's rule, the mother should provide care as long as the benefit to the infant is twice the cost to her own future fitness $(C < B \times 0.5)$. The infant also has an interest in protecting its present and future siblings. From its perspective the mother should provide care as long as the benefit to the infant is simply greater than the cost to her $(C < B)$. The infant's ideal equation would lead the mother to give a dispropor-tionate share of her parental effort to it. From the mother's perspective, she should allocate parental care more evenly across all of her present and future offspring. This leads to selection of mothers who reserve some of their energy and set limits on the amount of effort they are willing to give to one infant, and to selection of infants who demand more care and attention than their mothers are eager to provide.

Conflict, to some degree, is a universal feature of human parent-offspring relationships, and yet most theories of parenting in mainstream psychology have no coherent way of explaining it except as dysfunctional behavior. Trivers's theory of parental investment not only predicts conflict in normal parent-offspring relationships but also helps us to understand its origins and its functions.

The developmental changes in mother-infant interactions described here and in Figures 6.1 and 6.2 could be interpreted in different ways. In one view, the mother could be seen as promoting infant independence by increasing her rate of leaving and rejecting ventral contact. The hidden assumption in this interpretation is that maternal behavior is designed to be good for the infant, without regard for its effects on the mother or on other family members. One problem with this view is that it does not predict the infant's persistent attempts to gain access to the mother, in opposition to her efforts to limit access, and provides no rationale for understanding weaning conflict. A second interpretation is that maternal rejection of the infant's attempts to gain access to her is a negative by-product of the stressful and unnatural conditions found in captivity. This interpretation is contradicted, however, by studies of mother-infant interactions in undisturbed primate populations in the field. Direct comparisons of mother-infant behavior between field and captive settings have reported higher rates of maternal rejection in the field (Hauser & Fairbanks, 1988).

The difficulties faced by primate mothers become even more apparent when the energetic costs of lactation and infant carrying are quantified for primates living in the wild (Altmann, 1980). Baboon mothers at Amboseli National Park in Kenya must travel several miles each day to find food. In the first few months after giving birth, the mother carries her infant everywhere she goes. The burden of infant care increases as the infant grows, in terms of both the energy required to carry it and the calories needed to produce enough milk to feed it. The amount of time that a mother must spend feeding herself increases steadily up to the time the infant can forage for itself. The mother must balance her need to maintain her own health with the growing costs of infant care. She does this by reducing carrying time, but this is complicated by the fact that when the infant walks instead of being carried, it burns more calories and requires more milk to sustain growth (Altmann & Samuels, 1992). Maternal expenditure peaks when the infant is about four to five months of age. At that time, the mother has depleted her available resting time to accommodate the increase in foraging (Dunbar & Dunbar, 1988). In the fifth month, the mother gradually restricts nursing time as the infant is able to meet more of its own caloric needs through independent foraging. Ideally, this course of maternal care produces a healthy infant and a mother with enough reserves to reproduce again.

An evolutionary biological perspective on the primate mother-infant data provides a basis for understanding both nurturance and conflict. Trivers's theory of parent-offspring conflict explains changes in the mother's receptivity in terms of optimal allocation of parental investment across offspring. It predicts that mothers will be responsive to the needs of their infants, but also that there will be conflict over the duration and amount of parental care. Infants have evolved to want more from their mothers than their mothers will be willing to provide.

The longitudinal ethological methods used in most nonhuman primate studies underscore the changing nature of the mother-infant relationship over time. Human parent-child relationships are also continually changing in concert with maturational development in the child. Developmental psychologists recognize the value of longitudinal research designs. Yet the logistical constraints of time, money, and respect for privacy cause most parenting research to collect data at a single time point, or at a few longitudinal points widely separated in time (Collins et al., 2000). This makes it difficult to quantify the shifting roles of the participants and the subtle changes in the parent-child "dance" that are illustrated here in Figures 6.1 and 6.2. The ethological method also illustrates the degree to which infants act to counteract their mothers' behavior. When mothers try to keep the infants near, they try to get away. When mothers push them away, they increase their efforts to reestablish contact. To the transactional analyst, this is a developmental process that is fine-tuned to promote the welfare of the infant. To the evolutionary biologist, it is evidence of a basic conflict between mother and infant, each acting in its own self-interest. The behavior of both mother and infant has evolved in this context, and both are equipped with strategies to counteract the other.

It can also be used to explain why it is so difficult to demonstrate lasting effects of major differences in early rearing experiences, such as nursing versus bottle feeding or day care versus home care, on child outcomes. The evolutionary biologist would argue that because of the ubiquity of parent-offspring conflict, infants have evolved the capacity to seek what they need and to thrive in a relatively wide range of circumstances.

Mothers and Juvenile Offspring

The exclusive nature of the mother-infant relationship changes when the mother gives birth to a younger sibling. Nonhuman primates have an extended juvenile period between weaning and puberty when they continue to live with their mothers in the social group into which they were born. Mothers in many monkey species are likely to give birth again when the previous infant is one year old (in developmental terms equivalent to four to five years of age for a human child). At this time, the mother shifts her attention to the newborn. She dramatically increases the rate of threats and aggressive behaviors directed toward the yearling, beginning in the first week after the birth (DeVinney et al., 2001). Nursing of the older offspring stops completely, and the amount of physical contact it receives with the mother is sharply reduced (Lee, 1984; Tanaka, 1992). Most yearlings respond to this transition with an increase in independence and development of relationships with other group members (Holman & Goy, 1988). Some show signs of distress, but this overt expression is not very effective in eliciting further care from the mother.

After the birth of a younger sibling, young primates still maintain close affiliative and supportive relationships with their mothers (Nakamichi, 1989; Pereira & Altmann, 1985). As juveniles, they typically spend more time near their mothers than near any other group member (Kurland, 1977). Juveniles and yearlings huddle with their mothers at night (Hammerschmidt et al., 1994). They also are more likely to be groomed by their mothers than by any other group member (Pereira, 1988). Juvenile vervets receive more than twice as much grooming from their mothers as from any other class of animals, including their older sisters and other adult female kin (Fairbanks, 2000a).

Mothers play an important part in protecting and defending their juvenile offspring in aggressive interactions with other group members. One observational study of free-ranging rhesus monkeys on Cayo Santiago documented 339 interventions by adults in conflicts between juveniles. Of these, 80 percent involved the mother of at least one of the participants (Janus, 1992). Similar high rates of aiding and defending juvenile offspring by their mothers have also been reported for other primate species (Fairbanks & McGuire, 1985; Pereira, 1989). When juveniles are sick or injured, they will often go to their mothers for comfort and assistance (Dittus & Ratnayeke, 1989).

In all of these affiliative and supportive interactions, it is the juvenile more than the mother who is responsible for seeking and maintaining the relationship. When juveniles and mothers are apart, it is usually the juvenile who reestablishes proximity, and the amount of time that juveniles spend with their mothers depends more on the rate of approaches by the juvenile than on the behavior of the mother (Fairbanks & McGuire, 1985; Pereira, 1988; Rowell & Chism, 1986). Juveniles solicit grooming from their mothers (Pereira & Altmann, 1985; Walters, 1981), and in the cases of special attention described above, it is the juveniles who seek out their mothers. The relationship between mothers and juvenile offspring is not without conflict, and mothers do not always respond to approaches and demands of their juvenile offspring with kindness. Young juveniles, in particular, are often rambunctious and intrusive in their interactions with their mother. They may jump on top of her, try to take food from her hand, or grab at a younger sibling. Mothers will threaten, hit, push away, or even bite their juvenile offspring to inhibit these activities. As a result, juveniles typically receive more aggression from their mothers than from any other group member (Bernstein & Ehardt, 1986c). Much of this aggression is directed toward inhibiting the juvenile from unacceptable behavior and limiting access to the mother's body. In conflicts between siblings, mothers will usually side with the younger against the older sibling (Horrocks & Hunte, 1983a).

Ethological descriptions of ongoing relationships of mothers with their immature offspring are consistent with Trivers's theory of parental investment described above. The care and attention that a mother gives to her juvenile offspring should be balanced by the fitness payoffs that she would receive by directing her efforts to other offspring. Mothers continue to provide relatively low-cost care to their juvenile offspring, while they shift the greater costs of lactation and carrying to a new infant. The fact that there is conflict between mother and offspring over the timing and amount of this transition is highlighted by the increase in rejection and aggressive interactions following the birth of a new sibling. The conflict reflects the discrepancy between the ideal amount of care for the offspring to receive versus the ideal amount of care for the mother to give (Trivers, 1974). In this view, conflict between parent and offspring is not an occasional result of unusual stresses on the parent, but instead is a fundamental feature of the relationship.

Species Differences in the Role of Fathers:
Socioecology and the Comparative Method

While most features of maternal behavior are quite similar across primate species, the contribution of fathers varies widely from species to species. In some species, the father plays a major role in infant care, and may even spend more time than the mother holding and carrying their infant offspring. In others, fathers take no interest in their infant offspring. The use of the comparative method, in conjunction with basic principles from behavioral ecology, helps to make sense of this variation.

Males, like females, must weigh decisions about parental investment in one offspring against opportunities for producing more offspring. For males these decisions are complicated by the addition of uncertainty about paternity. A female can be certain that she is the parent of her infant, but a male is never quite as sure. The degree of ambiguity or certainty about paternity is influenced by the species-typical social organization. A brief synopsis of social organization and male paternal care for some of our closest living relatives illustrates this point.

Gibbons and siamangs are lesser apes that live in the tropical evergreen forests of Southeast Asia (Leighton, 1987). In these species, males and females form monogamous pairs when they are about eight years old. Together, the pair stakes out a territory that they defend against incursions by other pairs. Dependent offspring live with their parents until they reach puberty, at which time they leave to find a partner and form their own family unit. In this type of social system, gibbon and siamang males have a relatively high degree of certainty that they are the fathers of offspring born in the group, and both males and females participate extensively in the care of the young. Mothers do most of the early infant carrying and feeding, but after weaning, adult males spend more time than mothers carrying, grooming, and playing with dependent infants (Chivers, 1974). Siamang males have been observed spending up to 75 percent of their time carrying and caring for yearling infants.

Mountain gorillas live in the mountain forests of central Africa in groups consisting of several adult females, their young, and one or two adult males (Stewart & Harcourt, 1987). One of the males, the "silverback," is clearly dominant over the other adolescent or adult males. Male relationships are relatively stable, and a silverback may maintain his dominance position for up to ten years. Females typically emigrate from their natal groups at puberty, leaving behind their female kin to form a close relationship with the silverback male in a new group. In this type of social system, the silverback male does almost all of the mating and is likely to be the father of most of the infants who are born in the group. When an infant is born, the mother provides most of the early feeding and carrying, but the silverback male is an important focus for protection, affiliation, and play. As juveniles, gorillas spend one-third of their feeding time and almost two-thirds of their resting time in proximity to the silverback male (Watts & Pusey, 1993). Young juvenile gorillas who are orphaned have a good chance of surviving because they form close associations with the silverback and share his night nest.

Common chimpanzees live in large communities which fission into small subgroups for days at a time (Goodall, 1986; Wrangham et al., 1996). Males tend to spend their time with other males, while females travel and forage alone with their dependent young. Females do not form close associations with particular males but instead will mate promiscuously with as many males as possible when they are in estrus. Male chimpanzees have a dominance hierarchy, but with this type of social system, the alpha male has a relatively low degree of confidence that he is the father of individual infants born in his community. After their infants are born, chimpanzee mothers do not associate with adult males. Adult males are tolerant but show no interest in particular infants.

These three species were selected because they are all in the ape family and are relatively closely related, but they differ dramatically in the role that fathers play in infant care (see Table 6.1). Male parental care is not more

Table 6.1   Social organization, paternity certainty, and paternal contribution to infant care

Species	Social organization	Mating system	Paternity certainty	Paternal care
Gibbon, siamang	Small, family groups	Monogamous	High	Extensive caretaking
Mountain gorilla	Multiple females, 1–2 males	One male dominates	Moderate	Affiliation, protection
Common chimpanzee	Large fission-fusion communities	Promiscuous	Low	Absent

common in species that are more closely related to humans. It appears intermittently throughout the Primate order and is related more closely to social organization and paternity certainty than to phylogeny (Whitten, 1987). Males are more likely to take an active role in infant care when they are more assured of being the infant's biological father.

Intensive paternal care is particularly common in tamarins and marmosets, which live in small family groups in the rain forests of South America (Snowdon, 1996). In these species, the role of the mother in infant care is reduced, and males do most of the infant carrying after the first few months of life. Males are strongly attracted to infants, and captive studies have demonstrated that they will carry, caretake, and defend infants who are not their own (Whitten, 1987). Under natural circumstances, however, most groups contain one adult male, one breeding female, and their immature offspring. Caregiving males are highly likely to be related to the infants they care for. Paternity certainty does not explain all of the variation in male-infant interactions in nonhuman primates, but it is a useful construct to account for the major differences in paternal care across monogamous and polygamous social systems.

The participation of fathers in child care in human societies also varies widely and is influenced by historical, cultural, and demographic factors (Lamb, 2000). A fundamental difference between human and nonhuman primate fathers is in the material provisioning of children (Lancaster et al., 2000). Because of the high cost of provisioning, it is particularly important for men to be able to differentiate their own children from offspring sired by other men. Men will knowingly and willingly invest in children who are not their own, but when residence is controlled, fathers spend more time with, provide more care for, and spend more money on their biological children than on stepchildren (Anderson et al., 1999; Zvoch, 1999).

The comparative method used by biologists is different from the animal model approach typically used in biomedical research. The chimpanzee, for example, is considered not as a "stand-in" for a human, but rather as a different species solving the problems of raising healthy, successful offspring within the context of its own social organization, ecology, and evolutionary history. In some regards, human are an outlying species, but in many aspects of life history, reproduction, and development, we fall along the same allometric curves as our nonhuman primate cousins (Harvey et al., 1987). An appreciation of variation among species in features such as growth rates, length of the period of dependence, and paternity certainty can lead to a better understanding of the factors that influence variation within a species, including our own.

Individual Variation in the Quality of Maternal Care

The previous two sections described patterns of parenting behavior that are typical for a given primate species. Studies of primate parenting have also noted marked individual differences in parenting style among members of the same species. To some extent these individual differences appear to represent stable aspects of temperament or personality. In other regards, they can be seen as adaptive responses to circumstances.

Research on primate mother-infant relationships using the ecological methods described above has consistently demonstrated that individual differences in maternal care fit a two-factor model, with maternal protectiveness representing one dimension and maternal rejection representing the other. In one of the earliest quantitative studies of maternal behavior of rhesus monkeys, Hinde & Spencer-Booth (1971) found that differences between mothers in contact initiation and caregiving behaviors were correlated with each other, while rejection was positively correlated with the percentage of time the mother and infant were out of contact. Contrary to expectation, these two dimensions were only weakly negatively correlated with each other. Later studies at the same colony confirmed that restrictive and rejecting behaviors vary independently (Hinde & Simpson, 1975).

Variation in maternal behavior along the two dimensions of protectiveness and rejection has been clearly demonstrated for vervet monkeys at the VMRC (Fairbanks, 1996a; Fairbanks & McGuire, 1987). Mothers who are high in protectiveness are more likely to approach their infants, initiate contact, and restrain them when they try to get away. Mothers who are high in rejection have higher rates of breaking contact, leaving, and rejecting their infants. Interestingly, these are orthogonal factors and not just opposite ends of the same dimensions. Numerous subsequent studies of this and other primate species have now found that the same two factors emerge from correlational or factor analysis of ethological data on mother-infant interactions (e.g., Berman, 1990a; Fairbanks, 1996a; Maestripieri, 1998a; Schino et al.,

Table 6.2 Two-factor model of individual differences in maternal style

	Protectiveness	
Rejection	Low	High
Low	Laissez-faire	Protective
High	Rejecting	Controlling

1995). Primate mothers who are low in both protectiveness and rejection are referred to as "laissez-faire," and those who are high in both traits as "controlling" (Table 6.2).

 Individual differences in maternal protectiveness and rejection tend to be consistent over time and across infants (Fairbanks, 1996a). Although the rates of protective and rejecting behaviors vary markedly across the first six months of the infant's life (see Figures 6.1 and 6.2), mothers with relatively high rates of restraint, rejection, or contact in the early months continue to have above-average rates of the same behaviors in later months (Berman, 1990a; Simpson & Simpson, 1986). Longitudinal studies of maternal behavior have also demonstrated that mothers use a consistent maternal style across infants (Berman, 1990a). For vervet mothers with two or more infants, maternal style with one infant can be predicted with a high degree of accuracy using the mother's behavior toward her other infants (Fairbanks, 1996a). This supports the idea that maternal style is a reflection of stable individual differences in temperament or personality of the mother.

Longitudinal research with nonhuman primates has also shown that maternal style along these dimensions tends to run in families. Adult daughters are similar to their mothers in the amount of time they spend in contact with their infants, in their degree of protectiveness and rejection, and in the likelihood of abuse (Berman, 1990b; Fairbanks, 1989; Maestripieri et al., 1997). These effects may be due to the direct influence of early experience or to genetic and physiological influences on personality and maternal style that are shared between mothers and daughters (Maestripieri & Carroll, 1998; Suomi, 1987).

In large primate populations, a certain percentage of normally reared individuals will exhibit maternal behaviors that can be considered pathological. Maestripieri and Carroll (1998) estimate that serious infant neglect and abuse are not uncommon among primates, and that 5–10 percent of infant primates are subjected to physical abuse by their mothers. They interpret the high levels of maternal rejection as undesirable products of variation in underlying genetic, physiological, and neurobiological systems. Just as some normally reared individuals have defective immune systems or the predisposition for mental illness, so do some females have inadequate maternal responsiveness. Research is aimed at elucidating the proximate mechanisms involved so that dysfunctional maternal behavior may be ameliorated or prevented (Maestripieri & Megna, 2000; Troisi & D'Amato, 1991).

Adaptive Variation in the Quality of Maternal Care

The influence of evolutionary biology and parental investment theory has led primate researchers to interpret some differences in the quality of maternal care within an adaptive framework. Several examples of variation in maternal behavior of vervet monkeys that are consistent with an adaptive interpretation are presented here. These examples were selected because they represent situations that have relevance to areas of concern in human parent-child relationships. In each case, maternal behavior of the monkeys varies according to the mother's social circumstances and opportunities for future reproduction. They exemplify the trade-offs between maternal care and infant survival, and demonstrate that investment of effort in one infant influences a mother's ability to invest in future infants.

"Teenage" Mothers

There is widespread recognition of the adverse effects of teenage pregnancy and parenting in modern society, both for the mothers and for their infants. Teenage girls who have babies are less likely to stay in school, have lower job attainment, and are less likely to marry than comparable teenagers who wait to become pregnant (Voydanoff & Donnelly, 1990). Rating of maternal behavior during home visits has indicated that teenage mothers are less verbal with their infants, less responsive and involved, and more restrictive and punishing compared to older mothers (Garcia Coll et al., 1987). Their children are more likely to have problems with behavior, school performance, and impulse control (Brooks-Gunn & Furstenberg, 1986). As with humans, adolescent vervet monkeys make bad mothers. Under conditions of food abundance, a vervet female is capable of giving birth at three years of age, before she is fully grown. Three is a transitional age for vervet females. They are still spending a considerable amount of time engaged in juvenile play activities (Fairbanks, 2000b), and they are also striving to take their position in the adult female dominance hierarchy (Horrocks & Hunte, 1983a). Play behavior is sharply truncated if an adolescent female becomes pregnant, and she spends most of her time resting and feeding. Even in the presence of abundant provisioned food in the captive colony, these young females have a hard time sustaining normal growth rates during pregnancy, and many of their infants are stillborn or die as a result of a difficult delivery (Fairbanks, 1996b). Those that are born alive are often neglected by their mothers. There is always considerable interest in infants in a vervet group, and juvenile and adult non-mothers will try to hold and carry newborn infants (Fairbanks, 1990). The mother usually resists this early attention, but three-year-old mothers readily allow caretakers to take their infants and show little interest in getting them back. This leads to a higher-than-normal mortality rate for live-born infants of three-year-old mothers. Interestingly, if their infants die, these young mothers return to being kids again, and begin playing at the same rate as three-year-old females who have never been pregnant.

If the infant survives, adolescent females continue to show ambivalence about their role as mothers. Figure 6.3 shows the rate of maternal rejection for vervet monkey mothers who had their first infant when they were three years old compared to females who had their first infant when they were fully mature, at four or five years of age. The adolescent mothers were significantly more likely to push their infants off the nipple or prevent them from establishing contact.

Several lines of evidence indicate that inadequacy in the maternal behavior of adolescent females is a response to circumstances and does not represent a persistent attribute of the mother. Females who are neglectful mothers at three years of age do not neglect their later-born infants. Nor can the differences be attributed to lack of experience in maternal care. Females who have their first infant when they are four are better mothers than females who have their first infant at age three. The poor maternal behavior of "teenage" mothers is more easily understood in terms of parental investment and lifetime

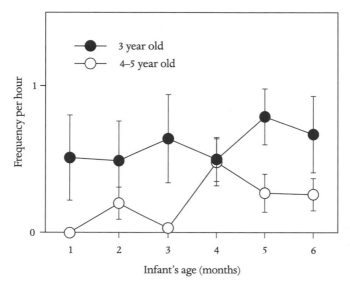

Figure 6.3 Mean (± s.e.) frequency per hour that vervet monkey mothers reject their first infants, by mother's age and infant's month of life.

reproductive success. A young female has her whole reproductive life ahead of her, and anything that she does to jeopardize her future welfare will have a larger impact on her lifetime reproductive success than a similar action by an older female. In many circumstances, an adolescent mother will produce more surviving offspring in her lifetime if she abandons her current infant and instead invests the time and energy in her own development. A female monkey who postpones maternal investment will reduce her risk of illness or death and reserve adequate energy for physical growth, play, and establishing favorable rank relationships. For humans, postponing child rearing also means that a woman will have more time for education to increase resource access and more opportunity to attract a mate who will invest in future offspring. From this point of view, inadequate teenage mothering is not a mistake that can be easily corrected by parenting education. Rather, it is the result of evolved contingencies in parental behavior that are adaptive when the mother's whole life history is considered.

Most primate mothers produce infants at an age and a time when they can provide adequate maternal care. The question is usually not whether a mother will care for her infant but how much care is the right amount. There is an implicit assumption in our culture that the ideal mother is completely nurturing and responsive to her infant's needs. Parental investment theory, by contrast, predicts that mothers will try to balance investment across infants to maximize their lifetime reproductive success. Research with vervet monkeys and other nonhuman primates provides evidence that variation in the quality of care given to one infant influences the mother's ability to invest in her next infant (Fairbanks, 1988a; Fairbanks & McGuire, 1995). Maternal rejection and weaning conflict is directly related to mating opportunities and the timing of the mother's next conception (Berman et al., 1993; Fairbanks & McGuire, 1987).

Grandmothers

Grandparents occupy a variety of roles in contemporary families depending on factors such as age, proximity, health, and family circumstances (Smith, 1995). During times of distress, such as divorce, grandparents can act as buffers and as an important source of support (Hetherington & Stanley-Hagen, 1995). Grandmothers provide substantial child care assistance in single-parent families, particularly those with young mothers (Burton & Dilworth-Anderson, 1991), and a grandmother is the most likely person to take responsibility for preschool children in parent-absent families (Kennedy & Keeney, 1988).

Grandmothers also provide support for their adult daughters and grand-offspring in vervet monkey society. Vervets, like most Old World monkeys, live in matrilineal family groups (Fairbanks, 2000a). Females maintain strong social ties with their mothers and other female kin throughout their lives. The presence of older adult females helps to maintain continuity in social relationships within and between matrilines. At the VMRC, the reproduction and maternal behavior of young adult vervet females is influenced by whether or not their own mothers are still alive and present in the group (Fairbanks & McGuire, 1986). A stable extended kin group provides a secure and predictable social environment for a young mother raising her first or second infant. In families that have lost the matriarch, young mothers tend to be more anxious and protective toward their infants, and they devote more time and effort to keeping the infant close (Fairbanks, 1988b). In contrast, if the infant's grandmother is available, young mothers take advantage of the security she provides and are more relaxed in the care of their infants.

The influence of the presence of the grandmother on maternal behavior of young adult mothers is summarized in Table 6.3. When the infant's grandmother is present, young mothers spend significantly less time in contact with their infants throughout the first six months of life. Mothers are less likely to initiate ventral contact with their infants when the grandmother is available, and are less likely to restrain their infants from leaving. Infants with grandmothers begin to leave and develop spatial independence from their mothers approximately one month earlier than infants without grandmothers in the group. They form special relationships with their grandmother and spend more time near her than near any other group member except the mother (Fairbanks, 1988c).

In evolutionary biological terms, the social support provided by grandmothers allows the young mothers to reduce the time and energy they invest in care of their infants without jeopardizing infant health and welfare. Mothers without this kind of family support must invest more effort to ensure infant safety. Maternal behavior is sensitive to this kind of variation in the social environment, and mothers are able to adjust the care they give to their infants to balance the infant's needs with their own lifetime reproductive success. Interestingly, the infants apparently try to thwart their mothers' attempts to regulate the amount of contact. Infants in the high-contact group (without grandmothers) attempt to break away from their mothers more often than infants in the low-contact group (with grandmothers).

Table 6.3   Influence of grandmothers on vervet monkey mother-infant relationships (summarized from Fairbanks, 1989)

Behavior	Grandmother present	Grandmother absent
Percent time in contact	Low	High
Mother restrains infant	Low	High
Mother initiates contact	Low	High
Infant breaks contact	Low	High
Infant carried by non-mother	High	Low

Social Status and Maternal Care

One of the most consistent predictors of poor outcomes in child development is socioeconomic status (SES). Children from lower-class families do worse in school and have more behavioral and emotional problems and lower economic achievement than children from middle- or upper-middle-class families. Researchers have looked at variations in parenting practices to identify specific sources of these differences (Hoff-Ginsberg & Tardif, 1995). Observational studies of mother-offspring interactions have repeatedly found that lower-SES mothers are more restrictive and controlling with their young children than higher-SES mothers (Hart & Risley, 1992; Hoff-Ginsberg, 1991). Consistent differences have also been found in the quality of verbal interactions between mother and child. Some of these differences in parenting style are believed to be a response to differences in external pressures on the parents and some to class-related differences in the parents' goals and expectations for their children.

The nonhuman primate analog of socioeconomic status is dominance rank. Dominance hierarchies are apparent in many primate species, with high-ranking animals enjoying advantages in resource access and social control. For females, dominance status is socially inherited through matrilines (Chapais & Schulman, 1980). Daughters of high-ranking females become high ranking themselves, while daughters of low-ranking females typically remain low ranking for life.

The influence of dominance rank on maternal behavior has been examined in a number of primate species. Low-ranking mothers, who face more threats from the social environment, tend to be more protective toward their infants (Berman, 1984; Maestripieri, 1993b). This is also true for vervet monkey mothers at the VMRC. Compared to high-ranking mothers, low-ranking females restrain their infants significantly more often (Figure 6.4a). High-ranking mothers, in contrast, are significantly more likely to reject their infants' attempts to gain contact (Figure 6.4b). As a result, infants in low-ranking families spend more time in close physical contact with their mothers, while infants in high-ranking families spend more time away with caretakers and peers (Fairbanks, 1990; 2000c).

Rather than using their greater command of social and material resources to increase the level of care they give to their infants, high-ranking mothers tend to do just the opposite. High-ranking mothers are better able to retrieve

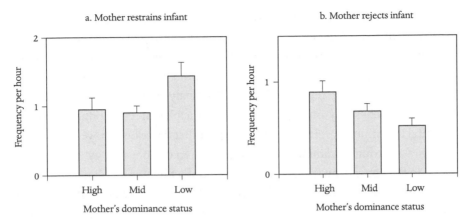

Figure 6.4 Mean (± s.e.) frequency per hour that vervet monkey mothers (a) restrain and (b) reject their infants, by mother's dominance status. Behavior averaged over the first six months of the infant's life.

their infants from caretakers when they choose, and to protect them from harm from other group members. They take advantage of the relative safety that their position provides to limit mother-infant contact time. This allows high-ranking females to begin cycling again sooner than low-ranking females without jeopardizing the health and welfare of their infants. Interbirth intervals are related to female dominance rank, with high-ranking females having shorter interbirth intervals and higher lifetime reproductive success than lower-ranking females (Fairbanks & McGuire, 1984).

The offspring of high-ranking females experience more maternal rejection as infants, but they also enjoy the advantages of growing up in a high-ranking family. Daughters will inherit their mother's high status as adults. Lifetime advantages in health, welfare, and reproductive success that have been documented for high-ranking females are more likely to be linked to this social transfer of rank than to the effects of early mother-infant contact or maternal rejection.

Evolutionary Psychology of Human Parenting

The evolutionary perspective used here to interpret nonhuman primate behavior has been applied directly to the study of human parenting by a small but bold group of evolutionary psychologists and anthropologists (e.g., Daly

& Wilson, 1995; Hrdy, 1999). Evolutionary anthropologists use life history data from human societies to determine how parenting practices have evolved to promote reproductive success in different socioecological contexts, while evolutionary psychologists look at how the mind has been shaped by natural selection to promote reproductive success (Daly & Wilson, 1995). Both of these approaches rely on the same basic principles of evolutionary biology, parental investment, and parent-offspring conflict described above.

In a thorough examination of human prenatal physiology, Haig (1993) came to the conclusion that mother-infant conflict begins in the womb, before the infant is even born. Parent-offspring conflict theory predicts that the "decision" to miscarry a pregnancy because of factors such as maternal malnutrition or fetal abnormality would be expected to differ for mother and fetus. The optimal feeding and growth rate for the fetus would also differ for mother and offspring. There is evidence for conflict in both of these areas in maternal-fetal interactions. For example, after implantation the fetal placenta taps directly into the mother's bloodstream and releases hormones to inhibit uterine contractions that could terminate the pregnancy. The fetus also produces hormones that raise the mother's blood sugar level. The mother counters with hormones that neutralize the effects of the fetal hormones, which leads to increased production by the fetus, and so on. The resulting compromise is physiologically wasteful and can best be understood as a kind of "arms race" with the mother and the fetus each trying to turn the interaction to its own advantage.

After birth, the decision to care for or abandon the infant can be predicted by effects on the parent's inclusive fitness. Daly and Wilson (1995) introduced the concept of discriminative parental solicitude to describe the contingent nature of parental care. They argue that humans have evolved to discriminate, consciously or unconsciously, based on attributes such as kinship and infant health that are related to the costs and benefits of infant care. The importance of kinship in parenting is illustrated in the high risk of serious maltreatment for children living in families with a stepparent (Daly & Wilson, 1996). Children living in families with one biological parent and one stepparent are hundreds of times more likely to die of child abuse than children living with both biological parents. In keeping with the importance of kinship in parental care, Daly and Wilson (1982) have shown that mothers are strongly motivated to point out father-infant resemblance.

Parental care decisions are also affected by nutritional and economic stresses on the family. Hrdy (1999) provides an extensive historical account that documents the high incidence of infant abandonment related to shifting economic circumstances in Western culture. For example, there is a remarkable correspondence between the price of rye and the number of abandoned children in Limoges, France, between 1726 and 1790 (Voland, 1989). Active infanticide is relatively rare, but mortality through neglect and abandonment has been exceedingly common in our history. In spite of good intentions, for centuries the majority of infants placed in orphanages and foundlings home did not survive (Hrdy, 1999).

The health of the child is another factor influencing parental solicitude. In a longitudinal study of 590 Hungarian mothers, Bereczkei (2001) found that mothers of high-risk infants, based on low birth weight and poor health status, terminated nursing sooner than mothers of healthy infants. They also conceived and gave birth to their next infant sooner than mothers of healthy full-term infants. In spite of cultural appeals to the contrary, parents are predisposed to limit investment in high-risk infants. This was also verified in a detailed study of home care by mothers of preterm twins (Mann, 1992). In a sample of fourteen twins, all seven mothers directed more attention to the healthier twin. Parents have even been shown to grieve more after the death of a healthy child than for an unhealthy one (Littlefield & Rushton, 1986).

Parenting practices can differ widely across cultures in accordance with socioecological principles. Blurton Jones (1993) describes major differences in the way children are treated in two hunter-gatherer societies. Kung mothers are indulgent and attentive and have never been observed to hit their children. Hadza mothers, in contrast, are more demanding, less responsive, and more likely to use physical punishment with their children. These differences are consistent with the view that parents are maximizing lifetime reproductive success in the context of the local ecology (Blurton Jones et al., 1989). For example, differences in mortality risks to children influence the role that children can play in foraging for themselves. The habitat of the Kung makes it difficult for children to gather food without traveling long distances and seriously risking their own survival. The Hadza habitat, in contrast, contains a range of foods close to camp that children can effectively exploit. The contribution that Hadza children make to their caloric intake, allows the Hadza mothers to have shorter interbirth intervals and produce more children in their lifetime. A model of the Kung birth spacing that incorporates travel distances, caloric intake and back load demonstrates that Kung mothers are limited by their local ecology to a four-year interbirth interval. Shortening the interbirth interval results in higher mortality and lower lifetime reproductive success. Thus, a Hadza mother can profit by weaning her infants early and putting her children to work, while a Kung mother cannot.

Cultural and economic factors also influence whether a parent will influence his or her lifetime reproductive success by investing more in sons versus daughters (Trivers & Willard, 1973). For example, in many societies, wealth and social class have a stronger influence on the reproductive success of sons than of daughters. Wealthy or high-status males have more opportunities than their sisters for mating success, while low-status males are less likely than their sisters to marry and produce children. An examination of one thousand probated wills demonstrated that these differences in expected fitness are reflected in inheritance patterns (Smith et al., 1987). Parents in the wealthiest quartile left twice as much money to their sons as to their daughters, while those in the lowest quartile left more money to their daughters than to their sons.

These and other studies by evolutionary psychologists and anthropologists have demonstrated the value of using evolutionary principles to understand variation in human parenting. This influence is also beginning to be felt in developmental psychology. For example, Belsky (1999) has argued that avoidant and ambivalent attachment patterns defined by Ainsworth's Strange Situation may represent different evolved strategies for increasing reproductive success in particular kinds of physical and social environments. In a major review of the etiology of child maltreatment, he acknowledged the importance of recognizing that the interests of parents and offspring are not always the same and delineated the ways that evolutionary theory can predict and guide our understanding of child neglect and abuse (Belsky, 1993).

Contemporary psychology generally views variation in parenting behavior as the result of a complex interplay between parent and child set in a social and cultural context (see Bornstein, 1995, for extensive reviews). Parenting experts recognize the contribution of the child in many dysfunctional parent-child relationships, and have developed educational programs to help parents find better ways to handle "difficult" children. Nevertheless, most research on parenting is influenced by the value judgment that some parents are better than others. Causative factors for poor parenting are sought in the parent's early experiences, current mental health, social support systems, and/or economic circumstances. In this regard, considerable attention has been paid to child abuse and neglect, teenage parenthood, and the role of parenting in perpetuating social class distinctions. Implicit in this research is the assumption that poor parenting is a dysfunctional product of deprived or inadequate conditions.

In contrast, the interpretive framework used here considers the effects of infant care on the parent's lifetime reproductive success as the primary factor influencing variation in the quality of parental care. This leads to recognition of circumstances that will cause parents to behave in ways that are clearly not in the best interest of an individual child. Consideration of maternal behavior within an adaptive framework, as presented here, can serve to complement research into the complex proximate factors that influence variation in parenting in humans and nonhuman animals. This approach poses different questions and provides alternate ways of looking at similar phenomena. The resulting evolutionary biological predictions do not always fit with our cultural conceptions of the ideal parent. Nevertheless, if evolved dispositions to withhold care or discriminate against certain children do exist, then it would be a disservice to the victims of this discrimination to avoid their consideration in explanatory models and social interventions. The evolutionary biological perspective used in primate parenting studies shifts the focus away from an almost exclusive attention to effects on the child toward a greater appreciation of the effects of caregiving behavior on parents and siblings. This has been a neglected perspective in developmental psychology (Ambert, 1992). Even when studies have specifically measured parental stress, the intention has been to use the information to explain variation in the effects of parenting practices on children. For example, studies of mothers working outside the home have used measures of maternal stress and role satisfaction as mediators of the influence of maternal employment on child outcomes (Gottfried et al., 1995).

Another contribution of the ethological and evolutionary biological approaches has been to highlight the responses of the offspring to different parenting behaviors. Both theory and empirical observations agree that primate infants are not passive recipients of maternal behavior. The ethological data clearly demonstrate many ways that infants act to extract parental care from their mothers. Evolutionary biological theories of parental investment predict a certain amount of conflict between parent and offspring, and also predict that offspring will evolve mechanisms to cope with expectable variations in parental care. These mechanisms may partly explain why it is difficult to demonstrate consistent effects of seemingly major differences in parenting practices on child outcome (NICHD, 1997).

I would like to thank Michael McGuire for establishing the vervet research colony, and Karin Blau, Glenville Morton, and Dan Diekmann for many years of assistance with data collection and animal care. I also thank Dario Maestripieri, Alan Steinberg, and Denise Gomez for helpful comments and assistance in preparing this chapter. The research reported here was funded in part by grants from the National Science Foundation and the Veterans Administration.

7

Social Development and Affiliation

James R. Roney and Dario Maestripieri

There have been major changes in the theoretical and empirical approaches to the study of socialization since the beginning of the twentieth century (Maccoby, 1992). The two research paradigms that dominated the first half of the century were psychoanalysis and behaviorism. Psychoanalysis viewed socialization as the process through which children gradually achieve control over their sexual and aggressive instincts and resolve their conflicts with their parents by identifying with the same-sex parent. Specifically, psychoanalysts argued that by identifying with their same-sex parent, children acquire sex-specific behaviors and prosocial values that allow them to function properly in their family and society. Behaviorism viewed socialization as the process by which children learn appropriate behaviors from their parents and society by means of rewards and punishments. Behaviorism shared with psychoanalysis the view of socialization as a process in which information flows unidirectionally from parents to children. This and other core assumptions and predictions of behaviorism and psychoanalysis, however, were not supported by empirical evidence. Therefore, both research paradigms underwent considerable criticism around the middle of the century and lost their prominence as frameworks for the study of socialization.

In the second half of the century, increasing emphasis was placed on the notion that children play an active role in their own social and cognitive development (e.g., Lewis & Rosenblum, 1974). Microanalytic behavioral observations of parent-child interactions revealed complex reciprocal influences between parents and children and continuous adjustment in their relationships across development (see, e.g., Trevarthen & Aitken, 2001, for a review). The study of relationships gradually acquired a prominent role in research on social development, and growing emphasis was placed on emotional and cognitive processes other than those involving associative learning (e.g., Bandura, 1989; Sroufe, 1996). Many developmental psychologists continue to believe that the parent-child relationship is unique in many respects and that parents play a unique role in the socialization of their children. An alternative view, however, is that social development is mostly affected by the experiences that children have outside the home (e.g, in peer groups) rather than inside it (Harris, 1995; but see Lowe Vandell, 2000).

Evolutionary approaches have been late additions to the study of socialization. Recent writings have emphasized the notion that different developmental trajectories may represent lifelong adaptations to particular environments or life history strategies (Belsky et al., 1991; Chisholm, 1993; Low, 1989; see below). Such evolutionary accounts of socialization attempt to provide adaptive explanations that complement the more proximate causal explanations pursued by traditional (e.g., learning-oriented) theories of socialization.

Systematic studies of nonhuman primate social development and socialization began in the 1960s and 1970s, at a time when behaviorism and psychoanalysis were already under scrutiny. Early studies were mostly descriptive and provided qualitative information on gross changes in behavior in relation to age (e.g., DeVore, 1963). There was a great deal of interest in primate socialization, but early attempts at studying this phenomenon lacked a well-defined theoretical framework, clear operational definitions of socialization processes, and standard techniques for behavioral observation and analysis. For example, the use of a broad definition of socialization as any influence of the social environment on the developing individual (Poirier, 1972) made it difficult for primate researchers to develop clear research questions and empirically testable hypotheses. As a result, although social development continued to attract the attention of primate researchers, the term "socialization" gradually disappeared from the primatological literature.

Attachment theory and the findings of human research on the effects of maternal deprivation stimulated a great deal of laboratory research on primate development in the 1960s and 1970s (see Chapter 5). Much of this research investigated the influence of the mother on behavioral development with deprivation studies, while the study of normative processes and individual differences in social development was relatively neglected. The quantitative study of normative processes and individual differences in mother-infant relationships and social development was pioneered by Hinde and collaborators (see Chapter 5 and 6). Hinde also provided a conceptual framework for the study of social behavior and development in primates by discussing how repeated interactions between individuals contribute to the formation of their relationship, and how many dyadic relationships contribute to the formation and maintenance of social structure (Hinde, 1976). This approach has had a great impact on primate research, and ever since Hinde's seminal contributions, the study of social development and affiliation in primates has been organized around the study of interactions, relationships, and social structure.

Parental Influences on Social Development

The study of parental influences on social development is an important area of research in both humans and primates. It is also a controversial area of research because it is often difficult to establish whether the correlation between parental and child characteristics reflects the causal influence of parental behavior on child development or the fact that parents and children share both their genes and their environment (e.g., Maccoby, 2000a). This is particularly true for examining variation in parental and child behavior within the normal range of the population rather than at the extremes.

Parenting Styles and Offspring Development

Research on the influence of normal variation in parenting styles (in particular, the degree of parental control and authority) on early child development was pioneered by Baumrind (1971), who distinguished among three qualitatively different types of parenting style: permissive, authoritarian, and authoritative. Baumrind's own research and other studies found that the children of authoritative parents tended to be mature, independent, and friendly, and had high self-esteem and good school performance. The children of authoritarian parents were less socially competent, had lower self-esteem, and were more aggressive. Finally, the children of permissive parents had the lowest levels of self-reliance and self-control and were likely to develop problematic behavior. Thus, the authoritative parenting style, which involved a combination of warmth and affection, prompt responsiveness to children's needs, and parental imposition of clear requirements for prosocial behavior, was regarded as optimal for socio-emotional development.

Baumrind's initial classification of parenting styles has been subsequently revised and expanded. For example, Maccoby and Martin (1983) proposed a two-dimensional model of parenting in which the interaction between the two orthogonal dimensions of demandingness and responsiveness results in four different parenting styles: authoritarian, authoritative, indulgent, and neglecting. Maccoby and Martin (1983), along with many other developmental psychologists, continued to regard the authoritative style as optimal, and much effort in this area of research has gone into differentiating the positive and negative developmental outcomes of optimal versus nonoptimal parenting styles (see Darling & Steinberg, 1993). Other researchers have suggested that some aspects of parenting style (e.g., high levels of parental responsiveness and protectiveness) can interact with particular temperamental types in children to produce high levels of behavioral inhibition (e.g., Kagan, 1998).

Research on parenting styles in nonhuman primates shares some conceptual similarities with human research, but the interpretation of parental influences on social development has often been quite different. In macaques and vervet monkeys, most variability in maternal behavior occurs along the two orthogonal dimensions of protectiveness and rejection; their interaction results in four different types of parenting styles: controlling, protective, rejecting, and laissez-faire (see Chapter 6). Most primate research has concentrated on the determinants of variability in parenting style, while the consequences of different parenting styles for offspring behavioral development have been investigated in only a few studies.

Early studies showed that whereas high rates of maternal rejection resulted in short-term increases in infant distress and clinging (Kaufman, 1974), infants of rejecting mothers eventually developed independence at an earlier age and were more enterprising and resourceful than infants of protective mothers (Hinde, 1974a; Schino et al., 2001; Simpson & Datta, 1990). Fairbanks and McGuire (1988, 1993) reported that vervet monkey infants and juveniles reared by laissez-faire mothers were more interested in the external environment and quicker to explore a novel environment and approach new objects than those reared by mothers with other parenting styles. Furthermore, when male adolescents were observed in proximity to a strange adult male, the individuals who experienced the highest rejection rates from their mothers during infancy approached more often and spent more time near the adult male than the adolescents who were rejected less often (Fairbanks, 1993). Therefore, moderate to high rejecting experiences during infancy may favor the development of independence and assertiveness during adolescence. Further evidence suggesting an influence of parenting styles on offspring development comes from studies showing cross-generational transmission of parenting styles from mothers to daughters (Berman, 1990a, b; Fairbanks, 1989; Maestripieri et al., 1997).

The correlation between parenting style and offspring behavior documented in monkeys could be due to similarities in temperament or might be the result of social learning processes. Fairbanks (1996a) argued that the relation between parenting style and juvenile responses to novelty in vervet monkeys was unlikely to be entirely accounted for by temperamental similarities between mothers and offspring. This is because environmentally induced changes in parenting style brought about concomitant changes in juvenile response to novel objects that were similar to those associated with naturally occurring differences in parenting styles. With respect to cross-generational consistencies in parenting styles, Fairbanks's (1989) data suggested that direct experience with their own mothers in infancy was a better predictor of cross-generational consistency in mother-infant contact than observations of maternal interactions with siblings. In contrast, Berman's data (1990a, b) suggested that daughters matched their rejection behavior to that of their mothers by attending to their mothers' behavior toward their younger siblings. The mechanisms underlying the relation between parenting styles and offspring development can be effectively investigated with cross-fostering experiments in which infants are swapped after birth between mothers with different temperaments and parenting styles (e.g., Suomi, 1987).

Despite some disagreements over the specific processes of primate socialization, there is a consensus among primate researchers that no parenting style should be viewed as intrinsically better than any other but that individual differences in parenting and development may represent adaptations to demographic and environmental circumstances (see Chapter 6). This reflects the belief that natural selection often produces flexibility in behavioral strategies in relation to different circumstances. This framework leads to testable predictions concerning the differential survival and reproductive success of individuals exposed to different parenting experiences in the same or different environments.

Other Parental Influences on Social Development

Parents can influence the social development of their children through their everyday interactions with them above and beyond their particular child-rearing styles. Both direct interaction between child and parent and the child's observation of the interaction between parents and other individuals are likely to play an important role in early social development. From very early on, parents display a typical pattern of interactive behaviors with their child including smiling, raised eyebrows and open eyes, and high-pitched speech patterns (e.g., Fernald, 1991). Synchrony in communication between parent and child and parental responsiveness to the child's demands for attention affect the quality of the parent-child bond and may have long-lasting consequences for the developing child (e.g., Ainsworth, 1979; Trevarthen & Aitken, 2001).

The relation between the quality of early parent-child relationships and social development has been thoroughly investigated by attachment theorists (Chapter 5). Belsky et al. (1991) have reinterpreted some of this research from an evolutionary perspective and incorporated its findings into a developmental theory of reproductive strategies (see also Chisholm, 1993).

The basic assumption of this theory is that parenting behaviors may provide cues regarding the type of social environment a child is entering and so shunt development toward behavioral strategies that are more effective within the environment in question. Thus, children who experience insensitive, intrusive, or unresponsive parenting (i.e., the conditions that foster insecure attachment) are more likely to embark on a developmental pathway that prepares them for a world in which adult social relationships (including reproductive pair bonds) are transient, unstable, and often exploitative. This pathway includes both higher thresholds for trusting and caring about others and accelerated pubertal timing and unrestricted sexual behaviors. According to this theory, variables such as father's absence or the quality of marital relationships between parents may be particularly important cues regarding the nature of sexual and social relationships in the child's developmental environment. Although some evidence is consistent with the idea that early social experiences may trigger distinct life history strategies, alternative explanations are also possible. Some research has indicated rather high heritability for variables such as empathic concern (e.g., Matthews et al., 1981), and studies of temperament raise the possibility that the correlation between parenting style and child outcomes reflects genetic similarity rather than causal effects of parenting behavior on social development (e.g., Kagan, 1998). Manipulative research with nonhuman primates may provide an avenue for disentangling these alternative explanations.

Early research with rhesus macaques by Hinde and Simpson (1975) suggested that the degree of meshing versus dissonance between mother and infant was an important aspect of the quality of their relationship. These authors also reported that mothers who had a well-meshed relationship with their infants were more likely to encourage their behavior and independence. Unfortunately, these early findings and suggestions have not been further investigated by subsequent research. Thus, with the exception of the research on parenting styles and behavioral development described in the previous section, most primate researchers interested in parental influences on development have concentrated on specific interactions between mothers and infants (e.g., involving proximity, grooming, play, or agonistic support) rather than on the overall quality of their relationship.

Many studies of primates have shown that immatures have proximity relationships similar to those of their mothers; that is, they tend to remain near and affiliate with members of the same matriline (e.g., in macaques: Berman, 1982a, b; Caine & Mitchell, 1979; de Waal, 1996c; Ehardt & Bernstein, 1987; Hanby 1980a, b; Nakamichi, 1989; Figure 7.1). The precise mechanisms through which infants develop social networks similar to those of their mothers, however, are not well understood. One possibility is that monkey infants develop kin-biased networks through their individual learning experiences with kin and non-kin. In this view, infants have greater opportunities to affiliate with kin than with non-kin simply by virtue of being close to their mothers (Gouzoules & Gouzoules, 1987). Alternatively, infants may acquire social preferences similar to those of their mothers by observing interactions between

Figure 7.1 Rhesus macaque mothers and infants.
(Photo: S. Ross)

their mothers and other individuals (Altmann, 1980; Evans & Tomasello, 1986). In this case, mothers play an important role in the development of their offspring's social networks, but their role is mostly passive. Finally, it is possible that mothers actively influence the social networks of their offspring by encouraging or constraining their choice of social partners (Berman, 1982a, b; Cheney, 1977).

Berman and co-workers (Berman, 1982a, b; Berman et al., 1997) argued that, in rhesus macaques, the mother initially regulates the infant's social interactions by selectively thwarting attempts by others to interact with the infant or vice versa. They also suggested that infants undergo a process of network differentiation in which they retain the maternal social network but develop independent relationships within it. Finally, they suggested that infants' relationships with their mothers are influenced by the size and composition of their group, and that infants autonomously control their social networks in response to demographic changes. Berman and Kapsalis (1999) found that similarities in the kin bias of rhesus macaque mothers and infants were predicted by similarities in social risk. Thus, although intense mother-infant interaction may contribute to the development of high degrees of kin bias in infants, the perpetuation of kin bias across generations is probably accounted for by shared environment and individual learning processes.

In addition to offspring social networks, another well-known example of parental influences on social development in primates concerns the transmission of dominance ranks. Since primate mothers support their offspring in fights against other individuals, the outcome of conflicts between immatures is determined by the dominance rank of their mothers (e.g., Chapais, 1992). Infants and other group members quickly learn the relationship between maternal interventions and conflict outcome so that infants eventually acquire dominance rank close to that of their mothers (see Chapter 2).

The issue of whether primate mothers intentionally promote social learning in their offspring (i.e., the issue of whether they are capable of teaching) is highly controversial. In nonhuman primates, mother-infant interactions suggestive of teaching have been observed in several domains such as tool use, alarm calls, food choice and processing, gestural communication, and infant independent locomotion (reviewed by Maestripieri, 1995b). Primates engage in elementary forms of teaching, but teaching is very rare and domain-specific (see Maestripieri et al., 2002). Although there have been suggestions that primate mothers may encourage their offspring to form social bonds with particular individuals (de Waal, 1990), convincing evidence that primate mothers intentionally shape the social development of their offspring is not available. Obviously, human children learn a lot about their social world through direct teaching and education. Thus, there are important limitations in the extent to which information on parental influences on offspring learning processes in nonhuman primates can be extrapolated to humans.

Sex Differences in the Development of Affiliative Behavior

Sex Differences in Human Social Development and Their Explanations

Behavioral sex differences in human infancy are not as marked as those occurring in childhood or adolescence. Nevertheless, there is some evidence that infant girls are generally more responsive to social cues than are infant boys. For example, some studies have shown that infant girls orient to other people's faces and gaze and show empathy to the distress of other individuals to a greater extent than infant boys do (e.g., Haviland & Malatesta, 1981; Zahn-Waxler et al., 1992). Sex differences in social interactions between children tend to appear at about two to three years of age, when children first show interest in social play. One of the most consistently reported differences in child social play is that boys engage in rough-and-tumble play more frequently than girls do (Maccoby & Jacklin, 1974). In contrast, girls are more likely than boys to engage in play parenting with dolls or younger siblings. Same-sex play groups begin to appear at about three years of age; during the ages of six to twelve years, friendships continue to be sex-segregated, and boys and girls virtually avoid interacting (Maccoby & Jacklin, 1987). In adolescence, same-sex groups gradually evolve into mixed-sexed crowds, which subsequently break down into heterosexual cliques and couples.

Children form same-sex peer groups and friendships on the basis of their mutual play interests and similar styles of social interaction. Friendships between boys are characterized by a higher degree of aggression and competitiveness than those between girls. In particular, school-age boys are more competitive with their friends than with non-friends, whereas the opposite tends to be true for girls. Friendships between girls are more intimate, cooperative, and supportive than those between boys. Boys' friendship groups tend to be large and accepting of newcomers, whereas girl friends tend to play in pairs or in smaller, more exclusive groups. Whereas boys play more outdoors and roam over a larger area during play, girls spend more time playing indoors or near home or school. Sex differences in the quality of early social relationships are similar to those of adults. Thus, similar to men, boys tend to have competitive and domineering styles of interaction, whereas, similar to women, girls tend to have more cooperative and supportive interactional styles (Maccoby & Jacklin, 1987). Differences in interactional styles are accompanied by differences in competitive tactics so that boys tend to establish dominance through physical assertion, whereas girls are more likely to use verbal competitive tactics such as gossip and ridicule.

As mentioned earlier, psychoanalytic theories of social development view sex differences in behavior as the product of identification with same-sex parents. Social learning theories interpret sex differences in social development as the product of learning processes, with a distinction between behavioristic theories that emphasize associative learning and modeling theories that emphasize observational learning and imitation (Maccoby, 2000b). It is unlikely that sex differences in behavior are the result of the children's imitation of the behavior of the same-sex parent, although it is likely that boys and girls selectively attend to different types of behaviors in adults. Parents in many cultures often treat sons and daughters in different ways, but the magnitude of these differences and their interpretation are controversial. In particular, the extent to which fathers and mothers use different parenting styles and interact differently with their sons and daughters is debated in the socialization literature (e.g., Russell & Saebel, 1997).

A meta-analysis of mostly North American studies conducted by Lytton and Romney (1991) failed to find a statistical interaction between gender of the parent and gender of the child. Thus, at least in Western societies such as the United States, it seems that mothers and fathers show similar interactions with sons and daughters in terms of general responsiveness, nurturance, and encouragement of independence. Some studies, however, have reported that mothers tend to use an authoritative or permissive parenting style, especially with their daughters, whereas fathers tend to use an authoritarian parenting style, especially with their sons (e.g., Conrade & Ho, 2001; Russell et al., 1998). Furthermore, in both Western and non-Western societies, parents encourage gender-typed activities in both their sons and daughters (Low, 1989; Lytton & Romney, 1991). Some authors interpret these differences in parental behavior in terms of cultural expectations concerning gender roles (e.g., Whiting, 1980). Others, however, argue that differences in parenting practices largely amplify or suppress preexisting sex differences in child behavior rather than create them (e.g., Geary, 1998).

Biological theories of sex differences in social development have traditionally been divided into those that address mechanisms and those that address adaptive function. The former view sex differences in social development mostly as the product of physiological differences (e.g., neurobiological and endocrine) between the sexes that are established prenatally and reinforced postnatally (e.g., at puberty). The latter view sex differences in social development as part of a complex set of adaptations to the life and reproductive histories of males and females, with particular attention paid to future patterns of mating and parental investment. Finally, evolutionary psychologists integrate the study of mechanisms and function by arguing that the adaptations are found at the level of the psychological mechanisms underlying sex differences in behavior rather than in behavior itself. Evolutionary psychologists, however, have for the most part concentrated on sex differences among adults and have not focused on issues of development.

Sex Differences in Primate Social Development and Their Explanations

In nonhuman primates, there are clear sex differences in affiliative behaviors such as grooming and play, and such differences emerge relatively early during development. In macaques, females start performing more grooming than males in the first year of life, and grooming behavior is fully differentiated by about two to three years of age, when females both give and receive more grooming than males (e.g., Eaton et al., 1985; Hadidian, 1979; Hinde & Spencer-Booth, 1967; Nakamichi, 1989). Female-female grooming, particularly between kin, increases steadily from the early juvenile period to adulthood. Juvenile females increasingly groom both older females and younger individuals, and grooming of males increases dramatically when females begin menstrual activity; juvenile females usually give younger individuals, older females, and adult males more grooming than they receive from them (Glick et al., 1986a, b; Goy, Kraemer, & Goldfoot, 1988; Hadidian, 1979; Kurland, 1977). Although in some macaque species females groom other females more than they do males at any age (e.g., Glick et al., 1986b; Lindburg, 1973), in others females tend to groom both females and males equally at all developmental periods (e.g., Hadidian, 1979). In one-male groups, adult females usually groom the male more than any other adult female in the group.

By the time macaque males are three to four years old, their grooming activity is considerably lower than that of females and no longer biased toward kin (Bernstein et al., 1993b; Hadidian, 1979). At this age, grooming is mostly directed to adult males and is largely unreciprocated. Male-female grooming increases in conjunction with puberty and emigration from the natal group (Goy, Kraemer, & Goldfoot, 1988; Hadidian, 1979; Lindburg, 1973). In other words, males mostly groom other males prior to departure from their natal group, and groom females after immigration into a new group. Males consistently receive grooming from females more than from males, but prior to emigration they receive grooming mostly from their mothers, while after immigration they receive grooming from unrelated females that are potential sexual partners. Grooming between adult males is generally much less frequent than grooming between adult females and mostly limited to the non-mating season (e.g., Drickamer, 1976; Hadidian, 1979; Oki & Maeda, 1973).

Sex differences in grooming behavior are unlikely to be the result of maternal influence. Macaque mothers frequently groom their infants during the

Figure 7.2 Grooming between adult female rhesus macaques. (Photo: S. Ross)

first two to three years of life without any appreciable difference between sons and daughters (Kurland, 1977; Missakian, 1974; Oki & Maeda, 1973). Maternal grooming of offspring tends to decrease after the first two to three years, and in the third year maternal grooming begins to be biased toward daughters (Kurland, 1977; Missakian, 1974; Muroyama, 1995; Oki & Maeda 1973). By this time, however, sex differences in offspring grooming behavior are already well established, and such differences are likely to be the cause rather than the consequence of sex-biased maternal grooming. Specifically, although both sons and daughters increasingly groom their mothers as they mature, daughters return grooming to their mothers more and earlier than sons do (Eaton et al., 1985, 1986; Hadidian, 1979; Kurland, 1977; Missakian, 1974; Muroyama, 1995; Figure 7.2). The rate at which sons groom their mothers does not change after the third year, whereas at this age, daughters begin to outperform their mothers (Hadidian, 1979; Kurland, 1977; Oki & Maeda 1973). Thus, the bias in maternal grooming of daughters is probably the result of the fact that daughters increasingly reciprocate their mother's grooming whereas sons do not. The mother-daughter grooming relationship persists through their lifetime, whereas mother-son grooming terminates with male emigration.

Sex differences in play are as dramatic as those in grooming and remarkably similar to those observed in children. Play behavior in macaques is first clearly observed in weeks seven to twelve and then increases rapidly throughout the first year of infant life (Eaton et al., 1985; Hinde & Spencer-Booth, 1967; Itoigawa, 1973; Lindburg, 1971). Initially, there are no clear sex differences in the time of appearance of play or in the age-related increase in play (e.g., Eaton et al., 1985; Hinde & Spencer-Booth, 1967). At about twenty weeks of age, however, the rate of female play reaches a plateau, whereas the rate of male play continues to rise. Sex differences in play have been observed in many other species of monkeys (e.g., patas monkeys: Rowell & Chism, 1986) as well as apes (e.g., Nadler et al., 1987).

In all macaque species, and probably in most primates, after the first six months of life and for the whole juvenile period, males engage in more rough-and-tumble play than females do (e.g., Burton, 1972; Eaton et al., 1985; Hinde & Spencer-Booth, 1967; Itoigawa, 1973). Juvenile males both initiate play and accept play invitations more frequently than females do; their social play is generally more vigorous than female play; and variability in the frequency of play is lower among males than among females (e.g. Symons, 1978). Male juveniles and sub-adults are the individuals most involved in social play (Figure 7.3). After puberty, the intensity of play gradually decreases for both sexes, but the sex differences persist into adulthood. Adult females show little or no play with other individuals, including their infants, whereas adult males often play with juveniles and sub-adults, and occasionally with other adult males as well (e.g., Fady, 1976; Symons, 1978).

Macaque immatures typically play with individuals of the same sex and age (e.g, Ehardt & Bernstein, 1987; Fady, 1976; Lindburg, 1971). Infants whose mothers are closely related and have similar rank also often play together (Caine & Mitchell, 1979). Play groups may range in size from two up to ten immatures and are typically larger for males than for females (Hinde & Spencer-Booth, 1967; Lindburg, 1971). Juvenile males initiate play with adult peripheral males or juveniles of other groups, whereas no female of any age plays with peripheral males.

Female infants engage in more solitary play than males do at any age (e.g., Deag & Crook, 1971; Ehardt & Bernstein, 1987). Females begin play parenting (e.g., they attempt to carry young infants) in the first year of life, whereas same-aged males show no interest in infants (Deag & Crook, 1971; Ehardt &

Figure 7.3 Rough-and-tumble play between juvenile male rhesus macaques. (Photo: S. Ross)

Bernstein, 1987; Kuyk et al., 1977; Lovejoy & Wallen, 1988; Symons, 1978). Female interest in infants peaks in the juvenile and sub-adult period and remains elevated throughout adulthood. In contrast, males do not show any significant interest in infants at any age, with the exception of a few species in which adult males use infants as agonistic buffers in interactions with other males (e.g., Deag & Crook, 1971). Sex differences in interest in infants may not be evident under particular laboratory conditions (e.g., Gibber & Goy, 1985), but sex differences in play parenting are striking and widespread in group-living monkeys.

Similar to grooming, sex differences in social play and play parenting are unlikely to be the product of socialization. Despite some early reports claiming that there were clear differences in maternal interactions with sons and daughters (e.g., Itoigawa, 1973; Mitchell, 1968), most studies have shown that there are few or no differences in maternal interactions with sons and daugh- ters from birth up to puberty (Brown & Dixson, 2000; Fairbanks, 1996a). Furthermore, there is no evidence that mothers have a different influence on the social networks of their sons versus their daughters. Evidence that male and female infants are treated differently by adults other than their mothers is also scanty. Some studies have reported that the daughters of low-ranking mothers receive more aggression than the sons (e.g., Silk, 1983), but other studies have failed to replicate these findings (e.g., Maestripieri, 1999b).

The sex differences in primate social development are more consistent with the biological theories than with socialization theories. Studies addressing mechanisms have provided evidence for hormonal regulation of sex differences in play, grooming, aggression, and sexual behavior (see Chapters 2 and 4). Studies emphasizing function have pointed out that early sex differences in social behavior are clearly adaptive in light of the different life histories and reproductive strategies of males and females. For example, the intense female-oriented grooming behavior exhibited by young females clearly preludes their integration and acceptance into the female core of the social group, whereas the grooming patterns of young males can be viewed as a prelude to emigration from the natal group. Early differences in play behavior are also likely adaptations to the different life histories of males and females, with a primary emphasis on the development of fighting skills for males and parenting skills for females (e.g., Meaney et al., 1990; Pereira & Altmann, 1985; see Maestripieri & Pelka, 2002, for humans).

Some of these sex differences in early social behavior can be viewed as species-typical adaptations, whereas others can be regarded as a more general primate or mammalian pattern. For example, sex differences in grooming behavior are expected to vary as a function of the social organization of the species and patterns of dispersal. Thus, direct extrapolation of grooming data and theory from a particular primate species to humans may not be warranted. Sex differences in play fighting and play parenting, however, are more likely to reflect patterns of mating competition and parental investment that are shared by most mammalian species. Thus, predominance of play fighting among males should be observed in most polygynous mammalian species in which males are larger than females and male-male competition is intense. Predominance of play parenting among females should be observed in all mammalian species in which maternal behavior has a strong learning component (i.e., Old World monkeys, apes, and humans). Thus, although general principles concerning the biological origin and significance of many sex differences in human social development can be obtained from comparative research with a wide range of species, in the case of behaviors such as play, data and theories from primate research can be more directly extrapolated to humans.

Determinants of Adult Affiliative Relationships and Social Structure

In both human and nonhuman primates, individuals bias their affiliative behavior toward particular classes of individuals. The developmental research reviewed above has shown that social preferences may emerge early in infancy or childhood. Such preferences play an important role in determining the distribution of affiliation among adults and the general structure of primate and human societies. The distribution of affiliation among adults and the structure of primate societies are in large part determined by ecological factors (e.g., van Schaik, 1989; Wrangham, 1980). The characteristics of the ecological environment in which humans evolved are not well known. Among the over three hundred extant species of primates there is great variability in social organization and mating systems. Thus, it would be difficult to draw many specific parallels between the distribution of affiliation in humans and other primates. It may be possible, however, to note some general principles that have validity across species and environments. In this section we attempt to discuss some such principles, with regard first to adult affiliative relationships and then to the study of social structure.

Determinants of Adult Affiliative Relationships: Kinship, Rank, and Reproductive Status

In addition to age and gender, kinship and social status play an important role in determining the distribution of affiliation within primate and human societies. Furthermore, in both primates and humans, changes in female reproductive status can often interact with or override the effects of kinship and status on affiliation.

Kinship probably played a very important role in affiliation patterns throughout human evolution. Studies of contemporary hunter-gatherer societies demonstrate that residence and activity patterns are primarily organized around extended kin groups (e.g., Chagnon, 1977; Hill & Hurtado, 1996). Furthermore, when group size becomes too large, fissioning patterns are such that the descendant groups tend to be more closely related than the original group (Chagnon, 1979, 1982). In modern societies, it is obvious that social relations are organized around kinship, given the division of most modern societies into nuclear family groups. Social structure in modern economies, however, likely underestimates the importance of kinship, given the dispersal of extended kin to seek employment opportunities where available. Humans follow a pattern of sex-biased dispersal that is typical of many vertebrate species and that results in kinship effects on affiliation being stronger for the non-dispersing (i.e., philopatric) sex. It has been argued that early humans had a pattern of female dispersal and strong affiliative bonds between related males similar to chimpanzees (e.g., Rodseth et al., 1991). Although patterns of sex-biased dispersal may have changed throughout human history and differ in relation to geographic areas, the principle that kinship results in stronger affiliative bonds within the philopatric sex has general validity (e.g., Wrangham, 1980).

Socioeconomic status is another major determinant of human affiliation patterns. It is without doubt one of the central theoretical tenets of sociology that human relationships are organized around social class (e.g., Weber, 1921/1968). Not only is social status a primary predictor of marriage patterns, for instance, but also when status is controlled in regression models, the similarity of spouses on variables such as IQ and personality is substantially reduced (e.g., Mascie-Taylor, 1990). Such findings suggest that much of the trend toward similarity in friends and mates (e.g., Rushton, 1989) may be attributable to the tendency for members of the same social class to associate preferentially with one another. Such a tendency is observed also in primate societies in which dominance hierarchies have a strong influence on access to resources (see below).

A prominent exception to the pattern of class endogamy is the relationship between high-status men and physically attractive women. Elder (1969) found that physically attractive women are more likely to achieve upward social mobility than less attractive peers because of their greater likelihood of marrying wealthy men. Among working-class women, physical attractiveness was a better predictor of marrying a high-status man than was educational attainment. Exceptions to endogamy are also found in caste systems in which men of higher castes are willing to have sexual relations with women who are otherwise considered unclean and hence socially excluded (Kamble, 1982, cited in Kurzban & Leary, 2001). Among the Ache of Paraguay, postmenarcheal teenage girls are always welcome at campsites, and "are fed abundantly wherever they go" (Hill & Hurtado, 1996, p. 225). In sum, it appears that female preferences for men of high status and male preferences for physically attractive women are means by which reproduction can alter the bias in affiliation toward kin and individuals of similar social status.

Similar interactions between kinship, social status, and reproduction have been reported in nonhuman primates. Kinship has long been recognized as a significant determinant of affiliative relationships in many species. Kinship effects are found for a variety of behavioral interactions, most notably grooming and proximity. In female philopatric species such as macaques, female grooming is more kin-oriented than male grooming is at any age, even in groups with natal males (Ehardt & Bernstein, 1987). Kinship effects on macaque affiliation almost invariably refer to matrilineal kinship. In a few cases, however, paternal kinship has been shown to influence social behavior (e.g., Alberts, 1999; Widdig et al., 2001). Proximity and grooming relationships are very stable over time, especially for adult females with young, but also for adolescent and juvenile females (Hanby, 1980a, b). Similar to what has been reported for some hunter-gatherers, group fissioning in macaques occurs along matrilineal lines so that fission increases the degree of kinship within descendant groups while increasing genetic diversity between groups (Chepko-Sade & Olivier, 1979; Cheverud et al., 1978).

The influence of social status on grooming has been reported in many primate species, with high-ranking individuals doing less grooming and receiving more than low-ranking individuals (de Waal & Luttrell, 1986; Mehlman & Chapais, 1988; Oki & Maeda, 1973; Rhine, 1973; Silk, 1982, 1992). In some macaque species, grooming is strongly correlated with dominance in females, but weakly in males, with the exception of the alpha male (e.g., Sade, 1972). Grooming is most common between adjacently ranked females and matrilines, whereas male-male grooming is not necessarily associated with adjacent ranks (Butovskaya & Kozintsev, 1994; Oki & Maeda, 1973; Rhine, 1973; Silk, 1982). Frequent grooming between closely ranked females is the result of joint effects of rank and kinship on affiliation (see below).

Changes in female reproductive status bring about a significant rearrangement of grooming distribution within a group. In seasonal species, the mating season is associated with increased grooming between adult males and females (e.g., D'Amato et al., 1982; Oki & Maeda, 1973), and in both seasonal and nonseasonal species, heterosexual grooming increases dramatically during the periods of estrus and consort formation (Hadidian, 1979; Lindburg,

1973; Mehlman & Chapais, 1988; Sade, 1965). Female estrus may alter the relationship between grooming and dominance so that dominant males may give more grooming to estrous females than they receive from them (Mehlman & Chapais, 1988). Many different hypotheses have been proposed to account for the association between female estrus and increased grooming (e.g., Carpenter, 1942; Drickamer, 1976), but the most likely explanation is that grooming reduces the risk of aggression and favors the maintenance of prolonged proximity necessary for mating to occur (e.g., Mehlman & Chapais, 1988).

In addition to mating, lactation can also alter grooming relationships. First, lactating females often maintain affiliative relationships ("friendships") with unrelated adult males, whose primary function may be protection from infanticidal males (e.g., Chapais, 1986; Maestripieri, 2000a; Palombit et al., 1997; Silk, 2002). Second, the presence of infants also alters affiliation between females, because female-female grooming intensifies and becomes unilateral, as lactating females groom far less than they receive. Interest in new mothers stems from interest in their infants, and the increase in proximity and grooming with new mothers takes place because it enables other females to interact with the infants (Henzi & Barrett, 2002). Thus, infant attractiveness can temporarily override the strong influences that dominance rank and kinship have on female affiliation patterns (e.g., Maestripieri, 1994b).

Determinants of Social Structure

Attempts to extrapolate information on primate social structure from behavioral data on affiliation have been made using many different conceptualizations and analytical procedures, some of which were directly imported from the social sciences. Role theory, as originally used in social psychology and anthropology, has often been invoked to understand the structure of primate societies. From this perspective, social structure was viewed as the outcome of the behavior of different individuals, each of whom performed a specific social role (e.g., Fedigan, 1976). Early studies equated social roles with the activity patterns of individuals from different sex, age, or rank classes (e.g., Bernstein & Sharpe, 1966; Fedigan, 1976). A somewhat different approach was used in describing the role of the alpha male in disrupting fights (Bernstein, 1964; Smith, 1973). Here role was defined as a consistent pattern of response occurring in specific contexts rather than a general activity profile (see also

Reynolds, 1970). Although some authors argued that learning strongly influenced the performance of social roles (e.g., Eaton, 1976), the concept of role was imported from human to nonhuman primate research without carrying along all of its cognitive and cultural connotations (e.g., the notion that an individual's role is often defined in relation to societal expectations). If social role in primates is simply equated with membership in a certain age, sex, or dominance class, it is little more than a descriptive tool. Thus, its usefulness in providing insight into social organization remains unclear. In fact, although popular in the 1960s and 1970s, analyses of primate social structure in terms of social roles have been subsequently abandoned.

In contrast to role theory, sociometric analyses of social structure recognize that the analysis must focus on dyadic relationships rather than on individual behaviors. The goals of this analysis are to create a matrix for the quantification of social exchanges among all potential dyads of individuals, identify preferential associations between individuals (i.e., strong relationships), and examine how relationships within dyads are connected to one another. Sociometric methods were used by Sade (1972) to identify grooming cliques, and by Hanby (1980a, b) to identify clusters of spatial associations with rhesus groups (see also de Waal, 1986b; Sade et al., 1988). A variant of this approach, the block-model approach, also imported from sociology, was used by Pearl and Schulman (1983) to compare the social structure of rhesus groups in Pakistan and on the island of Cayo Santiago. Sociometric and matrix analyses have been particularly useful in the studies of dominance hierarchies (e.g., Schulman, 1983). Studies of spatial or grooming relationships with such methods, however, have at best replicated findings already obtained with more traditional analytical approaches.

The approach to the study of primate social structure that has proved to be heuristically most valuable is probably the view that social structure can be studied and understood through the study of its components, that is, social interactions and relationships (Hinde, 1976). In this view, social relationships are conceptualized as a series of repeated interactions between individuals over a certain period of time, where early interactions influence subsequent ones. Social structure can be understood by studying how relationships are patterned and how they affect other relationships. This approach is best exemplified by the "grooming models" proposed by various investigators, in which the structure and distribution of grooming within a primate group are reconstructed from a study of grooming interactions and relationships within the group.

In an early paper, Rhine (1973) hypothesized that frequent grooming between adjacently ranked females was the result of interplay between attracting and repelling forces. This idea was subsequently articulated by Seyfarth (1977), who argued that the distribution of female grooming can be understood in terms of the attractiveness of high-ranking females as grooming partners and competition for grooming between females. Attractiveness of high-ranking females as grooming partners would stem from the fact that subordinates might receive benefits in proportion to their grooming efforts (see below). As competition for grooming increases with the rank of the partner, females compromise by directing more grooming to intermediate-ranking females. Seyfarth (1983) subsequently proposed a cumulative effect of kin- and rank-based attractiveness to take into account the fact that closely ranked females are often genetically related. Seyfarth's original (1977) and revised (1983) models of grooming distribution generated a great deal of research, which provided evidence both in favor of and against the model (reviewed by Schino, 2001).

On the basis of patterns of interactions within male peer networks, Colvin (1982) proposed an alternative grooming model that emphasized attraction to similar rank rather than to high rank. Colvin discussed at length the reasons why partners of similar rank should be most attractive, and the predictions of his model coincide with those of a model based on attraction to kinship/familiarity (e.g., de Waal & Luttrell, 1986). Simply put, attraction to similar individuals is a self-reinforcing mechanism that acts as a conservative force maintaining stability within the social system. Colvin's (1983) data did not fully support the Seyfarth (1977) or the Colvin (1982) model, leading him to integrate these two models into a new model. His own empirical tests revealed that the new model was a better predictor than the Seyfarth or the Colvin models alone (Colvin, 1983). Unfortunately, no one attempted to replicate Colvin's (1983) findings or further test his integrated model.

Overall, a focus on social relationships has allowed researchers to make significant progress in understanding social structure. A focus on relationships has also proved indispensable to understanding the dynamics of affiliative exchanges between individuals. In fact, in order for affiliative interactions between two individuals to be fully understood, they must be viewed in relation to the history of their previous interactions, the expectation for future interaction, and their context of occurrence with regard to other relationships.

Dynamics of Altruistic Interactions

Research on altruism in human and nonhuman primates has focused on different levels of analysis. Work on humans has predominantly addressed the proximate causes of altruistic and prosocial behaviors by examining what developmental, cognitive, affective, and situational factors predict these behaviors. Nonhuman primate researchers, by contrast, have focused on ultimate explanations by examining whether behavioral patterns conform to predictions drawn from evolutionary theory and game theoretic models. Thus, each of these research literatures has addressed important questions that the other has left relatively neglected. In this section we briefly review each of these literatures and also present an evolutionary framework for the integration of proximate and ultimate causes of altruistic behaviors.

Human Altruism and Its Explanations

Psychologists broadly define as prosocial any behavior that benefits another individual, such as giving, sharing, cooperating, protecting. In some cases prosocial behavior may be motivated by selfish interests so that the actor gains an immediate benefit from the behavior. Forms of prosocial behavior that benefit the recipient while imposing a cost to the actor are considered altruistic.

Social psychologists have employed two basic approaches for explaining altruism and prosocial behavior (Batson, 1998). The first is a variance-accounted-for approach that attempts to use as many variables as necessary to predict outcome behaviors as accurately as possible. Researchers have proposed countless dispositional and situational variables that predict altruistic and prosocial behaviors under various conditions. For example, research on bystander intervention has led to complex models in which altruistic intervention is seen as the outcome of a number of variables including the expected benefits to the altruist, the benefits to the victim, the cost of intervention and nonintervention, the feelings of responsibility, and the potential availability of help from other people (e.g., Batson, 1998). Batson (1998) criticized the variance-accounted-for approach on the grounds that successful prediction requires such specificity of circumstances, behaviors, predictors, and subject populations that useful generalizations are impossible.

The second approach attempts to subsume helping behaviors within broader social psychological theories. Social learning theories have emphasized the role of experience and, specifically, either reinforcement/punishment or observational learning mechanisms. For example, some studies have shown that altruistic behaviors in children can be increased with parental encouragement (e.g., rewards) or punishment (Grusec, 1991; Rushton, 1980). Children are also well known to imitate other children or adults who behave altruistically (e.g., Eisenberg, 1986). Social learning approaches may also explain various mood effects on helping behaviors. If children learn to internalize the external rewards (including social praise) that have been associated with past helping behaviors, then altruistic acts may be used to reduce negative moods. Various studies have implicated negative mood reduction as a motivation for helping (e.g., Cialdini & Kenrick, 1976). At the same time, however, numerous studies have shown that positive mood induction can increase helping behaviors (for review, see Batson, 1998). The explanation for the latter finding is unclear, but may have to do with the avoidance of a negative mood caused by distress over perception of a person in need. Although the emphasis on experience placed by social learning theories may be useful to account for changes in altruistic tendencies over time or differences among individuals, such theories do not fully explain why altruistic behavior occurs in the first place and continues to be exhibited even in the absence of reinforcement, punishment, or modeling.

A more cognitive approach to helping behaviors implicates the importance of norms and social roles. Within this tradition, for example, developmental psychologists have focused on children's acquisition of norms concerning social responsibility and reciprocity, and on socialization practices within the family that may foster altruistic behavior (e.g., Eisenberg, 1986). Cultural psychologists and anthropologists have addressed both developmental and normative (e.g., the social responsibility and reciprocity norms) aspects of altruism within the context of variation across cultures (e.g., Shweder, 1982a). In general, broad social norms appear to have limited ability to predict specific individuals' helping behaviors (Batson, 1998). This has led some to focus on more personal norms as predictors of behavior. Staub (1978), for instance, proposed that individuals possess networks of beliefs that define expectations for prosocial behaviors under certain conditions (i.e., personal norms) and that cause feelings of guilt given failure to live up to one's own values and beliefs. Kohlberg's (1981) stages of moral reasoning might also be seen as cognitive structures that define personal norms for moral conduct under various circumstances.

A more affective approach to prosocial behavior concerns the link between empathy and altruism. Empathy is defined here as "an other-oriented emotional response congruent with the perceived welfare of the other" (Batson & Moran, 1999, p. 911) and is indexed by items assessing how sympathetic, warm, compassionate, softhearted, tender, and moved one feels toward another. A large number of studies have demonstrated a causal relationship between feelings of empathy and altruistic behaviors directed toward those with whom one empathizes (for a review, see Batson, 1998). The effects of experimental induction of feelings of empathy can be so powerful that subjects are moved to "cooperate" in one-trial Prisoner's Dilemma games even when they know the target of their empathy has already defected (Batson & Ahmad, 2001). It is important to point out, however, that feelings of empathy are often artificially induced in altruism research, and it is thus largely unclear what sorts of ecological cues prime empathic feelings outside of the laboratory (but see Preston & de Waal, 2002).

Altogether, the social psychological literature on altruism and prosocial behavior suggests that one may act altruistically because he or she has been rewarded for such behavior in the past, possesses cognitive structures that instantiate norms for altruism under specific circumstances, feels good, feels bad, or feels empathy. Altruistic behavior is also viewed as the product of countless situational and cultural factors. Thus, it seems that no coherent, general theoretical framework for the interpretation of human altruism has emerged from social psychological research. Such a general theoretical framework, however, could be provided by evolutionary theory.

An evolutionary psychological perspective would argue that prosocial behavior and altruism evolved by natural selection and that distinct psychological mechanisms are likely to have been shaped by distinct selection pressures for prosociality. Such a perspective is capable of integrating proximate and ultimate explanations for altruism by using analyses of selection pressures to predict the processing characteristics of psychological mechanisms. A summary of these selection pressures and corresponding mechanisms is presented in Table 7.1.

Table 7.1 Hypothesized selection pressures for altruism and correspondent psychological adaptations

Selection pressures for altruism	Psychological mechanisms
Kin selection	Solicitude toward kin; nepotism
Reciprocal exchange	Cognitive adaptations for social exchange
Indirect reciprocity	Reputation-maintenance mechanisms; moralistic aggression
Friendship/Banker's Paradox	Dyadic alliance mechanisms
Honest signaling	Status mechanisms; hierarchy negotiation
Group selection	Coalition maintenance mechanisms

Kin selection. Kin selection theory predicts higher rates of altruism toward kin versus non-kin (Hamilton, 1964). Although relatively few studies on human altruism have examined differential solicitude toward kin (e.g., Burnstein et al., 1994; Essock-Vitale & McGuire, 1985; Hill & Hurtado, 1996), such effects are more or less common knowledge. The available studies do suggest the existence of psychological mechanisms that favor the expression of altruistic behavioral tendencies in response to cues of kinship, such as familiarity or phenotype-matching.

Reciprocal exchange. Consistent with the theoretical formulations of Trivers (1971) and Axelrod (1984), many studies have shown that people are motivated to help those who have previously helped them (e.g., Wilke & Lanzetta, 1970, 1982). Specific psychological mechanisms have been proposed to underlie reciprocal exchange. Cosmides and Tooby (1992; Cosmides, 1989) argued that since reciprocal exchange requires the ability to detect cheating (i.e., receiving a benefit without reciprocating), humans have evolved specialized reasoning structures for detecting defection in exchange relationships. Their experiments demonstrate facilitated performance on the Wason selection task when selection problems are framed as social contracts in which one party could cheat by taking particular actions (see also Gigerenzer & Hug, 1992).

Indirect reciprocity. The notion that altruism and cooperation can evolve if altruists receive benefits from third parties who preferentially assist other cooperators is known as indirect reciprocity (Alexander, 1987; see Nowak & Sigmund 1998, for a formal test of this idea via computer simulations). The dynamics of indirect reciprocity suggest the importance of a number of psychological proclivities. The first is a concern for one's own reputation since the reception of benefits from others relies on how cooperative one is perceived to be. Likewise, people should be interested in the reputations of others to the point of being willing to invest time and effort to obtain such information through mechanisms such as gossip (e.g., Dunbar et al., 1997). Finally, the enforcement of systems of indirect reciprocity may have selected for emotional proclivities for moral outrage and moralistic aggression against those who cheat or otherwise defect against the system of indirect exchanges. Indirect reciprocity constitutes a route to altruism that is distinct from kin selection and reciprocal altruism. An individual may be motivated to help others out of reputational concerns (conscious or unconscious) or the desire to reward other cooperators even if the helper and recipient are unlikely to meet again in the future. Direct reciprocity, however, requires the possibility of future interactions and prospects for repayment from the recipient.

The "banker's paradox." Tooby and Cosmides (1996) pointed out that relationships based on reciprocity are subject to the "banker's paradox": the idea that banks are most eager to loan money to those who need it least. If close relationships with non-kin are really based on reciprocal exchanges, then one should be least likely to receive help at exactly those times when one most needs it (because the ability to reciprocate is compromised under difficult circumstances). Close friendships may instead be based on something more like a social insurance model. When two individuals receive mutual benefits from their relationship (which may not be based on exchange), each has an interest in the continued well-being of the other; this fact, in turn, increases the value of each to the other even more, since losing the other person costs one not only the benefits of the relationship but also someone who has an interest in one's own well-being. Thus, a type of runaway friendship process may ensue until each person has such a strong stake in the other's well-being that they are especially likely to help when the other is in trouble (Tooby & Cosmides, 1996). This model of friendship may account for findings that are anomalous for accounts of altruism based on reciprocity. Research has shown that individuals who desire a close (or "communal") relationship with another person are disappointed when that person quickly and directly reciprocates favors, as if paying off the debt and so discharging any obligation (e.g., Mills & Clark, 1982). Likewise, exact repayments for benefits received are often seen as signs of a lack of intimacy and may even be insulting among friends. These facts seem difficult to reconcile with the idea that all non-kin cooperative behaviors are based on reciprocity.

Honest signaling. Individuals may sometimes commit altruistic acts as costly signals of their fitness or social status. Zahavi and Zahavi (1997) reviewed otherwise inexplicable examples of birds competing with one another to perform altruistic actions such as sentinel duties and feeding other adult birds. Evidence that nonhuman primates share food as a means of enhancing or solidifying their social status is reviewed by Flack and de Waal (2000a). Similar arguments have been applied to food sharing in human hunter-gatherer groups and large donations to charity by wealthy individuals (e.g., Hawkes, 1990; Hill & Hurtado, 1996). Although one could argue that status signaling fails to meet the definition of altruism (e.g., Zahavi, 2000), this may be yet another explanation for various prosocial behaviors.

Group selection. Individuals may be willing to incur costs to help other group members if doing so enhances group success to the point that payoffs to altruistic individuals in the long run are greater than if they had not committed the altruistic acts. Wilson and Sober (1994) argue that strategies like "tit for tat" are successful only because they increase between-group fitness (where "group" is defined as the interacting dyad), and thus reciprocal and even kin altruism are really special cases of group selection. Even if all altruistic tendencies are group-selected, however, there is no reason to believe that the same types of psychological mechanisms underlie distinct prosocial behaviors. Altruistic acts committed out of allegiance to a group without expectation of reciprocity (see Caporael et al., 1989), for instance, are almost certainly motivated and triggered by different factors than behaviors undertaken out of the expectation of future reciprocation.

The theoretical framework sketched above suggests a roadmap for systematically investigating the nature of psychological mechanisms that may predict altruistic behavior or mediators of altruism such as feelings of empathy. The basic idea is that each type of selection pressure is likely to have fashioned distinct psychological mechanisms that operate according to distinct decision rules. The empirical delineation of these decision rules would thus link ultimate explanations based on selection pressures with proximate explanations based on processing characteristics. Since these mechanisms may have evolved by natural selection, furthermore, similar mechanisms may occur in nonhuman primates, as the product either of adaptive convergence or of common evolutionary history.

Primate Altruism and Its Explanations

Altruism and cooperation permeate many aspects of primate social life and have been extensively studied both in the laboratory and in the field. Altruism may manifest itself in many different aspects of primate behavior including grooming, agonistic support, food sharing, cooperative hunting or cooperative defense of resources, alloparenting, or food or alarm calls (see Dugatkin, 1997). In many cases, specific parallels can be drawn between the dynamics of reciprocation in some primate and human activities. For example, the type of cost-benefit analyses used by humans in bystander intervention studies are very similar to those exhibited by nonhuman primates in the context of decisions about whether or not to support another individual or join a coalition (e.g., Chapais, 1995). Similarly, the dynamics of food sharing and cooperative hunting investigated in some hunter-gatherer societies share several similarities with like activities in chimpanzees (Stanford, 1995). Rather than review all aspects of altruism in primates, we focus on one type of altruistic activity, namely grooming, and attempt to extract general principles that can be extrapolated to other altruistic activities or across species.

Social grooming fits the definition of altruistic behavior in that it entails costs to the actor and benefits to the recipient. The costs of grooming include expenditure of time and energy and reduced opportunity to engage in other activities such as vigilance or feeding (Dunbar & Sharman, 1984; Maestripieri, 1993c). The recipients of grooming benefit in terms of hygiene and tension reduction (Dunbar, 1991). In some cases, groomers can be thought of as gaining an immediate benefit from their behavior (e.g., protection or reinforcement of their own dominance rank; Silk, 1982). In most circumstances, however, it is assumed that the benefits to the actor are lower than the costs, and therefore grooming is considered altruistic.

In most primate species, grooming is heavily biased toward kin (Gouzoules & Gouzoules, 1987). Altruistic behavior toward kin can evolve and be maintained in a population through kin selection (but see Chapais, 2001), and therefore the primate data on kin-directed grooming are consistent with evolutionary theory. Grooming between non-kin can evolve and be maintained through mechanisms of reciprocity, where reciprocation occurs in the same currency (referred to as reciprocity by Hemelrijk, 1990b) or in a different currency (referred to as interchange by Hemelrijk, 1990b).

Studies of grooming reciprocity at the dyadic level have generally been consistent with the predictions of evolutionary theory. An early study of stumptail macaques by Goosen and Ribbens (1977) showed that unrelated individuals were often in conflict about initiation of grooming and refused one another's persistent invitations. Once affiliative interactions had started, grooming was generally reciprocal, but individuals attempted to maximize grooming received while doing as little grooming as possible (e.g., by grooming their partner for very short bouts and immediately soliciting grooming). Maestripieri, Schino, and Scucchi (unpublished data) found that interactions in caged pairs of familiar individuals were characterized by a high degree of grooming reciprocity. In contrast, individuals who had not met before were reluctant to initiate or reciprocate grooming. The interactions between these individuals were characterized by a high number of grooming solicitations not followed by grooming. These findings suggested that, under the experimental conditions of this study, familiarity was used as a predictor that two individuals would meet again in the future, so that familiar individuals tended to cooperate whereas unfamiliar individuals tended to defect. These findings are consistent with those reported by Muroyama (1991a, b; 1996), showing that most grooming sessions between unrelated female macaques involved only one short bout followed by grooming solicitations. Therefore, grooming between non-kin appears to be consistent with strategies of reciprocation such as "tit for tat," in which the first move is cooperative and then the individual will continue to cooperate or defect depending on its partner's behavior (Axelrod, 1984). If unrelated individuals interact repeatedly, for example, because they live in the same group, some degree of grooming reciprocity between them can be expected. In fact, when appropriate statistical techniques are used (e.g., Hemelrijk, 1990a, b), a correlation between grooming given and received by individuals usually emerges, although the correlation is not strong (e.g., Barton, 1987; Borries et al., 1994; Hemelrijk & Ek, 1991; Rowell et al., 1991; Watts, 2002).

Asymmetries in grooming between unrelated individuals are often associated with substantial differences in their dominance ranks. For example, grooming interactions may begin with a short bout of grooming by the dominant individual followed by a grooming solicitation, and one or more long bouts of grooming by the subordinate (e.g., de Waal & Luttrell, 1988). In these circumstances, it must be assumed that the subordinate is willing to accept the exploitative behavior of the dominant because he or she expects that grooming will be reciprocated in a different currency. The benefit that primates most often obtain from grooming is probably social tolerance. Thus, by grooming high-ranking individuals, low-ranking individuals are able to stay near them and benefit from their protection or from sharing resources with them (Silk, 1982). The findings of research on the exchange between grooming and agonistic support, however, are not easy to interpret. Some of these studies found a positive correlation between giving grooming and receiving support, others found no correlation, and a few even found a negative correlation (e.g., Chapais et al., 1995; Watts, 2002). The causal relationship between grooming and support may be difficult to establish with observational/correlational studies alone, in part because both grooming and agonistic support occur at relatively low rates and may be separated by long time intervals (Chapais et al., 1995). Some evidence for a correlation between grooming and support, however, has been provided by experimental studies in which one or both variables were manipulated (e.g., Chapais et al., 1995; Hemelrijk, 1994; Seyfarth & Cheney, 1984). In addition to tolerance or agonistic support, there is evidence that grooming can also be exchanged with sexual favors, food, or access to infants (e.g., de Waal, 1997; Henzi & Barrett, 2002; Stammbach, 1988).

Whenever grooming is reciprocated with a different currency, assessing whether the data fit the predictions of reciprocity theory or game theory has not been easy for many reasons. When different altruistic behaviors are involved, it is difficult to compare their costs and benefits and therefore assess whether symmetries or asymmetries in reciprocity occur. Furthermore, a positive correlation between giving grooming and receiving support or food does not necessarily constitute evidence of a causal relationship between these two variables, because both could be independently correlated with a third variable such as proximity or the act of grooming itself. For example, upon reinterpreting Seyfarth and Cheney's (1984) data on grooming and support, Hemelrijk (1990a) found that being groomed and providing support were each independently correlated with the act of grooming itself, and when this correlation was statistically controlled for, the correlation between grooming and support disappeared. This raises the possibility that grooming may not function to establish and maintain strong social bonds, but that it simply reflects such bonds (e.g., Oki & Maeda 1973). For example, grooming may function to "test" the quality of social relationships or to create an atmosphere of friendliness between two individuals, which in turn might facilitate tolerance or support (Dunbar & Sharman, 1984; Henzi & Barrett, 1999). These different hypotheses, however, are not incompatible with reciprocity hypotheses, can be empirically tested, and do not pose a fundamental challenge to the application of evolutionary theories of altruism to primate behavior, although some of their limitations must be acknowledged (e.g. Chapais, 2001; Chapais et al., 2001).

Overall, primate research on altruism has emphasized the functional and evolutionary aspects of this phenomenon, whereas only a few studies have addressed its proximate determinants. Given the available evidence, learning is not viewed as a major developmental determinant of altruistic behavior, although it has been argued that punishment of cheaters does occur (Hauser, 1992). Similarly, the ability to take another individual's perspective and empathize with his or her feelings is not viewed as a requirement for altruistic behavior. Rather, some degree of cognitive complexity is viewed only as necessary to keep track of reciprocal exchanges (de Waal & Luttrell, 1988; Seyfarth & Cheney, 1988).

One specific area of possible convergence between primate and human research is the investigation of the neural substrate for altruistic behavior. Recent research has shown that distinct brain structures are activated given different outcomes in Prisoner's Dilemma games. For example, Rilling and colleagues (2002) demonstrated activation of the posterior fusiform gyrus in subjects who had played to cooperate when their opponent had defected (the p. f. gyrus is involved in face identification and may be activated as a means of facilitating memory for cheaters; see Chapter 12). In addition, while mutual defection produced activation in phylogenetically conserved structures such as the striatum, mutual cooperation produced activation in the phylogenetically more recent orbitofrontal cortex. This raises the question whether humans evolved unique or elaborated brain structures for mutual cooperation that are largely absent in nonhuman species. Any definitive judgment on this question would seem to require much more knowledge of the cognitive and neuro-biological mechanisms that underlie cooperative behaviors in both human and nonhuman species.

Primate and human research on social development and affiliation share numerous similarities in terms of conceptual background, analytic proce-dures, and basic empirical findings. Both primate and human researchers studying the influence of parenting on social development must disentangle the relative contributions of genes, early experience, and shared current envi-ronment in accounting for cross-generational similarities in behavior. The information produced by cross-fostering experiments with primates could be very useful to human research in this area. Primate research can also make an important contribution by providing information on the neuroendocrine, neurochemical, and neuroanatomical characteristics of individuals exposed to different parenting experience. Finally, evolutionary research can encourage developmental psychologists to investigate further the possible adaptive func-tion of individual differences in parenting and development, including the origins and functional significance of behavioral sex differences. Primate re-search can gain many benefits from integration with human research, includ-ing the development of standardized procedures for behavioral testing as well as increased sophistication at both the theoretical and empirical level.

Our current understanding of the determinants of affiliation and social structure in both human and nonhuman primates has benefited from the development of a science of relationships that has cross-disciplinary and cross-species foundations. Although the distribution of affiliation among adults and the characteristics of social structure vary across species in relation to ecologi-cal factors, there are nevertheless several broad principles with general valid-ity. Thus, it is likely that new knowledge produced by nonhuman primate research about the dynamics of social interactions, relationships, and social structure could be extrapolated to humans and vice versa.

Primate and human studies on altruism have addressed different but complementary aspects of this phenomenon. Thus, there is great potential for integration in the study of both the proximate mechanisms and the evolutionary aspects of altruism. Primate research on altruism is firmly grounded in evolutionary theory and has made much progress in the use of sophisticated game theory models to predict the occurrence and distribution of altruistic acts among individuals and over time. Little is known, however, about the cognitive and emotional aspects of altruism in primates, and this is an area of research that could benefit considerably from importing both conceptual models and relevant empirical information from human research.

Human research on altruism and prosocial behavior could make significant progress by focusing on whether altruistic tendencies are governed by experience, specifically designed cognitive mechanisms, or more general emotional states. Although some anthropologists have consistently studied human altruism from an evolutionary perspective, such an approach is still vastly underused in much research within social psychology. Thus, the use of theoretical models imported from evolutionary theory and primate research (as well as other disciplines such as economics) in the study of human altruism could strengthen the conceptual foundations of this research and allow investigators to assess whether the empirical findings obtained with human and nonhuman primates have cross-species validity.

8

Comparing Cognitive Development

Jesse M. Bering and Daniel J. Povinelli

People who study the intellectual abilities of chimpanzees or other great apes are frequently asked some version of the following, somewhat impatient question: "Yes, yes, all of your experiments and data are very interesting, but let us cut to the chase: How smart, exactly, are your apes? Are they are smart as a two-year-old child? A three-year-old? A four-year-old?" This question, graphically presented in Figure 8.1, naturally tempts an answer—especially because it is framed in terms of a notion already common among students of cognitive evolution; namely, that mental evolution invariably proceeds through the progressive addition of new stages to the end of a monolithic, domain-general ancestral pathway of cognitive development. Even Jean Piaget, the famed Swiss epistemologist, was tempted by the apparent allure of answering this question. "[Chimpanzees] are superior to one-year-old babies," he once replied during an interview, "but they don't progress much beyond that."

By the end of this chapter, we hope to show why the question depicted in Figure 8.1 may be unanswerable: not because it is too difficult but, rather, because it is incoherent. By way of foreshadowing, the reason is simple. In addition to whatever quantitative modifications may have occurred in the various cognitive systems that were present in the common ancestor of humans

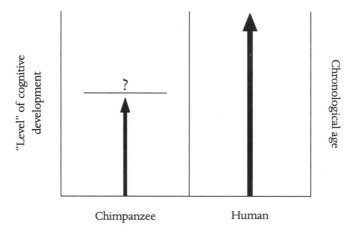

Figure 8.1 "How smart is your ape compared to a human child?" This question assumes that cognitive evolution is best depicted as the addition of new "stages" or "levels" to the end of a monolithic linear pathway of the common ancestor. In the tradition of Darwin and Romanes, notions about cognitive evolution are even more divorced from modern biological thinking in that species can be arranged in an ordered fashion from the "lowest" worm to the "highest" ape in a kind of "phylogenetic scale" (see Hodos & Campbell, 1969). Compare to Figure 8.8.

and chimpanzees, humans appear to have evolved additional, qualitatively new abilities (ones not found in other species). More important, the system or systems that support such abilities were not tacked onto the end of the general developmental pathway found in the common ancestor but rather were woven into development early on, so that they now develop in parallel to the systems we share in common with other species. We shall argue that one of the most important features (or consequences) of these systems is the human ability to form concepts about purely abstract things: concepts about things that cannot be directly observed by the senses; concepts about the "hidden" world—the world of forces and causes that lie behind the surface appearance of things. In the domain of animate beings, we are referring to things like emotions, intentions, perceptions, and beliefs; in the domain of inanimate, physical objects, we are referring to things like gravity, force, mass, physical connection, and the like. Indeed, we suspect that this core difference between humans and chimpanzees may have such sweeping effects on our common-sense understanding of the world that it may mask our very ability to recognize its existence.

Continuity and Discontinuity:
From Bridging Gaps to Accepting Diversity

Our starting assertion, then, is that humans may be the only species on this planet that possesses the natural ability to reason about unobservable entities, and furthermore that the systems which support these differences are already manifest during the earliest moments of infancy. Curiously, the possibility of the existence of a fundamental difference such as this has only rarely captured the attention or interests of researchers who have compared the cognitive development of humans and apes, for at least two reasons. First is the idea that cognitive evolution occurs exclusively by gradual, quantitative modification— by a periodic swelling and receding of psychological competencies within particular phylogenetic lineages. Even Darwin himself, otherwise a champion of evolution as a branching process of diversification, caricatured psychological evolution as having produced a kind of phylogenetic scale in which animals could be arranged along a continuum from "lower" to "higher," with humans anchoring the highest (most psychically developed) point (for discussions of Darwin's gradualist and scale-like views on the evolution of the brain and mind, see Povinelli & Giambrone, 1999; Preuss, 1995). This intellectual framework emphasized commonality and continuity among species, with each adjacent species up the scale possessing just a little more of the same basic faculties as the ones just below it (see Spencer, 1887, for a historically influential version of the idea of the phylogenetic "scale"). There can be no cognitive traits, according to this logic, which exist in a fully functioning form in one species (e.g., humans) but not in another, closely related one. There must at least be some trace, some vestige, some meager but nevertheless present semblance of the particular cognitive trait in the sister species. Indeed, this logic has led to something of a recrudescence of recapitulation theory, with apes seen as progressing farther than monkeys, humans getting farther than apes, and human ontogeny reflecting the same general pattern of emergence (e.g., Langer,

1998, 2000; McKinney, 1998; Parker & Gibson, 1979; Parker & McKinney, 1999; Suddendorf, 1999; Whiten, 1996).

In principle, of course, there is nothing inherently problematic about the idea of some kind of psychological continuity among species (or even, in some cases, recapitulationist patterns of development). Indeed, it follows as a virtual truism of modern biology that sibling species such as humans and chimpanzees will share many behaviors in common, and that these shared behaviors (and at least some aspects of the psychological systems generating and attending them) will be genuinely homologous (see also Povinelli et al., 2000, 2002; Povinelli & Giambrone, 1999). What is problematic is the simplistic misapplication of such notions. Particularly problematic is the centerpiece of the agenda laid out by early philosophers such as Hume and the founders of comparative psychology, Darwin and Romanes, that the presence of homologous behaviors necessitates strong similarity in the underlying cognitive systems which produce and/or represent the behavior (for a detailed discussion of this history, see Povinelli & Giambrone, 1999). This, of course, is an extreme form of continuity theory. It is a view that promotes the idea that the agenda of studies of comparative cognitive development should be a focus on understanding how "far along" certain species get with respect to the "most evolved" species, *Homo sapiens*. Further, it is an idea that is still alive and well among researchers who compare the psychological systems of humans and chimpanzees (see Chapter 9).

In this chapter we outline a different, more pluralistic idea. We argue that the existence of massive homology in the behavioral and psychological systems of humans and chimpanzees in no way precludes the possibility that humans have evolved either one or many pathways not found in other species—pathways that develop in parallel to these ancestral systems and that now reside alongside and interact with the more ancient ones. Further, we show how this idea seems increasingly supported by mounting empirical evidence. Note that there is nothing about this view which rejects profound similarities between humans and chimpanzees, nor anything which rejects the idea that each species may tweak ancestral systems in certain quantitative ways depending on the particular socioecology of its evolutionary history. Nor, for that matter, is there anything in this account which denies the possibility that chimpanzees have evolved their own, peculiarly chimpanzee-like cognitive systems.

Similarity and Difference:
Is One More Important than the Other?

At this point one might reasonably ask why we appear to be so concerned with the possible differences (as opposed to the similarities) between chimpanzees and humans. More directly, one might ask, "Aren't the similarities profoundly important as well—perhaps even more important than any minor differences that might exist between the species?"

The first part of the question demands an unequivocal yes. Indeed, many will be startled by how similar the spontaneous, natural behavior of chimpanzees is to our own. A short, nonexhaustive list includes the following: complex alliance formations (de Waal, 1982), conflict resolutions (Baker & Smuts, 1994; de Waal, 1989a), tool-using technologies (Boesch & Boesch, 1990; McGrew, 1992), subtle regional behavioral differences (i.e., local "dialects" of behavior: Whiten et al., 1999), political maneuvering and fluid social hierarchies (de Waal, 1982), sex differences in group-living tasks (Boesch & Boesch-Achermann, 2000; Wrangham, 1986), juvenile play (Mendoza-Granados & Sommer, 1995), and strong maternal attachment (Goodall, 1986). Thus, the one thing we certainly do not wish to do is minimize the extent of overlap between human and chimpanzee behavior. It is important and immense.

Unfortunately, the enormous similarity in the spontaneous behavior of humans tosses up blinding, and at present almost intractable, distractions from what we believe is the core task: the project of formulating a genuinely evolutionary science of other minds—a science dedicated to understanding both the similarities and the differences between the two species. This is because the very mind (the human one) that seeks to analyze objectively the behavior of other species in order to determine the nature of their cognitive systems is already wired to interpret their behavior from a human standpoint—regardless of the objective reality. Put another way, here is one thing of which we can be sure: the human mind is extremely adept at seeing the world through its own lens, and indeed, the more that the things it sees physically resemble human beings, the more powerful and complete that transformation will be. This means that the work of comparative psychology must be converted from the easy task of simply cataloguing behavioral similarities among species to the very hard task of grappling with the ways in which fundamental patterns of behaviors can interact with multiple systems which have evolved for representing and interpreting them. Elsewhere we have shown analytically that when it comes to certain classes of conceptual cognitive systems, the mere presence of a given spontaneous behavior can never reveal, without careful experimental analysis, whether such systems are actually present (see Povinelli et al., 2000; Povinelli & Giambrone, 1999).

Perhaps by now the answer to the second part of the question—whether the similarities between humans and chimpanzees are actually more important than the minor differences—should be obvious. If aspects of the reorganization of the human cognitive system over the past 5 to 7 million years have been so colossal, so jarringly dissimilar from anything the natural world has yet known, then it would stand to reason that this state of affairs should demand at least some attention from all of those who wish to understand the true nature of the minds of humans and chimpanzees.

Reasoning about the Hidden World

Our laboratory has conducted an extensive amount of investigation designed to explore the psychological systems of a cohort of seven chimpanzees that we have followed from infancy to adulthood. The results of these studies have emphasized two themes. First, they have underscored the already well-established conclusion that chimpanzees share with humans an impressive ability to represent and reason about the observable contingencies that exist in the world. Second, and far more interesting, they have pointed to the possibility that, unlike humans, chimpanzees may not impose on observable events explanations for why they exist in the first place. That is, the search for underlying, unobservable causal mechanisms may be a uniquely human cognitive specialization.

This hypothesis has profound implications for understanding what is fundamentally human about the human mind. Because both humans and other closely related species share vast networks of homologous psychological mechanisms for uncovering and representing the observable regularities in the world, it is virtually guaranteed that both humans and chimpanzees will possess many of the same behaviors for coping with similar problems they encounter in the social and physical domains. Less obvious, perhaps, is that the same spontaneous behavior, whether produced by a human and a chimpanzee, two different humans, or even the same human at different time points, may have very different psychological causes. This new system (one undoubtedly tied up with the evolution of the universal human capacity for language) did not replace the operation of these ancestral systems but rather resides alongside them, both modulating and being modulated by them (see Povinelli et al., 2000).

We shall explore this idea further, but for now we turn to an examination of some of the empirical data which, we believe, strongly support the existence of precisely the mosaic pattern to which we have been alluding: the existence of profound similarity in the cognitive developmental pathways of humans and other species (see, for example, contributions in Antinucci, 1989) right alongside profound differences. In particular, we examine what can be thought of as the quintessential case of the human capacity to reason about things which cannot be directly perceived—namely, the human ability to conceive of internal, unobservable mental states.

Social Understanding in Chimpanzees and Children

At some point humans develop the ability to reason about the mental states of themselves and others. This capacity has been referred to as "theory of mind" (Premack & Woodruff, 1978). The exact age at which various aspects of these abilities emerge, and the mechanisms responsible for their emergence, is a matter of considerable controversy (for review, see Mitchell, 1997). Proposals for how children get from their initial understanding of the world to some more mature state abound, and they differ in important and subtle ways in (1) the nature of the starting state of the system that is present at birth, (2) which aspects of the system are not reducible, and (3) the mechanisms by which new aspects of the systems are produced (Carey & Spelke, 1994; Gopnik, 1993; Gopnik & Meltzoff, 1997; Gopnik & Wellman, 1992; Harris, 1991; Leslie, 1994; Perner, 1991). Regardless of the outcomes of these ongoing debates, however, two statements seem uncontroversial: large aspects of this "intentionality system" are cross-cultural (e.g., Avis & Harris, 1991; Povinelli & Godfrey, 1993; Lillard, 1998; Vinden & Astington, 2000), and its application is very general indeed, with humans attributing emotions, desires, thoughts, and feelings to a dramatic range of other animals and even objects (Eddy et al., 1993). In short, whatever the evolutionary forces that sculpted it, the human theory of mind is not particularly sensitive to the particular animal before it.

Humans may therefore simply be built in such a way that we will attribute mental states to chimpanzees, and further, we will attribute to chimpanzees the ability to do the same. This fact, of course, has no bearing on whether chimpanzees really possess a system for attributing mental states to others; it only bears on the far less interesting claim that humans possess such a system. Thus, the difficult, nontrivial empirical problem still looms large and unsolved: Do chimpanzees actually possess such a system?

Knowledge about Visual Perception

In what follows, we focus on the question of whether chimpanzees reason about mental states (such as perceptions, beliefs, desires, intentions, and emotions) by asking whether chimpanzees understand one of the earliest emerging aspects of social understanding in young children: the understanding of "seeing"—that is, the understanding that other individuals have unobservable visual experiences.

We focus on what chimpanzees know about seeing for several reasons. First, it is the most widely explored facet of nonhuman primates' understanding of the mental states (e.g., Call et al., 2000; Cheney & Seyfarth, 1990a; Hare et al., 2000, 2001; Kummer et al., 1996; Povinelli & Eddy, 1996a, b; Povinelli et al., 1990, 1991, 1999, 2002; Povinelli, Theall, et al., in press; Premack, 1988; Reaux et al., 1999; Theall & Povinelli, 1999; Tomasello et al., 1999). Furthermore, substantial enough research has been conducted with chimpanzees, in particular, to take stock of the database, allowing us to make some meaningful assessments. Several reliable findings have emerged, and several other, still controversial findings crisply illustrate the main theoretical point we wish to make in this chapter. Second, an understanding that others "see" things emerges fairly early in human development—somewhere around the child's second birthday—and a mentalistic understanding of visual reference or attention may emerge even earlier. A related point is that this kind of understanding seems fundamental to our mature representation of others as psychological agents. And finally, based on our assessment of the current empirical evidence, we believe that this is an excellent example of how profoundly similar humans and chimpanzees can be in their spontaneous, everyday behavioral interactions while still remaining radically different in their interpretation of such behaviors.

Sensitivity to the Eyes

A variety of birds, reptiles, fish, and mammals have been shown to be sensitive to the presence of eyes or eye-like stimuli (e.g., Burger et al., 1991; Burghardt & Greene, 1988; Gallup et al., 1971; Ristau, 1991b; review by Argyle & Cook, 1976). For example, Blest (1957) showed that birds were less likely to prey on moths with eyespots than on those without. In general, of course, such sensitivity is to be expected. After all, from the potential prey's perspective, what could be more ecologically relevant than a pair of eyes looming in your visual field? But such sensitivities would not seem to qualify as unambiguous evidence that the bird is reasoning about "seeing."

A moment's reflection will reveal that the same logic applies in the context of highly social organisms for which vision is an important sensory modality. Nonhuman primates, in particular, appear quite sensitive to the movements of the head and eyes of others (see Figure 8.2). The basic capacity to follow the gaze direction of others, for example, has now been demonstrated in a wide range of nonhuman primate species (e.g., Emery et al., 1997; Ferrari et al., 2000; Itakura, 1996; Povinelli & Eddy, 1996b, c; Povinelli et al., 1999, 2002; Tomasello et al., 1999). Chimpanzees have been examined most thoroughly, and have exhibited the same range of complexity of components of the gaze-following system that is present in human infants aged eighteen to twenty-four months (see Povinelli, 2001). Aspects of these findings suggest that the neuropsychological system controlling these behaviors is a shared primitive feature of the chimpanzee-human clade (and, most likely, an even larger clade).

Inferences about Seeing

But does this tell us anything about whether chimpanzees understand that others have unobservable internal perceptual states of "seeing" things? An intensive longitudinal investigation of a group of seven chimpanzees conducted by our research group has provided convergent evidence that despite their remarkable gaze-following skills, they do not understand the perspectival, subjective experience associated with the orientation and movement of the head and eyes of other individuals, despite the fact that these very same subjects have robustly exhibited the most complex aspects of gaze-following for which this species (or human infants, for that matter!) has been tested.

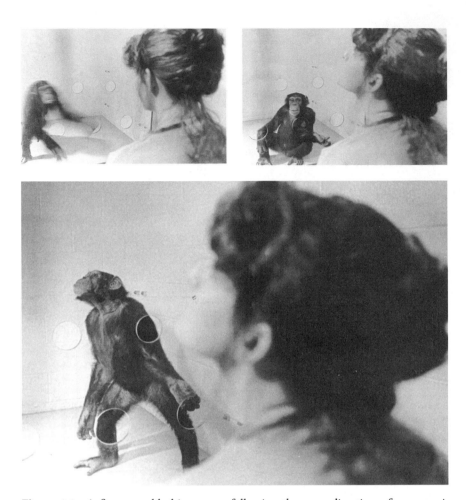

Figure 8.2 A five-year-old chimpanzee following the gaze direction of an experimenter who turns her head and eyes in concert with a predetermined target above and behind the chimpanzee.

For example, in one series of experiments, we probed whether, when faced with two familiar human experimenters, our chimpanzees would selectively deploy their visually based, species-typical begging gesture to the person who could see them (see Figure 8.3). Assessments were made when the apes were five to six, seven, and eight to nine years of age (for results, see Povinelli & Eddy, 1996a; Reaux et al., 1999). The results of nearly twenty experiments showed that although our chimpanzees actively used their communicative gestures, they did not seem to appreciate that only one of the two people

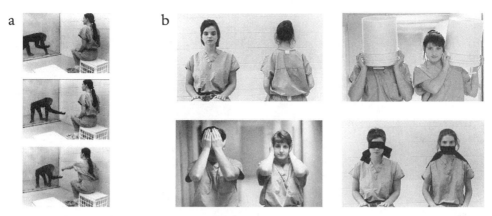

Figure 8.3 Do chimpanzees understand "seeing"? (a) On standard (background) trials, Mindy uses her natural begging gesture to "request" food from a familiar experimenter. (b) On probe trials, various conditions instantiating the seeing/not seeing distinction require the chimpanzee to choose to whom it will gesture. Although the chimpanzees are correct from trial 1 forward on back versus front, they do not appear to understand any of the other conditions. Follow-up tests revealed that their understanding was about the observable postures of the experimenters, not who could see them.

could see them. It is essential to note, however, that in virtually every instance, the chimpanzees learned the contingencies involved quite rapidly. Thus, with enough experience and feedback, the animals learned to gesture to the person who could see them. But it is equally important to note that follow-up tests consistently revealed that the hypotheses that best predicted which experimenter the ape chose were about the postures, not the mental states, of the people involved (see Figure 8.4). Furthermore, in a longitudinal project with these same animals, we assessed their understanding of seeing when they were juveniles, adolescents, and young adults. The results consistently yielded the same pattern described above: they were reasoning about the postures, not the perceptual states, of their communicative partners. Other research with these same animals, using quite different methodologies, has converged on a similar interpretation (review by Povinelli & Giambrone, 2001).

Other researchers have questioned this conclusion, and have suggested that tests involving competition may reveal the presence of an ability to reason, at some primitive representational level at least, about "seeing" (e.g., Chapter 9). Perhaps the most direct evidence for this view comes from a study by Hare et al. (2000), who placed subordinate chimpanzees in one-on-one

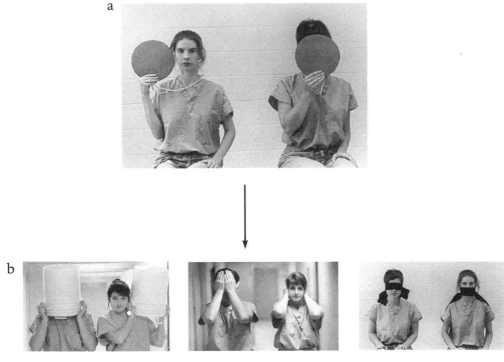

Figure 8.4 An important example of what our chimpanzees learned in the seeing/not seeing tests. After many trials of screen over the face (a), the subjects learned to choose whichever experimenter was holding the screen beside the face. Had they learned that that person could see them, or simply "Gesture in front of the person whose face is visible"? In (b) we tested them on several of the original conditions. If, on the one hand, they had learned something about seeing, they should have transferred their understanding of the screen condition equally to all of these old conditions. On the other hand, if they were just using the rule about the presence or absence of a face, they should have performed well in the buckets and hands-over-the-face conditions but not in the blindfold condition (because an equal amount of the face was visible)—which was exactly the pattern of results we obtained.

competitive situations with dominant rivals over two food items, in which one item was visible to both participants but the other was visible only to the subordinate (e.g., food placed behind an opaque barrier). These tests were designed to determine if the subordinate animals were capable of reasoning about which food items their dominant rival could and could not see. The most compelling of their tests involved positioning the chimpanzees directly across from each other in holding cages with a testing arena between them (see Figure 8.5). With the doors to the two holding cages closed, two food

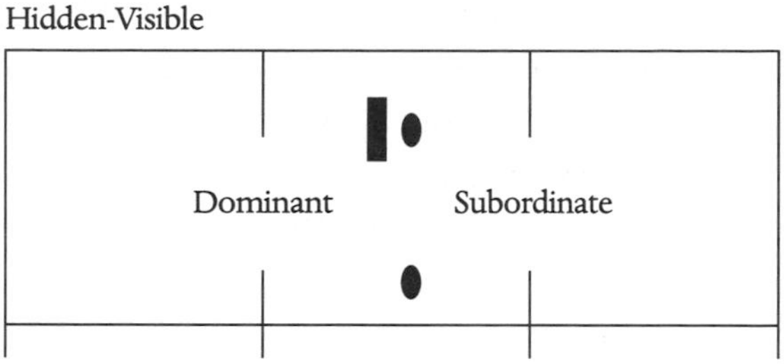

Hidden-Visible

Figure 8.5 The hidden-visible condition used by Hare et al. (2000) to test subordinate chimpanzees for their understanding of what a dominant rival can or cannot see. (Redrawn from Hare et al., 2000)

items were placed on the floor of the testing arena an equal distance from both animals. One of the food items was in the open (and therefore would be visible to both the dominant and the subordinate animal when the doors were opened), whereas the other was behind an opaque barrier (so that the subordinate but not the dominant could see it). Next, the doors were opened slightly, allowing both animals to look into the enclosure. Finally, the subordinate was released and allowed to enter the testing arena. As soon as he or she took a couple of steps toward one of the food items, the dominant's door was opened as well. The logic of this procedure was that the subordinates would use their perspective-taking abilities to infer that the dominant did not see the food behind the barrier (and therefore did not know that it was there). Thus, according to Hare et al. (2000), they ought to prefer to take the hidden food.

Hare et al. (2000) reported that the subordinates tended to approach the hidden items first, and obtained more hidden items than visible ones by the end of each trial. Although both measures are of some interest, it should be noted that with respect to the question of visual perspective-taking, the crucial question is whether the subordinates approach the hidden item first, because only this finding supports the idea that they are reasoning about what their rival can or cannot see. After all, the subordinates might obtain more hidden items by the end of the trial simply because the dominant typically takes the visible one, leaving only the hidden one for the subordinate.

How are we to integrate these findings with our extensive previous work? Should all previous research be considered "overturned by an elegant experiment more intuitive for chimpanzees," as Whiten (2001, p. 133) has claimed?

Although there are a number of a priori concerns about the logic of the study, we were nonetheless prepared to reconsider the generality of our conclusions about what chimpanzees know about seeing. First, however, we concurred with Hare (2001) that "if an experiment is controlled well, a positive result (rejection of the null hypothesis) reflects the ability of the test subjects and should be replicable," and thus we sought to satisfy ourselves that the results could, in fact, be replicated. Second, we wanted to determine if they would hold up in some simple variations of the procedures that could tease apart the visual perspective-taking model from some rather obvious alternative interpretations.

So that our animals would have the same experiences as the ones used by Hare et al. (2000), Karin-D'Arcy and Povinelli (in review) initially attempted to replicate several studies they reported which they acknowledged were difficult to interpret. Interestingly, we completely replicated those effects. When the diagnostic "hidden-visible" tests (described above) were conducted, however, a very different pattern of results emerged. First, the end-of-the-trial effect was consistently replicated. That is, in a series of studies, our subordinates consistently obtained more of the hidden food than the visible food by the end of the trials. Recall, however, that this effect may simply be due to the fact that once the dominant is released and takes the food in the open, the only food left for the subordinate is the hidden one.

Strikingly, in each experiment, the first-choice effect was consistently not replicated. Despite several variations, there was no evidence that the subordinates were reliably selecting the hidden food first. Thus, the initial studies consistently found patterns of results that were inconsistent with the idea that the subordinates were reasoning about what the dominant could or could not see. Further studies revealed that even the subordinates who showed a marginal tendency to approach the hidden food first did not differentiate between identical occluders which were simply turned in ways that did and did not result in obscuring the dominant's view (Figure 8.6; see Karin-D'Arcy & Povinelli, in review, experiments 6–7). Indeed, in some cases our subordinates showed a statistically significant preference when the barrier was turned to the side so that the food was equally visible to both the dominant and subordinate!

In summary, the only reliable finding to emerge from the Hare et al. research is one which has no real bearing on the question of visual perspective-

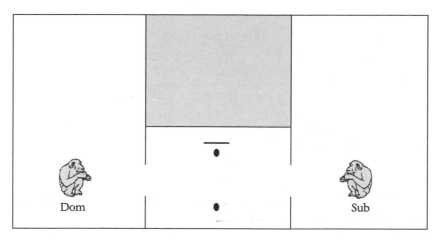

Figure 8.6 Modification of the Hare et al. (2000) design used by Karin-D'Arcy and Povinelli (under review) to determine whether subordinates' choices were due to reasoning about what the dominant could see or a preference for the food near the structure.

taking. Clearly, additional studies are needed to assess further the robustness of the effects reported by Hare et al. (2000).

"Pointing" out the Differences:
How Chimpanzees and Children Understand Gestures

Our initial studies on chimpanzees' understanding of the visual perspectives of human experimenters relied on our apes' use of their natural begging gestures. But how, exactly, were our subjects representing these gestures?

Even if one were to accept uncritically the findings from our seeing/not seeing tests described earlier, one should still wonder if the robust use of directed begging gestures by our apes might reflect some understanding of the communicative intent behind them. In other words, they might deploy their gestures in a "proto-declarative" fashion, in which they are directed at the actual internal, representational states of the communicative partner, without any immediate, instrumental function (Bates et al., 1975; Camaioni, 1991). For instance, if Kevin and Mary are sitting by the lake, and Kevin suddenly taps Mary on the shoulder and then points behind her at a rare black swan swimming by, it would be because he wants her to share in the experience with him. His representation of her representation does not consist of the perceptual experience of seeing the black swan. Thus, the referent of the pointing gesture in this case is not, per se, the swan itself, but rather it subsumes and is "about" Kevin's entire communicative intent (i.e., "I want you to look at that, Mary"). In humans, the first pointing gestures emerge around nine to twelve months of age (Desrochers et al., 1995; Franco & Butterworth, 1996). But at this age there appear to be some dissociations between the production and comprehension of the pointing gesture (Baldwin, 1993; Franco & Butterworth, 1996). The infants may be deploying the gesture without truly understanding its referential significance, perhaps acquiring it through imitation or some form of ritualization (e.g., Tomasello, 1999). Thus, young infants (and, we hypothesized, perhaps other species) may understand and use pointing in a starkly "proto-imperative" fashion (*sensu* Bates et al., 1975), in which gestures are used to manipulate actors in the environment to perform certain activities in the external world. The complete absence of proto-declarative gesturing of any kind whatsoever among wild populations of chimpanzees that have been studied for over forty years is simply so striking and overwhelming that the ambiguity of the one published report of what might or might not be a single, isolated instance of pointing (see Vea & Sabater-Pi, 1998) to our minds simply further highlights the robust nature of this difference in the natural gestural systems of humans and chimpanzees.

But what about conditions in which chimpanzees interact frequently with humans and are exposed to their caregivers' pointing behaviors on a fairly regular basis? Might an understanding of the referential nature of the gesture emerge? Findings from various researchers, in fact, confirm the appearance of spontaneous "pointing" gestures in captive apes, and many researchers have interpreted these gestures as evidence of apes' understanding of the mental states of the humans with whom they are interacting (see Chapter 12; Gómez, 1998; Krause & Fouts, 1997; Leavens et al., 1996; Miles, 1990; Whiten, 2000b). In most of these cases, however, it is difficult to rule out the possibility that these gestures are proto-imperative in nature; indeed, this is difficult in the case of human infancy as well. This interpretive difficulty is underscored by one final point about chimpanzees' production of such "pointing" gestures: despite years of reliably using them in (mostly food- or grooming-related) situations with humans, to our knowledge they have never been reported to use them with one another. This raises the distinct possibility that they have no general understanding of their proto-declarative function, but only a limited understanding of how they affect the behavior of human beings (creatures that invariably respond as if the apes did mean them in some more mentalistic manner).

Because of the difficulty in disentangling proto-imperative and proto-declarative pointing (both conceptually and practically), in our own laboratory we have instead investigated what chimpanzees understand about the referential nature of the gestures of others. That is, when observing the intentional communicative attempts of other agents, do chimpanzees comprehend these attempts as such, or do they merely learn, over time, how these actions "tag" important aspects of the environment? We conducted a series of experiments with our seven chimpanzees that was explicitly designed to determine whether they understood the referential significance of the human pointing gestures which they had been exposed to since birth (see Povinelli et al., 1998). To begin, we simply trained our apes to pick a box to which an experimenter pointed. Over time, they learned that only the box to which the experimenter pointed contained food. For these initial training trials, the experimenter placed his or her hand a distance of 5 cm from the correct box (see Figure 8.7a). Interestingly, our apes did not appear to grasp the referential nature of the experimenter's actions from the start, initially choosing at random between the two boxes. Eventually, however, all seven of our animals succeeded in learning to select reliably the box to which the experimenter pointed.

But what, precisely, did this tell us about how the chimpanzees comprehended the gestures? One possibility was that even though they required some experience in the testing conditions, they did in fact come to understand the referential intent of the gesture, thus demonstrating an appreciation of the proto-declarative function of the experimenter's pointing gesture. Alternatively, perhaps the animals did not understand the referential function of the gestures, but instead had learned to exploit certain contingent relations associated with the experimenter's actions. That is, maybe the apes were employing some heuristic in the form of "Open the box closest to the experimenter's hand" (distance-based rule) or "Open the box + finger/hand configuration" (local-cue rule), thus obviating any need for representing the experimenter's actual communicative intent of conveying information.

To tease apart the subtleties inherent in these conflicting models, we devised several test conditions designed to identify which strategy would be used by both chimpanzees and two- to three-year-old children in retrieving rewards. In one study we merely moved the experimenter's hand so that instead of having it placed 5 cm from the correct container and 75 cm from the incorrect one (as it had been in the training and standard trials), it was now 120 cm from the correct container and 150 from the incorrect one in the probe trials (Figure 8.7b). If the subjects were relying on the local-cue rule, their performance should be crippled by this new configuration. Indeed, for five of the seven animals this proved true: the likelihood of these apes' choosing the box to which the experimenter pointed was at chance, despite the fact that on standard trials they continued having no difficulty. Two of the apes, Apollo and Kara, however, continued to make the correct choice on the majority of the new probe trials. Did this mean that Apollo and Kara, unlike their peers, understood the referential nature of the gestures? Recall that we had hypothesized that there were two heuristic strategies relying on readily observable spatiotemporal patterns that the apes could be employing: the distance-based rule and the local-cue rule. While we had demonstrated that five of the apes had in fact been using the local-cue heuristic, we could not be certain whether the remaining two animals were relying on the distance-based rule or the communicative intent model. Although the experimenter now sat away from the correct box, there was still a marked difference in the distance between his extended hand and the correct box and his extended hand and the incorrect box. Thus, Apollo and Kara might just have been better at judging which box was closer to the experimenter's hand—better at connecting the observable dots, as it were.

After we introduced additional configurations in which the experimenter's hand was closer to the incorrect container but referencing the correct container with the index finger (Figure 8.7c), and the tip of the index finger was equidistant between the two boxes but clearly referencing the correct container (Figure 8.7d), all of the animals (including Apollo and Kara) chose at random between the boxes. In direct and striking contrast, twenty-six-month-old human children were virtually at ceiling on even the most difficult of these conditions.

Similar experiments (which have carefully dissected the variables influencing apes' performance on object-choice tasks) involving intentional communi-

where is
experimenter
looking?

a
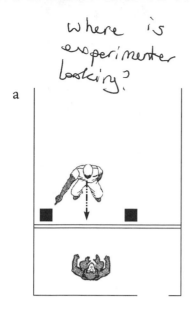

b
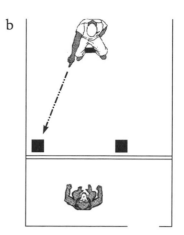

c
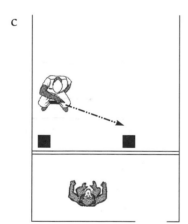

d
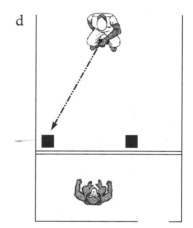

Figure 8.7 Conditions (a–d) used to test chimpanzees for the understanding of the communicative intent and referential significance of pointing gestures.

cation have found converging evidence of a lack of referential comprehension, regardless of the particular communicative device used. Povinelli and colleagues (1999) showed that chimpanzees do not appear to understand the intentionally communicative referential aspect of gaze (see also Call et al., 2000), while three-year-old children have no trouble understanding that an individual's gaze directed toward an external referent is "about" that object. And Tomasello, Call, and Gluckman (1997) reported that, while children two to three years of age were immediately able to infer novel signals (e.g., a marker placed on top of the correct box) as communicative symbols, their apes' performance was, at the beginning, dramatically unimpressive for the same conditions, only appearing to improve across trials as a result of the apes' using them as discriminative cues.

Comparative Investigations of the Attribution of Goals, Intentions, Knowledge, and Belief

In humans, the conception of others as animate, goal-directed agents emerges surprisingly early in development. Researchers have discovered that infants as young as nine months appear to appreciate, in some fashion of another, the goal-directed nature of human action, and even abstract computer animations engaging in sequential movement patterns (e.g., Gergely et al., 1995; Leslie, 1984; Phillips et al., 2002; Woodward, 1998; for review, see Flavell, 1999). Thus, early in life, infants appear sensitive to the goal-directed nature of action, an aspect of the human cognitive system that may be built into the way we perceive certain classes of object motion (e.g., Premack, 1990). Infants seem to treat other agents not merely as objects jetting about in haphazard ways but as beings with intentional states. While few researchers will claim that infants this age are necessarily explicitly representing goals and intentions as such, most conclude that at the very least the existing data demonstrate that infants are "on the right track," and are using precisely the right information from which later-developing, explicit understandings of goals and intentions will develop (e.g., Wellman & Phillips, 2001; Woodward et al., 2001). Additional evidence of such competence can be found in infants' ability to parse the behavior stream at its intentional joints (Baldwin & Baird, 1999). At around eighteen months of age, children appear able to appreciate that others have wants and desires that are different from one's own (e.g., Repacholi & Gopnik, 1997). Later, during the preschool years, the notion of belief emerges (e.g., see review and meta-analysis by Wellman et al., 2001).

Hauser (1998b), and to a lesser extent Tomasello and Call (1997), have proposed that other primate species may understand agents in a somewhat more primitive manner, reasoning that, at a very basic level at least, other species clearly segregate the animate from the inanimate world on the basis of superficial properties such as self-propelledness and irregular movement (see

Premack, 1990). Hauser (1998b) found support for this in a modified "looking time" paradigm, in which tamarins were observed to spend more time looking at an inanimate object (e.g., cereal) that moved between two box chambers than they did at an animate one (e.g., live mouse). According to Hauser, this demonstrated that the monkeys were "surprised" by this breach of agency laws. Of course, this tells us little about how the animals were reasoning about the mediating cognitive forces generating the behavior of the inanimate objects, only that it violated some set of expectations they possessed about the things they observed.

What about chimpanzees' and other nonhuman primates' understanding of intentions? Some researchers have attempted to use imitation to determine whether chimpanzees, for instance, reason about the intentions underlying behavior (e.g., Adams-Curtis & Fragaszy, 1995; Bjorklund et al., 2002; Call & Tomasello, 1994, 1996; Myowa & Matsuzawa, 2000; Nagell et al., 1993; Tomasello, Savage-Rumbaugh, & Kruger, 1993; Visalberghi & Fragaszy, 1990; Whiten et al., 1996). Although the conceptual issues are notoriously slippery, Tomasello (1990, 1996) has argued that in order for an organism to engage in true imitation, it must take the perspective and represent the intentions of the model from whom it learns how to perform an action (for critical considerations of the animal imitation literature, see Galef, 1992; also Whiten & Ham, 1992; Zentall, 1996). That is, the organism must demonstrate that it understands what the model is trying to do, not simply that the model is doing something to achieve some goal. Alternatively, an organism can be said to engage in emulation whenever the means for achieving a goal are essentially ignored and the organism only reproduces, through a different set of actions (perhaps through trial and error), the same end state. Emulation is envisioned as occurring through a form of stimulus enhancement in which the model's achieving some goal (e.g., getting termites from under a log) captures the attention of the observer and brings the observer to discover the same goal on its own, using different behavioral means to get there (e.g., rolling a log instead of poking through it). Unfortunately, the distinction between true imitation and emulation is sharper in theory than in practice, and has itself been the subject of considerable disagreement.

Nonetheless, the emulation-imitation distinction has led to experiments that have been somewhat effective in pulling apart these forms of social learning, and which have produced findings converging on the conclusion that nonhuman primates do not view others as mental agents. When properly controlled to rule out lower-level processes of social learning, no monkey species has been found either to imitate or emulate (Adams-Curtis & Fragaszy, 1995; Galef, 1992; Visalberghi & Fragaszy, 1990; for review, see Whiten & Ham, 1992), and while chimpanzees may be proficient emulators (see Tomasello, 1996), and may actually be more efficient (at times) in their social learning than human children, they appear mostly to ignore the behavioral mechanisms used to get there. According to Tomasello, they ignore the underlying reason these behaviors are performed, precisely because they do not reason about intentions per se. The best evidence of imitation in apes comes from several human-reared subjects (e.g., Bering et al., 2000; Tomasello, Kruger, et al., 1993), a topic we shall discuss shortly.

Myowa and Matsuzawa (2000) used a procedure originally designed for human infants (e.g., Meltzoff, 1995) in an attempt to show that chimpanzees could infer the intentions of a human model when observing actors unsuccessfully attempting goal-directed tasks (e.g., removing a lid from a tube). Eighteen-month-old human infants successfully accomplished the goal, thus demonstrating that they could read through the surface behavior (the literal, unsuccessful act) down to the intended (undemonstrated) act. Although Myowa and Matsuzawa (2000) report preliminary evidence of this form of intentionality attribution in their chimpanzees, the majority of the apes' "successful" attempts occurred at baseline (before witnessing the demonstrations), precluding any definitive statements on what the animals actually learned by watching the model.

In another study, Bjorklund et al. (2002) exposed their group of three human-reared chimpanzees to a series of generalization of deferred imitation tasks, which included four phases. In phase 1, the chimpanzees were given two sets of objects (e.g., a pair of cymbals and a pair of trowels) and were permitted six minutes with the objects to determine if they would spontaneously exhibit the target behavior associated with either set of objects. In phase 2, the animals were shown six demonstrations of the target actions with one of the sets of objects from the baseline (e.g., clanging the cymbals together by holding the outside knobs). A ten-minute delay followed the demonstration. In phase 3, the animals were given either the set of objects witnessed in the demonstration (standard deferred imitation) or the other set of objects from baseline not demonstrated in the previous phase (generalized deferred imitation). Finally, in phase 4, the set of objects not used in phase 3 was given to the subject. Bjorklund et al. (2002) argued that successful generalization of deferred imitation (e.g., clanging the trowels together by holding the outside wooden handles) required the apes to represent the actual purpose, or goal, of the model's actions in the demonstration phase, because instead of simply reproducing the actions on an identical task, the animals were asked to translate what they had learned to an entirely different set of materials that could be used to generate a similar outcome. All three chimpanzees displayed evidence of this kind of generalized deferred imitation. Interpreting these results is difficult, however, because no control was established for the perceptual similarity of the objects used; the apes may only have been mapping what they had witnessed with the objects from the demonstration phase onto the new set of objects—objects that bore the same general affordances of the original set.

Unfortunately, the exact relationship between social learning (e.g., imitation) and an understanding of intentions and goals remains elusive. Some authors (e.g., Bjorklund et al., 2002; Tomasello, 1999) continue to argue that evidence of true imitation, in which actions are reproduced with fairly high degrees of fidelity, is symptomatic of the ability to represent explicitly the intentions and goals of others' mental state attributions, whereas others (e.g., Heyes & Ray, 2000) maintain that imitation has no bearing on the question of mental state attribution whatsoever. Heyes (1998), for instance, has noted that the best evidence for nonhuman animal imitation comes not from primate species but rather from rats and budgerigars—species that very few researchers have claimed possess a theory of mind. In short, it is not at all apparent to us that the ability to reproduce another agent's actions, at any level of precision, necessarily hinges on an ability to represent intentions explicitly. This is not to say, however, that organisms with the capacity to reason about intentions (e.g., humans) will not use this ability during some of their attempts to imitate the behavior of others; indeed, in species with a theory of mind, such attributions should occur regularly.

Other studies, not involving imitation, which have attempted to assess chimpanzees' ability to attribute intentions, beliefs, and knowledge have overwhelmingly found support for the hypothesis that chimpanzees do not represent the mental states of others. While space prevents us from providing accurate descriptions of all of these, some of the more relevant findings include the following: chimpanzees do not appear to distinguish between accidental and intentional actions (Povinelli et al., 1997; but see Call & Tomasello, 1998), do not instruct ignorant others how to perform novel cooperative tasks

(Povinelli & O'Neill, 2000), and fail to differentiate between a knowledgeable and a naïve experimenter (Call et al., 2000; Call & Tomasello, 1999).

Physical Causality

It is important to touch at least briefly on a related project with our chimpanzees which was designed to map their understanding of unobservable forces in the physical world (see Povinelli, 2000). The initial round of nearly thirty studies, conducted over a five-year period, was centered on the widely celebrated ability of chimpanzees to make and use simple tools. Inspired largely by some work by Elisabetta Visalberghi and her colleagues, we asked not whether chimpanzees could learn to make and use tools, nor even the level of complexity that such tool use and construction might achieve, but whether they reason about more than the mere appearances of the objects as they make and use simple tools (e.g., Limongelli et al., 1995; Visalberghi et al., 1995; Visalberghi & Limongelli, 1994; Visalberghi & Trinca, 1989). Of specific interest to us was whether chimpanzees delve into the unobservable causal structure of the objects and events they observe, and whether their understanding of the physical world is mediated by concepts about unobservable phenomena such as gravity, force, shape, physical connection, and mass—an understanding that seems robustly in place by about three years of age in human children, if not earlier (for a review, see Povinelli, 2000).

The results of these studies consistently converged on a finding strikingly analogous to what we have uncovered about chimpanzees' understanding of the social world: they are excellent at extracting from the statistical regularities about what objects do and how they behave, but appear to have little or no understanding that these observable regularities can be accounted for, or explained, in terms of unobservable causal forces. Indeed, we have speculated that for every unobservable causal concept that humans may form, the chimpanzee will rely exclusively on an analogous concept, constructed from the perceptual invariants that are readily detectable by the sensory systems (see Table 8.1). Of course, humans will rely on these same spatiotemporal regularities much of the time as well, perhaps relying on systems that are homologs of those found in chimpanzees and other primates. But unlike apes, humans, we have speculated, evolved the capacity to form additional, far more abstract concepts that posit unobservable phenomena to explain observable

Table 8.1   Examples of perceptual invariant analogs of causal concepts

Causal concept	Perceptual invariant analog
Gravity	Downward trajectories
Transfer of force	Motion/contact movement sequences
Strength	Propensity for deformation
Shape	Perceptual form
Physical connection	Degree of contact
Weight	Muscle/tendon stretch sensations

events. Indeed, we have begun to explore whether this difference between human and apes can be described more simply in terms of the widely celebrated human capacity for explanation (e.g., see contributions to Keil & Wilson, 2000)—a capacity that may be lacking in chimpanzees (see Povinelli & Dunphy-Lelii, 2001).

Thus, we have speculated that a core difference between humans and chimpanzees may be that humans have evolved a unique capacity to develop representations about unobservable causes—a difference that manifests itself equally in the two species' understanding of the social and physical worlds.

The Immersion of Apes in Human Culture

The conclusions that we have drawn above primarily concern captive chimpanzees raised by their mothers or in nursery-peer groups with human caretakers. But what about chimpanzees raised with human beings and immersed in human culture? One possibility is that although chimpanzees do not normally develop the ability to reason about unobservable causal forces in either the social or physical domains, they do have the innate capacity to develop such abilities if they receive more of the experiences that human children receive. The idea that such intimate contact with human culture might bootstrap the development of certain cognitive structures that do not normally develop in apes is an old one (e.g., Hayes, 1951; Kellogg & Kellogg, 1933), and one that was played out more recently in the context of attempts to teach home-raised chimpanzees and other great apes certain language systems (Gardner & Gardner, 1971; Miles, 1994; Patterson & Linden, 1981; Premack,

1976; Savage-Rumbaugh & Lewin, 1994; Terrace, 1979). More recently, a considerable amount of speculation has been devoted to whether human rearing reorganizes the cognitive systems of apes, causing them to develop or elaborate on core systems that they do not normally develop or express strongly (e.g., Bjorklund & Pellegrini, 2002; Call & Tomasello, 1996; Donald, 2000; Tomasello, Kruger, & Ratner, 1993). Aspects of this debate pivot on the issue of how these systems normally develop in humans (see Povinelli, 2000; Chapter 12). To be sure, apes raised in human homes have exhibited more human-like performance in a variety of relevant areas, including imitation (Hayes & Hayes, 1952; Russon & Galdikas, 1993), deferred imitation (Bering et al., 2000; Bjorklund et al., 2002; Tomasello, Savage-Rumbaugh, et al., 1993), joint attention (Carpenter et al., 1995; Gómez, 1990), referential comprehension (Call & Tomasello, 1994), knowledge attribution (Call et al., 2000), and even pretense (Gardner & Gardner, 1971; Hayes, 1951; Temerlin, 1975; but see Bering, 2001). Whether we should conclude, as has Donald (2000), that "this demonstrates convincingly that the enculturation process can successfully uncover and exploit cognitive potential that had remained untapped for millions of years" (p. 29) is another matter entirely.

At present the scattered nature of the findings with apes raised in human homes prevents us from commenting intelligently on the subject. That some differences exist between home-raised versus other apes should not be surprising. At the very least, chimpanzees will bring their extraordinary intellectual resources to bear on the human culture in which they are immersed. But are these changes superficial, or do they reflect deeper changes in their core cognitive systems? This is a question that simply cannot be answered by the current strategies of post hoc retrodiction of experimental results (e.g., Call et al., 2000; Gómez, 1996), assessments with apes raised in human homes (e.g., Bering et al., 2000), or comparisons of such apes to other animals who do not have the requisite background familiarity with comparable testing situations (e.g., Tomasello, Kruger, et al., 1993). In this context, it is worth pointing out that there has never been a systematic test of the enculturation hypothesis; namely, no one has ever reared an appropriate number of apes in different ways and then experimentally assessed the effects of this experience on their cognitive development. In brief, a project of daunting scope would be needed. Perhaps a minimum of six to eight chimpanzees would need to be raised in human home settings for the first four years or so of their lives, with adequate safeguards in place to ensure that they were indeed brought into maximum contact with human social and material culture. Further, a control group of the same number of animals reared primarily with other chimpanzees would be needed to assess the effects of this massive enrichment intervention. We recognize that such a project would be extraordinarily costly and time-consuming—perhaps impossibly so—but if it were conducted properly, such an undertaking might stand as one of the most important achievements in the history of humanity's attempt to define its very nature.

An Evolved Conclusion: Getting Used to Psychological Diversity

In recent years, several researchers have advocated abandoning the global question of whether chimpanzees have a "theory of mind" as too broad to be useful (e.g., Povinelli & Eddy, 1996a; Tomasello & Call, 1997; Whiten, 2000b; see Chapter 9). The nature of the question that will replace it is not yet universally agreed on, however. Given the difficulty they have with tests of their understanding of complex epistemic states such as knowledge and belief, it seems undeniably reasonable to ask whether chimpanzees possess a better understanding of other, perhaps "less complex" mental states such as intentions, perceptions, goals, and desires (see Povinelli & Eddy, 1996a; Chapter 9), or whether the ability can be elicited only in certain kinds of situations (e.g., competitive ones: Hare et al., 2000).

We have in addition, however, pursued a third possibility, a possibility which, in the interests of fairness to the chimpanzees and other nonhuman species, sets aside the recapitulationist undercurrents inherent in the idea that there is some monolithic vector of cognitive development along which humans and apes and other species can be universally compared (for fuller descriptions of this approach, see Povinelli, 2001; Povinelli et al., 2000; Povinelli & Giambrone, 1999; Povinelli & Prince, 1998). In the face of parallel sets of converging experimental findings, we have come to give serious consideration to the possibility that humans may possess unique, specialized capacities when it comes to representing mental states and other unobservable phenomena, and that these systems appear early in development, entangling themselves into ancestral systems we share in common with other systems. In

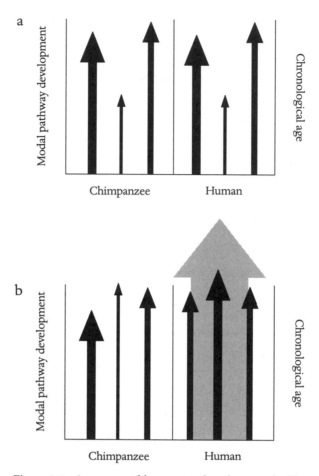

Figure 8.8 A more sensible question than the one asked in Figure 8.1 is: "How, at every stage of development, are chimpanzees and humans both similar and different?" In the example here, both humans and chimpanzees have added new systems to their development pathways (represented in "b" by novel arrows not present in "a") and expanded or contracted the functions of systems present in the common ancestor (represented by changes in the thickness of the arrows from "a" to "b"). We graphically represent the possibility that humans have woven in a new system or systems (represented by the large shaded arrow) that operates in parallel to ancestral systems, allowing for use of the older systems as input into new conceptual systems which "reinterpret" ancient behaviors in new ways.

humans, the very same action pattern—for example, following someone's gaze—may often be prompted by the mere detection of observable regularities (e.g., Driver et al., 1999; Kingstone et al., 2000; Langton & Bruce, 1999), whereas at other times it is prompted by a specialized system dedicated to representing why an event occurred in terms of unobservable variables (e.g., Adam wants me to think there's a bear behind me, so I'm going to play along)—especially when an event deviates from some canonical routine (Bruner, 1990). If true, the uniquely human system for representing unobservable causal states is parasitic on other, ancestral psychological systems that we share in common with our closest living primate relatives, and it imbues the ancestral representations of particular behaviors with psychological and causal content. For this reason, we have labeled it the "reinterpretation" hypothesis (see Povinelli et al., 2000).

By now, the incoherence of the question we graphically presented at the outset of this chapter (see Figure 8.1)—"What is the intellectual age of an adult chimpanzee in human terms?"—should be obvious. If the reinterpretation hypothesis is correct, then it is possible that there is simply no age at which humans and chimpanzees share a completely overlapping set of cognitive developmental pathways (see Figure 8.8). From birth forward the two species will share a suite of homologous systems, but also from birth forward specialized systems in the human species (and perhaps in the chimpanzee as well) will reside alongside these systems, and interact and influence them in ways that are complex and difficult to identify (see Povinelli & Giambrone, 1999). Thus, if something like the reinterpretation hypothesis turns out to have substantial merit, then the quest to find the "rudiments," the "simpler forms," the "less complex aspects" of an ability to reason about mental states in chimpanzees and other species may be nothing short of a fool's quest—one driven by the very faculty we sought to understand in the first place.

This writing was supported by a Centennial Fellowship from the James S. McDonnell Foundation to DJP. We thank Dario Maestripieri and one anonymous reviewer for helpful comments.

9

Social Cognition

Josep Call and Michael Tomasello

Arguably the most active and exciting area of research in contemporary developmental psychology, and perhaps psychology and cognitive science as a whole, is the study of social cognition. Many interesting and seemingly important phenomena have turned up as researchers have concerned themselves less with how people understand space, objects, tools, mathematics, and mechanical gadgets, and more with how they understand other people. Understanding how other people work as intentional and mental agents is at the heart of many of the most essential cognitive and social skills of human beings, such as imitative learning, language, teaching, theory of mind, and many forms of collaborative and competitive interactions (Tomasello, 1999).

In children's social-cognitive development there seem to be two key moments. First, the best-known moment is at age four, when children are said to attain something resembling an adultlike "theory of mind." They evidence this by reasoning correctly about false beliefs; that is, they can understand and reason about what other persons believe to be the case (e.g., that a box contains candy) even when they themselves know that it is not the case (e.g., the box really contains pencils). Children before this age seem to be unable to differentiate the state of the world as they know it to be from what other people may believe about the state of the world based on their own personal experience and knowledge. "Mind-reading" skills of this type are required for sophisticated forms of deception, teaching, and meta-cognitive monitoring, and some forms of role-taking and collaboration (Perner, 1991).

Second, and less well known, is a fundamental change in children's social cognition that takes place at about nine months to one year of age. It is at this age that young children first become able to conceptualize the psychological states of other persons, albeit at first much simpler states than knowledge and beliefs. Key components in this so-called nine-month revolution are the ways in which young children understand the visual perception and intentional actions of other people. Thus, at this age young children begin to show evidence of an emerging understanding that other people see things and that this affects their behavior in various ways. For example, infants begin to follow the gaze direction and pointing gestures of other persons and even to direct the attention of others by themselves pointing to outside events or by holding up objects to show them to others. It is at this age as well that their earliest linguistic skills for following into, directing, and sharing attention with other persons first emerge (Carpenter et al., 1998). One-year-old children can also tell the difference between a behavior that another person performed intentionally and one that she performed unintentionally, and they make use of this discrimination in deciding what they will and will not imitate (Carpenter et al., 1998; Meltzoff, 1995). Indeed, the forms of imitation that emerge at this age enable children not just to copy the body movements of others, but also to reproduce goal-directed (intentional) actions—a key skill in learning many cultural activities, such as the use of tools and linguistic symbols, which are directed at outside goals and referents (Tomasello, Kruger, & Ratner, 1993).

After a splashy start with a 1978 article claiming that chimpanzees have a theory of mind—based on a study investigating chimpanzees' ability to attribute goals to human actors (Premack & Woodruff, 1978)—the serious study of what nonhuman primates and other nonhuman animals share with humans in the domain of social cognition has only recently begun. There have been some well-publicized observations of human-raised apes doing humanlike things (e.g., Savage-Rumbaugh & Lewin, 1994) and some published anecdotes of ape and monkey behavior in the wild that are claimed to indicate all kinds of mind-reading skills for nonhuman primates such as deception (e.g., Whiten & Byrne, 1988). But the fact is that there have been very few well-controlled experiments with nonhuman primates focused on these issues, and they have not been set up to specify and interrelate primate understanding of different types of psychological states, from visual perception to false beliefs. Indeed, on the basis of our thorough review of literature—both observational and experimental—we concluded (Tomasello and Call, 1997) that there was very little evidence that nonhuman primates understand any of the psychological states of others at all. That is, we found no solid evidence that nonhuman primates understand the perception, intentions, knowledge, or beliefs of other individuals.

More recently, however, some new findings have emerged that require a modification of this conclusion. These findings demonstrate that nonhuman primates do indeed have some understanding of the psychological functioning of other animate beings, both humans and conspecifics. This is most clearly true for chimpanzees, and especially with regard to their understanding of the visual perception of others. Chimpanzees still do not have human social-cognitive skills, of course, and this poses a serious theoretical challenge to the field of cognitive ethology: Precisely which psychological states do nonhuman primates (or other animals) understand? To answer this question, many of the old methods will not be sufficient; researchers are going to have to turn up the microscope. There is no longer the luxury of saying that nonhuman primates either do or do not have a theory of mind, but we must now attempt to specify precisely what kinds of psychological phenomena they do and do not understand. This review is meant to be a first step in that direction, focusing on our new knowledge of chimpanzees' understanding of perception, intention, and knowledge.

Understanding Visual Perception

Eyes and vision play a very important role in the social lives of many animal species. Human infants can discriminate faces from other similarly complex objects soon after birth, mainly on the basis of the spatial organization of the eyes and the rest of the face (Fantz, 1963). But whether or not animals understand that there is someone behind those eyes, perceiving through them, is a different question. Until recently there was no solid (i.e., experimental) evidence that any nonhuman species understood this. Although the research is still very new and open to different interpretations, there is now some fairly convincing evidence that chimpanzees do indeed understand what other chimpanzees can and cannot see. Whether this is a humanlike understanding, or something else, is still to be determined.

Understanding Seeing

The conclusion that chimpanzees do not understand that others see things is mostly based on Povinelli and Eddy's (1996a) experiments involving seven four- and five-year-old chimpanzees' understanding of how humans must be bodily oriented for successful communication to take place (see also Reaux et al., 1999). They found that chimpanzees were sensitive to the body orientation of a human and consistently gestured toward the human who was facing toward them, thus confirming the naturalistic findings of Tomasello et al. (1994; Tomasello, Call, Warren, et al., 1997). In other experiments, however, chimpanzees did not seem to distinguish between more subtle cues. For example, they did not gesture differentially for a human who wore a blindfold over his eyes (as opposed to one who wore a blindfold over his mouth), or for one who wore a bucket over his head (as opposed to one who held a bucket on his shoulder), or for one who had his eyes closed (as opposed to one who had his eyes open), or for one who was looking away (as opposed to looking at the subject). A number of control experiments ruled out possible artifactual explanations for these results, and the basic findings have been replicated by Reaux et al. (1999) with the same chimpanzees at an older age.

We have been engaged in another series of studies, however, that give a much more positive picture of what chimpanzees understand about seeing (Hare et al., 2000). The basic set-up was as follows (see Figure 9.1). In each of five experiments a subordinate and a dominant individual were placed in rooms on opposite sides of a third room. Each had a guillotine door leading into the third room which, when cracked at the bottom, allowed the subjects to observe a human placing two pieces of food at various locations within that room—and to see the other individual looking under her door. After the food had been placed, the doors for both individuals were opened and they were allowed to enter the third room. The basic problem for the subordinate in this situation is that the dominant will take all of the food it can see, and indeed in all the studies in which dominants had good visual and physical access to the two pieces of food, they took the food on most occasions (i.e., they went for

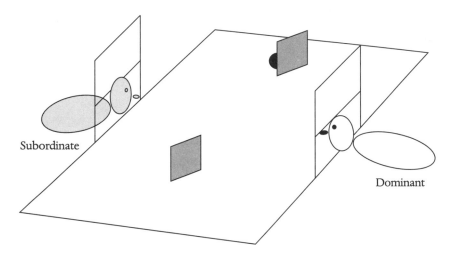

Figure 9.1 Experimental set-up for Hare et al. (2000). (Redrawn from Hare et al., 2000)

one piece while staring at and so intimidating the subordinate from taking the other piece). In some cases, however, we arranged things so that the subordinate could see a piece of food that the dominant could not see, for example, by placing it on the subordinate's side of a small barrier. The question in these cases was thus whether the subordinates knew that the dominant could not see a particular piece of food and so it was safe for them to go for it.

The basic finding was that the subordinates did indeed go for the food that only they could see much more often than they went for the food that both they and the dominant could see. In some cases, the subordinate may have been monitoring the behavior of the dominant, but in other cases this possibility was ruled out by giving subordinates a small headstart and forcing them to make their choice (to go to the food that both competitors could see, or to go to the food that only they could see) before the dominant was released into the area. Moreover, we also ran two other control conditions. In one, the dominant's door was lowered before the two competitors were let into the room (and again the subordinate got a small headstart), so that the subordinate could not see which piece the dominant was looking at under the door (i.e., it is possible that in the first studies the subordinate saw that the dominant was looking at the food out in the open and so went for the other piece). In the other control study, we followed the same basic procedure as before (one piece of food in the open, one on the subordinate's side of a barrier), but in this case we used a transparent barrier that did not prevent the dominant from seeing the food behind it. In this case, chimpanzees chose equally between the two pieces of food, seeming to know that the transparent barrier was not serving to block the dominant's visual access (and so her "control" of the food) at all. The findings of these studies thus suggest that chimpanzees know what conspecifics can and cannot see, and further, that they use this knowledge to devise effective behavioral strategies in food competition situations.

contrast w/ previous chapter

Although this study is silent regarding the role of the eyes in this situation, there is some evidence suggesting that chimpanzees and orangutans may indeed be sensitive to the role of the eyes in communicative situations. Gómez (1996) found that six chimpanzees requested the attention of a human by touching her when she had her back turned or her eyes closed or was looking in the other direction. A comparison with a condition in which the experimenter was staring at the subject with her eyes open indicated that the eyes-closed condition was hardest and looking in a different direction was the easiest condition for chimpanzees to discern. Similarly, we found (Call and Tomasello, 1994) that one of two orangutans pointed more often to request food when the experimenter who was facing him had his or her eyes open compared to eyes closed or the back turned. As in the case of the chimpanzees, the eyes-closed condition was harder to discern than the back-turned condition. Interestingly, those subjects who discriminated better between the attentive and non-attentive conditions had received more human contact during their upbringing.

All these studies have dealt with what individuals understand of the visual perception of others. Until recently, no attention had been devoted to the question of what individuals understand about their own visual perception. Call and Carpenter (2001) investigated this question in chimpanzees, orangutans, and children aged two and a half by presenting a finding game in which a reward was hidden in one of two or three hollow tubes. Whether subjects saw the baiting of the tubes and whether they could see through the tubes was varied, as was the delay between baiting and presentation of the tubes to subjects. Dependent measures were (1) whether subjects chose the correct tube, and (2) most important, whether they spontaneously looked into one or more of the tubes before choosing one. Most apes and children appropriately looked into the tubes before choosing one more often when they had not seen the baiting than when they had seen the baiting—suggesting that they knew something about the function of their own looking behavior. In general, they used efficient search strategies more often than insufficient or excessive ones.

Gaze-Following into Distant Space

Two previous studies have suggested that several primates follow gaze from humans to locations above and behind them (Itakura, 1996; Povinelli & Eddy, 1996b). More recent studies have reinforced and extended this finding in several directions. First, these initial studies have been replicated with various species such as chimpanzees (Call et al., 1998; Povinelli & Eddy, 1997), various macaques (but not with lemurs: Anderson & Mitchell, 1999), and dolphins (Herman et al., 1999). Several studies have used conspecifics as the source of information and have found identical results. Chimpanzees, macaques, mangabeys, and even goats follow the gaze of conspecifics to locations to the side of, above, or behind subjects (Emery et al., 1997; Tomasello et al., 1998). Second, some of these studies have provided additional details about the nature of gaze-following in these species. For instance, Call et al. (1998) found that when a chimpanzee tracked the gaze of another individual to a location and found nothing interesting there, the chimpanzee quite often looked back to the individual's face and tracked the gaze direction a second time—"checking back" in this way being a much-used criterion in assessing human infants' understanding of the visual experience of others (Bates, 1979). Moreover, some authors have even shown that macaques, baboons, chimpanzees, and orangutans are sensitive to eye direction (not just head orientation) under experimentally controlled situations (Ferrari et al., 2000; Itakura & Tanaka, 1998; Lorincz et al., 1999; Povinelli & Eddy, 1997; Vick & Anderson, 2000).

One possible explanation for simple gaze-following behavior is that individuals learn through experience that when they look in the direction toward which another individual is visually oriented, they often find something interesting or important. The cognitive process might thus be: turn in the direction in which others are oriented and then search randomly until you find something interesting. This is what Povinelli (1999) calls the "low-level" explanation, since it is based on an individual's learning what amounts to a conditioned discriminative cue: when another individual turns in a particular direction, you will quite often be rewarded for looking in that direction yourself. The "high-level" explanation may be variously characterized, but in a loose formulation it consists simply in understanding that the other one is having a visual experience (perhaps similar to one's own).

Following a suggestive finding of Povinelli and Eddy (1996b), Tomasello and colleagues (1999) effectively disproved the lower-level explanation of chimpanzee gaze-following with two experiments. In the first experiment, a chimpanzee watched as a human experimenter looked around various types of barriers (see Figure 9.2). If chimpanzees simply look in the same direction as others and then search randomly, as claimed by the low-level explanation, then they should either look at the barrier (if they find it interesting) or else look for something else in that direction. They did not do this, however. Instead, the chimpanzees actually moved themselves to new locations so that they could look around each of the barriers—seemingly to see what was behind it, where the experimenter was looking (they did this much more than in a control condition in which the experimenter looked in another direction, so it was not just natural curiosity about what was behind the barrier that caused this behavior). In the second experiment, chimpanzees watched as a

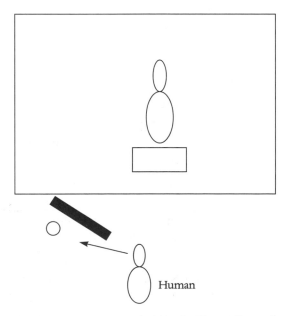

Figure 9.2 Experimental set-up for Tomasello et al. (1999). (Redrawn from Tomasello et al., 1999)

human experimenter looked toward the top and then the back of their cage. As they turned to follow the human's gaze, a distractor object was presented. Again in this case, if chimpanzees simply look in the same direction as others and then search randomly, as claimed by the low-level explanation, then they should look at the distractor (if they find it interesting) or something else in that direction. What they did, however, was to follow the experimenter's gaze all the way to the back of the cage, and they did this even though they clearly noticed the distractor. Together, these two studies effectively disconfirm the low-level model of chimpanzee gaze-following, supporting instead the hypothesis that chimpanzees follow the gaze direction of other animate beings geometrically to specific locations in much the same way as human infants (Butterworth & Jarrett, 1991). These studies provide no information, however, about the degree to which chimpanzees interpret the gaze of others mentalistically.

Gaze-Following in Object Choice

A great deal of attention has been devoted to investigating the ability of several species to spontaneously use gaze (head) direction from either conspecifics or humans to locate hidden food in the so-called object choice paradigm (Anderson et al., 1995). In a typical situation, subjects were presented with two opaque containers, only one of which contained food. A human experimenter then looked continuously at the container with food inside and, after a variable delay, allowed the subject to select one of the containers. With the exception of one chimpanzee (Itakura and Tanaka, 1998) and four gorillas (Peignot & Anderson, 1999), the only positive results in this paradigm have come with several human-raised chimpanzees in various studies (Call et al., 2000; Itakura & Tanaka, 1998). Several species of primates including macaques, baboons, chimpanzees, orangutans, and capuchin monkeys from different laboratories have shown a very inconsistent ability to spontaneously use the gaze direction of humans to help them locate hidden food (Anderson et al., 1995, 1996; Call & Tomasello, 1998; Call et al., 1998, 2000; Itakura & Anderson, 1996; Vick & Anderson, 2000). This failure to use gaze consistently to find hidden food contrasts with the ability to follow gaze to distant locations. This discrepancy seems particularly puzzling because some of the chimpanzees that failed this test in our studies were the same ones that were able to follow gaze to distant locations (e.g., Call et al., 1998). Itakura and colleagues (1999) used a trained chimpanzee conspecific to give the gaze direction cue, but still found mostly negative results. Failure to use gaze cues extends into other more overt cues such as pointing gestures or touching the stimuli (Call et al., 2000; Tomasello, Call, & Gluckman, 1997). In contrast, several studies have shown that other mammals such as dogs and dolphins are capable of spontaneously using gaze and other cues in object choice situations (Hare et al., 1998; Hare & Tomasello, 1999; McKinley & Sambrook, 2000; Miklósi et al., 1998; Tschudin et al., 2001).

Despite this lack of spontaneous use of gaze direction, several studies have shown that a number of species can learn to use cues such as head direction and gaze after multiple trials (Itakura & Anderson, 1996; Vick & Anderson, 2000). Additional tests have shown, however, that this was for them only a learned behavioral cue, not an indicator of the visual experience of others. For example, when the experimenter turned his head in the direction of the baited container but looked to the ceiling and not at the baited container, chimpanzees chose the correct container just as often as when the experimenter looked directly at it (Povinelli & Eddy, 1997). Other studies have investigated the conditions that help subjects improve their performance. For instance, Call et al. (2000) found that vocalizations (and other noises) and various behavioral cues such as touching, approaching, or lifting and looking under the correct container facilitated the performance of a minority of chimpanzees.

Overall, and with regard specifically to chimpanzees, it is safe to say that studies have revealed that in some situations chimpanzees know much more about seeing than is apparent in the classic gesture choice and object choice experimental paradigms. Discrepancies among the findings in the different situations could conceivably be due to methodological issues not related to chimpanzees' understanding of seeing per se, but this is extremely unlikely since in several different paradigms the findings are extremely robust and replicable across variations of task design, subjects, and in some cases laboratories. And so the theoretical challenge is to explain the apparently reliable yet different chimpanzee behaviors that emerge in the different observational settings, and to do so in a way that is revealing about the nature of chimpanzee social cognition.

Understanding Intention

Premack and Woodruff (1978) reported that a chimpanzee could identify the intentions of humans, that is, she could choose a photograph of the end state toward which a human was directed in familiar activities. The problem is that Savage-Rumbaugh et al. (1978c) demonstrated that this experiment had some very plausible interpretations in terms of associative learning: the subject could easily have solved the task simply by recalling familiar sequences of human behavior. Since this seminal study, the notion of understanding intentions has received little attention (much less than seeing or knowing, for instance) and the only evidence in support of intention is weak. There have also been some specifically negative findings, for example, Povinelli et al. (1998) found that chimpanzees did not discriminate between intentional and accidental actions. But there are now some new findings on chimpanzees' understanding of intentions.

First, we (Call and Tomasello, 1998) investigated the understanding of others' intentions using a task in which juvenile and adult orangutans, chimpanzees, and two- and three-year-old children had to discriminate between the intentional and accidental actions performed by the experimenter. During training, subjects learned to use a discriminative cue (marker placed on top of box) to select the baited box. During testing, however, the marker was not initially situated on any of the boxes, and the experimenter in each trial had to place it on top of the baited box to inform the subject of the food's location. During the course of a trial, the experimenter also "accidentally" dropped the marker on top of an unbaited box. In this way, during any given trial the experimenter marked two boxes, one intentionally and the other accidentally. Subjects of all three species preferentially selected the box the experimenter had marked intentionally, thus supporting the hypothesis that they understood something about the experimenter's intentions. This preference was especially pronounced for all three species during the initial trials. Three-year-old children presented the most robust performance of all groups of subjects.

Second, Myowa and Matsuzawa (2000) used the reenactment procedure pioneered by Meltzoff (1995) with eighteen-month-old children. In one condition (unfulfilled intention), chimpanzees observed a human model trying and failing to effect a desired manipulation, thus demonstrating an appropriate action without demonstrating the desired outcome (e.g., pulling on the tube's ends to no avail). In another condition (fulfilled intention), chimpanzees observed the human model successfully opening the object. Results showed that chimpanzees performed at the same level in both conditions. Myowa and Matsuzawa (2000) interpreted these results as evidence that chimpanzees understood intentions. The number of successes was very low in both conditions, however, and most of the successes occurred during the baseline. Call, Carpenter, and Tomasello (unpublished data) found very similar results in a study with chimpanzees. Subjects were allowed to witness different demonstrations by another chimpanzee of how to open a tube. In different experimental conditions, demonstrations consisted of (1) the actions necessary to open the tube without actually opening it (action only); (2) the open tube, without showing any actions (end state only); (3) both of these components (full demonstration); or (4) none of these components (baseline). In the first three conditions subjects saw two different ways that the tube could be opened (break in middle; caps off ends). As in the previous study, there were no differences between the full demonstration and the action-only conditions, and also there were no differences between conditions, with a high baseline. We also tested children with the same paradigm and confirmed previous studies showing that children performed better on both the full and action-only conditions compared to the baseline and end-state conditions. These two studies on social learning are inconclusive. The main problem is that the baseline is too high; future studies should use more difficult tasks.

At this time there is not enough solid research to come to any firm conclusions on whether chimpanzees and other nonhuman primates understand the intentions of others, as distinct from their behavior in general. This would seem to be an important question, because having goals as distinct from the behavioral means for pursuing those goals would seem to be an especially useful cognitive skill for predicting the behavior of conspecifics in novel situations—and so competing and cooperating with them more effectively (Tomasello & Call, 1997).

Understanding Knowledge and Beliefs

In developmental psychology and some areas of cognitive science, understanding perception and intentions is thought by many researchers to be only preparation for the real thing: children's understanding of thoughts and beliefs. Indeed, the most specific meaning of the term "theory of mind" refers only to children's understanding of false beliefs (understanding false beliefs is especially important because only they require the child to distinguish beliefs from reality). The question of whether nonhuman primates understand that others have thoughts and beliefs that may differ from their own and from reality has yet to receive a definitive answer. But a number of studies represent important progress.

Initially Povinelli and colleagues (1990; see also Premack, 1988) reported that chimpanzees made the distinction between knowing and guessing, but further analyses (Heyes, 1993; Povinelli, 1994) and failed attempts at replication (Povinelli et al., 1994) cast some doubt on those results. Call and colleagues (2000) used the same paradigm with some modifications. These authors capitalized on the ability of chimpanzees to follow gaze, and particularly two chimpanzees who satisfactorily used gaze information in object choice situations. Chimpanzees were confronted with two experimenters giving conflicting cues about the location of the hidden food. One of them baited the container (thus knew the food location), while the other experimenter remained ignorant about the food location (he did not either bait or witness the baiting). Two of the eight chimpanzees tested significantly followed the individual who had baited the food (i.e., the knower), thus suggesting that the chimpanzees were making this distinction. When a third experimenter did the hiding, however, while one of the others simply watched and the remaining one did not, neither of the two chimpanzees that had passed the previous test succeeded. It is interesting to note that the two individuals who passed the initial tests were human-raised. Touching rather than seeing is related to knowing; or perhaps they had learned that handling the food (rather than witnessing the food's location) was the cue they needed to follow.

Another set of experiments originated in the paradigm designed by Gómez (1996). In this experiment, a researcher hid food in one of two boxes, locked the boxes, and hung the key in a predetermined location in the room. Another experimenter entered the room, retrieved the key, and waited until the subject, an orangutan, pointed to one of the boxes. In some trials, the experimenter who hid the food also moved the key to a location other than the usual one. Gómez's results showed that the orangutan pointed preferentially to the key when the human had not seen its current location but would point to the food if the human had seen the location of the key. We (Call and Tomasello, 1994) used a similar paradigm and found that two orangutans pointed to the location of the hidden key. Unfortunately, this study did not include a condition in which the human had witnessed the current location of the key. In both studies, however, orangutans pointed to the key location after a few trials.

Whiten (2000b) used a modified version of the original procedure with three language-trained chimpanzees. Chimpanzees were presented with three conditions. In the baseline condition, subjects witnessed an experimenter (the hider) hiding food in one of the two boxes, and another experimenter (the helper) picking up a key (placed at a fixed location) and opening the box indicated by the chimpanzee. In the experimental condition, the helper baited the box, the hider then hid the key, and the helper reentered the room to help the subject obtain the food. Results showed that one of the chimpanzees (Panzee) preferentially pointed to the key in experimental trials while she pointed to the baited box in baseline trials. In a third control condition in which the helper (rather than the hider) had hidden the key, Panzee pointed to the box containing food. Panzee's results are remarkable because she pointed to the correct location from the first experimental trial. The other two subjects also pointed to the key in experimental trials but only after a few trials, as was the case with the orangutans in previous studies.

There have been two other approaches to adapting methods used in developmental psychology for use with apes. One such approach consists of using more implicit measures such as looking time. Following Clements and Perner (1994), Hauser (1999a) tested the ability of cotton-top tamarins to pass an implicit false belief situation. Hauser (1999a) presented tamarins with the following basic setup. An experimenter (E1) entered the room, took a few bites from an apple (and gave some pieces to the tamarin), placed it inside one of two boxes, and left the room. While E1 was outside the room, another experimenter (E2) moved the apple to the other box. When E1 returned, he searched either in the box where he had left the apple or in the other box. There were three conditions corresponding to the following variations on the basic situation. In the opaque box–leave condition, both boxes were opaque. In the opaque box–stay condition, both boxes were opaque, but E1 stayed inside the room while E2 transferred the apple from one box to the other. Results showed that in the opaque box–leave condition, tamarins looked significantly longer when E1 searched in the box currently containing the food than in the box that had originally contained the food. This means that the subjects expected E1 to look in the currently empty box, the box where E1 had deposited the apple. In contrast, when E1 stayed in the room, tamarins showed the opposite pattern of results, since they looked significantly longer when E1 searched in the box that had originally contained the food compared to the box that currently held the food. This means that the subjects expected E1 to look in the box containing the food, presumably because E1 had a chance to observe the transfer. In a follow-up experiment, Hauser (1999a) probed further the conditions that may lead subjects to create certain expectations about their behavior. Using the basic set-up of the opaque box–stay condition, he reran the following two variations. In one variation E1 closed her eyes while E2 transferred the food between the boxes, whereas in the other variation E1 turned her back while the transfer was taking place. Unlike the differences observed in the original conditions, the differences between the expected and unexpected conditions disappeared in both variations. Hauser interpreted these results in terms of the expectations about behavior that tamarins form rather than explicit false belief. Moreover, these expectations, Hauser argued, are based on rough cues such as presence versus absence in the room rather than more precise cues such as whether the eyes are open or closed.

Another avenue for exploring the ability to understand false belief in nonverbal organisms is to develop a nonverbal analog to the traditional false belief tasks along the lines of previous attempts by researchers such as Premack (1988). We (Call and Tomasello, 1999) investigated the question of understanding others' knowledge in adult orangutans, juvenile and adult chimpanzees, and four- and five-year-old children using a novel nonverbal version of the "location change" false belief test (Wimmer & Perner, 1983). In one condition, one experimenter (the hider) baited one of two identical boxes out of view of the subject while a second experimenter (the communicator) witnessed her actions. Then the communicator indicated the baited box to the subject by placing a marker on top of it. In the critical condition, the hider baited the box while the communicator observed her, and then the communicator left the room before marking the box. The hider then switched the location of the boxes in full view of the subject. The communicator returned and marked the location where he had originally seen the food being placed (although this location was now incorrect owing to the hider's box switch). If subjects understood that the communicator was wrong (i.e., had a false belief) because the hider had changed the location of the box containing the reward without the communicator's knowledge, they would choose the box *not* marked by the communicator. Additional control tests were conducted to ensure that subjects possessed the various cognitive skills (e.g., memory) necessary to solve the task. Human children were also given a verbal version of the false belief task so its results could be compared with the nonverbal task. All three species passed the control tests, but only five-year-old children and some four-year-old children (but not the apes) consistently passed both the verbal and nonverbal false belief tasks. Results from the verbal and nonverbal versions of the false belief task were highly correlated, with both tasks being of comparable difficulty. These results suggest that this new nonverbal false belief task is equivalent to its verbal counterpart, with the advantage that it can be used to test nonverbal children and nonhuman animals.

Finally, there has been a new paradigm specifically designed to test chimpanzees. Hare and colleagues (2001) used the same competition paradigm previously described but added some complexity. The experiments previously described dealt with chimpanzees competing over food when all the relevant information was available at the time of choice. In the next series of experiments (Hare et al., 2001), we investigated whether chimpanzees were also able to take into account past information such as whether the dominant had seen the baiting. For these experiments we used two barriers and one piece of food, and manipulated what the dominant saw. In experimental trials, dominants had not seen the food hidden, or food they had seen hidden was moved to a different location when they were not watching (whereas in control trials they saw the food being hidden or moved). Subordinates, by contrast, always saw the entire baiting procedure and could monitor the visual access of their dominant competitor as well. Subordinates preferentially retrieved and approached the food that dominants had not seen hidden or moved, which suggests that subordinates were sensitive to what dominants had or had not seen during baiting. In an additional experiment, we switched the dominant who had witnessed the baiting for another dominant who had not witnessed the baiting and compared the results with a situation in which the dominant was not switched. We found that subordinates retrieved more food when the dominant had been switched than when she was not switched, thus demonstrating their ability to keep track of precisely who had witnessed what. This result also ruled out the possibility that subordinates were using just the sequence of door opening and closing to decide which food to take.

The interesting question is thus whether this new competitive paradigm enables chimpanzees to display more sophisticated social-cognitive skills than other paradigms, as it seems to, or whether there is some other explanation for chimpanzees' sophisticated behavior in these experiments. Even if we take the results at face value, however, exactly how to interpret them is still an open question. For example, we could either say that the subordinate chimpanzee knows what the other one knows, or we could say—following the interpretation of Hare et al. (2001)—that each knows what the other one has seen in the immediate past. This latter characterization is thus much more concrete than one expressed in terms of thoughts and beliefs. Only future research can tell us the level of flexibility and abstractness with which chimpanzees can use their skills of gaze-following and monitoring.

Recent years have produced a wealth of new information on the social cognition of nonhuman primates that has changed the way we think about what and how animals think. At least for chimpanzees, there is good evidence that individuals know what others can and cannot see. They follow the gaze of conspecifics and humans, follow it past distractors and behind barriers, "check back" with humans when gaze-following does not yield interesting sights, use gestures appropriately depending on the visual access of their recipient, and select different pieces of food depending on whether their competitor has visual access to them.

Chimpanzees may even know what others know, and tamarins may even be sensitive to false beliefs, although the most direct tests of false belief with nonhuman primates have produced only negative results. And the interpretation of all of these results is far from straightforward. For instance, even putative examples of attribution of knowledge or false belief (Hare et al., 2001; Hauser, 1999a; Whiten, 2000b) may be explained as cases of sophisticated understanding of perception in others. For instance, the chimpanzees in Hare et al. (2001) could have solved the problem by referring to what others were able to see or not see rather than their knowledge states. In order to ascertain whether chimpanzees rely on what others know rather than what others have seen, future studies will have to investigate how dependent these results are on the visual modality. If chimpanzees can use other sources of information besides visual access to solve the problem, this would support the notion that they rely on something more abstract than simply visual perception.

The observed progress in both understanding of visual perception and knowledge contrasts with the little progress experienced in the area of understanding intentions in others. Although some results suggest that chimpanzees and orangutans may be sensitive to the distinction between accidental and intentional actions, the evidence is still too weak to make any strong claims. The second fiddle status of intentions in comparative research mirrors the state of affairs in developmental psychology, in which research on the attribution of knowledge and beliefs has received a disproportionate amount of attention in comparison with research on the attribution of intentions and attention.

To summarize, then, there is good evidence that chimpanzees know what others can and cannot see, but the evidence on intention and knowledge attribution, although suggestive, is still too fragmentary and vulnerable to alternative explanations. This may not seem to be significant progress, but our view is that real progress has been made in answering the seminal question, "Does the chimpanzee have a theory of mind?" Several theorists (Gómez, 1991; Whiten, 1994; Whiten & Byrne, 1988) have argued that a more gradual and differentiated response to this question is needed, not simply a yes/no answer. Recent work has started to produce the evidence to support this more gradual and detailed approach so that it is no longer easy to answer the key question with a single word. The question is not whether chimpanzees have a theory of mind but rather which of the many psychological states of others that chimpanzees might possibly understand do they really understand.

It is premature to give a full answer to this question because the evidence available is still very fragmentary; most of what we know is about understanding visual perception, mostly regarding chimpanzees. We can, however, present various alternatives that have been offered to explain chimpanzee (and primate) social cognition. In general, there have been three such alternatives. First, there is the meta-representational account in which individuals attribute mental states such as knowledge or beliefs to others. Within this position various authors have distinguished different degrees of mental attribution. Whereas some authors argue that apes may understand false beliefs (Byrne, 1997; Dunbar, 2000), others argue that apes may simply be capable of re-representation or secondary representation (Suddendorf & Whiten, 2001; Whiten, 2000b). The key distinction between these two meta-representational accounts lies in the fact that the former is based on representing a representation as a representation whereas the latter is based on representing a representation but without understanding the representational nature of the representation (Perner, 1991). Second, there are those authors who argue that chimpanzees are simply good at reacting to cues provided by others. Povinelli and his colleagues have championed this view in recent years (Povinelli, 1994; Povinelli et al., 1994; Povinelli & Eddy, 1996a; Reaux et al., 1999). According to them, chimpanzees learn rules with little understanding of those rules—a concept that Povinelli has extended beyond social cognition into physical cognition (Povinelli, 2000). Heyes (1993, 1998) has also pointed out that the current putative evidence on mental state attribution can be easily explained as a case of learned contingencies between stimuli and responses.

Finally, there is a third option between the two previous alternatives. We have called it an explanation of the third kind (Call & Tomasello, in press; Tomasello & Call, 1997) or representational account (Call, in press), to differentiate it from various types of cue-based explanations (nonrepresentational) on the one hand and various types of meta-representational explanations on the other. This third way, like the meta-representational account but unlike the cue-based explanation, maintains that social problem solving is based on using knowledge. The two knowledge-based mechanisms differ, however, in the nature of this knowledge. While the representational account accepts that individuals may have a notion of seeing in others, it does not postulate any insight into the subjective experience of seeing in others. In contrast, this insight into others' subjective experience is a key component of the meta-representational account. One way to describe this third way is to present it as the social counterpart of insight into physical problems. Some animals have insight into some social problems, which enables them to develop intelligent problem-solving strategies, especially in novel situations. But this does not necessarily mean that they also have the ability to imagine the visual experience of others or to understand that others have beliefs about things that may differ from their own and from reality.

We believe that the current evidence suggests that at least some animals such as chimpanzees have social knowledge as a sort of social insight. We have discussed social knowledge, but there is still the issue of receptive (reactive) versus productive (proactive) knowledge. By far the evidence gathered here is about reacting to certain situations rather than producing or coming up with clever solutions. An example will illustrate the point. Upon finding food that is hidden from dominants, chimpanzees may choose to approach and take it, but not if the dominant can see. It is less clear whether chimpanzees would also engage in more proactive strategies such as actively hiding food or driving others away. The first case is reactive (whether to go or not to go) while the second case is productive (what to do to change the situation). There is some anecdotal evidence that chimpanzees may engage in proactive strategies. For instance, some individuals may lead others away from a food site, direct their attention toward nonexistent predators, or cover parts of their anatomy (see Whiten & Byrne, 1988, for a compilation). More systematic studies on the prevalence of these strategies and mechanisms underlying them are sorely needed. The prevalence of reactive compared to proactive strategies is also seen in other domains such as coalition formation in which most of the episodes are reactive in nature, that is, a coalition is formed after the initial bout of aggression (i.e., as a reaction to aggression), not before.

It is encouraging that so much progress is occurring. We now have a much better idea about the understanding of psychological states in chimpanzees, especially with reference to visual perception. Progress, however, often brings costs as well as benefits. A positive consequence of the parallel growth of the areas of understanding of visual perception and knowledge is that they have become tightly intertwined. Instead of two separate areas, they can be viewed as different aspects of a single phenomenon, with distinctions between representational and meta-representational processes falling into a graded continuum. We do not view this as a bad thing, since a continuous view strikes us as more fruitful that a polarized view, which in many cases simply reflects dogmatic positions rather than the complexity of the findings in a given field.

Nevertheless, progress in some areas has produced the undesirable effect that other areas such as understanding intention and attention have not received much research. Another cost associated with progress is that the efforts have been directed at only a very few species, most notably chimpanzees. If the field of comparative cognition is to flourish, more species should be added to the mix. Research with other animals such as dogs, goats, birds, and dolphins shows that social cognition is not the exclusive realm of chimpanzees or even primates. Future studies should include extensive research on other non-primate species. This more broadly comparative approach will not only help to complete the current fragmentary picture but also aid in discovering new phenomena in comparative social cognition.

10

Personality

Samuel D. Gosling, Scott O. Lilienfeld,

and Lori Marino

People who have worked closely with primates have long been struck by individual differences among these animals (Crawford, 1938). "When observers spend hours recording behavior, they end up not only with behavioral data, but also with a clear impression of individuals" (Stevenson-Hinde, Stillwell-Barnes, & Zunz, 1980a, p. 66). Although this view is not always discussed explicitly, it is widespread among primate researchers. Yet few of them have actually published research on primate personality, perhaps because they believe that such work will be viewed as unscientific or anthropomorphic, or because their central interests lie elsewhere. Nonetheless, a relatively small but growing published literature on primate personality is emerging. This literature ranges from theoretical essays (e.g., Hebb, 1946) and case studies of single animals (e.g., Buirski & Plutchik, 1991) to large-scale investigations (e.g., Capitanio, 1999; King & Figueredo, 1997).

In this present chapter we review, summarize, and attempt to integrate the research on personality in nonhuman primates. Unlike many other areas reviewed in this book, personality research in primates has yet to develop into an established subfield in its own right. No investigator could be identified primarily as a "primate personality researcher." Instead, a diverse group of researchers has contributed to the fifty or so studies that have examined primate personality, with the result that the literature has no central focus or unifying research agenda. To provide standards against which the primate research can be evaluated, we draw on the human research literature.

By virtue of its recent history, the literature on human personality provides a clear set of criteria for evaluating personality findings. These criteria have been articulated in response to a series of influential critiques aimed at the field over the past several decades. These critiques attacked the theoretical and methodological bases of the field and appeared to call into question many fundamental assumptions regarding human personality. As a consequence, a good deal of research in personality has focused on responding to these criticisms (Pervin, 1999). As suggested by the title of an article that summarized one debate—"Profiting from Controversy"—much has been learned in the course of these interchanges (Kenrick & Funder, 1988). Indeed, the consensus is that the responses prompted by the criticisms have actually strengthened the field. We use the criteria that emerged from the defense of human personality research as standards or benchmarks against which to evaluate research on primate personality.

In contrast to the field of human personality, which traces its roots to the earliest days of scientific psychology, the study of personality in nonhumans is relatively new. Despite a few early attempts to examine personality in nonhuman primates (Hebb, 1946, 1949; Yerkes, 1939) and other animals (Pavlov, 1955), the field is too short-lived to have gone through many cycles of profitable debate. Nonetheless, because many of the issues that confronted human personality research now confront animal research, primate researchers should reap whatever profits they can from the hard lessons learned in human research. In this chapter we use the controversies from human personality research as a launching point for critically examining the issues that confront primate personality researchers. These issues suggest a set of criteria that must be met before the field can advance meaningfully. In addition, we ask whether these criteria have been met. Where evidence from the primate literature pertains to the controversies, we draw on these findings. Thus, the challenges that confront the field of primate personality serve as a conceptual framework within which to summarize findings from studies of primates.

Definitional Issues

Before we examine specific studies, it is necessary to lay the groundwork for what follows by posing some key definitional questions. The first concerns the constituent elements of personality. In other words, what is the best way to conceptualize an individual's personality? For example, if we want to know what individuals are like, do we want to know about their goals and motives, about their behaviors, or about abstract themes that imbue their lives with meaning? Human personality psychologists adopt a variety of orientations and often differ in the personality constructs they emphasize. The phenomena studied include temperament and character traits, dispositions, goals, personal projects, abilities, attitudes, physical and bodily states, moods, and life stories (Angleitner et al., 1990; John & Gosling, 2000; Little, 1996; McAdams, 1995, 1996). A review of more than 150 studies of personality in nonhuman species showed that the personality phenomena studied in animals constitute only a subset of the personality phenomena studied in humans (Gosling, 2001). The overwhelming majority of animal studies focused on traits, behaviors, and abilities, but no animal research examined personal projects, identity, attitudes, or life stories. Presumably this discrepancy between the domains of human and animal personality stems largely from the nature of the latter concepts, which require individuals to articulate their internal motives, feelings, and beliefs. Clearly, any phenomena dependent on self-reports cannot be examined in nonhuman populations (with the possible future exceptions of language-trained animals). More fundamentally, such constructs as projects, identity, and life stories may depend on complex mental representations of the self, some of which may be largely or even uniquely human (Call & Tomasello, 1999; Hampton, 2001; Hare et al., 2000; Povinelli & Giambrone, 2001). It is therefore not surprising that research on animal personality is composed essentially of observational studies of traits.

But how should traits themselves be defined? Broadly speaking, personality traits are temporally and situationally consistent patterns in the way individuals behave, feel, and think (John & Gosling, 2000). Although traits are meant to characterize consistencies across time and situations, modern dispositional conceptualizations of traits acknowledge that behaviors are not independent from situations. In other words, traits can be thought of as tendencies to behave in certain ways given certain situational triggers (see Tellegen, 1991, for a discussion).

A second definitional issue concerns temperament, a construct closely related to personality. In human research, temperament has been defined by some researchers as comprising the congenital, early-appearing tendencies that continue throughout life and serve as the developmental foundation for personality (Buss, 1995; Goldsmith et al., 1987). Although this definition is not uniformly embraced by human researchers (McCrae et al., 2000), animal researchers agree even less about how to define temperament (Clarke & Boinski, 1995). In some cases the word "temperament" appears to be used purely to avoid using the word "personality," which some animal researchers associate with anthropomorphism. To further complicate matters, some human researchers use "temperament" to refer to higher-order (superordinate) personality dimensions and reserve the term "personality" to refer to lower-order traits. This distinction quickly becomes fuzzy, however, because personality can be conceptualized at many hierarchical levels, ranging from broad superfactors (e.g., activity level) to specific characteristics or "habits" (e.g., dynamic as a lecturer). Moreover, in the context of the present discussion, the distinctions between personality and temperament are not directly relevant. We therefore use the term "personality" to refer to both constructs.

Criteria for Evaluating Personality in Primates

Mischel's (1968) influential book was the first of a series of direct challenges to the assumptions that personality (1) exists and (2) predicts meaningful real-world behaviors. Based on a review of the personality literature, Mischel (1968) pointed to the lack of evidence that individuals' behaviors are consistent across situations (Mischel & Peake, 1982). The logical implication of this claim is that situations, not traits, are the prime determinants of behavior. According to Mischel, measures of personality traits are rarely predictively useful. Instead, Mischel contended that we must carefully specify the situational context in order to predict behavior. Mischel's arguments engulfed the field of personality psychology in intense controversy for well over a decade (Epstein, 1979).

A number of authors marshaled a broad array of evidence to challenge

Mischel's attacks. Kenrick and Funder (1988) summarized this evidence by systematically addressing seven challenges to the existence and predictive utility of personality traits. It is beyond the scope of the present chapter to review how Kenrick and Funder addressed each of these challenges. Their paper is useful, however, because it delineated the kinds of evidence that are needed to establish the viability of personality traits. Essentially, three major criteria must be met. First, ratings must be shown to reflect genuine attributes of the individuals rated, not merely the observers' implicit theories about how personality traits covary. Second, assessments by independent observers must agree with one another. Third, these assessments must predict behaviors, real-world outcomes, and non-test criteria such as biological variables.

Criterion 1: Ratings Must Reflect Attributes of Targets', Not Observers', Implicit Personality Theories

The Issue

When Passini and Norman (1966) compared personality trait ratings of complete strangers with ratings of well-known friends, they found something extraordinary. Specifically, they found that the patterns of correlations among the traits for strangers were strikingly similar to the patterns of correlations for friends. That is, the observers were apparently perceiving and distinguishing personality dimensions in people they had met only fifteen minutes earlier. One interpretation of this finding is that personality traits are mere fictions, projections of our cognitive structure onto the interpersonal world. According to this interpretation, the correlations among trait ratings reflect not the actual co-occurrence of these traits in strangers but rather the shared meaning of trait terms. Proponents of this view maintain that ratings of assertive and active, for example, are correlated not because assertive individuals also tend to be active, but because assertive and active have similar meanings. This explanation has been termed the "systematic distortion hypothesis" (Shweder, 1982b; Shweder & D'Andrade, 1980) and concerns the accuracy of our implicit theories about the extent to which traits tend to co-occur. This "what goes with what" form of accuracy is distinct from a "what goes with whom" form of accuracy, which refers to the degree to which traits are related to what an individual actually does (Borkenau, 1992; Norman & Goldberg, 1966). (The "what goes with whom" form of accuracy is the subject of Criteria 2 and 3 below.)

Evidence from the Literature on Humans

In humans, several lines of evidence have been invoked to refute the systematic distortion hypothesis (Block et al., 1979; Romer & Revelle, 1984). One line of evidence is based on personality ratings obtained using procedures designed to prevent raters from relying on semantic similarities among words (Weiss & Mendelsohn, 1986). Instead of a single rater being asked to rate a target on multiple traits, multiple raters were used, each rating the target on different traits. If under these conditions assertive correlated with active, it could not be due to raters' inferring one trait from another, because raters rated either assertive or active, not both. Studies that have used this approach have yielded similar patterns of correlations among traits, as have studies using traditional methods. These findings and others suggest that the correlations reflect genuine attributes of the target individuals, not merely semantic relations among traits. Although Weiss and Mendelsohn's (1986) findings do not conclusively refute the systematic distortion hypothesis (because raters may nevertheless have implicitly invoked correlated traits when asked to rate a single trait), they provide one important piece of evidence against this hypothesis. In general, the systematic distortion hypothesis has been discredited as a primary explanation for personality descriptions of humans (Borkenau, 1992). Researchers now interpret Passini and Norman's (1966) findings to indicate that individuals learn the true structure of personality from their observations of the real world and simply "fill in the blanks" when presented with incomplete information about individuals.

Evidence from the Literature on Nonhuman Primates

But what about personality descriptions of nonhuman primates? Is it possible that observers are not detecting the true structure of personality traits in primates and are instead simply filling in the blanks using their knowledge of human personality structure? Four forms of evidence argue against this primate version of the systematic distortion hypothesis. The first piece of evidence derives from studies that have examined personality structure in animals; these studies identify the broad dimensions that characterize various species.

If the relations among traits reflect only shared word meanings, we should find the same personality structures in all species. Comparisons across different species, however, often reveal different structures. For example, the traits fearful and aggressive are part of a single personality dimension in rhesus monkeys (Stevenson-Hinde, Stillwell-Barnes, & Zunz, 1980a), whereas these traits load on two separate dimensions in gorillas (Gold & Maple, 1994). Because these findings derive from studies using exploratory, rather than confirmatory, factor analytic techniques, however, it will be necessary to corroborate them further with statistical techniques that lend themselves more directly to hypothesis testing.

The second form of evidence also derives from studies of personality structure. Although most animal studies of personality structure are based on personality ratings, a small number of studies are based on behavioral tests and carefully recorded ethological observations. For example, van Hooff (1973) meticulously observed the naturally occurring expressive behavior of chimpanzees. A social play factor was marked by such behavior patterns as "grasp and poke" (boisterous but relaxed contact), "pull limb" (playful social contact), and "gymnastics" (exuberant locomotory play, such as swinging, dangling, rolling over, turning somersaults), and an affinity factor was marked by behavior patterns indicating social closeness, such as touching (gentle contact, such as stroking another over the head), grooming, and embrace. These behavior-based factors cannot be explained solely in terms of semantic similarity. Yet the factors obtained from behavioral codings resemble factors obtained from rating data, suggesting that both methods assess the same underlying construct (Gosling & John, 1999).

The third form of evidence against a semantic similarity explanation for personality descriptions in primates is provided by studies showing that trait ratings are correlated with independently coded behaviors. For example, Capitanio (1999) found that ratings of sociability were significantly correlated with independently assessed affiliative behaviors, while ratings of confidence were significantly correlated with independently assessed aggressive behaviors. These and other findings again suggest that the ratings reflect genuine attributes of individuals, not merely semantic relations among traits.

The fourth form of evidence comes from studies showing that at least some primate personality traits (e.g., dominance) are heritable (Weiss et al., 2000), suggesting that biological mechanisms, not semantic similarity, account at least partly for intra-subject variance in personality dimensions.

All four sets of evidence suggest that similarity in the meaning of the terms cannot explain the empirical relations among the traits. The findings refute the systematic distortion hypothesis and suggest that the structure of personality ratings is based, at least partially, on aspects of the individuals being rated.

Criterion 2: Independent Assessments Must Agree

The Issue

Agreement among observers about target individuals reflects observers' ability to differentiate among individuals. If individual differences exist, and if they can be detected, then independent observers should agree that baboon X is more anxiety-prone than baboon Y. Consensus among observers has been considered a crucial criterion of accuracy by many personality researchers (e.g., Funder, 1987; Kenny, 1991, 1994; Kenny & Albright, 1987; Kruglanski, 1989; McCrae, 1982; Wiggins, 1973).

Evidence from the Literature on Humans

There is now overwhelming evidence that humans agree strongly in their ratings of other humans. Studies of humans typically elicit inter-observer agreement correlations in the region of .50 (e.g., Funder et al., 1995; McCrae, 1982), and provide a benchmark by which primate ratings can be judged.

Evidence from the Literature on Nonhuman Primates

Table 10.1 summarizes the inter-observer agreement correlations among observers in primate studies in which these correlations were reported. The studies are further divided into those that computed reliability in the conventional manner, across subjects, and those that computed reliability within subjects. The first column shows the species, and the second through fourth columns show the mean levels of inter-observer agreement for each study and the 95 percent confidence intervals around the mean. The fifth through eighth columns show correlations and labels (where appropriate) for the indicators with the maximum and minimum levels of agreement obtained in each study. The last three columns list the number of animals studied, the number of indicators used, and the relevant citations.

Table 10.1 Inter-observer agreement and test-retest reliability for personality ratings of primates

| Species | Reliability correlations | | | | | | | Sample size | No. of indicators | Study | Retest interval |
| | Mean correlation | 95% CI | | Maximum | | Minimum | | | | | |
		Lower	Upper	Item label	Correlation	Item label	Correlation				
	INTER-OBSERVER AGREEMENT (COMPUTED FOR EACH VARIABLE ACROSS SUBJECTS)										
Chimpanzee	.51[a,b]	.45	.57	Dominance	.70	Cleanliness	.36	9	16	Crawford (1938)	
	.86	.78	.92	Friendly to staff	.90	Friendly to "timid man"	.75	30	6[c]	Hebb (1949)	
	.33[d]	.29	.37	Dominant	.61	Erratic	.10	100	43	King & Figueredo (1997)	
Vervet monkey							.55[a,d]	97	12	McGuire et al. (1994)	
Japanese macaque	.69	.52	.81	Aggressive	.86	Gregarious	.38	14	8[c]	Martau et al. (1985)	
	.61	.43	.74	Aggressive	.82	Distrustful	.24	12	8[c]	Martau et al. (1985)	
Rhesus monkey	.53[e]	.49	.58	Dominant	.70	Sensitive	.17	45	33	Stevenson-Hinde et al. (1978, 1980a)	
	.08[d]	-.01	.18	Sociable	.42	Opportunistic	-.45	42	25	Capitanio (1999)	
Baboon	.75[d]	.66	.81	Rejection	.84	Exploration	.54	7	8[c]	Buirski et al. (1973)	
	.68	.46	.83	Dyscontrolled	.85	Gregarious	-.01	6	8[c]	Martau et al. (1985)	
Gothic-arch squirrel monkey	.71	.35	.88	Timid	.92	Depressed	.01	5	8[c]	Martau et al. (1985)	
Roman-arch squirrel monkey	.76	.61	.85	Aggressive	.86	Dyscontrolled	.35	8	8[c]	Martau et al. (1985)	
Median	.68				.84		.30	13	8		

INTER-OBSERVER AGREEMENT (COMPUTED WITHIN SUBJECTS, ACROSS VARIABLES)

										Source
Chimpanzee	.75	.57	.86	.98	n/a	.20	n/a	14	8[c]	Buirski et al. (1978)
	.57[f]	.44	.67	.78	n/a	.33	n/a	8	15	Crawford (1938)
Japanese macaque	.72	.59	.81	.90	n/a	.07	n/a	14	8[c]	Martau et al. (1985)
	.51	.29	.69	.81	n/a	–.14	n/a	12	8[c]	Martau et al. (1985)
Rhesus monkey	.81							12	28	Locke et al. (1964a, b)
Baboon	.71	.33	.89	.90	n/a	.20	n/a	6	8[c]	Martau et al. (1985)
Gothic-arch squirrel monkey	.78	.66	.86	.85	n/a	.67	n/a	5	8[c]	Martau et al. (1985)
Roman-arch squirrel monkey	.82	.55	.93	.97	n/a	.15	n/a	8	8[c]	Martau et al. (1985)
Median	*.74*			*.90*	*n/a*	*.20*		*10*	*8*	

TEST-RETEST RELIABILITY (COMPUTED FOR EACH VARIABLE ACROSS SUBJECTS)

										Source	
Chimpanzee	.71	.62	.78	.85	Confidence	.21	Cleanliness	9	16	Crawford (1938)	4 months
	.81	.72	.87	.98	Cheerfulness	.12	Destructiveness	8	22	Crawford (1938)	10 month

TEST-RETEST RELIABILITY (COMPUTED WITHIN SUBJECTS, ACROSS VARIABLES)

										Source	
Japanese macaque	.92	.83	.96	.98	n/a	.17	n/a	12	8[c]	Martau et al. (1985)	12 months

Note: 95% CI = 95% confidence intervals. "No. of indicators" refers to the number of items or scales on which the summaries are based; some studies report reliabilities for single items and other studies report reliabilities for multi-item scales. Some item/scale labels have been slightly abbreviated. The studies' authors' definitions of personality have been used, so we have not excluded items that would not ordinarily be considered personality constructs (e.g., masturbation). Within-subject correlations are computed across items, and the maximum and minimum reliability correlations refer to individual animals, not items; therefore, the item labels are denoted "n/a" (not applicable) for the within-subject studies. All means have been computed or recomputed using Fisher's r-to-z transformation where appropriate.

a. Pairwise correlations computed from the alpha reliability using the Spearman-Brown prophecy formula.
b. Ratings were made on three occasions. The three occasions did not yield similar patterns of inter-observer agreement, only correlating .11, on average, across the 16 traits. To obtain the most stable inter-observer agreement estimates, we took the mean of the three sets of agreement correlations, and these combined correlations are the data reported here.
c. Data reported for scales (i.e., aggregates of multiple items), not single items.
d. Inter-observer agreement computed using intraclass correlations.
e. Mean of three annual sets of ratings.
f. These figures probably offer inflated estimates because Crawford (1938) reported agreement only among the three observers showing strongest agreement.

One can obtain a rough estimate of the magnitude of inter-observer agreement by computing the grand mean of the eleven mean correlations (weighted by the number of estimates on which each mean is based) in the first section of Table 10.1. This estimate is approximate because it summarizes a set of agreement correlations that were (1) computed at both the item and scale level, (2) based on a variety of methods, (3) derived from observers who varied in their acquaintance with the animals, (4) based on animals observed across diverse situations, and (5) often computed using suboptimal sample sizes. Nevertheless, the grand mean of the eleven means reported in Table 10.1 is composed of 183 reliability estimates and is probably the best estimate to date of the level of pairwise (dyadic) agreement between observers making personality ratings of primates. The weighted grand mean correlation is .48. This correlation is substantial and compares favorably with equivalent inter-observer correlations in the human personality literature (e.g., Funder et al., 1995).

Within-subject agreement correlations adopt an "ipsative" (within-individual) approach by computing the inter-observer agreement correlations across traits within an animal. This approach asks, for example, whether observers A and B agree that a particular baboon is more trustful than she is timid. Within-subject correlations, which have been computed for six primate species, are shown in the second section of Table 10.1. Again, the mean of these correlations is substantial: the weighted grand mean of the within-subject mean correlations is .72. This grand mean, based on ninety-one reliability estimates, again suggests that observers agree in their appraisals of the personalities of the animals they are rating. These within-subject correlations should be interpreted cautiously, however, because they do not show that observers can discriminate among the individuals being rated. Strong within-subject reliabilities would be obtained even if observers based their judgments on stereotypical views of the animals, with each observer giving each animal the same stereotype-based ratings. To show that the observers discriminate among individuals, it is necessary to compare the within-subject correlations with the "off diagonal" cases. For example, is the correlation between observer A's rating of animal X and observer B's rating of animal X higher than the correlation between observer A's rating of animal X and observer B's rating of animal Y?

It is reasonable to posit that the longer two observers have known an animal, the more strongly they will agree about its personality. This supposition has been widely supported by research on humans (Funder & Colvin, 1988;

Kenny et al., 1994). For example, Funder et al. (1995) found that targets elicit stronger agreement about their personality traits when judged by their acquaintances than when judged by strangers, who had only viewed them on videotape. Nevertheless, only one animal study has directly tested the effect of acquaintanceship. Martau et al. (1985) compared personality ratings made by observers who had watched Japanese macaques for two hours per day for a month with ratings made by observers who had less than five hours of experience with the monkeys. Agreement improved slightly with increased acquaintance; the average inter-observer correlation (computed at the scale level) was .69 for relatively acquainted observers and .61 for the relatively unacquainted observers. Although suggestive, these findings should be interpreted cautiously because Martau et al. (1985) (1) reported only within-subject inter-observer agreement correlations, (2) examined only twelve animals, and (3) confounded familiarity with the animals with collaboration among observers (i.e., the observers with greater amounts of familiarity with the monkeys also had more opportunities to discuss their ratings with one another); in addition (4) the difference between the two levels of agreement was neither statistically significant nor substantial. We strongly encourage personality researchers to examine familiarity further as a potential moderator of inter-observer agreement in nonhuman primates. If possible, such studies should use observers who are differentially acquainted with the animals but who have had few or no previous opportunities to share their impressions of these animals with one another.

Overall, the accumulated evidence suggests that observers tend to agree strongly in their appraisals of primates' personalities. According to the correlations obtained from studies of humans as a benchmark, Criterion 2 has been met.

Criterion 3: Assessments Must Predict Behaviors and Real-World Outcomes

The Issue

To have any value, personality traits must predict behaviors and real-world outcomes. Personality differences should reflect differences in behavior. Thus, one of Mischel's (1968) most pointed criticisms of personality was to argue that personality traits rarely predict behaviors or real-world outcomes at meaningful levels, with trait-behavior correlations rarely exceeding .30.

Evidence from the Literature on Humans

Mischel's critique prompted two major responses. First, some authors argued that Mischel's review of the literature was biased and ignored studies with stronger predictive correlations (Block, 1977; Hogan et al., 1977; Moskowitz, 1990). Moreover, some authors pointed out that trait-behavior correlations are considerably stronger when aggregates (composites) of behavior rather than single instances are used (Epstein, 1979). As we later explain in more detail, aggregates typically increase correlations because they reduce measurement error, making them more reliable than estimates based on single behaviors. As Epstein (1979) noted, isolated behaviors almost always tend to possess a high component of "situational uniqueness." Second, correlations in the region of .30 are not as weak as they may appear. Funder and Ozer (1983) showed that the effect size represented by a correlation of .30 is comparable to the effect sizes in some of the most widely known studies in social psychology. Moreover, Rosenthal and Rubin (1982) showed that such correlations correspond to quite substantial and practically significant outcomes. Rosenthal and Rubin developed a method for converting a correlation coefficient between two variables into a metric that reflects the percentage of cases with which one variable predicts the other. This "binomial effect size display" is computed as $.50 + r/2$ where r is the correlation coefficient. According to this method, a correlation of .30 between two variables is equivalent to being able to predict one variable from the other 65 percent of the time (where chance prediction is 50 percent), a 30 percent improvement over chance. To illustrate this point with an example from the primate literature, consider the correlation of .36 between the personality dimension of social competence and grooming behavior in vervet monkeys (McGuire et al., 1994; see below). This correlation is equivalent to predicting grooming behaviors from social competence scores for 68 percent of the cases across which the correlation was computed. Although far from perfect, this predictive capacity represents a substantial (36 percent) improvement relative to chance levels.

Evidence from the Literature on Nonhuman Primates

Evidence for concurrent and predictive validity is quite strong in the few primate studies in which personality measures were tested (see Gosling, 2001). In a study of vervet monkeys, McGuire et al. (1994) predicted three overt behavioral measures (exploration, grooming, and initiating submissive gestures) from personality dimensions that had been derived from personality ratings and behavioral codings of monkeys. Monkeys that performed many exploratory behaviors were high on the dimension labeled playful/curious ($r = .80$) but neither high nor low on the socially competent ($r = -.02$) and opportunistic ($r = .16$) factors. Similarly, the dimension labeled social competence positively predicted grooming ($r = .36$) and negatively predicted initiations of submissive behaviors ($r = -.53$), but these behaviors were not predicted by the playful/curious ($r = .12$ and $-.20$, respectively) and opportunistic ($r = .14$ and $.15$, respectively) dimensions. Thus, personality dimensions predicted conceptually relevant behaviors (i.e., convergent validity) but not unrelated behaviors (i.e., discriminant validity), supporting the construct validity of the dimensions.

Studies of rhesus monkeys have also predicted behavioral codings from personality measures at the trait and dimension levels (Stevenson-Hinde, 1983; Stevenson-Hinde, Stillwell-Barnes, & Zunz, 1980a). For example, ratings of aggressive predicted occurrences of hits, threats, and chases toward other group members ($r = .49$), and ratings of fearful predicted occurrences of fear grins and avoidance of others ($r = .57$). All six of the ratings assessed by Stevenson-Hinde and colleagues (1980a) yielded significant predictive correlations ranging from .45 for excitable to .73 for effective, which are well above Mischel's .30 barrier. At the dimension level, monkeys' personality scores predicted their behaviors in test situations. For example, male infants' confidence scores predicted the duration that they voluntarily spent in a cage out of reach of their mothers ($r = .68$). Other correlations ran counter to expectation, however; for example, infants with high excitability scores made fewer distress calls than infants with low excitability scores when mother and infant were removed from their colony ($r = -.72$).

The most systematic study to predict behavioral codings from personality ratings was performed on forty-two captive male rhesus monkeys (Capitanio, 1999). The goal of this investigation was to examine the predictive validity of personality ratings, that is, the degree to which the personality measures could predict subsequent, conceptually related outcomes. Specifically, Capitanio (1999) attempted to determine whether personality dimensions, identified in adult monkeys living in half-acre cages, predicted behavior in situations different from that in which the dimensions were originally derived at time points up to four and a half years after the original assessments. Although the results were somewhat mixed, they tended to support the predictive validity of the personality dimensions assessed. For example, sociability measured when the monkeys were in their natal groups was positively associated with earlier codings of the number of approaches initiated ($r = .38$) and received ($r = .27$), and was negatively associated with codings made approximately three years later of agonistic signaling (e.g., threats, $r = -.39$, and lipsmacks, $r = -.29$) in response to affiliative stimuli (videotapes of affiliative monkey behavior).

Personality traits have also been shown to predict important biological outcomes in primates. Capitanio and colleagues (1999) used personality dimensions to predict physiological measures associated with survival in monkeys who had been experimentally inoculated with the simian immunodeficiency virus (SIV). Corroborating research on extroversion in humans (e.g., Miller et al., 1999), Capitanio et al. (1999) found that sociability in animals was related to immune parameters. Specifically, they found that after inoculation with SIV, animals higher in sociability showed a greater immune response (i.e., a more rapid decline in plasma cortisol concentrations, elevations in the anti-RhCMV IgG response, and a decline in viral load) than did animals lower in sociability. These findings are important because they demonstrate that behaviorally based personality dimensions can predict criterion measures that are not based on behavior. As a consequence, it cannot be argued that the relations between personality traits and outcome variables are merely a tautological consequence of content overlap (e.g., see Nicholls et al., 1982).

Although the evidence is far from complete, the findings to date suggest that personality traits predict theoretically related behaviors and real-world outcomes. We therefore conclude that the third criterion has provisionally been met, although additional research examining laboratory and biological outcome variables is warranted to buttress this conclusion.

In summary, personality research in primates fares relatively well when held to the standards expected of human personality research. As we have already noted, much work remains to be done to fortify the foundations of this emerging area of research. With the preliminary groundwork complete, however, it is safe to proceed to our next set of questions, which focus on assessment methods and the personality constructs that characterize primates.

How Should Personality Be Measured in Primates?

Two main methods have been used to assess personality in primates: codings of the animals' behaviors and subjective ratings of traits. These two methods reflect different resolutions to the trade-off between quantifying personality in terms of objective behaviors and using humans to record and collate information more subjectively (Block, 1961; Stevenson-Hinde, 1983).

Some researchers have tried to adopt an objective stance by coding narrowly defined behaviors and assessing individual animals over a series of behavioral tests, such as coding an animal's response to a novel object or a new environment (e.g., Spencer-Booth & Hinde, 1969). These coding studies are listed as "coding" in the fourth column of Table 10.2. Other researchers have chosen to sacrifice the objectivity supposedly gained from such detailed behavioral codings in favor of obtaining ratings by people who are familiar with individual animals on such traits as confident, curious, and playful (e.g., Buirski et al., 1978); these rating studies are identified as "rating" in the fourth column of Table 10.2. Are these two methods equally good at assessing personality, or is one superior to the other?

One criterion by which a method can be judged is its reliability. Most questions about reliability have been raised in the context of trait ratings (Gosling, 2001), suggesting that primate researchers are concerned more about observer ratings than about behavior codings. With a few exceptions, researchers have largely ignored reliability issues when using behavioral-coding methods, perhaps because they assume that such codings lend themselves to high levels of agreement across observers. It would seem that many animal behavior researchers regard behavioral tests as intrinsically superior to global personality ratings (Gosling, 2001). Yet human research would suggest that animal researchers are incorrect, and that behavioral codings actually deserve the closest scrutiny. Human studies have shown that codings of behavior yield unstable estimates (Borkenau, 1992) and can be difficult to assess reliably (Gosling et al., 1998). Specific behaviors tend to have low cross-situational consistency and are notoriously difficult to assess and predict. The consistency of

Table 10.2 Summary of research on primate personality: species, sample size, variables, method, and focus of study

Species	Study	Sample size	Method	Focus of study
Chimpanzee	Bard & Gardner (1996)	29	Coding	Influence of early maternal environment on temperament and cognition
	Hebb (1949)	30	Coding	Determine feasibility of measuring complex individual differences in temperament
	van Hooff (1970, 1973)	25	Coding	Identify broad dimensions of social behavior
	Yerkes & Yerkes (1936)	29	Coding	Describe fear responses to novel objects
	Lilienfeld et al. (1999)	34	Rating and coding	Develop psychopathy measure and relate it to sex, personality variables, and behavior codings
	Buirski et al. (1978)	23	Rating[a]	Assess reliability of subjective ratings of personality; relate personality to dominance rank and sex
	Buirski & Plutchik (1991)	1	Rating	Test personality instrument's ability to detect psychological well-being
	Crawford (1938)	9	Rating	Assess reliability of subjective ratings
	Dutton et al. (1997)	24	Rating	Compare structure underlying personality ratings made by different observers
	King & Figueredo (1997)	100	Rating	Assess reliability of subjective ratings; identify broad personality dimensions
	Murray (1998)	59	Rating[b]	Identify broad dimensions of personality and relate them to group size and rearing conditions
	O'Connor et al. (2001)	145	Rating	Examine relations among personality dimensions, subjective well-being, and psychopathology
	Weiss et al. (2000)	145	Rating	Examine heritability of personality dimensions
Gorilla	Gold & Maple (1994)	298	Rating[b]	Identify broad personality dimensions; provide normative personality ratings
Vervet monkey	Fairbanks & McGuire (1993)	83	Coding	Relate early maternal protectiveness to protectiveness subsequent responses to novel objects

Species	Reference	Method	N	Purpose
	McGuire et al. (1994)	Rating	97	Assess reliability of subjective ratings; identify broad personality dimensions; relate personality to sex, age, social status, and group composition
Japanese macaque	French (1981)	Coding	3	Identify individual differences in play behavior
	Martau et al. (1985)	Rating[a]	14	Assess reliability of ratings
Stumptail macaque	Figueredo et al. (1995)	Rating[b]	13	Assess reliability, stability, and validity of personality dimensions
	Nash & Chamove (1981)	Rating	13	Identify broad personality dimensions; relate personality to dominance rank and self-aggression
Longtail macaque	Heath-Lange et al. (1999)	Coding	3	Describe developmental trends and cross-species differences in temperament
	Clarke & Lindburg (1993)	Coding	5	Describe cross-species differences in behavior
Liontail macaque	Clarke & Lindburg (1993)	Coding	5	Describe cross-species differences in behavior
Pigtail macaque	Heath-Lange et al. (1999)	Coding	7	Describe developmental trends and cross-species differences in temperament
	Reite & Short (1980)	Coding	21	Relate behavioral codings to physiological measures
	Westergaard et al. (1999)	Coding	30	Relate behavioral codings to biochemical measures
	Caine et al. (1983)	Rating[b]	10	Assess reliability of ratings; relate personality ratings to dominance rank and early separation experience
Rhesus macaque	Capitanio (1984)	Coding	12	Relate early rearing experience to subsequent social ability
	Chamove (1974)	Coding	91	Develop a device and system for scoring social behavior; identify broad behavioral dimensions
	Chamove et al. (1972)	Coding	168	Identify broad personality dimensions
	Freedman & Rosvold (1962)	Coding	7	Relate aggression and anxiety to sexual behavior
	Kalin et al. (1998)	Coding	50	Relate fearfulness behaviors to brain activity and cortisol levels
	Locke et al. (1964a, b)	Coding	12	Develop series of behavioral tests and assess their reliability; identify broad dimensions of social behavior
	Maestripieri (2000b)	Coding	10	Develop behavioral measure of emotionality and examine stability over time and across situations

Table 10.2 (continued)

Species	Study	Sample size	Method	Focus of study
Rhesus macaque (continued)	Spencer-Booth & Hinde (1969)	16	Coding	Develop series of behavioral tests and assess their reliability; examine relations between tests and temporal stability
	Stevenson-Hinde, Stillwell-Barnes, & Zunz (1980b)	25–31	Coding	Examine consistency in behaviors over time and across situations
	Suomi (1987)	12	Coding	Describe behavioral and physiological characteristics of reactivity; examine environmental and genetic influences on reactivity
	Suomi (1991)	24	Coding	Describe behavioral and physiological characteristics of reactivity; examine environmental and genetic influences on reactivity
	Suomi et al. (1996)	8	Coding	Examine consistency in behaviors over time and across situations
	Westergaard et al. (1999)	31	Coding	Relate behavioral codings to biochemical measures
	Stevenson-Hinde, Zunz, & Stillwell-Barnes (1980)	25	Rating[b] and coding	Examine consistency in behaviors over time and across situations; relate behaviors to broad personality dimensions
	Capitanio (1999)	42	Rating[b] and coding	Identify broad personality dimensions; relate personality dimensions to behavior codings over time and across situations
	Schneider et al. (1991)	23	Rating[b] and coding	Relate temperament ratings to behavioral codings over time and to environmental enrichment in infancy
	Bolig et al. (1992)	22	Rating[b]	Assess reliability and validity of subjective ratings of reactivity and personality; identify broad personality dimensions; relate personality to reactivity, dominance rank, age, and sex
	Capitanio et al. (1999)	18	Rating[b]	Relate personality to physiological indexes of disease progression
	Clarke & Snipes (1998)	48	Rating	Relate temperament to rearing experience, sex, and age

Species	Citation	Method	N	Purpose
	Stevenson-Hinde, Stillwell-Barnes, & Zunz (1980a)	Rating[b]	45	Assess reliability and validity of subjective ratings of personality; identify broad personality dimensions; examine temporal stability of dimensions
	Stevenson-Hinde & Zunz (1978)	Rating[b]	45	Assess reliability of subjective ratings of personality; identify broad personality dimensions
Baboon	Heath-Lange et al. (1999)	Coding	4	Describe developmental trends and cross-species differences in temperament
	Sapolsky & Ray (1989)	Coding	30–45	Identify individual styles of dominance; relate dominance style to endocrine levels and dominance tenure
	Virgin & Sapolsky (1997)	Coding	26	Identify individual styles of subordinance; relate subordinance style to endocrine levels and rise to dominance
	Buirski et al. (1973)	Rating[a]	7	Assess reliability of subjective ratings of personality; relate personality to dominance rank
	Martau et al. (1985)	Rating[a]	6	Assess reliability of ratings
Tufted capuchin	Byrne & Suomi (1995)	Coding	17	Examine development of social and exploratory behaviors and activity
Gothic-arch Squirrel monkey	Martau et al. (1985)	Rating[a]	5	Assess reliability of ratings
Roman-arch squirrel monkey	Martau et al. (1985)	Rating[a]	8	Assess reliability of ratings
Small-eared bushbaby	Watson & Ward (1996)	Coding	45	Assess reliability of behavioral codings; examine consistency of behaviors across situations; identify broad dimensions of behavior; relate dimensions to age, sex, handedness, and problem-solving

a. Rating instrument wholly or partially based on the Emotions Profile Index (Plutchik & Kellerman, 1974).
b. Rating instrument wholly or partially based on Stevenson-Hinde & Zunz (1978) instrument.

aggregated scores is usually much higher because the aggregation process tends to cancel out the random errors associated with measures of single instances (Epstein, 1979, 1983). Global ratings effectively rely on the benefits of aggregation by using human observers to implicitly combine specific behaviors across time and situations. Historically, data obtained from observers have been derided as subjective and inappropriate for the objective requirements of scientific measurement. Block (1961) argued convincingly, however, that aggregated observations composed of ratings by several independent observers meet the standards required of any measurement instrument; that is, aggregate scores composed of ratings by multiple observers are reliable and are largely independent of the idiosyncrasies of individual observers. With the psychometric odds stacked in favor of global ratings, substantial effort and resources are typically required to code behaviors as reliably as trait ratings (Moskowitz & Schwarz, 1982). It should be acknowledged, however, that the practical challenges raised by assessing behaviors may be less acute for animals than for humans. This is because animals' behavioral repertoires can be controlled (e.g., by keeping animals in enclosures) and recorded using ethological coding procedures (e.g., van Hooff, 1973) and new technologies (i.e., Noldus "Observer" and "Ethovision"; see *www.noldus.com*) developed especially to capture animal behavior (see also Martin & Bateson, 1993).

One primate researcher who struggled with the advantages and disadvantages of codings versus ratings was Donald Hebb (1946, 1949). Hebb (1946) described attempts at the Yerkes Primate Laboratories to implement a behavioral recording system in which specific behavioral acts of chimpanzees were recorded. After two years it became clear that the behavioral recording system was incapable of capturing distinctions that were important for describing interactions with chimpanzees. For example, before a researcher entered a cage with a chimpanzee, it was more useful to know simply whether his animal was "aggressive" than it was to be presented with a series of detailed behavioral recordings. Nevertheless, behavioral codings gained a strong foothold in the primate personality literature.

How reliable are behavioral codings? Many studies have used behavioral coding methods (see Table 10.2), but very few have reported reliability data (Byrne & Suomi, 1995; Chamove, 1974; McGuire et al., 1994; Watson & Ward, 1996). To make matters worse, the few studies that do include reliability data often report different reliability indices, making it virtually impossible to perform a meaningful quantitative summary. The findings are generally promising, however. For example, in a study of tufted capuchins (Byrne & Suomi, 1995), successive scorings of the same videotapes by the same observer elicited 90 percent agreement across the two scorings.

What about the reliability of ratings? One way to gauge reliability is in terms of inter-observer agreement. As noted earlier in our discussion of Criterion 2, global ratings of personality in primates show acceptable levels of reliability, in some cases exceeding those obtained in human personality research. Although the variability across studies suggests that reliability is not guaranteed, the findings are sufficiently consistent across traits, species, and independent studies to suggest that personality traits can be rated reliably.

As in all assessments, the strongest evidence is obtained from converging sources of data. Therefore, where resources permit, we recommend using both rating and coding methods. As discussed under Criterion 3 earlier, several studies have found evidence for convergence between ratings and behavioral codings. Where resources are limited, however, our review suggests that rating methods, performed with care and rigor by observers who know the targets well, provide the most efficient means of assessing personality in primates.

Nevertheless, the use of such ratings is not without potential hazards. In particular, such ratings almost inevitably raise the specter of anthropomorphism. Although the accumulated evidence we have reviewed strongly suggests that anthropomorphism (i.e., the imposition of observers' human-derived implicit models of personality onto animals) cannot fully account for either the co-variation among ratings or their external (e.g., biological) correlates, it is incumbent on investigators—particularly those who examine personality traits in primate species that have not been extensively studied—to demonstrate that ratings exhibit meaningful external correlates.

What Aspects of Normal and Abnormal Personality Have Been Identified and What Should We Measure?

Studies of Normal Personality

As noted in the introduction, the vast majority of nonhuman primate personality studies have focused on traits. But of the thousands of possible constructs in the domain of normal personality (Allport & Odbert, 1936), which should be examined? The field of animal personality currently resembles the early stages of human personality research, when there was widespread disagreement concerning which trait terms to adopt. More recently, however, some consensus has emerged in the human domain regarding a unifying framework: the Five Factor Model (FFM; McCrae & Costa, 1999). The FFM is a hierarchical model including five higher-order factors, which represent personality at the broadest level of abstraction. Each factor (e.g., extroversion versus introversion) summarizes several more specific facets (e.g., sociability), which in turn subsume a large number of even more specific traits (e.g., talkative, outgoing). The FFM framework suggests that most individual differences in human personality can be classified into five broad, empirically derived domains (John & Srivastava, 1999). The FFM has provided a common language for human personality researchers and has helped to unify a field that was in danger of disintegrating as each researcher invented his or her own system for classifying personality. We use the FFM here as a framework for organizing the literature on primate personality, although our adoption of this model should not be taken to imply that we endorse it over other personality taxonomies. We use the FFM for convenient heuristic purposes because this framework is (1) robust across samples and cultures, (2) widely accepted and researched, (3) readily applicable to primates, and (4) reasonably comprehensive in its coverage of major personality traits. It should be noted, however, that the FFM has not been universally accepted as the ultimate model of personality (see Block, 1995, and Loevinger, 1994, for criticisms of the FFM), and that other broad models have been proposed (e.g., Almagor et al., 1995; Cloninger, 1987; Eysenck, 1967; Gray, 1981; Watson & Tellegen, 1985; Wiggins, 1979), many of which share commonalities with the FFM (e.g., Watson et al., 1994).

Although it is difficult to make quantitative comparisons across primate studies, a number of dimensions have appeared repeatedly across multiple studies. Several studies identified a dimension reflecting an individual's characteristic reaction to novel stimuli or situations. This dimension has been referred to as emotionality, fearfulness, or reactivity (Higley & Suomi, 1989). Note, however, that the commonly used term "emotionality" is somewhat of a misnomer, as the construct refers only to negative, not positive, emotions. In the human FFM this dimension is most similar to the FFM dimension of neuroticism (versus emotional stability). In nonhuman primates, the dimension has been measured by behavioral measures such as scratching (Maestripieri, 2000b) and rated on traits such as insecure (Caine et al., 1983). Much of the work on rhesus monkeys by Suomi and colleagues (e.g., Bolig et al., 1992; Suomi, 1987, 1991) has examined this dimension, focusing on reactivity, in which "up-tight" and "laid-back" individuals are distinguished. According to Bolig et al. (1992), up-tight monkeys were apprehensive, excitable, insecure, and tense, whereas laid-back monkeys were confident, understanding, and equable. In gorillas, fearful animals were rated as high on measures of fearful, apprehensive, insecure, and tense and low on a measure of confidence (Gold & Maple, 1994).

A second recurring dimension, identified in several articles as exploration, is the propensity to seek out novel stimuli or situations, and resembles some elements of the human FFM openness dimension. This dimension has been measured by behaviors such as approach to novel objects (Hebb, 1949) and rated on traits such as inquisitive (Buirski et al., 1978). Watson and Ward (1996) identified four dimensions in small-eared bushbabies, one of which, curiosity, was characterized by a tendency to examine novel objects placed in an arena and to investigate a clear Plexiglas cage containing a live snake. Similarly, in one study of chimpanzees, one dimension was defined by the traits of inventive and inquisitive (King & Figueredo, 1997).

A number of studies also identified an aggression dimension. This dimension would be placed on the low pole of agreeableness in the human FFM. In primates it has been assessed by behavioral measures such as attacking another individual (Capitanio, 1984) and ratings of traits like aggressive (Martau et al., 1985). In ratings of rhesus monkeys, Bolig et al. (1992) identified an aggression dimension marked by high ratings on aggression and irritability. Chamove and colleagues (1972) identified a similar dimension, using behavioral coding techniques, whereby aggression was marked by hostile behaviors such as biting or grabbing another animal.

As noted earlier, the FFM factors consist of higher-order dimensions that subsume several lower-order facets. We next describe three domains that have been identified as independent dimensions in the primate literature, although in humans they would all be considered facets of extroversion.

The first of the extroversion-like dimensions differentiates individuals who seek out social interactions from those who prefer to remain solitary. In primates, this dimension, usually referred to as sociability, has been measured by behaviors such as proximity to other group members (Byrne & Suomi, 1995) and frequency of social encounters (Chamove et al., 1972) and rated on traits like solitary (Clarke & Snipes, 1998). In gorillas, extroverted animals were rated as high on traits such as sociable, playful, and popular and low on solitary (Gold & Maple, 1994).

Another dimension that in humans would be largely subsumed by FFM extroversion refers to an animal's general activity level. In primates it has been measured by behaviors such as the amount of enclosure covered by the animal's roaming (Reite & Short, 1980) and rated on traits such as active (McGuire et al., 1994). In chimpanzees, an activity dimension was defined in terms of four behavioral-coding categories tapping general levels of body movement and activity (Bard & Gardner, 1996).

Several studies also identified a dominance or assertiveness dimension. In humans this dimension would also be a facet of FFM extroversion. In primates, this dimension was usually related to the individual's rank in the dominance hierarchy (e.g., King & Figueredo, 1997; Sapolsky & Ray, 1989). In rhesus monkeys, a dimension labeled confident was marked on the positive pole by traits such as confident, effective, and strong, and on the negative pole by apprehensive and subordinate (Stevenson-Hinde, Stillwell-Barnes, & Zunz, 1980a).

Studies that focus on characterizing particular dimensions are crucial to establishing the construct validity of these dimensions. To understand which dimensions are to be found in which species (i.e., personality structure), however, exploratory studies with a broader focus are needed. Gosling and John (1999) reviewed nineteen such studies (eight based on primates), using the FFM as an organizing framework. The FFM dimensions of extroversion, neuroticism, and agreeableness showed considerable generality across species. Of the eight primate studies, seven identified a factor closely related to extroversion, capturing dimensions ranging from surgency in chimpanzees to sociability in rhesus monkeys. Naturally, the overt manifestations of these personality dimensions differ across species. For example, whereas the human scoring low on extroversion stays at home on Saturday night, or tries to fade into a corner at a large party, the rhesus monkey scoring low on sociability initiates few fear grins and receives grooming presentations (Capitanio, 1999).

Factors related to neuroticism appeared almost as frequently, capturing dimensions such as fearfulness and excitability. Factors related to agreeableness appeared in six of the eight studies, with dimensions like affinity representing the high pole and aggression and hostility representing the low pole. Separate openness dimensions were identified in only two studies and combined components of curiosity-exploration (interest in new situations and novel objects) and playfulness (which is associated with extroversion when social rather than imaginative aspects of play are assessed). Chimpanzees were the only species characterized by a separate conscientiousness factor, which was defined more narrowly than in humans. Nevertheless, this dimension included the lack of attention and goal-directedness, and the erratic, unpredictable, and disorganized behavior typical of the low pole. Dominance emerged as a clear separate factor in five of the eight studies, and a separate activity dimension was identified in only one study.

Studies of Abnormal Personality

Up until now we have focused largely or entirely on studies of normal-range personality traits in primates. Nevertheless, one potentially important but heretofore largely neglected domain concerns the assessment and correlates of psychopathology in primates (see Chapter 16). Some forms of psychopathology (e.g., generalized anxiety disorder; see Ruscio et al., 2001) most likely represent extremes of normal-range personality traits (including the dimensions of the FFM), whereas others (e.g., schizophrenia; see Golden & Meehl, 1979; Lenzenweger, 1999b) may instead represent categorical (qualitative) differences from normality. In either case, studies of psychopathology in primates may elucidate the similarities and differences between the correlates of normal and abnormal personality traits across species.

Lilienfeld and colleagues (1999) examined the applicability of the construct of psychopathic personality (psychopathy) to chimpanzees. In humans, the construct of psychopathy refers to a constellation of personality traits including superficial charm, guiltlessness, callousness, dishonesty, manipulativeness, risk-taking, and poor impulse control (Cleckley, 1941/1988). Lilienfeld et al. (1999) developed a twenty-three-item Chimpanzee Psychopathy Measure (CPM) that consisted of items believed to represent analogs of human psychopathy in chimpanzees (e.g., confident, deceptive, greedy, opportunistic, risk-taking). The CPM was completed by six observers on a sample of thirty-four chimpanzees at the Yerkes National Primate Research Center Field

Station. Lilienfeld et al. found that the CPM exhibited moderate to high inter-observer reliability, suggesting that observers tend to agree on the personality features ostensibly constituting psychopathy in chimpanzees. In addition, the CPM correlated positively and significantly with measures of extroversion and agreeableness from the FFM (the correlation with neuroticism was positive but nonsignificant; measures of conscientiousness and openness were not administered) and with independently rated observational measures of agonism, sexual activity, daring behaviors, teasing, and silent bluff displays derived from a behavioral ethogram. The CPM also correlated significantly and negatively with observer-rated generosity. All of these findings accord broadly with the conceptualization of psychopathy in humans. In addition, paralleling findings in humans (Lykken, 1995), males scored higher on the CPM than females, although this difference was only marginally significant. The findings of Lilienfeld et al. provide preliminary evidence that the construct of psychopathy is applicable to nonhuman primates, although further validation of the CPM using laboratory and biological indices is clearly necessary.

O'Connor and colleagues (2001) examined the relevance to chimpanzees of several psychopathological constructs in the fourth edition of the *Diagnostic and Statistical Manual of Mental Disorders* (*DSM-IV*; American Psychiatric Association, 1994), namely, major depression and generalized anxiety disorder (which are classified as Axis I disorders, i.e., major mental disorders) and antisocial, borderline, and schizoid personality disorders (which are classified as Axis II disorders, i.e., more trait-like disorders). In addition, O'Connor et al. obtained ratings on a measure of subjective well-being (SWB), which exhibited high inter-observer reliability. Observers completed measures of these characteristics on 145 chimpanzees in the ChimpanZoo project. The authors found that the measures of major depression, generalized anxiety disorder, and borderline personality disorder were negatively and significantly correlated with SWB. Interestingly, antisocial and schizoid personality disorders were not significantly correlated with SWB, consistent with observations that humans with these personality disorders tend to be relatively immune from high levels of negative affect (e.g., American Psychiatric Association, 1994; Lykken, 1995). O'Connor et al. also reported provocative sex differences on several of these measures. For example, adult females scored higher than adult males on the measures of major depression, generalized anxiety disorder, and borderline personality disorder, whereas adult males scored higher than adult females on the measure of antisocial personality disorder. All of these sex differences are consistent with those found in humans (American Psychiatric Association, 1994). Interestingly, there were no sex differences on the major depression measure among juveniles, which is consistent with findings indicating that in humans the rates of depression among prepubertal boys and girls are essentially equivalent (American Psychiatric Association, 1994). Finally, O'Connor et al. found that their measures of psychopathology exhibited several theoretically meaningful ethogram correlates. For example, the measures of major depression and generalized anxiety disorder were positively correlated with independently rated submissive behaviors, while the measures of antisocial and borderline personality disorder were positively correlated with independently rated agonistic behavior.

The systematic study of abnormal personality among nonhuman primates is clearly in its infancy, and a great deal more research incorporating diverse external correlates (especially laboratory and biological measures) is required before the applicability of these findings to the human literature can be ascertained. In other words, the extent to which constructs such as psychopathy and major depressive disorder in nonhuman primates are isomorphic with those in humans remains unclear (see also Chapter 16). In addition, as in the human literature, it is not entirely clear which forms of psychopathology might be adequately captured within an existing dimensional model of normal personality traits (e.g., the FFM). Moreover, it is not known how these disorders might manifest themselves differently across species. For example, whereas depression often manifests itself as reduced social contact in humans, it may manifest itself in a highly specific way (e.g., as reduced grooming frequency) in many primates. To establish these potential correspondences, investigators will need to adopt a construct validational approach (Cronbach & Meehl, 1955), which, as noted earlier, involves comparing personality measures in terms of their situational, behavioral, and physiological correlates across species.

A Caveat

We end this section with a note of caution regarding cross-species comparisons of both normal and abnormal personality. The aforementioned comparisons were made on the basis of the rating or behavioral-coding items that characterize each dimension. Comparing dimensions across species is a perilous task, however, because it is often difficult to determine whether apparently similar manifest behaviors reflect similar underlying constructs. For example, is a given capuchin monkey's gesture, which resembles a gesture that would be considered friendly in chimpanzees, really a friendly gesture? By virtue of being in the same phylogenetic group, primates are closely related, so it is reasonable to posit that similar mechanisms link latent constructs to manifest expressions across primate species. There are also wide variations among primate species, however, in terms of social structures, habitat, climate, predation risk, and so on, so it is also reasonable to posit that at least some links between underlying constructs and manifest expressions vary across species. The challenge of establishing the cross-species equivalence of normal and abnormal personality characteristics in primates is somewhat analogous to the challenges encountered in research on the universality of human emotions (Russell, 1994). For example, a cross-cultural emotion researcher may be faced with the task of establishing whether apparently similar emotional expressions in two cultures are associated with equivalent underlying emotions. Likewise, primate researchers are faced with the question of whether superficially similar behaviors in two species are associated with isomorphic personality characteristics in these species. It may be profitable for animal personality researchers to adopt the methods employed by cross-cultural emotion researchers and compare personality traits across species in terms of their situational, behavioral, and physiological correlates. Although one can never demonstrate with absolute certainty that two different behaviors in two different species correspond to the same latent trait, findings demonstrating that these behaviors exhibit similar correlates across multiple independent domains lend considerable weight to this possibility.

Implications and Recommendations

Instrument Development/Item-Generation and Measurement Guidelines

It is important that the nascent field of primate personality develop a common language rather than repeat the youthful mistakes made in the field of human personality. Researchers should carefully select which variables they study to ensure that they are both comparable and comprehensive.

To ensure comparability, primate researchers should use a standardized set of characteristics translated into species-typical behaviors. Researchers should not render their interesting findings incommensurate by using idiosyncratic terms. This issue is of particular concern in animal research, which is more vulnerable than human research to the danger of using inconsistent trait concepts because there is typically much more variability across species (e.g., between chimpanzees and black lemurs) than across groups of humans (e.g., between French people and Americans). Nevertheless, researchers should make every effort to ensure that their item pool is as comparable as possible with that of other investigators by using variables that have been used in other studies.

To ensure comprehensiveness, the range of personality traits studied in a species must fully represent the behavioral repertoire of that species. Research on animal personality often neglects to use systematic item-generation procedures, thereby exposing animal studies to the danger of overlooking important features of a species' personality. Too often, item pools are generated on the basis of the intuitions of one or two people who are familiar with the animals. Care should be taken to account for all the relevant domains of individual differences. For example, interactions with humans often play a large part in the lives of captive animals, but this realm of behavior is frequently overlooked in research on captive populations. Therefore, in addition to the within-species social interactions that are typically studied in animal personality research, studies should also examine facets of human-animal interactions.

Inevitably, a natural tension exists between the demands of comparability and comprehensiveness. In capturing the idiosyncrasies of a particular species, researchers may be forced to adopt traits that are not applicable to other species. A balance should ideally be reached in which a basic set of standardized descriptors (operationalized in species-appropriate terms) is supplemented by important species-specific descriptors. We concur with Loevinger's (1957) recommendation that test developers should be overinclusive in the early stages of item development. Only in this fashion can the investigator be reasonably certain that important content domains are not omitted. Moreover, this approach ultimately permits the investigator to learn not only which traits are relevant for a species but also which traits are irrelevant (see also Clark & Watson, 1995), with the result that discriminant validity is built into the test-construction process and the boundaries of still poorly understood constructs can be clarified. Because the field of animal personality is still in its infancy, we urge researchers to err of the side of overinclusiveness at this stage.

In addition, we advise researchers to follow several basic measurement guidelines to optimize the reliability of ratings (John & Benet-Martinez, 2000). Specifically, studies of animal personality should (1) use sufficient numbers of observers to obtain reliable estimates of the constructs being rated and, where inter-observer agreement is low, add more observers (Block, 1961); for example, if one conservatively estimated that pairwise inter-observer agreement would be around .30, then (using the Spearman-Brown prophecy formula) one should obtain an alpha reliability of .72 for a composite of six judges; (2) where possible, examine easily observable traits or behaviors (Funder & Dobroth, 1987; Gosling et al., 1998; John & Robins, 1993) and, where low-observability traits must be examined, add observers to improve reliability (according to the Spearman-Brown prophecy formula); (3) use rating scales with a range that allows observers to express the full breadth of variability observed, but since Likert-type scales with too many responses can be susceptible to response biases involving extremity of responding (e.g., Clark & Watson, 1995), the range of the rating scales should not exceed the range of detectable variability; (4) ensure that there are sufficient items to provide a reliable estimate of each anticipated dimension (i.e., start with at least three or four items per dimension, preferably more; Goldberg & Digman, 1994); and (5) use item-analytic or factor-analytic techniques to ensure that the items selected assess a common construct. In addition, to obtain stable reliability estimates, researchers should collect ratings on as many animals as is feasible. In short, animal personality researchers should adopt the principles of personality assessment that have been developed over many years of research on human participants (see, e.g., Kenrick & Funder, 1988, for further discussion).

Distinguish between Within-Species and Cross-Species Differences

There are two ways to view personality differences in comparative research: within-species differences and cross-species differences. It is important to distinguish between these two views because they are based on different sources of information and have radically different research uses and implications. Within-species (or intraspecies) comparisons are what people normally mean when they talk about personality comparisons, such as comparing individual

X's level of friendliness to individual Y's, where X and Y are members of the same species. Cross-species (or interspecies) comparisons compare the personality characteristics of one species as a whole with the personality characteristics of another species. Cross-species comparisons can be made in terms of a trait common to both species (e.g., Are lion-tailed macaques more curious than long-tailed macaques? Clarke & Lindburg, 1993) or present in only one (e.g., Is curiosity present in chimpanzees but not in orangutans?). In human personality research, the distinction between within-species and cross-species comparisons does not arise because studies are performed only on a single species. In the realm of animal personality, however, the distinction is important because an animal's personality could be characterized one way using a within-species framework but another way using a cross-species framework. For example, using a within-species framework one might characterize a bonobo as unsociable (i.e., it is low for a bonobo), but using a cross-species framework one might characterize the same animal as sociable (i.e., by virtue of being a member of a sociable species).

Note that these two frameworks are independent. That is, one could discover cross-species differences in the absence of within-species differences and vice versa. We draw attention to this distinction because the choice of comparison determines which kind of research question can be addressed. A within-species framework would be appropriate to examine how biological, social, and environmental factors shape an individual's personality. In contrast, a cross-species framework would be appropriate for examining the origins and adaptational significance of specific traits (Hodos & Campbell, 1969; Lesch et al., 1997).

Evolutionary Implications

Scientists assume that natural selection is the mechanism underlying the evolution of most anatomical, physiological, and now, with increasingly accepted evidence, cognitive abilities in humans and other species. Nevertheless, two of the domains that investigators have traditionally been reluctant to include among these dimensions are personality and emotion. Of these two, there appears to be a greater acceptance of continuity between humans and non-humans for emotions than for personality traits. This may, on the one hand, be because the external signs of many emotions are so overtly similar across humans and other species, especially other primates (see Chapter 11). On the other hand, personality is typically thought of as a rather complex dimension that is attributable to humans but seems a farther stretch for other species. Anthropomorphism appears to be a major threat in the realm of personality. Moreover, the nature of personality seems to be a central component of "who we are" as individuals. Therefore, it may be difficult to acknowledge the role of evolutionary processes in this intimate aspect of our identity. Thus, the insights that can come from understanding the evolution of personality in both humans and nonhumans have not always been readily accepted. Nevertheless, there is no inherent reason to accept the premise that some traits, such as morphological characteristics, are evolved by natural selection but that others, such as personality traits, are not. Darwin (1872) argued for continuity between humans and other species in emotions, setting the stage for the realization that emotional, cognitive, and social evolutionary processes play a role in the evolution of personality in humans and other species. The field of evolutionary psychology has used evolutionary arguments and theories to explore the origins of human psychology, including behaviors well within the realm of personality (Barkow et al., 1992; Buss, 1984, 1991; Daly & Wilson, 1999). In this chapter we have demonstrated the important contributions that studies of human and primate personality can make to each other.

Perhaps the most compelling argument for the validity and usefulness of comparing human and nonhuman primate personality traits is the principle of evolutionary parsimony. According to this principle, we should avoid postulating very different mechanisms for very similar behaviors and traits in closely related species unless compelled by strong evidence (Flack & de Waal, 2000b). It is more likely than not that similar behaviors and traits across closely related species are based on similar rather than different underlying cognitive, emotional, and social processes. Evolution tends to be economical in its use of mechanisms of change in morphological and physiological traits. Therefore, there is every reason to accept evolutionary economy in the realms of cognition, emotion, and personality. This position does not imply that every behavior or trait that appears to be similar across closely related species is identical in every dimension. As Flack and de Waal (2000b) have argued, however, the burden of proof lies with the less parsimonious explanation of dissimilarity in mechanisms. It is probable that the similarities (and differences) that exist in personality traits across human and nonhuman primates are underwritten primarily by similar mechanisms that are expressed in species-specific ways at various levels and to various degrees. We propose that this view is the most parsimonious and plausible for explaining the bases of personality (and other psychological) traits across species.

Personality research in primates is still at a relatively nascent stage of development. The relatively few studies that have been conducted thus far are promising, however. The research literature suggests the following conclusions:

1. Personality research in primates is beginning to meet the basic standards expected of personality research in humans. That is, personality ratings in primates do not seem merely to reflect the implicit theories of observers projected onto animals (i.e., anthropomorphism). In addition, these assessments are reliable, at least as gauged by inter-observer agreement. Finally, these assessments show evidence of validity in terms of predicting behaviors and real-world outcomes such as susceptibility to disease progression. Nevertheless, considerably more research on the external (e.g., laboratory, physiological) correlates of personality ratings in primates is clearly necessary.

2. It is feasible to measure personality traits in nonhuman primates using either rating or coding methods, though rating methods generally seem to be more efficient. Although this finding may appear counterintuitive to some primate researchers, it reflects the importance of aggregation over time and situations, a principle that has been well established in the psychometrics literature.

3. Studies of normal personality characteristics have identified a number of recurring dimensions, many of which appear similar to those identified in human research (including dimensions identified from the FFM). The very few studies that have examined abnormal personality have identified promising primate analogs of several forms of human psychopathology. In both normal and abnormal domains much work remains to be conducted, and great caution must be exercised in determining the cross-species comparability of constructs.

4. When devising assessments of personality in primates, researchers should profit from the measurement guidelines developed by the older field of human personality. In particular, researchers should be certain to use trait measures that broadly represent the behavioral repertoire of each species, assess each trait with multiple items that are then aggregated to enhance reliability and validity, ensure that these items assess the same latent construct, and use multiple observers whose ratings are then aggregrated.

5. Researchers should be certain to distinguish between within-species and cross-species comparisons, since they are based on different sources of information and have radically different research uses and implications.

6. Like the fields of cognition, morphology, physiology, and emotion, personality should be addressed within (and informed by) an evolutionary framework.

Taken together, the approximately fifty studies of primate personality demonstrate the viability of assessing personality in primates and hint at the potential benefits of doing so. We anticipate that work in this area will continue to flourish and provide new insights into personality in both human and nonhuman primates.

11

Emotions and Behavioral Flexibility

Filippo Aureli and Andrew Whiten

Emotions are at the core of people's experience. Sometimes they are simply in the background; at other times they seem to pervade our entire consciousness. But regardless of whether we are aware of them or not, emotions affect the way we behave. Their pervading role in our lives is reflected by the tremendous number of words that we use to describe emotions. Johnson-Laird and Oatley (1989) were able to gather a corpus of no fewer than 590 English terms describing and differentiating the complexities of human emotions.

Several influential psychologists and biologists of the late nineteenth and early twentieth centuries considered emotion of great importance to understanding behavior. The empirical study of emotion, however, has been infrequently pursued until recently. Reasons for this neglect are multiple, ranging from behaviorism's rejection of studying internal states to the scarcity of appropriate methods, from lack of an evolutionary perspective to disregard for the notion of homeostatic regulation in the study of brain and mind (Damasio, 1998, 2000; Davidson & Cacioppo, 1992).

The situation has dramatically changed since the early 1980s. The number of scientific articles published on emotions has grown enormously during this period. Further evidence for the explosion of research on emotion can be found in the many books published on the topic, the creation of at least two specialized journals, the organization of several conferences and symposia to cover various aspects of emotions, and the establishment of at least one international society devoted explicitly to the study of emotion.

The explosive volume of recent research on emotion is partly due to the eventual acceptance within psychology of the significance of internal states in explaining behavior, the adoption of an evolutionary perspective, and the development or refinement of methods to study emotional expression (Cacioppo & Gardner, 1999; Davidson & Cacioppo, 1992). Emotion is now recognized to be of importance in many subdisciplines of psychology, including developmental, comparative, evolutionary, social, health, and neuropsychology. Research effort in the neurophysiological basis of emotion has even resulted in the emergence of "affective neuroscience" as a new discipline (Davidson & Sutton, 1995; Panksepp, 1998).

The popularity of emotion research is nevertheless biased taxonomically. The majority of research effort has focused on humans. Terms with emotional connotation are still rarely found in studies of animal behavior. It remains uncommon, for example, to find descriptions of the emotional repertoire of a species, whereas it is customary to publish detailed lists of behavior categories. A notable exception is Goodall's (1986) study of wild chimpanzees, which lists terms such as *apprehension, fear, distress, annoyance, anger, enjoyment,* and sexual and social *excitement.* Another is found in Smuts's (1985) study of olive baboons, in which she uses terms with emotional connotation such as *jealousy, ambivalence, flirtation, trust, affection,* and *grief* in describing the special relationships between adult males and females and their kin. Although the recent development of affective neuroscience has produced a boost in research using animal models, relatively few studies have systematically investigated behavioral phenomena related to emotions in nonhuman animals, including nonhuman primates.

We suspect that the scarcity of research on animal emotions is largely due to the fact that emotions have traditionally been described in terms of subjective experiences. Most students of animal behavior have thus preferred to avoid emotional terms like *afraid* and *jealous,* seeing them as difficult or impossible to substantiate, and often overly anthropomorphic. They have attempted instead to relate behavioral sequences to contextual contingencies and motivating factors (van Hooff & Aureli, 1994). The principle was that only behavior could be seen, so only behavior could be valid subject matter for science. The closing decades of the twentieth century, however, witnessed a major reappraisal of this position, one of the most obvious symptoms being an increasing willingness to study seriously the animal "mind" and investigate cognitive phenomena underlying surface behavior (Ristau, 1991a; Shettleworth, 1998; cf. Kennedy, 1992).

We suspect that the impossibility of ascertaining whether nonhuman animals have subjective emotional experiences has mistakenly been seen as preventing the scientific investigation of animal emotion (cf. Panksepp, 1989). The degree of subjective feeling associated with emotion may vary, depending on the extent to which a species possesses the capacity of conscious awareness (LeDoux, 1995). The fact that we may never be able to measure the subjective experience of other animals should not keep us from studying animal emotions. After all, we have difficulty in knowing whether our own subjective emotional experience is similar to that of other human beings, yet solid scientific research on human emotions has nonetheless progressed greatly in recent times (see references above and below). Although some (most notably Griffin, 1981, 1992) argue that aspects of animal consciousness should be taken as a main object of scientific study, we think that there are fundamental obstacles to achieving this; instead we urge optimism about tackling other aspects of animal emotion.

Subjective feeling, though important, is only one aspect of emotion (Davidson & Sutton, 1995; LeDoux, 1995, 1996; Öhman, 1993; cf. Damasio, 2000). According to MacLean's (1952) evolutionary approach to brain anatomy, emotions cannot be assumed to be uniquely human traits because most of the brain structures involved in emotions are essentially the same in all mammals and perhaps in all vertebrates (LeDoux, 1996; Panksepp, 1998). To take the example of fear, emotional mechanisms functionally link the perception of danger with the response of finding protection in both humans and other animals. MacLean's conceptual framework has been strongly supported by neurophysiological evidence (Brothers, 1990; Davidson & Sutton, 1995; LeDoux, 1996; Levenson, 1992; Panksepp, 1989, 1998; Rolls, 1999). Such inferences about emotion need not rely on conscious introspective report (Davidson & Sutton, 1995); we can use the term *emotion* without implying the conscious feeling that humans associate with it. Other animals may experience it as well, but it remains unclear if, or how, we shall ever know.

In this chapter we suggest the principal ways in which the topic of animal emotions can be approached as a valid and rigorous subject for scientific enquiry. We note that when we talk of emotion in the familiar human case, we find that our emotion terms tend to refer to both behavior and states of mind: mention of the emotion "x" (e.g. "anxiety," "anxious") can typically refer to both *being x* (being anxious) and *acting x* (appearing anxious). Our perspective here is that we can legitimately set to one side the deeply problematic question of how the animal feels, yet nevertheless study with good scientific rigor its emotional expression (*acting x*, i.e., acting anxiously) and even its emotional state (*being x,* i.e., being anxious). (Note that it could be objected that our setting aside of the issue of what emotions feel like to other animals, in order to deal with questions we believe more scientifically tractable, appears to abandon responsibility for animal welfare. In other words, one could argue that if we cannot establish if, and what, an animal feels, we have no basis for informed decision making about its welfare. We suggest taking a conservative position for practical purposes when we need to make decisions related to animal welfare. The morally responsible attitude should be to assume that animals may feel the emotions that on all the other objective grounds [i.e., the correlates of emotion states] we ascribe to them, even if we cannot confirm those feelings in the same scientific manner.)

Theories of Emotions

The topic of emotion has been approached from many different angles by scientists with very different backgrounds, so that theoretical issues and empirical data have been interpreted in a variety of ways. This variation has led to the development of a rather large number of theories of emotions. Our brief review focuses on those theories that have particularly important implications for the comparative study of emotions in human and nonhuman primates.

Given his interest and the historical context, it is not surprising that Darwin (1872) was the first to present a substantial evolutionary analysis of emotions. He focused mainly on emotional expression. Darwin was convinced that animals and humans express emotion in similar ways. Their facial, vocal, and postural expressions convey the underlying emotion, and by doing so they increase the chance of survival and ultimately individual fitness. Darwin's perspective has led to research on the phylogeny of human facial expressions which has found similarities with the displays of nonhuman primates, thus reconstructing various precursors in our common ancestors (reviewed in more detail later in this chapter).

Over two thousand years ago, Greek physicians already linked human emotions with physiological processes. They believed that bodily fluids were associated with specific emotions. In more recent times, James (1884) emphasized the role of physiology. He argued that peripheral physiological changes in response to certain stimuli produce emotions instead of being a product of emotions. According to James, distinct emotions are experienced through the perception of unique physiological changes. This theory generated an extensive research effort in search of physiological discriminators of emotions. Early negative results prompted Cannon (1929) to criticize James's theory (most often known as the James-Lange theory) by admitting that the same physiological change (e.g., an increase in heart rate) occurs during very different emotional experiences and also during non-emotional events (e.g., while running for sport). Although no single measure can distinguish each emotion from another, experiments have provided substantial evidence for emotion-specific physiological patterns (Collet et al., 1997; Ekman et al., 1983; Levenson, 1992). Other scientists have pointed out that evidence for emotion-specific physiological activity is more likely to be found in brain circuits than in the peripheral nervous system (LeDoux, 1994; Panksepp, 1994).

Another controversy concerns the relative importance of the components of the peripheral nervous system in the emotional process. Although Darwin (1872) emphasized the role of bidirectional communication between brain and peripheral nervous system via what is now called the vagus nerve, most research followed Cannon's assumption that emotions reflect activation of the sympathetic nervous system for "fight or flight" responses. The vagus nerve, however, is the major component of the parasympathetic nervous system. Evidence now shows that the latter system is at least as important as the more widely studied sympathetic nervous system in regulating emotional experience (Porges, 1991, 1997). The parasympathetic nervous system is involved in the control of facial expression, breathing, and vocalizations, as well as of organs that are involved in emotional experience such as the heart, bronchii, and salivary and lacrimal glands. Building on this and other evidence, the polyvagal theory proposes that the evolution of the nervous system provides neural substrates for three main emotion systems in mammals. In any one emotional episode each system may be involved in turn, following a hierarchical response strategy in which the most evolutionarily recent modifications are employed first (Porges, 1997). The polyvagal theory does justice to Darwin's intuition of parasympathetic involvement in emotional experience while maintaining a role for the sympathetic branch of the nervous system.

Alternative perspectives to physiology-centered theories have held that discrete emotional experiences derive from cognitive appraisals of the perception of undifferentiated physiological arousal (Mandler, 1975; Schachter & Singer, 1962). The ambiguity of the sensory information motivates a mental search for an emotional label for the perceived physiological state. There is therefore a cognitive evaluation of the perceived physiological change. The appraisal process is likely to combine characteristics of the individual and the situation to produce a variety of emotions (Frijda, 1986; Lazarus, 1991; Smith & Pope, 1992). Debates have continued about the actual role of appraisal and the degree of cognitive involvement in processes that lead to emotional experience (Frijda, 1993; Izard, 1993; Zajonc, 1980). The level of complexity of the attributional processes needed for appraisal is of course of great relevance for the study of animal emotions because it questions directly whether animals other than humans possess the cognitive abilities required for such appraisal (cf. Cacioppo et al., 1992). A position on this issue fully in line with the view that nonhuman animals experience emotion is the one proposed by Frijda (1993). While fully acknowledging the possible involvement of complex cognitive elaboration, Frijda suggested that emotion could adequately result from non-conscious, simple appraisal bound to basic processes of information intake and action monitoring. Similarly, Öhman's (1993) information-processing model views the role of cognitive involvement in the production of emotion as variable and critically influenced by a non-conscious expectancy system that biases stimulus appraisal depending on individual personal memories. Furthermore, Izard (1993) proposed four types of emotion-activating systems, three of which involve non-cognitive information processing.

There is also still open debate about the degree of specificity in emotion-linked physiological responses and the subsequent implications for theories of emotion (Cacioppo et al., 1992, 1993; Davidson & Sutton, 1995; LeDoux, 1996; Levenson, 1992). There has been at least an attempt to reconcile the various positions into a single general model. The Somatovisceral Afference Model of Emotion specifies the mechanisms and conditions for (1) how the same pattern of physiological changes leads to discrete emotional experiences and (2) how different patterns of physiological change lead to the same emotional experience (Cacioppo et al., 1992). According to the model, a stimulus leads to variable changes in physiological activity with ambiguous activation (i.e., partially differentiated activation specific to multiple emotions) falling between emotion-specific patterns of activation and undifferentiated activation along a continuum of physiological patterns. Similarly, the extent of cognitive elaboration of the somatovisceral afference required to produce emotion ranges from simple informational analyses such as pattern recognition to much more complex attributional analyses. The more complex the analysis, the longer it takes to be completed and produce the emotional experience (Cacioppo et al., 1992).

The emphasis on brain systems underlying emotional experience (LeDoux, 1996; Panksepp, 1998; Rolls, 1999) can be viewed as complementary to the theories that instead stress the importance of peripheral physiological changes and cognitive appraisal. Progress in affective neuroscience has allowed the linking of certain brain regions and certain neurotransmitters to particular emotions (see below). In addition, this progress has prompted a debate about the nature of the involvement of separate brain subsystems in the production of certain emotions and the role of reinforcing stimuli and their appraisal in such processes (Gray, 1975, 1990; Rolls, 1990, 2000; and commentaries below).

Emotions as Intervening Variables

All the theories reviewed above, though developed mainly to explain human emotions, pertain in general to nonhuman animals as well. In practice, however, not all of them can be easily applied to the study of animal emotions. We can distinguish two main approaches in which animals' emotional states and emotional expressions can be investigated.

The first is based on proposals developed by one of us for thinking about when it becomes appropriate to attribute a "state of mind" (be it emotion, thought, or belief) to another individual, whether animal or human. Whiten (1993a, b, 1996) suggested that such states have the status of 'intervening variables," a construct used by earlier researchers to explain complex webs of causation, such as the way in which many different yet related aspects of drinking behavior (e.g., effort to obtain drink, amount drunk, tolerance of impurities) can be caused by many different factors (e.g., time since last drink, salt load). The causal linkage between a multiplicity of such independent variables and the many dependent variables they may influence can be explained most economically by positing a central "intervening variable"—in this case called thirst (Miller, 1959; Figure 11.1). Although we may not be able to see an animal's state of thirst directly, we can infer it on the basis of observable phenomena, and having done so we can well predict how the animal will behave in a variety of contexts. Note that we can do so without having to decide whether thirst *feels* the same in ourselves and the animal of interest. This kind of approach to identifying phenomena that are not directly observable is normal and common scientific practice; we are simply suggesting that the same logic should apply to the recognition of animals' emotional states, such as anger and fear (cf. Hinde, 1972).

The second approach is to rely instead on an objectively measurable variable, such as a physiological correlate, known to be intimately linked to a specific emotional state in humans. For example, if a psychoactive drug influences observable phenomena labeled "anxiety" in humans, a similar effect of the drug in another species offers evidence of a state most aptly labeled "anxiety" as well.

These two approaches to the objective study of emotions—inferring intervening variables and identifying a specific physiological correlate—seem at first sight to illustrate a clear dichotomy. They may be better thought of, however, as extremes of a continuum, and both can gain from our combining aspects of the other approach. Thus, on the one hand, physiological phenomena may be incorporated into the web of identifiable causal relationships that justifies ascription of a particular state of emotion, that has the explanatory power of an intervening variable. On the other hand, direct inferences from physiological variables become the more powerful when an intercorrelated array of them is identified rather than a single one (cf. Cacioppo et al., 1993), and this really reflects the logic of the "intervening variable" perspective.

The integration of the two approaches is also urged from other theoretical grounds. Economical representations of events (such as intervening variables), though elegant, may not reflect reality if too simplistic. Biological solutions do not emerge ex novo as the most parsimonious possibilities, but rather develop from preexisting structures and therefore are constrained by their evolutionary past. This is a further reason why biologically relevant

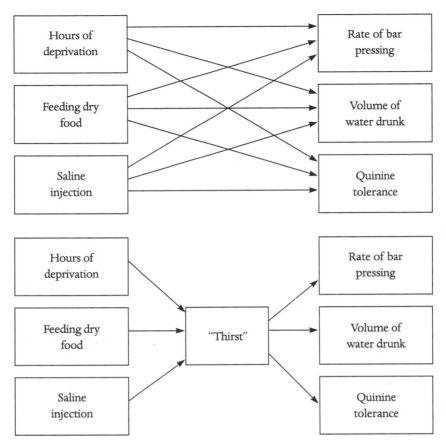

Figure 11.1 Thirst as an intervening variable. Top graph: the relationship between three independent variables (on the left side) and three dependent variables (on the right side) in the case of rats' drinking. Bottom graph: the recognition of "thirst" as an intervening variable permits a more economical representation of the causal linkages, in this case reduced from nine shown on top graph to six. (After Miller, 1959; redrawn from Whiten, 1996)

explanations of emotions as intervening variables should aim to incorporate information about the underlying neurophysiological processes (cf. Zupanc & Lamprecht, 2000, for motivation as an intervening variable). The integration of the two approaches may lead to a more powerful explanation of emotional phenomena. In the case of anxiety, for example, identifying this emotion as an intervening variable linking a variety of alternative inputs (e.g., social insecurity, past stresses, lack of support) with an array of different outputs (e.g.,

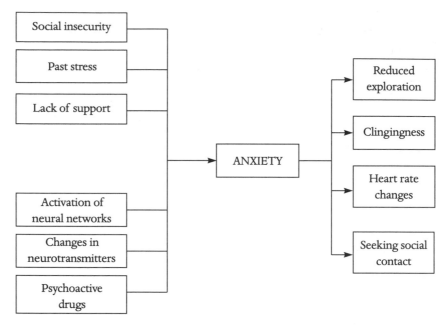

Figure 11.2 The integrated view of anxiety as an intervening variable, including effects of neurophysiological mechanisms. Note that linkages may be more intertwined than those indicated here, with some variables (e.g., social insecurity, psychoactive drugs) acting through others (e.g., changes in neurotransmitters), but the essential role of anxiety as the central intervening variable remains.

reduced exploration, clingingness, seeking social contact) might be strengthened by examining neurophysiological inputs such as neural substrates, neurotransmitters, or anxiety-eliciting drugs, and outputs such as sweating and heart rate (Figure 11.2).

Functions of Emotions

Emotions, being psychobiological phenomena, require two sets of explanations, proximate and ultimate. Tinbergen's (1963) influential "four whys" provide an excellent framework for a comprehensive explanation of emotions. At the proximate level, emotions need to be explained in terms of the neurophysiological, behavioral, and cognitive mechanisms underlying them and of the development process in each individual from young to adult. At the ultimate level, the explanations refer to the functions of emotions in terms of adaptive significance and to their evolutionary history (or phylogeny) in terms of how they were shaped by natural selection. We review evidence on mechanisms, development, and evolutionary origin in the next section; we focus here on functional explanations.

A basic and widely recognized concept is that each emotion is useful only in certain situations. Its function can be understood by an analysis of the characteristics of the situations. In general, emotions can be viewed as shaped by natural selection to adjust the physiological, psychological, and behavioral parameters of an organism in order to increase its capacity to respond to the threats and opportunities characteristic of particular situations (Nesse, 1990). For example, fear prepares the organism for appropriate responses in situations of danger (e.g., escaping a predator, finding protection, attacking in defense). Such situations, occurring repeatedly in the course of evolution, have selected for certain characteristics of emotional states that are particularly significant for coping with the challenges involved.

Various specific functions can be ascribed to single emotions (e.g. Nesse, 1990; Rolls, 1990, 1999), and such a list of functions could potentially be extremely long. More important for our overview is the growing consensus among researchers from various theoretical backgrounds on the general function of emotions in terms of adaptive response to environmental demands, preparing the individual to cope with them and increasing survival (e.g., Damasio, 1998, 2000; Ekman, 1984; Frijda, 1986; Izard, 1977; Johnson-Laird & Oatley, 1992; Lazarus, 1991; LeDoux, 1996; Öhman, 1993; Plutchik, 1980; Smith & Pope, 1992; Tomkins, 1963). Several aspects of the adaptive functions of emotions (e.g., homeostatic regulation, motivation, communication) are especially important for the scientific investigation of animal emotions.

First, at the most basic level, emotions are part of homeostatic regulation that ensures the survival of the organism (Damasio, 2000). This basic function is probably achieved by relying on a relatively simple evaluative system that differentiates between hostile and hospitable stimuli and allows rapid orienting responses of approach or withdrawal (Cacioppo & Gardner, 1999; Cacioppo et al., 1999; cf. Davidson, 2000). This characteristic can explain why emotions are inseparable from the concepts of reward or punishment, of pleasure or pain, of advantage or disadvantage (Damasio, 2000) and why brain systems critical for reward-punishment evaluation are involved in emotional experience (Rolls, 1999). Even where the behavioral response is governed by a single dimension of approach or withdrawal, it is probably based on at least two evaluative channels that process information in parallel from the flow of sensation: one that derives threat-related (i.e., negative) information and the other that derives safety and appetitive (i.e., positive) information (Cacioppo et al., 1999). The information processed in the two channels is subsequently integrated for the production of emotions. This conceptualization of the initial phase of the emotional process may explain why the dichotomous classification of "positive" and "negative" emotions, though potentially arbitrary, appears intuitively grounded: it is actually based on very basic adaptive responses.

Another important functional issue is the specific role of emotions in motivating organisms to act (LeDoux, 1996; Rolls, 1990). Emotions can be viewed as interfacing between sensory inputs and motor outputs in a way that allows flexibility in the response, in contrast to (genetically programmed or learned) fixed action patterns as a reaction to a stimulus (Gray, 1975; Panksepp, 1989; Rolls, 1995). We suggest that the essential function of emotions is *to gear a particular type of motivational control to the perception of critical circumstances*. This leads the individual to take a particular motivational stance, which severely constrains its further behavior, that is, it constrains decision making for some time appropriate for its referent (e.g., a longish period in the case of fear through sighting a leopard) (cf. Damasio, 1994; Johnson-Laird & Oatley, 1992). Although decision making is constrained, this is only for an appropriate and temporary period. Overall flexibility of response is not hampered but is achieved by the change in motivational stances elicited by different circumstances.

To conclude this brief review, we suggest that the essential function of emotions in increasing the organism's survival is applicable to humans as well as to other animals, regardless of the level of cognitive abilities possessed (even though such level has implications for the complexity of the underlying process). In addition, the view of the state of an emotion as an intervening variable is highly compatible with the concept of flexible responses derived from emotional interfacing between sensory inputs and motor outputs, involving a process of canalization of motivational stance and decision making. Furthermore, the concept of positive and negative emotions, so pervasive in human psychology, can be confidently applied to the study of animal emotions because it derives from a basic evaluative system leading to adaptive orienting responses of approach or withdrawal to different stimuli.

Emotions in Human and Nonhuman Primates

Which emotions might be comparable across primate species? Starting from Johnson-Laird and Oatley's (1989) list of the 590 English terms describing human emotions would be a daunting task. Fortunately, a number of scientists studying human emotions claim that such complexity can be reduced to a list of "basic" emotions which can be counted on our fingers. The emotions included in such a list vary somewhat among scholars, but overall there is great consistency (Ekman, 1984; Frijda, 1986; Izard, 1977; Johnson-Laird & Oatley, 1992; Plutchik, 1980; Tomkins, 1962; but see Ortony & Turner, 1990, for a critical position). Scholars, however, have taken different positions about the reasons for considering certain emotions as basic, as well as the existence and nature of non-basic emotions; specialized journals and landmark volumes have dedicated special issues and sections to such debates (Ekman & Davidson, 1994; Stein & Oatley, 1992). Relevant to the study of emotions in nonhuman primates is the fact that according to most theorists, basic emotions have been shaped by natural selection to enable humans to deal with fundamental life tasks, and most of their characteristics are shared with other animals (Ekman, 1992; Izard, 1992; LeDoux, 1996).

There is thus theoretical justification for the study of basic emotions in nonhuman primates; yet the systematic investigation of emotions such as happiness/enjoyment, sadness, or surprise has been a difficult task. Most research has focused on anger and fear, for which behavioral expressions are more obvious, or on anxiety, for which pharmacological manipulation has provided additional tools (see below). Our bias in reporting studies on negative emotions reflects this situation. It remains to be seen if this is due to greater difficulty in operationally defining positive emotions and developing standardized indicators for these in animals other than humans.

Developmental processes play a role in emotional states and expressions because previous experiences certainly influence our emotions, but little is known about their role in nonhuman primates. There is indication of similarities in the emotional responses of infants to separation from their mother in human and nonhuman primates. Disturbance of the mother-infant attachment can elicit separation anxiety (Coe & Levine, 1981; Suomi et al., 1981; see Chapter 5). These emotional responses appear early in life, but similar responses are also found during separation or other disturbance of close relationships among adults (Feeney, 1998; Mason & Mendoza, 1998). In addition, it is known that there are developmental differences between the production and the appropriate use of emotional expressions. For example, during the first two weeks of life rhesus macaques are emotionally expressive, but only by nine to twelve weeks of age do they use their defensive emotional responses in the appropriate contexts (Kalin et al., 1991).

Primates are regularly faced with challenges in their environments which are expected to elicit emotional responses, given the primary function of emotion in increasing the organism's survival. According to this expectation, the findings of several studies on nonhuman primates have been interpreted in terms of emotional responses to a variety of natural challenges: predators (Joslin et al., 1964; Vitale et al., 1991), unfamiliar conspecific intruders (French & Inglett, 1991; Mendoza & Mason, 1986), introduction to an unfamiliar conspecific group, mimicking the challenge posed by group transfers (Suomi et al., 1981), and exposure to novelty (Buchanan-Smith, 1999; Clarke et al., 1988; Fairbanks & McGuire, 1993).

In the remainder of this section, we review examples of research on emotional states and emotional expressions in nonhuman primates, comparing them with empirical evidence on human emotions. Our review necessarily emphasizes certain areas that have received more research effort than others. The review cannot be exhaustive, given the recent growth of the field. Instead we illustrate achievements and limitations through a representative sample of studies. We start with evidence of neurophysiological correlates of emotions, followed by evidence of behavioral expressions.

Neurophysiological Correlates

The causation of psychobiological phenomena is always linked to the anatomical structures involved. For emotions, one critical structure is certainly the brain. MacLean (1952) introduced the term "limbic system" to identify areas of the forebrain involved in emotional experience. He later proposed the "triune brain theory," according to which the evolution of the forebrain has passed through three main recognizable stages corresponding to the reptilian brain present in all vertebrates, the paleomammalian brain present in all mammals, and the neomammalian brain characteristic of primates (MacLean, 1990). Research has shown that MacLean's theory is flawed because some areas included in the original limbic system are not involved in emotional experience and because there is no single system for emotional experience; different systems are involved in different emotions (Damasio & van Hoesen, 1983; LeDoux, 1996). MacLean's evolutionary approach has been instrumental, however, in guiding the interpretation of the growing number of findings on brain structures involved in emotions in different species (e.g., LeDoux, 1996; Panksepp, 1998).

Various methods have been used to explore the role of structures of the central nervous system and different neurotransmitters in the regulation of emotion. Traditionally these methods have been invasive and include electrical stimulation, chemical manipulation, cerebrospinal fluid sampling, and lesions of specific brain areas. Through the use of these techniques much has been learned, for example, on the role of the amygdala in regulation of fear and other emotions (Aggleton & Young, 2000; Davis & Whalen, 2001; LeDoux, 1996).

Noninvasive methods may also be utilized to study the role of the central nervous system in emotion. Scientists have examined the emotional and behavioral changes associated with human brain damage occurring from natural causes or trauma (Damasio & van Hoesen, 1983; Robinson & Downhill, 1995). The activity of specific brain areas and its possible asymmetric nature has been successfully investigated with electroencephalography (EEG) in both human and nonhuman primates, and then related to emotion and temperament (Davidson, 1995; Kalin et al., 1998). Important advances in investigating the neural substrates of emotion have been made through positron emission tomography (PET) and functional magnetic resonance imaging (fMRI). These techniques allow scientists to monitor cerebral blood flow, thereby obtaining images of brain physiology and function, and to relate particular neuroanatomical areas and brain activity with emotional responses and emotional memory (Davis & Whalen, 2001; Dolan & Morris, 2000; Kalin et al., 1997; Reiman et al., 2000).

Early emphasis in the study of the neural substrates of emotion focused on subcortical areas, influenced by MacLean's (1952) framework. It has become clear, however, that in primates the neocortex plays a critical role in aspects of emotional experience (Kolb & Taylor, 1990). As in other aspects of brain functioning, cortical and subcortical asymmetries are involved in emotion processing (Davidson & Hugdahl, 1995). Research findings have suggested that positive emotions are associated with greater left hemisphere activity and negative emotions are associated with greater right hemisphere activity (Canli et al., 1998; Davidson, 1995). For example, in human subjects, EEG measurements revealed left-sided activation in the anterior temporal regions during the experience of happiness (elicited by the viewing of video clips), whereas right-sided activation in the same regions occurred during the experience of disgust (Davidson et al., 1990). In addition, human subjects that were impaired in recognizing facial expressions of negative emotions had brain damage in the right hemisphere (Adolphs et al., 1996). There is also suggestive evidence from studies on nonhuman primates supporting the lateralization of emotion (Hauser, 1993; Kalin et al., 1998; Parr & Hopkins, 2000). Particularly relevant for comparison with the human findings described above is a study on chimpanzees that found an increase in right brain temperature during the viewing of negatively emotional videos (Parr & Hopkins, 2000).

Identification of brain areas involved in different emotions is only a first step in unfolding the mechanisms underlying emotional states and expression. To understand the neural network underlying emotional experience we need to study the regulation of the electrical communication between nerve cells both in the brain and in the peripheral nervous system. The plasticity underlying the flexibility of emotional states is achieved by the actions of neuromodulators. These are substances released by neurons in the hypothalamus, basal forebrain, and brain stem which transform the working mode of neural circuits (Damasio, 2000).

An important example of neuromodulators relevant to our comparative review is provided by brain opioids. These are substances that produce pleasant sensations and that are usually released during "positive" emotions associated with rewarding situations (Dum & Herz, 1987; Panksepp et al., 1997). Studies on nonhuman primates have shown that such neuromodulators play an important role in the regulation of friendly interactions with conspecifics. In talapoin monkeys the receipt of grooming increases the concentration of endogenous brain opioids (Keverne et al., 1989). More important, pharmacological manipulation produces changes in patterns of grooming and other affiliative behaviors in various monkey species. The administration of opiate agonists decreases the motivation to seek contact with others and to be groomed, whereas opiate antagonists increase the same motivations (Fabre-Nys et al., 1982; Keverne et al., 1989; Martel et al., 1995; Schino & Troisi, 1992).

These results suggest that neuromodulators such as brain opioids may be a basis of "positive" emotions that monkeys may experience during friendly interactions.

Much progress has been made in the identification of subcortical and cortical areas involved in the regulation of different emotions in human and nonhuman primates, and promising insight has been gained into the neural networks and neuromodulators at the root of such regulation (Damasio, 1998; Davidson & Sutton, 1995; LeDoux, 1996; Panksepp, 1998; Rolls, 1999). In the coming years, higher-resolution, noninvasive techniques complemented by sophisticated research designs will certainly increase the understanding of brain mechanisms of emotional experience in human and nonhuman primates from a comparative perspective. At the same time, it is important to continue the study of the peripheral nervous system that has been the focus of scientific investigation since James's (1884) emphasis on the critical role of physiology in emotional experience.

The monitoring of peripheral physiological parameters generally relies on relatively noninvasive techniques. Common techniques include the measurement of physiological functions such as heart rate, blood pressure, hand temperature, skin resistance, muscle tension, and hormonal levels (reviewed in Bauer, 1998; Cacioppo & Tassinary, 1990; see also Cacioppo, Tassinary, & Berntson, 2000). Some studies use only one of these measures, whereas others monitor several physiological functions simultaneously. An example of a multi-measure study is the classical work by Ekman et al. (1983). The authors measured heart rate, hand temperature, skin resistance, and forearm flexor muscle tension of human subjects while eliciting six target emotions by two different tasks. The use of multiple measures provided evidence for the differentiation of physiological responses not only between "positive" and "negative" emotions but within "negative" emotions as well (see also Collet et al., 1997). Skin temperature patterns, reflecting peripheral vasoconstriction or vasodilation, have proved to be particularly consistent in differentiating among states with different emotional valence (reviewed in Rimm-Kaufman & Kagan, 1996). In human subjects, positive emotions are commonly associated with an increase in skin temperature, whereas negative emotions are accompanied by cooling.

Physiological changes depending on the quality of the elicited emotion have also been reported for nonhuman primates. In chimpanzees, video scenes depicting darts and needles or conspecifics being injected with needles elicited a decrease in skin temperature, suggestive of negative emotional experience (Parr, 2001). Another study monitored cardiac activity of infant chimpanzees to differentiate their emotional responses to two conspecific vocalizations (Berntson et al., 1989). Heart rate decreased following playback of chimpanzee screams but increased after chimpanzee "laughter" (the rhythmic panting that accompanies the playface in situations such as being tickled). Analysis of another parameter of cardiac activity (vagal tone as indexed by the magnitude of respiratory sinus arrhythmia) suggested that the deceleration in heart rate resulted from an increase in parasympathetic activity and heart rate acceleration from sympathetic activation. The direction and nature of these physiological changes allowed the authors to interpret the chimpanzees' emotional reaction to conspecific screams as an orienting response and that to laughter as a defensive response (cf. Graham, 1979). Interestingly, the physiological response to laughter decreased with age, indicating a developmental change. In addition, the cardiac response to laughter was similar to the one evoked by threat vocalizations (Berntson & Boysen, 1989). One possible interpretation is that young individuals may not discriminate perceptually between the two vocalizations; alternatively, infant chimpanzees' defensive response to laughter at an early age may be adaptive because such vocalization is primarily given during intense play interactions that may be threatening to vulnerable young individuals (Berntson et al., 1989).

Increase in heart rate is commonly associated with acute experience of anxiety in humans (Berntson et al., 1998; Fredrikson, 1989; Öhman, 1993) and is also produced by treatment of rhesus macaques with anxiety-eliciting drugs (Ninan et al., 1982). Such an association was exploited in a study of emotional responses of rhesus macaques during social interactions (Aureli et al., 1999). The study used biotelemetry to monitor the heart rate of free-moving individuals, recording their response to being approached by different group members while sitting still. The approach of a dominant individual produced an increase in heart rate, whereas the approach of a subordinate or kin did not cause consistent heart rate changes (Figure 11.3). These results are in agreement with the view that the strict dominance hierarchy of this species creates risky situations when animals interact with more dominant individuals. The increase in heart rate may be taken as an indication of anxiety, probably due to the increased risk of being attacked when a dominant individual is in close

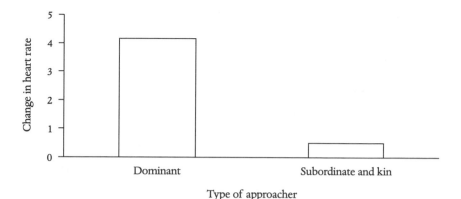

Figure 11.3 Mean changes (after/before the approach) in heart rate (beats per minute) of adult female rhesus macaques during approaches by dominant individuals and by subordinate individuals or kin. (Redrawn from Aureli et al., 1999)

proximity. After all, anxiety is an adaptive response to uncertainty and antici-pated threat (Lazarus, 1991).

Heart rate changes associated with emotional experience can also be found in dominant individuals. Although activity transitions are usually re-flected in a change in heart rate (Smith et al., 1993), cardiac response before the initiation of a movement can be investigated to assess whether any change could be linked with the emotional content of the behavior associated with the following movement. Another biotelemetry study found that dominant male baboons showed an anticipatory increase in heart rate in the seconds pre-ceding their locomotion with aggressive posture toward a group member compared to locomotion without such a posture (Smith et al., 2000). Different degrees of heart rate changes were found between dominant and subordinate individuals in response to varying proximity with group members during the initial phases of group formation (Manuck et al., 1986).

Different heart rate responses are found when friendly interactions, possi-bly associated with positive emotions, are examined. In humans, for example, receiving gentle touching reduces heart rate (Drescher et al., 1980). A similar form of skin stimulation in nonhuman primates is allogrooming. The impres-sion of relaxation of primates receiving allogrooming has inspired the tension-reduction hypothesis (Terry, 1970). Biotelemetry studies in various monkey species have provided support for this hypothesis. Heart rate decelerated faster during receipt of grooming than in control periods (Aureli et al., 1999; Boccia et al., 1989; Smith et al., 1986).

There are thus many similarities in brain mechanisms and peripheral physiological activation involved in emotions in human and nonhuman primates. To give a final example, a very coherent picture can be put together for the role of the amygdala in the experience of fear from comparative studies of humans and other mammals, using evidence from a variety of sources (e.g., lesions, chemical or electrical stimulation, neural activity recording, neuroimaging; see Davis & Whalen, 2001; Ono & Nishijo, 2000). Similarly, studies using peripheral physiological parameters, such as skin temperature and heart rate, have revealed consistent patterns of emotional responses to comparable stimuli in human and nonhuman primates (see above). Interspecific and developmental differences may exist, however, in the effectiveness of mechanisms for the inhibition of emotions in relation to perseverative errors and the involvement of the prefrontal cortex (Hauser, 1999b). The similarities reported so far and the possible differences should inspire an increasing effort in the comparative study of the neurophysiology of primate emotions.

Facial and Vocal Expressions

Since Darwin's (1872) original emphasis, facial expressions have been among the emotional indicators primarily studied. It has been claimed that nowhere in the body are human emotions so clearly differentiated as in facial muscle movements (Rinn, 1984). Even though there is no consensus among emotion theorists, many investigators believe that different emotions are characterized by distinctive facial muscle responses (Ekman & Oster, 1979; Izard, 1977; Lanzetta & McHugo, 1989; Tomkins, 1962). In particular, based on cross-cultural and experimental evidence, Ekman (1993) claimed that there are a limited number of basic facial expressions of emotions that are "universal," occurring in all human populations (but see Russell, 1994). Following this perspective, such facial expressions provide true cues for discrete emotional states serving communicative function and being predictive of the sender's future actions (see also Chapter 12).

Facial displays have been studied in animals, especially in nonhuman primates, in order to trace the evolutionary history of human facial expressions (Andrew, 1963a; Chevalier-Skolnikoff, 1973; Preuschoft & van Hooff, 1995;

Redican, 1982; van Hooff, 1976). These studies analyzed expressions by using careful descriptions of the various movements of the anatomical components involved in each facial expression. Such comparative analyses have suggested, for example, that human facial expressions such as laughter and smiling, which are generally considered to be functionally similar, derive from two independent facial displays (Preuschoft & van Hooff, 1997; van Hooff, 1972). According to van Hooff and Preuschoft, the smile derives from the "silent bared-teeth display" that is typical of most species of Old World monkeys and apes and also found in several New World monkeys. Such a display is usually associated with fearful situations and is a sign of submission and appeasement in social contexts (Miller, 1971; Preuschoft, 1992). In some species, however, the same display is associated with the initiation of positive social interactions, as is the human smile (Preuschoft & van Hooff, 1997). The evolutionary origin of laughter has been traced back to the "relaxed open-mouth display," or "playface," observed in all primates studied and therefore likely a heritage from ancestral forms (Preuschoft & van Hooff, 1997). Although such a display is strictly confined to the play context in the majority of species, it is used in a broader affiliative context in those same species that exhibit the silent bared-teeth display as a friendly expression. Interestingly, the function of possibly homologous displays varies between species in ways not tied to phylogenetic relationships; the silent bared-teeth display and the relaxed open-mouth display are used by Tonkean macaques in ways more similar to human smiling and laughter than those by chimpanzees. The available evidence suggests that species characterized by tolerant societies use the two displays in highly overlapping contexts, whereas species living in societies with strict dominance hierarchies keep the two displays clearly separate (Preuschoft & van Hooff, 1997).

A different hypothesis for the evolutionary origin of human laughter and smiling is based on the consideration that nonhuman primates have a limited repertoire of facial and vocal expressions of positive emotions. Owren and Bachorowski (2001) suggest that laughter and smiling arose as mechanisms that allowed early hominids to form and maintain cooperative relationships. According to this hypothesis, laughter and smiling evolved as reliable indicators of the sender's positive emotional state, generating positive emotions in the receiver, who likely responded with similar facial expressions. The feedback loop involving communication and generation of positive emotions in the sender and the receiver is seen as instrumental in fostering the building of trust between partners, critical for cooperative actions, and providing inherent protection from exploitation (see Chapter 13 for discussion of the evidence supporting this hypothesis).

The production of facial expressions of emotions appears at an early age in both human and nonhuman primates. Human infants reliably smile at the mother's voice by the fourth week of age and laugh appropriately at about four months (Sroufe & Waters, 1976). Chimpanzees exhibit adult-like facial expressions, such as the silent bared-teeth display and the relaxed open-mouth display, during the first weeks of life (Bard, 1998). Laughing and smiling are shown by blind and deaf children from early infancy (Eibl-Eibesfeldt, 1989), suggesting that spontaneous laughing and smiling are not behaviors that human infants learn from observation of others. Similarly, there is evidence that rhesus macaques reared in isolation perform species-typical facial expressions in the right context (Miller, 1971; Sackett, 1966), suggesting a strong genetic component in the production of such displays. The same monkeys, however, were unable to respond appropriately to facial expressions of conspecifics, suggesting that the comprehension of such expressions requires social learning (see Chapter 12 for further discussion). Human infants show signs of comprehension of the mother's facial expressions by ten weeks of age (Haviland & Lelwicka, 1987). Similarly, young chimpanzees respond differentially and appropriately to expressions of happiness and fear by human caretakers early in life (Russell et al., 1997).

The study of facial expressions of emotions has employed a range of behavioral and physiological methods to discriminate among the expressions themselves and their emotional valence. Sophisticated coding techniques, such as Ekman and Friesen's (1978) Facial Action Coding System, specify all minimal units of facial movements. Whereas these techniques provide reliable inferences about the muscular basis of any expression, not all muscle activation is expressed in overt facial action (Cacioppo et al., 1993). Recent improvement of facial electromyography (EMG) provides a more sensitive method to record facial responses throughout the entire range and to monitor the dynamic nature of emotion (Cacioppo et al., 1990). These techniques are not easy to use in free-moving individuals and pose difficulties for application to the study of nonhuman primates during social interactions.

Whereas the ability of people to attribute consistently certain emotional states to particular facial expressions has been widely documented and is at the basis of the concept of universality of facial expressions (reviewed in Ekman, 1993), the study of such ability in nonhuman primates is intrinsically more difficult because language cannot be so easily used to gather the needed answer. The ingenuity of researchers has, however, circumvented such difficulty, and there is evidence for the perception of the emotional significance of facial expressions in nonhuman primates. As part of a series of experiments involving pairs of rhesus macaques in the role of "informer" and "responder," Miller (1971) reported that the view of a silent bared-teeth display exhibited by the informer was sufficient for the responder to press a lever and avoid an electric shock. The informer responded with this facial display when seeing a warning signal indicating the upcoming electric shock, but the warning signal was not visible to the responder, who had to base its response only on the fear conveyed by the display of the partner. In another elegant experiment, chimpanzees were required to categorize emotional video scenes presented on a computer monitor by matching them to conspecific facial expressions with similar emotional valence, so that matching was based on emotional similarity instead of perceptual features (Parr, 2001). Without previous training, chimpanzees spontaneously matched positive or negative video scenes to conspecific facial expressions according to their shared emotional valence, suggesting an ability to discriminate facial expressions on the basis of their underlying emotional states.

Darwin (1872) also pointed to vocal signals as primary carriers of emotional valence. There is substantial evidence that emotional experience produces changes in respiration, phonation, and articulation, which affect acoustic features, and that the acoustic features of vocal emotional expression are similar in human and nonhuman primates (Jürgens, 1998; Scherer & Kappas, 1988). These likely deep-reaching phylogenetic roots may underlie the ability of naïve human listeners to identify the emotional content of vocalizations of stumptail macaques (Leinonen et al., 1991). Detailed acoustic analyses of human speech and primate vocalizations have revealed that vocal parameters not only index the degree of emotional intensity but also differentiate emotional valence (Bachorowski & Owren, 1995; Banse & Scherer, 1996; Fichtel et al., 2001; Jürgens, 1979).

The vocal expressions of emotion that have been intensively studied across the primate order are distress calls (Todt, 1988). Like the human cry, species-specific distress calls appear to increase in frequency during social separation in various species. The acoustic structure and rate of these calls can be readily manipulated through administration of exogenous neuropeptides and anxiety-related drugs (Newman, 1991; Newman & Farley, 1995; Panksepp et al., 1997). These properties make distress calls particularly useful as emotional indicators. Distress calls appear from the first day of life in most species. Developmental patterns of fussiness and screaming/crying are comparable in human and chimpanzee infants, and similar patterns are also found in nursery-reared chimpanzees, but their frequencies are much higher (Bard, 2000).

The similarities in distress vocalizations across species may be due in part to the direct relationship between the properties and the immediate function of such vocalizations. It is likely that general properties of the mammalian auditory system make acoustic characteristics such as overall amplitude and noisiness particularly unpleasant in listeners, and this negative effect would start off a response that reduces the unpleasantness (Owren & Rendall, 1997; Todt, 1988). Some forms of crying in human infants are in fact particularly aversive to human listeners, who are prompted to sooth the crying infant (Ostwald, 1972; Zeskind & Lester, 1978).

An elegant, although invasive, study of the emotional valence of different calls in a nonhuman primate species was carried out by Jürgens (1979). Squirrel monkeys were implanted with intracerebral electrodes, and species-specific vocalizations were elicited by electrical stimulation of different brain areas responsible for their control. The experiments were set up to give the monkeys the opportunity to switch on or off the vocalization-eliciting stimulation themselves, so that a quantitative measure of whether the monkeys avoided or sought the stimulation could be obtained. The monkeys' choices were then interpreted as reflecting the degree of aversiveness or pleasantness of the emotional state linked to each vocalization.

Perception of emotion appears to involve different brain areas depending on the mechanisms the individual uses to extract the emotional information while listening to the vocal expression. A study measured cerebral blood flow through positron emission tomography in human subjects listening to utterances and identifying their emotional content (George et al., 1996). When comprehension of emotion was achieved by word meaning, the activity of both frontal lobes increased, the left more than the right. By contrast, comprehension of emotion from tone of voice increased the activity of only the right prefrontal cortex.

Research on vocal expressions has traditionally emphasized the emotional experience of the producer of the vocalization. One proposal for the evolution of nonhuman primate vocal signaling has suggested that vocalizations are also used to elicit emotional responses in the recipients in order to modify the latter's behavior (Owren & Rendall, 1997). For example, by having produced in the past individually distinctive vocalizations prior to attacking a certain subordinate group member, a dominant individual has conditioned it to experience a negative emotion in response to the simple sound of such vocalizations. A threat signal by the dominant individual would then be sufficient to promote avoidance or submissive behavior by the subordinate as a way for the latter to reduce its negative emotional experience. This could apply to positive emotions as well. Coo calls of stumptail macaques and grunts of chacma baboons are given before they engage in friendly interactions (Bauers & de Waal, 1991; Silk et al., 1996). These calls are viewed as eliciting positive emotion in the receiver because they had been previously associated with pleasant sensations such as those produced in receiving grooming (see above). The positive emotion promotes tolerance and facilitates the friendly interaction (Owren & Rendall, 1997).

Our brief review shows that there are several lines of evidence for the expression of emotions in facial and vocal signals and that there are many similarities in this regard between human and nonhuman primates. There is certainly a need to integrate research on these two forms of emotional expression and to evaluate the degree of their overlap versus their independent contributions (Scherer, 1994). After all, vocal production often occurs during facial expressions, and facial movements are always present during vocal expressions.

We must not forget, however, that there are several controversial issues regarding the interpretation of vocal and facial displays as expressions of emotions (see Chapters 12 and 13 for more discussion). One of these issues, common to both human and nonhuman research, is whether the expressions convey only information about the emotional state of the sender, whether they communicate intent without being readouts of underlying emotions, or whether they are functionally referential about objects and events in the environment (Buck, 1994; Cheney & Seyfarth, 1990b; Fridlund, 1991; Hauser,

1996; Marler, 1978; Marler & Evans, 1997). It is likely that these various communicative functions coexist in many signals. For example, screams produced during aggressive interactions in primate species in which other group members may join the action have been shown to provide information about the severity of the conflict and the type of opponent, but they are also a reflection of the caller's emotional disposition and communicate its conditional future behavior (Gouzoules et al., 1995; Rowell & Hinde, 1962; cf. Hinde, 1985; Marler & Evans, 1997; Owren & Rendall, 1997).

Self-Directed Behavior

A less conventional set of behavioral indicators of emotional experience has been used, especially in nonhuman primates. Self-directed behaviors such as self-scratching and self-grooming obviously have a hygienic function, but they have also been considered displacement activities, that is, activities which are apparently irrelevant to an individual's ongoing behavior and which reflect motivational ambivalence or frustration (McFarland, 1966; Tinbergen, 1952). Ethological studies on nonhuman primates have documented an increased frequency of these behaviors in situation of uncertainty, social tension, or impending danger (Maestripieri, Schino, et al., 1992; Troisi, 2002). Similarly, high levels of self-directed behavior are characteristic of human subjects experiencing anxiety (Fairbanks et al., 1982; Troisi et al., 1998; Waxer, 1977).

In addition to the ethological evidence, pharmacological manipulations have demonstrated that self-directed behavior is strongly associated with anxiety in various primate species (Barros et al., 2000; Cilia & Piper, 1997; Maestripieri, Martel, et al., 1992; Ninan et al., 1982; Schino et al., 1991, 1996). For example, Schino et al. (1996) explored the acute effects of two benzodiazepine receptor ligands known to have opposite effects in humans: the anxiolytic lorazepam and the anxiogenic FG 7142. These drugs were given to seven sub-adult male long-tailed macaques in low doses to avoid side effects such as excitation or sedation, along with the administration of a placebo in a counterbalanced order. As expected, lorazepam caused a dose-dependent reduction in the frequencies of self-directed behavior such as scratching and self-grooming, whereas the anxiogenic FG 7142 caused a dose-dependent increase. The drugs did not change the rates of aggression and allogrooming received by the subjects, so the changes in self-directed behavior were unlikely to be a consequence of the modification of their emotional state resulting from the receipt of such behaviors (Schino et al., 1996).

Self-directed behavior has been used in several studies investigating emotional responses. For example, proximity to a dominant group member may be potentially risky because reduced interindividual distance increases the likelihood of aggression. Under these circumstances subordinate individuals may be tense or anxious, especially in species with relatively strict dominance hierarchies (Figure 11.4). Studies of captive long-tailed macaques have shown that rates of self-grooming and self-scratching of adult females are higher while they are in proximity to a higher-ranking male than when alone (Pavani et al., 1991; Troisi & Schino, 1987). Similar results were found for adult female rhesus macaques when near more dominant males and females (Maestripieri, 1993a). Further evidence for differential anxiety depending on the dominance relationship with neighbors was provided by a study of wild olive baboons. When the animal nearest to an adult female was dominant over her, she increased the rate of self-directed behavior by about 40 percent over that

Figure 11.4 Self-scratching by a female pigtail macaque as a sign of anxiety while in proximity with a more dominant female. (Photo: M. Seres)

observed when the nearest neighbor was subordinate to her (Castles et al., 1999). In addition, intermediate-ranking rhesus macaques scratched themselves most at feeding time probably because they experienced higher uncertainty about how to behave than did dominant individuals, who easily monopolized the food, and subordinate individuals, who waited for access to the food (Diezinger & Anderson, 1986).

Differential anxiety may reflect not only the risk associated with interaction with dominant individuals but also the degree of uncertainty in dominance relationships. For example, when two long-tailed macaque females were paired in a small pen, the nature of their relationship was reflected in the rate of self-directed behavior (Schino et al., 1990). When the females were familiar with each other, the rate of self-directed behavior was low. When the females were unfamiliar with each other, the rate of self-directed behavior depended on whether a dominance relationship was established or not. In the pairs in which one partner displayed submissive signals at the beginning of the experimental session, the rate of self-directed behavior did not differ from that of familiar pairs. When the unfamiliar females delayed the establishment of dominance relationships, the rate of self-directed behavior was much higher than that of the other two types of pairs, in which dominance relationships were clear-cut.

The risk of aggression is also present in other situations, and various studies used self-directed behavior to document the emotional responses. Two studies of macaques documented maternal anxiety as a correlate of maternal behavior. The first study found that the Japanese macaque mothers that showed higher levels of maternal possessiveness displayed higher rates of self-scratching (Troisi et al., 1991). This finding suggests that the mothers who are most apprehensive about their infants adopt a highly protective mothering style. A second study of rhesus macaques made the case for an anxiety-eliciting situation when a young infant was away from its mother and in close proximity with dominant adult males and females, where it was at high risk of harassment by them (Maestripieri, 1993a). The perceived danger for the infant elicited signs of anxiety in the mother such as increased rates of scratching.

In captive chimpanzees, vocalizations of neighboring groups elicit excitement in the group that may lead to intragroup aggression (Baker & Aureli, 1996). Such impending risk of aggression is reflected in an increase in self-directed behavior by group-living chimpanzees when they hear neighbor vocalizations (Baker & Aureli, 1997). Similarly, an increase in self-directed behavior has been reported to occur in macaques and baboons soon after receiving aggression, a period in which further attacks are likely (Aureli, 1992; Aureli & van Schaik, 1991b; Aureli et al., 1989; Castles & Whiten, 1998b). The increased risk of aggression is unlikely to be the only cause of uncertainty in post-conflict situations because former aggressors also display elevated rates of self-directed behavior, although they are not likely to be attacked (Aureli, 1997; Castles & Whiten, 1998b; Das et al., 1998). The post-conflict uncertainty is likely to be due to disturbance of the relationship between the former opponents (Aureli & Smucny, 2000). Supporting this interpretation, post-conflict scratching rates are higher after conflicts between more strongly bonded partners, who have more to lose from the disturbance of their relationship in terms of mutual tolerance and cooperation (Aureli, 1997).

Changes in the patterns of self-directed behavior have also been used to investigate the reduction of anxiety and social tension after friendly interactions. Post-conflict anxiety is reduced by friendly interactions, but the effect depends on the identity of the partner. Whereas friendly post-conflict reunion between opponents, or "reconciliation" (de Waal & van Roosmalen, 1979), reduces rates of self-directed behavior to baseline rates (Aureli & van Schaik, 1991b; Castles & Whiten, 1998b; Das et al., 1998), friendly post-conflict interactions with third parties fail to do so (Das et al., 1998). Similarly, heart rates of opponents increase following a conflict and are reduced faster following reunions between opponents than following friendly post-conflict interactions with third parties (Aureli & Smucny, 2000; Smucny et al., 1997). In addition, comparisons of the occurrence of self-directed behavior support the tension-reduction function of allogrooming (see above). Rates of self-directed behavior of long-tailed macaques were lower after receipt of allogrooming than at other times (Schino et al., 1988).

The review presented above focuses on emotional responses in social contexts. Studies using self-directed behavior have also provided evidence for emotional experience during non-social cognitive challenges. Self-directed behavior of chimpanzees increases during more difficult tasks and decreases after auditory feedback signals, suggesting an association with the degree of uncertainty of the situation (Itakura, 1993; Leavens et al., 2001). The effectiveness of changes in self-directed behavior to index emotional responses in a variety of social and non-social situations is very promising for further comparative research on the conditions associated with anxiety, uncertainty, and tension in human and nonhuman primates.

Behavioral Flexibility and Emotional Mediation

As we hope has been apparent in our review of the evidence for human and nonhuman primates, emotions are not dichotomous phenomena (i.e., present or absent), but they come in shades. They are modulated by many factors, including the type and intensity of the stimuli, the previous internal state, and individual characteristics. This intrinsic plasticity makes them suitable candidates to mediate the behavioral flexibility typical of human and nonhuman primates, especially during social interaction.

Individual primates behave in different ways depending on the circumstances. They also interact differently with various group members. For example, not only do nonhuman primates distribute their investment in a particular behavior, such as grooming, differently across group members (Castles et al., 1996; Cheney, 1992; Di Bitetti, 2000; Henzi, Lycett, & Weingrill, 1997; Sambrook et al., 1995), but also they exchange different types of behavior depending on the identity of the partner. For example, individuals that spend more time in proximity with one another usually engage more frequently in friendly interactions and less frequently in aggression (de Waal & Luttrell, 1988; Fairbanks, 1980; O'Brien, 1993), even when the number of interactions is corrected for time spent in proximity (Bernstein et al., 1993a). Thus, this variation is not random, and can be summarized conceptually by the differences in the quality of the relationships with other group members (cf. Hinde, 1979).

A good example to illustrate the relevance of relationship quality in affecting behavioral expression is the patterning of reconciliation, an interaction that promotes conflict resolution (see Aureli & de Waal, 2000a, and Chapter 3). In macaque species in which kin are powerful allies, for example, reconciliation occurs more often after conflicts with kin than with non-kin, whereas in those macaque species where kin and non-kin are treated more similarly reconciliation patterns are less kin-biased (reviewed in Aureli et al., 1997). This means that the same individual engages in a particular interaction depending on the quality of the relationship with the former opponent. Furthermore, reconciliation frequency between two individuals may change over time; in fact, by experimentally increasing their relationship value, pairs of macaques increase their reconciliation rates (Cords & Thurnheer, 1993). This is just one example; there is more evidence for great flexibility in the interaction of single individuals with different group members or with the same partner over time.

A key issue is to understand the mechanisms that underlie an individual's ability to modify its behavior depending on the quality of the relationship with the potential partner in a particular context. This can be accomplished by relationship assessment based on the information contained in the various interactions that the partners exchange (Aureli & Schaffner, 2002). Depending on the partner, the exchanges differ in the relative amount of different types of behavior, and the context of occurrence may change the quality of the same interaction (e.g., a grooming bout would communicate different information depending on whether it occurs soon after an aggressive conflict or during a resting period). Therefore, assessment appears to require bookkeeping of the various interactions, computation of their relative frequencies, and conversion of their quality and associated information into a common currency.

Emotion could play a role in such processes and provide the individual with a timely relationship assessment to guide its social decision (Aureli & Schaffner, 2002). This can be achieved by considering the mediating role of emotion implicit in the concept of intervening variable (see above). There has been growing attention to the mediating role of emotions in the human literature (Frijda, 1994; Panksepp, 1989; Rolls, 1995) which is paralleled by new perspectives in animal research that consider emotions as mediators between an animal's perception of the environment and its responses (Crook, 1989; Lott, 1991; Owren & Rendall, 1997; Pryce, 1996; Whiten, 1996). The emotional experience of an individual is certainly affected by the frequency and quality of previous interactions with group members (see above). Emotional states may express a crucial (and, in many contexts important to primates, urgent) integration of the information contained in the various interactions between two partners. The emotional experience can then be functionally equivalent to the processes of bookkeeping, frequency computation, and quality conversion of the interactions with a partner needed for updated relationship assessment. The resulting emotional experience is partner-dependent. Thus, emotional differences can be at the core of the observed variation in social interactions reflecting the variation in relationship quality across partners (Aureli & Schaffner, 2002).

Our explanation for behavioral flexibility based on emotional mediation shares various aspects with hypotheses put forward to describe the functioning of human emotions. The role of emotion in mediating social interactions is based on our integrated view of emotion as a biologically functional intervening variable, realized physiologically (see above). According to this view, emotion leads the individual to take a particular motivational stance, which constrains its further behavior. In other words, a certain emotional state constrains decision making for some time appropriate for the situation eliciting such a state. This perspective shares several critical aspects with the Somatic Marker Hypothesis proposed by Damasio (1994, 1996) to explain how emotions play a role in human reasoning and decision making. According to this hypothesis, emotions, even at an unconscious level, are viewed as crucial components of human decision making because previous experiences relevant to the current situation trigger the reactivation of the emotional state associated with the previous experience. Such emotional reactivation qualifies the behavioral option related to that previous experience, alerting the individual about the positivity or negativity of such an option as a response to the current situation. The hypothesis thus considers emotions as somatic markers that help constrain the decision-making process by rapidly rejecting or endorsing certain options and permitting the individual to make a decision efficiently.

A crucial point in our hypothesis of emotional mediation of relationship assessment is that the emotional experiences linked to previous interactions with a partner are integrated according to the concept of biologically relevant intervening variables to influence the individual's motivational stance and future interaction with that partner. Such integration of emotional experiences is similar to what Öhman (1993; see above) proposed for his nonconscious expectancy system. Such a system relies on networks of emotional memories that, when activated by matching input information, bias the assessment of the current situation. Furthermore, the notion of emotional mediation of behavioral flexibility based on relationship quality finds encouraging parallels with the concept of attitudes viewed as evaluative categorizations of stimuli and action predispositions toward future encounters with such stimuli (Cacioppo & Berntson, 1994). As explained earlier, such attitudes are organized in terms of a positive-negative dimension at the level of behavioral responses, but probably derive from two evaluative channels, one for negative and one for positive information (Cacioppo et al., 1999). Applied to our case, this suggests that the quality of future interactions with a partner will depend on the individual attitude toward such a partner.

Considering the topic of animal emotion by setting aside the problematic question of conscious feeling has allowed us to pursue a constructive approach to the comparative study of emotional states and their expressions. Our review of research in both human and nonhuman primates, though not exhaustive, shows, on the one hand, many points of similarity with respect to the empirical evidence and, on the other hand, great potential for fruitful integration of theoretical approaches. This is an encouraging conclusion that we hope can promote a more consistent comparative perspective in further research.

From a theoretical point of view, the concept of emotion as an intervening variable is widely applicable to a variety of circumstances and, when combined with knowledge of the underlying neurophysiological processes, can provide a powerful tool for biologically relevant explanations of the role of emotional states. We have especially emphasized the role of mediation of behavioral flexibility. In particular, our view for the role of emotion in mediating social interactions depending on the quality of the social relationship between the partners shares important features with theoretical constructs that other authors have proposed for the role of emotion in stimulus assessment, evaluative categorization, and decision making (Cacioppo & Berntson, 1994; Damasio, 1994, 1996; Öhman, 1993).

From an empirical point of view, Tinbergen's (1963) "four whys" framework provides guidance for reaching a more complete explanation of any emotional state. We have presented evidence for proximate explanations, such as the neurophysiological, behavioral, and cognitive mechanisms underlying different emotional states and the circumstances that trigger them as well as the relative role of genetic and learning components during their development. We have also focused on ultimate explanations in terms of functional significance and evolutionary history of certain emotions and their expressions. The overall picture is rather patchy, with a general tendency to investigate only a subset of the fundamental "four whys" and often only certain aspects within a single "why." This unsatisfactory outcome is probably due to the challenging task of setting up a research program with the goal of providing all the explanations at the various levels. It is conceivable, however, that such a goal can be achieved by the coordinated effort of various research teams focusing on different aspects of the same or similar emotional states. In the synthesis of such research efforts, explanations at different levels can be integrated for a better understanding of various emotional states.

A promising example of such synthesis is the integration of proximate and ultimate explanations of empathy presented by Preston and de Waal (2002). Empathy viewed as an emotional change that promotes social interaction, such as the comforting of distressed individuals, fits our proposal of emotional mediation of behavioral flexibility. The synthesis by Preston and de Waal (2002) is not based on findings of a coordinated research effort, but nevertheless provides a powerful example of how comparative evidence can be used to unveil the basic functions and mechanisms underlying an emotional phenomenon that can be applied to humans as well as to other animals.

There are other issues that future comparative research on emotions should address. Our proposal of a central role of emotion in social relationships finds interesting parallels with attachment theory (reviewed in Chapter 5) that are worth exploring further. After all, the responses to separation differ depending on the identity of the attachment figure. In addition, similarities also exist with theoretical work in microsociology (Collins, 1981, 1993), which proposes that assessment of social relationships in group-living species (including humans) is based on emotions generated during previous interactions that affect subsequent interactions. Integrating the theoretical constructs from different disciplines could represent a productive development for comparative empirical research on the function of emotions.

Furthermore, the relative lack of voluntary control which humans (and probably other animals) have over emotions points to two areas of promising investigation. On the one hand, emotions could be used as reliable signals to build the trust required for long-term relationships and provide repeated signs of the commitment to the partner where rewards are remote in time (Frank, 1988; cf. Trivers, 1971). On the other hand, the inherent honesty of emotions could have promoted selection for sophisticated emotion-reading capacities from subtle cues (e.g., self-directed behavior: Bradshaw, 1993), but also selection for counterstrategies of dissimulation of emotional experience (Whiten, 1993a, 1997; Whiten & Byrne, 1988). We believe that these intriguing hypotheses as well as the more general hypothesis for emotional mediation of behavioral flexibility could inspire a new wave of interest in the comparative study of emotions, supported by sophisticated noninvasive techniques and well-planned research designs.

❖

We thank Darlene Smucny for sharing information on the neurophysiological correlates and Dario Maestripieri and Colleen Schaffner for discussion and comments on an early draft. Andrew Whiten was supported by a British Academy Research Readership during preparation of this chapter.

12

Nonvocal Communication

Lisa A. Parr and Dario Maestripieri

Face Recognition

Faces are undoubtedly one of the most important classes of stimuli involved in social communication, providing viewers with a wealth of information about other individuals, including their age, sex, individual identity, and even their underlying emotional state and impending behavior (Brothers, 1990; Buck, 1988; Diamond & Carey, 1986; Tomonaga et al., 1993). The distinctiveness and long-term memory of faces is reflected in their visual complexity: the face contains both repetitive and distinctive patterns of visual information (Benton, 1980; Bruce et al., 1994; Ellis, 1992). Each face, for example, contains the same set of features: two eyes, one nose, and one mouth. The general pattern of these features is, likewise, repeated in every face: the position of the eyes, nose, and mouth is fixed in a triangular pattern. It is the precise size, shape, and overall configuration of these individual facial features that vary across individuals, making each face unique (Friedman et al., 1971). Therefore, the individual quality and configuration of facial features create a face's saliency and maximize rapid and accurate recognition of social peers in both human and nonhuman primates.

Over the last several decades, research on face recognition and facial expression has been on the rise (see Ekman, 1998; Fridlund, 1994; Russell & Fernandez-Dols, 1997). Important advances have been made in understanding the neural basis of face recognition, at both the cellular and domain-specific levels. These studies have been aided by the innovation of neuroimaging and electrophysiological recording techniques, including PET (positron emission tomography) and fMRI (functional magnetic resonance imaging), single-cell recording studies, and EEG (electroencephalography) (Allison et al., 1999; George et al., 1993; McCarthy et al., 1999; O'Scalaidhe et al., 1997; Perrett et al., 1992; Puce et al., 1999; Sergent & Signoret, 1992; Tarr & Gauthier, 2000). Electrophysiological recordings in macaque monkeys, for example, have demonstrated that cells in the inferior temporal cortex and superior temporal sulcus preferentially respond, in the form of spikes per second, to faces over other classes of visual stimuli (Desimone, 1991; Gross, 1992; Perrett et al., 1992; Tovee et al., 1994). Perhaps one of the most interesting and elegant contributions of these studies has been the identification of a face area in the brain. The fusiform gyrus, or FFA (fusiform face area), is an area of the medial temporal lobe that responds selectively to the presentation of faces compared with other objects (Gauthier & Logothetis, 2000). In addition, it has been shown that this area is particularly sensitive for processing stimuli for which subjects have considerable expertise (Gauthier et al., 2000). Since most humans have been exposed to conspecific faces their entire lives, we may consider faces to be the most extreme example of a stimulus for which people have developed expertise, as opposed to people who have learned expertise in an area, such as bird experts (Tarr & Gauthier, 2000).

The ability to use the information present in faces and respond to it discriminatively has undoubtedly had important consequences for the evolution of social animals, particularly social mammals. Social mammals rely heavily on their ability to recognize one another in order to form individual alliances, identify kin, and maintain social cohesion. Although some mammals, such as dolphins, elephants, and carnivores, have elaborate individual recognition systems that involve echolocation and/or olfactory cues, the face remains the most distinctive feature for visual recognition among nonhuman primates.

Face Recognition in Monkeys and Apes

One of the seminal studies on face recognition in monkeys was performed by Rosenfeld and Van Hoesen (1979). They trained four rhesus macaques on a series of simultaneous two-choice discriminations using the Wisconsin General Testing Apparatus (WGTA). The stimuli consisted of geometric patterns and full-size portraits of unfamiliar conspecifics' faces. A correct discrimination was to select the pre-trained target monkey's face from a variety of distractor stimuli. Examples of typical discriminations included selecting the stimulus monkey from the distractor monkey when it was presented at three-quarter profile, full-face profile, or as a mirror image. All subjects learned the discriminations fairly quickly (within four hundred trials), but the major finding supporting evidence of face recognition was that once the full-face discrimination had been acquired, all other facial discriminations were acquired with less difficulty. Therefore, manipulations of posture, orientation, and overall size of the stimuli did not impair the subjects' ability to respond to the faces as discriminable stimuli. Qualitative observations of the monkeys' behavior revealed that they withdrew from the stimuli, displayed submissive facial expressions, and reacted as though the image was an unfamiliar conspecific.

Numerous studies by Hamilton and Vermeire have demonstrated superior performance by the right hemisphere for face discriminations in a population of split-brain macaques, that is, monkeys with a severed corpus callosum and optic chiasm, a process that functionally separates the two hemispheres (Hamilton & Vermeire, 1988; Ifune et al., 1984; Vermeire & Hamilton, 1998a, b). These monkeys learned and remembered discriminations of conspecific faces better when they were presented to the right hemisphere (Vermeire et al., 1998). Monkeys were trained first to discriminate the face of one individual from those of other monkeys presented in several different photographs, in addition to several facial expressions of the same individual. This involved a go/no-go task similar to that described above. Once these discriminations had been learned, subjects were required to recognize the same individual and expressions from novel examples (Vermeire & Hamilton, 1998b). Thus, these studies were substantially more controlled and more pertinent to face recognition than the studies previously described because the subjects were actually required to recognize one individual presented in several different photographs and then to generalize this performance to novel examples.

The monkeys were much better at discriminating individual identity than facial expressions. To date there are no published accounts reporting the ability of monkeys to discriminate facial expressions (see, e.g., Kanazawa, 1996). These studies are particularly informative because of the ability to separate the relative contribution of each hemisphere to face-processing tasks, and because the sample size (n = 26) is quite large for this type of study.

Parr and colleagues (2000) tested individual recognition in four rhesus monkeys that had been trained on a computerized joystick-testing paradigm (see below). These monkeys, as in other studies using macaques, required considerable training to learn matching-to-sample (MTS) discriminations of unfamiliar conspecifics' faces. Once they learned the discriminations, however, they generalized their performance after only a few repetitions. This was true whether the discrimination involved recognizing identical photographs of conspecifics' faces, or whether the task was to recognize the same individual presented in two different photographs (Parr et al., 2000). Thus, monkeys seem competent in treating faces as discriminable stimuli and performing tasks of individual recognition that, as in humans, appear to be specialized to the right hemisphere.

Bauer and Philip (1983) performed a cross-modal (visual/face to auditory/vocal) matching paradigm in three chimpanzees trained to use American Sign Language (ASL). Subjects were trained to sign the names of six familiar cage mates when shown their photographs. During testing, a prerecorded vocalization of a pant-hoot, a typical chimpanzee greeting, was played and the experimenter asked, using ASL, "Who is that?" A correct response was for the chimpanzee to select the photograph of the individual whose vocalization it had just heard. All subjects are reported to have made the transfer from naming a photograph of a familiar conspecific to associating that photograph with a taped recording of that individual's vocalizations. The individual data were presented only as a graph of percent of correct responses per trial session, with multiple trials given per session. Therefore, it is impossible from the data presented to interpret the subject's performance on the first transfer trial after training. The data do suggest, however, that facial cues provide a more salient stimulus than vocal cues alone, but both together may provide the most reliable information about individual identity.

Boysen and Berntson (1989) measured heart rate responses in a chimpanzee as she viewed photographs of familiar chimpanzees who had a history of being either friendly or aggressive toward her. Heart rate is a physiological index of arousal and can be used to assess the psychological and emotional meaning of a stimulus (see Chapter 11). Unfamiliar chimpanzees were also presented. Heart rate increased when the subject viewed the hostile chimpanzee, suggestive of a defensive response, but decreased when she viewed the photograph of the unfamiliar chimpanzee, reflective of an orienting response to a novel individual. No significant changes were observed in her response to the friendly chimpanzee. Thus, the heart rate changes in response to the different categories of individuals reflected her relationship history with those chimpanzees and thus demonstrated individual recognition.

Studies on face recognition in chimpanzees conducted at the Yerkes Primate Center have used a computerized joystick-testing paradigm. Six adult chimpanzees were trained to manipulate a joystick to control the movements of a cursor displayed on a computer screen using a matching-to-sample (MTS) paradigm. According to this paradigm, subjects were first presented with a sample image on the top portion of the computer monitor. They had to orient to this image by contacting it with the joystick-controlled cursor, after which the screen cleared and two comparison images were presented at the lower left and right positions on the monitor. The correct image (target) was one that looked similar to the sample, on some perceptual dimension, while the nonmatching image (foil) was different (see Figure 12.1). If subjects selected the correct image, they were rewarded with a squirt of fruit juice and the next trial was presented. If they selected the incorrect image, there was no reward and they had to wait through a longer intertrial interval (10 s) than if a correct response was made (3 s).

Testing began by presenting subjects with the task of discriminating two identical photographs of an unfamiliar conspecific's face. The stimuli presented in these experiments were all high-quality digitized black and white portraits. With no prior experience discriminating social stimuli or with digitized images, subjects performed above chance on their first exposure to this task—twenty-five trials each representing a different pair of individuals (Parr et al., 2000). Since the correct pair in this experiment were identical photographs, we could not conclude that the subjects were viewing the stimuli as faces per se, only that they saw the images as being salient and discriminable. This problem represents the major limitation of the monkey studies described above. The discriminations were then changed to address the specific issue of

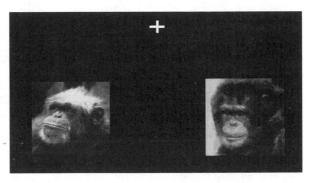

Figure 12.1 An illustration of an MTS trial. On the left is the sample image featuring the cursor and an unfamiliar conspecific's face, and on the right are two comparison faces. The comparison face on the bottom right is the correct response, the sample individual depicted in a different photograph. (Photo: F. de Waal)

individual recognition by presenting subjects with two different photos of the same individual as the correct pair. Thus, this task required that subjects recognize the identity of the individuals in the photograph and not rely on the photographic features. Subjects also performed very well on this task with few trial exposures. It is important to note that the goal of the task, match-the-sample, had not changed between this task and the first face-matching task. The only difference was in the dimension of similarity applied to the stimuli in the correct pair: two different photographs of the same individual, as opposed to two identical photographs. Chimpanzees showed remarkable skill in moving from one task to another, and performed above chance on these individual discriminations on the first testing session. These studies clearly demonstrate that chimpanzees recognize unfamiliar individuals using facial cues, thus making face recognition a highly salient aspect of chimpanzee social cognition. This is especially striking in light of the fact that the chimpanzees did not require any training in order to perform the initial face discriminations, as is the case for all the monkey studies.

The fact that monkeys require training to recognize faces should not necessarily be seen as evidence against the importance of face recognition in their social interactions. It may be more reasonable to assume that the laboratory testing procedures used to address social cognition and face recognition are not as salient for monkeys as they are for chimpanzees and humans, or that their behavior in these contexts is not as flexible.

Specific Feature Recognition and the Face Inversion Effect

Previous studies on facial feature detection in nonhuman primates have suggested that monkeys focus primarily on the region of the eyes when visually scanning faces of other monkeys and humans. Keating and Keating (1993), for example, examined the visual scanning patterns of monkeys viewing photographs of conspecifics. These studies revealed that the region of the eyes received more visual scanning than other areas of the face and head. This implies not that those areas of the face are attended to more, but that the eyes follow a predictable pattern when viewing a face. Kyes and Candland (1987) masked various combinations of facial features on a stimulus photograph of one baboon and recorded slide preference and viewing duration of four cage mates to this familiar individual. Their results showed that all subjects had a viewing preference and looked for a longer duration at the slides in which the eyes of the stimulus animal were visible, supporting the contention that the eyes are the most important feature for maintaining visual attention in this species. Using a technique more specialized for detecting specific eye movements, Nahm and colleagues (1997) demonstrated that monkeys scan the eyes in more detail than other areas of the face, regardless of whether the conspecific is presented in a still photograph or in a short video. Finally, Parr and colleagues (2000) reported that chimpanzees' ability to discriminate unfamiliar conspecifics based on individual identity was impaired when the target's eyes were masked, also suggesting that the eyes are the most salient facial feature in great apes.

The Gestalt psychologist Wolfgang Köhler was one of the first scientists to describe what is now known as the face inversion effect, or an impaired ability to discriminate faces when they are presented in their inverted orientation (Köhler, 1925). Since Köhler's work, the face inversion effect has been widely studied in humans and is often used as a marker of a specialized neural face-processing mechanism (Carey & Diamond, 1977; Diamond & Carey, 1986; Farah et al., 1995; Yin, 1969). When faces are inverted, their typical configuration of features is disrupted and recognition becomes impaired. The face inversion effect provides evidence that faces are not recognized simply by their specific features, but rather individuals are sensitive to the precise configuration and orientation of these features which are unique in every face. Only a few studies have investigated the inversion effect for faces in nonhuman primates, yielding inconsistent results.

Vermeire and Hamilton (1998a) reported a right hemisphere advantage for both learning and discriminating conspecific faces in split-brain macaques. These individuals showed impaired recognition of inverted conspecific faces presented to the right hemisphere but no deficit in their discrimination of the same stimuli presented to the left hemisphere. This finding reinforces the notion that face processing occurs primarily in the right hemisphere. These studies may be flawed, however, with regard to the amount and quality of information that was received by each eye. Severing the optic chiasm, for example, can lead to various forms of visual neglect in which portions of the visual field are eliminated and, therefore, process no visual information. If one eye sustained more visual neglect than the other, this may explain the observed findings without the need to involve differences in the ability of each hemisphere to process facial stimuli.

Wright and Roberts (1996) found that monkeys showed a face inversion effect for viewing human faces only. The nonhuman primate face stimuli consisted of species ranging from prosimians to apes; therefore, no stimulus consistency was maintained. Others have failed to find any evidence of a face inversion effect in monkeys (Bruce, 1982; Overman & Doty, 1982). Tomonaga et al. (1993) failed to find evidence of an inversion effect for human faces, but not for conspecific faces, in one chimpanzee. Therefore, although the face inversion effect represents an intriguing way to investigate the underlying neural mechanisms that may be involved in face processing, the nonhuman primate data are inconsistent and contradictory. If the face inversion effect is representative of a specialized face recognition process in humans, then it is expected that chimpanzees will also demonstrate impaired recognition of inverted conspecific faces.

Parr and colleagues have tested this assumption comparatively in chimpanzees (Parr, Dove, & Hopkins, 1998) and rhesus monkeys (Parr et al., 1999). Furthermore, these studies addressed the issue of the expertise effect, that is, whether the inversion effect in monkeys and apes is sensitive to the subject's expertise with a stimulus category, as reported in humans (Diamond & Carey, 1986). We presented chimpanzees with unfamiliar conspecific faces and human faces, two categories for which the chimpanzees had considerable expertise: capuchin monkey faces, a non-expert face category, and automobiles, a non-face category. The expertise effect was supported in that the chimpanzees showed the inversion effect for the human and chimpanzee faces but not for the unfamiliar capuchin monkey or the automobiles. This finding implies that stimuli that are highly familiar are processed by means of configurational cues, which is the critical feature that is disrupted during inversion. No evidence of the expertise effect was found in rhesus monkeys (Parr et al., 1999), however. Rhesus monkeys showed the inversion effect for most stimulus categories (also tested with human faces, conspecific faces, capuchin monkeys, automobiles, and abstract shapes), suggesting that their ability to recognize and discriminate visual images is highly dependent on the way those stimuli were originally presented (i.e., upright). This finding again suggests that monkeys are not as efficient or skilled in performing cognitive tasks as chimpanzees. The data also suggest that monkeys may process faces differently from chimpanzees, an intriguing but still unconfirmed hypothesis.

In a follow-up experiment, Parr (unpublished data) manipulated the likelihood that chimpanzees would use configurational information when processing similar categories of stimuli. To this end, chimpanzees were presented with human faces that contained highly salient features such as jewelry, facial hair, odd expressions, hats, glasses, and so on. Additionally, the same capuchin faces and automobiles that were used in the previous experiment were blurred so as to eliminate the detection of any specific feature. The hypothesis was that if the inversion effect is the result of highly familiar stimuli being processed configurationally, then altering this type of processing (by blurring or increasing the distinctiveness of specific features) should increase or decrease the likelihood that those stimuli will subsequently elicit the inversion effect. Consistent with the hypothesis, this study showed that chimpanzees process highly familiar categories of faces using configurational cues, thus providing evidence of the expertise effect.

Kin Recognition

The ability to recognize the relationship between individuals, especially kin relations, plays an important role in primate societies. In one of the best-known studies of kin recognition in monkeys, Dasser (1987) trained two long-tail macaques to discriminate slides of familiar mother-offspring pairs. After a predetermined criterion was reached, novel transfer trials were presented to test whether subjects would generalize the kin relationship to other known mother-offspring pairs. Generalization performance was above chance when the slides of the offspring had been taken recently, meaning that the ages of the offspring in the group and as depicted in the slide were similar. Performance was not above chance, however, when there was a discrepancy between the current age of the offspring in the group and the age of the offspring when the slide was taken. In a second study, the same two subjects were trained to discriminate only one example of either a familiar or an unfamiliar group member. They were then tested on their ability to generalize this "knowledge" to novel slides of those monkeys in different postures, or by body parts other than the face (Dasser, 1988). The results showed that generalization was above chance for the familiar group mates only, regardless of whether the novel slides depicted the face or another body part of the same individual. This suggests that more than just the face may be involved in discriminations of familiar individuals by rhesus monkeys, but that personal experience is needed with an individual before these discriminable cues are learned.

A study by Parr and de Waal (1999) tested whether chimpanzees may recognize kin relationships among unfamiliar conspecifics using facial similarity, that is, phenotypic matching (Lacy & Sherman, 1983). Five chimpanzees were presented with portraits of unfamiliar mothers and their sons (MS) or daughters (MD). Also presented were individual recognition trials (IR) presenting two different photos of the same individual and unrelated individuals (UC). These controlled for whether subjects were simply learning the correct responses by reinforcement history, or whether they actually recognized the facial similarity between mothers and their offspring. Subjects were given several hundred trials, and then their performance was compared across the four stimulus categories. As expected, subjects performed well on the IR trials, but they also performed above chance on the MS trials, and this result did not differ significantly from the IR trials. Also, as expected, subjects did not perform above chance on the UC trials. But, contrary to what we expected, the chimpanzees also performed poorly on the MD trials; this result did not differ significantly from the UC trial performance, suggesting that the chimpanzees saw no more similarity between portraits of mothers and daughters than between those of two unrelated individuals.

Facial Expressions and Emotions

Building on Darwin's (1872) classification of facial expressions as consisting of either inherited habits or learned simulations of those habits, Ekman and collaborators have proposed that there are two basic types of facial expressions: "innate" and "learned" facial expressions of emotions (Ekman & Friesen,

1975). Ekman and others maintain that humans possess a limited number of basic facial emotions (i.e., happiness, fear, disgust, anger, surprise, and sadness) and that these facial emotions are recognized across all human cultures, both literate and preliterate (Ekman, 1973). Humans, however, can also learn to display these facial expressions without their accompanying emotion (e.g., the smile of politeness versus the smile of enjoyment), and this use of nonverbal behavior shows considerable variability across cultures. Others have argued that true versus posed expressions are readily discernible, for example, the social versus the Duchenne smile (Frank et al., 1993; Surakka & Hietanen, 1998).

Experimental studies of facial expressions in nonhuman primates have also examined the extent to which these signals are innate or learned. In one of the earliest studies, Sackett (1966) demonstrated that socially naïve rhesus monkeys raised in isolation produced threatening facial expressions in response to photographs of threatening conspecifics. When the isolation-reared monkeys were reintroduced into a social group, however, they were unable to respond appropriately to the range of facial displays and complex emotional messages produced by their conspecifics. This was interpreted as evidence that the production of facial expressions may be the result of innate mechanisms, whereas the comprehension of facial displays made by other individuals in a social context requires a period of normal social and emotional development (see also Chevalier-Skolnikoff, 1974a; Kirkevold et al., 1982; Mason, 1985).

To investigate further the development of facial expression recognition, Mendelson and colleagues (1982) assessed how one-, three-, and seven-week-old rhesus infants responded to slides depicting faces of conspecifics with direct or averted gaze. No differences in responses were detected in the one-week old infants. By three weeks of age, however, infant monkeys showed social responses to, and looked for a longer time at, the faces with the direct gaze compared to those with averted gaze. This increased viewing was accompanied by negative emotional responses such as submissive squealing, grinning, and lip-smacking at the staring faces. By seven weeks of age, the infants avoided looking at the direct-gaze faces altogether. Instead, these infants preferred to look at the faces with the averted gaze and no longer produced emotional responses in their presence. These results suggest that by three weeks of age, rhesus monkey infants understand the social implications of the direct stare (i.e., that it is a threat) and respond submissively to this stimulus. These infants, however, do not as yet have the social competence or knowledge to avoid looking at these faces. By seven weeks of age, infants learn to respond to this socially threatening stimulus by looking away, a response that is critically important for integration in a rhesus macaque social group. It is important to note that these monkeys lived in social groups before the experiments were conducted. Therefore, they could have learned the appropriate response to facial expressions within their social group.

A series of intriguing studies conducted in the 1960s by Robert Miller and colleagues demonstrated that rhesus monkeys could obtain information on the emotional states of other individuals through their facial expressions (Miller, 1971). In one of these studies, Miller and colleagues (1959) exposed monkeys to a photograph of a familiar group mate with a neutral facial expression. The presentation of this conditioned stimulus (CS) was paired with an electric shock that subjects could avoid only by pressing a lever in their test cage. Subjects quickly learned to press the lever when this face appeared, thereby avoiding the shock. Next, subjects were shown another photograph of the same CS individual but this time making a fearful facial expression. When subjects were shown this new stimulus, they spontaneously produced significantly more avoidance responses than during the acquisition of the conditioned response when the CS was the same monkey with a neutral face. This suggests that a negative facial expression is much more effective in communicating an upcoming aversive event (i.e., the shock) than a neutral face, illustrating that at some level the monkeys understood its inherent negative emotional meaning.

Together, these studies indicated that early social experience is less important for the production than for the comprehension of facial expressions of emotions, and that monkeys that have been raised with conspecifics are able to obtain information on the emotional states of other individuals through their facial expressions. Some of these studies also suggest that threatening expressions are better for communicating negative events than neutral faces. More experimental evidence that primates can discriminate between different facial expressions and recognize their emotional content has been provided by studies of chimpanzees.

Facial Expression Discrimination in Chimpanzees

Studies conducted at the Yerkes Primate Center have confirmed the ability of chimpanzees to categorize facial expressions of emotions typical of their own species. These studies were done by means of a similar computerized MTS paradigm as that described above. In one study (Parr, Hopkins, & de Waal, 1998), subjects were first presented with a sample image on the computer monitor showing an unfamiliar conspecific making one of five facial expressions: hoot face, bared-teeth display, relaxed open-mouth display, scream face, and relaxed-lip face (see Figure 12.2). The correct comparison was always another individual making the same type of facial expression, while the nonmatching face was a neutral portrait of a third individual. Therefore, these experiments extended the MTS paradigm so that the basis of similarity between the sample and the correct comparison was the type of expression being made, not the identity of the individuals making them. Three of the five facial expressions—the bared-teeth display, scream, and relaxed open-mouth display—were discriminated above chance levels on the first session (day 1). This makes any explanation of performance based on the subjects' learning which stimulus was correct from their prior history of reinforcement extremely unlikely. More interesting, however, is the fact that the relaxed-lip face was never discriminated above chance levels from the nonmatching neutral face. This finding seems to indicate that although subjects could have used the distinctive feature of the relaxed-lip face, the droopy lower lip, to discriminate it from the plain neutral face, they did not. Considering the fact that both of these expressions are functionally neutral in a social context, and communicate no overt emotional or motivational predisposition, these results raise the possibility that subjects may have relied on the emotional content or intensity of the faces to perform the discriminations instead of using distinctive facial features.

In a second experiment, the faces were arranged more systematically so that each of the five expressions was combined with every other expression as the nonmatching comparison. The result was twenty directional dyadic stimu-

Figure 12.2 An illustration of six facial expressions in chimpanzees. From left to right these include the neutral portrait, pant-hoot, bared-teeth display, relaxed open-mouth display (play face), scream, and relaxed-lip face. (Photo: F. de Waal)

lus sets in which every expression was paired with every other expression. The stimulus sets that resulted from the dyadic combination technique could be divided into two main types: those in which the nonmatching comparison shared features in common with the sample expression, and those in which the two expressions were featurally distinct. An example of each of these trial types can be seen in Figure 12.3. The hypothesis under investigation was that if the chimpanzees were relying predominantly on specific facial features to discriminate their facial expressions, then they should perform much worse on the trials in which the two expressions were similar in appearance compared to the trials in which the two expressions appeared distinct. The data supported this hypothesis, but only for a select few expressions (the bared-teeth display, the hoot face, and the relaxed-lip face). Two expressions showed the reverse pattern; that is, performance was actually better when the sample and nonmatching comparison shared similar features (the scream face and the relaxed open-mouth display). So, while the data support a trend for a reliance on salient features in the categorizing of facial expressions, this does not seem to be true for all expressions.

In summary, our research has shown that chimpanzees do not seem to be relying on the distinctiveness of facial features to discriminate facial expressions, although there was a trend for distinctive facial features to be important in the dyadic expression-matching task. One might speculate that subjects were relying on one type of feature more than another (i.e., emotional content and

 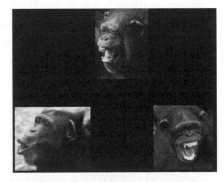

Figure 12.3 An example of the two trial types in the dyadic expression matching task. On the left is a trial in which the two expressions (scream and play face) share similar features, and on the right is an example in which the two expressions (scream and pant-hoot) are featurally distinct. (Photo: F. de Waal)

intensity), and only if this feature was not easily discernible did they move to another discriminative cue. In addition, subjects may have perceived the facial expressions differently owing to morphological variability in the way different expressions look when made by different individuals; that is, a correct pair in these experiments showed two different individuals making the same facial expression. It also raises the possibility that the expression types that did not seem to be affected by an overlap in facial features (i.e., the scream face and the relaxed open-mouth display) may be perceived as highly salient categories not prone to confusion with any other expression type. Within a social group, these expressions may be easy to recognize regardless of social context, changing environmental conditions, or the individual performing the expression. Future studies should try to isolate these possible cues and test each of them systematically, in addition to examining the relative contribution of facial and vocal channels in recognizing and categorizing dynamic social signals. Indeed, the presentation of facial expressions as static, black and white stimuli, as in our experiments, is hardly analogous to the way in which these dynamic social displays are normally encountered. The following section addresses some of these concerns by describing how chimpanzees process dynamic social information.

Dynamic versus Static Facial Expressions

Our studies on social cognition in chimpanzees have adopted a more ecologically salient format in that we are no longer presenting static photographs for discrimination but rather short video clips. These clips contain images of unfamiliar conspecifics and are inserted into the basic MTS format as the sample stimulus. To our knowledge, this is the first demonstration of MTS in nonhuman primates in which subjects are required to match photographs to video. Detailed data on emotional processing using this MTS format have been published (Parr, 2001), but here we will describe the results of unpublished experiments in which we first presented four of the original five chimpanzees with dynamic videos in a matching task.

After a brief training session to acclimate the subjects to the new video format, subjects were presented with short videos of unfamiliar conspecifics to which the correct matching photograph was a still frame from the sample video. The nonmatching photograph was another unfamiliar individual. Subjects performed considerably better on this task than on the training task, which was to match familiar objects. The average number of trials to reach the performance criterion of two sessions above 85 percent correct was 195. Considerable variability was present in that one subject met this criterion in the first two sessions while the slowest subject required ten sessions. These were the same two subjects who represented the group extremes during the object training task.

Next we attempted to replicate the first expression matching task described above, in which the nonmatching comparison was always a neutral portrait. The expression categories this time included short videos of each expression type (i.e., bared-teeth display, pant-hoot, the play face, scream, and relaxed-lip face), which were combined with neutral portraits as the non-matching comparison photograph. There were three examples of each of the six categories, so eighteen different trials were presented. Subjects required an average of 422 trials to reach criterion on this dynamic version of the facial expression matching task, with one subject reaching criterion in three sessions while another required twenty-two sessions. The performance on this dynamic expression matching task and the data from Parr, Hopkins, and de Waal (1998) for the static expression matching task are presented in Table 12.1.

According to Table 12.1, all categories were discriminated better when the sample was presented in a dynamic format, although this result is inherently confounded with subjects' experience. Some notable exceptions are apparent,

Table 12.1   Mean performance of subjects on the video expression matching task

Subject	Bared-teeth	Play face	Scream	Hoot	Relaxed-lip	Neutral
Jarred	42.10	77.90	77.30	78.90	95.50	98.90
Katrina	84.75	72.25	86.25	62.50	83.50	94.50
Lamar	43.09	74.32	57.50	65.68	95.68	69.95
Scott	45.00	91.50	76.25	77.00	85.38	84.75
Mean	53.74	78.99	74.33	71.02	90.01	87.03
Mean static expression matching task	63.72	70.29	60.57	62.81	39.32	N/A

however, the most prominent being that subjects had no problem distinguishing the relaxed-lip faces from the neutral portraits in the dynamic version of the task. This is interesting in light of the original findings which showed that subjects were unable to make this discrimination (Parr, Hopkins, and de Waal, 1998). In addition, subjects performed well on the play face in both tasks, better on the scream in the dynamic version, and worse on the bared-teeth category when it was presented dynamically. This expression is probably the most ritualized primate expression, and is observed in many species.

The data become more interesting when the static and video versions of the dyadic expression matching task are compared. Overall, subjects showed a significant advantage in discriminating dyads of expressions when there was little feature overlap, compared to dyads in which the two expressions appeared similar (see Figure 12.3). In the new dynamic task format, however, this performance advantage disappeared. Therefore, while it appears that chimpanzees are able to categorize their own expressions, the type of information that is most critical for these discriminations remains unclear, though it seems related to the dynamic nature of the information. For some expression types, the salient cue appears to be distinctive features, as for the play face and bared-teeth face in the static version of the expression-matching task, but this does not appear to hold true when the task presents short video clips of these expressions. Future studies should examine the role of multimodal features, such as visual and auditory characteristics, in greater detail. For example, how would chimpanzees discriminate facial expressions when the corresponding vocalization is altered or mismatched to the visual image? Are there some expressions for which the vocalization is the more salient feature than the visual information and vice versa? Studies under way at the Yerkes Primate Center are attempting to understand further how chimpanzees process and categorize the emotional information conveyed by facial expressions.

The Communicative Use of Nonvocal Signals

Ekman's view that facial expressions necessarily reflect underlying emotional states has been criticized by Fridlund (1994), whereas others, such as Russell (1994), have expressed doubts about Ekman's universality hypothesis. Fridlund (1994) maintains that facial expressions are signals that evolved by natural selection to elicit specific responses in other individuals. In his view, similar to many animal signals, human facial expressions convey context-specific information about an individual's intent or motivation and co-evolved with receiver psychology. Clearly, the two views are not incompatible but differ in their focus; that is, Ekman emphasizes the proximate control of facial expressions, whereas Fridlund focuses on their adaptive function. Thus, it is very likely that many facial expressions convey information about emotions while at the same time have evolved to elicit specific responses from other individuals.

Both Ekman and Fridlund recognize that some nonvocal signals serve a communicative function that is strictly associated with our use of language. Most facial expressions and hand gestures occur during speech and can be interpreted as a type of facial paralanguage. Ekman and Friesen (1969) distinguished four types of facial (or gestural) paralanguage: emblems, illustrators, regulators, and self-manipulators. Emblems are conventional, symbolic gestures such as nods, head-shakes, or thumbs-up, which we use to replace words or remark upon ongoing speech. They appear to be strictly related to words, and, similar to words, they exhibit contrast in their expression, are learned, and vary cross-culturally (but see Fridlund, 1994). Illustrators are facial expressions and gestures used to place accents on words (e.g., raising eyebrows), describe the shape and size of objects (iconic gestures), or indicate objects in space or time (indexical gestures). Unlike emblems, illustrators mimic the forms of the objects they stand for, have a graded nature, and lack arbitrariness. The third category of paralanguage includes regulators, that is, expressions or gestures that are used to regulate conversation (e.g., turn-taking, pauses, etc.). Finally, self-manipulators are self-directed activities such as self-grooming or scratching. They probably reflect emotional states associated with motivational conflict, and there is little or no evidence that they are functionally related to language (Maestripieri, Schino, et al., 1992). Following Darwin (1872) and Andrew (1963b), Fridlund (1994) suggested that facial paralanguage evolved through the emancipation and ritualization of physiological reactions or facial movements initially associated with vocalizations.

While both Ekman and Fridlund draw a sharp distinction between facial expressions of emotions and paralanguage, Burling (1993) emphasizes the distinction between symbolic and nonsymbolic facial expressions and gestures. According to Burling (1993), all symbolic signals including emblems and sign language are strictly related to language, whereas all nonsymbolic signals belong to what he calls the "gesture-calls" system of communication, which has little to do with language. In his view, nonsymbolic signals convey a tiny amount of information relative to that conveyed by language, and such signals seem better suited for communicating emotions and intentions than for describing the world. For example, Burling (1993) argued that there are only a limited number of nonsymbolic signals and these are characterized by immediacy; that is, they cannot be used to refer to the past or the future, or to things that are far away or out of sight. The gesture-call system also lacks syntax and is less subject to deliberate control than language. Finally, nonsymbolic gestures are indeterminate and extensively graded. Although Burling's view is shared by many linguists, it probably underestimates the importance of nonsymbolic gestures in human communication. In fact, in addition to conveying subtle information about emotions and intentions, nonsymbolic gestures also provide a necessary context for interpreting speech.

Similar to Fridlund (1994), Burling (1993) argued that primates produce no symbolic gestures and few or no iconic signals. Thus, both authors seem to agree that parallels can be drawn between primate gestures and nonsymbolic human gestures, but not between primate gestures and human symbolic gestures or language. According to Burling (1993), primate gestures and human nonsymbolic gestures share several similarities including the fact that (1) they both convey information about emotions, individual identity, intentions, or impending behavior; (2) their vocabulary is restricted, and new signals cannot be added through learning because their use is narrowly determined by inheritance; and (3) they are both graded rather than discrete systems. Thus, human nonsymbolic gestures are similar to primate nonvocal signals, but both are very different from language.

While linguists have emphasized the distinction between symbolic and nonsymbolic facial expressions and gestures, cognitive psychologists have emphasized the role of intentionality in primate and human nonvocal communication. According to Tomasello and co-workers, "non-intentional" gestures are signals that have evolved by natural selection to regulate social interactions involving sex and aggression (Tomasello & Call, 1997). For example, among chimpanzees, non-intentional signals are "more or less involuntary postural and facial displays that express mood, for example, piloerection indicating an aggressive mood, penile erection indicating a sexually receptive mood, and 'play face' indicating a playful mood" (Tomasello & Camaioni, 1997, p. 9). Many non-intentional signals do not seem to be dependent on individual or social learning, and they are not used flexibly. This is in contrast to the process of intentional communication, in which alternative means may be used toward the same end, and the same means may be used toward alternative ends (Bruner, 1972). According to Tomasello and colleagues, intentional gestures are not genetically determined and have not evolved by natural selection. Rather, they are learned during ontogeny. In some cases the learning involves not the signal itself but the appropriate communicative use of the signal. Thus, evidence that either gestures or their usage are learned would constitute evidence that they are used intentionally.

In the following sections we review and discuss some of the research on the communicative use of facial expressions and gestures in monkeys and apes to assess whether some of the above distinctions (e.g., between innate and learned signals, expressions of emotions and communicative displays, symbolic and nonsymbolic gestures, and intentional and nonintentional signals) can be successfully applied to primate nonvocal communication.

Nonvocal Communication in Old World Monkeys

Socially living primates interact with conspecifics on a daily basis, and communication plays an important role in both cooperative and competitive interactions. Two basic functions of communication are to bring individuals together when there is need for cooperation and to keep them apart whenever competition arises (Shirek-Ellefson, 1972). Many of the complexities of primate nonvocal communication result from the elaboration of this simple approach-avoidance system.

Most quantitative studies of nonvocal communication in Old World monkeys have been conducted in macaques. In this section we review and discuss how macaque nonvocal signals are used and what social and communicative functions are accomplished through them in the contexts of competition, mating, affiliation, and parental care. In many cases the precise meaning of macaque signals has not been unequivocally assessed, and therefore the interpretation of their functional significance is only tentative.

Communication in the context of competition allows individuals to negotiate access to resources while reducing the probability of costly fights. Macaques use virtually all of their body parts to communicate their intention to engage in or to avoid a fight. Facial expressions of threat typically involve staring at the opponent with eyes wide open, mouth open without showing the teeth, eyebrows raised, and ears flattened (Altmann, 1962; Hinde & Rowell, 1962; Kaufman & Rosenblum, 1966; van Hooff, 1967). The number of facial elements present in the threat varies with its intensity: simple staring signals a threat of low intensity, but as the intensity of the threat increases, the eyebrows, the ears, and the mouth are progressively recruited into the signal (Zeller, 1980, 1996). Competition over feeding and mating or simple proximity to another individual can elicit a threat. The threat signals the individual's potential or motivation to engage in a conflict. The relationship between threat and aggression, however, need not be necessary. Aggression may not be preceded by threats, and in most cases threats are not followed by aggression. "Defensive" threats are often accompanied by recruitment screams (Gouzoules et al., 1984), or by rapidly alternating the gaze between the target of the threat and a potential ally (de Waal et al., 1976; Maestripieri, 1996a).

In most cases, macaques respond to threats with submissive signals. These include facial expressions and postures that expose vulnerable regions of the body, or behavior patterns belonging to the mating or infantile repertoire. The most common submissive signal in macaques is the silent bared-teeth display, also referred to as "fear grin" or "grimace" (Figure 12.4). The bared-teeth display may also be accompanied by scream vocalizations, and the function of this display appears to differ depending on whether the bared-teeth is a silent or vocalized display (de Waal & Luttrell, 1985). In most macaque species, the bared-teeth display occurs primarily in response to threats or aggression or the approach of a dominant individual (Maestripieri, 1997). In this latter context, the most likely meaning of the bared-teeth is "I am afraid" or "Do not attack me" or a combination of both (Maestripieri, 1996c). The function appears to be to reduce the likelihood of future aggression. The signal, however, may or may not be effective in preventing aggression depending on the circumstances. In some species of macaques, and also in other primates, the bared-teeth display may occur without any prior interaction between two individuals and be followed by affiliation or mating (Petit & Thierry, 1992; Thierry et al., 1989). Therefore, the way in which the bared-teeth display is used seems to vary across species.

Another common submissive signal in macaques is the hindquarter presentation. As in the bared-teeth, subordinates present to dominants upon receiving aggression or in situations with high risk of aggression (Chadwick-

Figure 12.4 Bared-teeth display by female Guinea baboon. (Photo: D. Maestripieri)

Jones, 1989; Maestripieri, 1996a; Maestripieri & Wallen, 1997). The presentation can also be displayed to initiate affiliative interactions. The bared-teeth display and the presentation probably have a similar meaning but are used in slightly different contexts (Maestripieri & Wallen, 1997). Bared-teeth and presentation can occur in conjunction with other submissive signals such as lip-smacking and teeth-chattering (Altmann, 1962; Dixson, 1977; Hadidian, 1979; Hinde & Rowell, 1962). Lip-smacking may follow a submissive signal upon receipt of aggression and perhaps communicates an intention to engage in reconciliation. Teeth-chattering also occurs in response to aggression or risk of aggression and is frequently displayed immediately before or after the bared-teeth display (Maestripieri, 1996b; Preuschoft, 1992). Teeth-chattering is also used during the course of affiliative interactions and seems to have a meaning similar to that of lip-smacking but a stronger emotional component, ranging from fear to excitement (Chevalier-Skolnikoff, 1974b; Maestripieri, 1996b).

Nonvocal communication plays an important role also in mating interactions. Male and female must communicate to each other their intention to mate and coordinate their behavior so that a sexual interaction can physically take place. In fact, aggression between potential mates is always latent. Females in estrus signal their readiness to mate by approaching males and presenting their hindquarters to them. Estrous females may also slap the male or climb on the male's back and engage in a series of pelvic thrusts. Macaque males use facial expressions such as the pucker, bared-teeth, lip-smack, or teeth-chatter while approaching an estrous female (Christopher & Gelini, 1977; Goosen & Kortmulder, 1979; Maestripieri, 1996a). Once the distance between males and females has been reduced, males use tactile signals such as hip-touches to induce the female to present her hindquarters. During copulation, the female often reaches back, grasping the male's flank or leg with her hand, and lip-smacks while the male displays bared-teeth, squeaks, or teeth-chatters (Maestripieri 1996a, b). In this context, facial expressions could simply reflect an underlying orgasm-related emotion (Goldfoot et al., 1980).

Mating interactions are often preceded or followed by affiliative behavior such as grooming. Prior to grooming, an individual may use the pucker, lip-smack, teeth-chatter, the bared-teeth, or the presentation while approaching or to induce another individual to come closer. Once distance is reduced, grooming is usually requested by lying on the ground and exposing the part of the body to be groomed. Postural changes are also used to signal the intention to terminate the interaction (Boccia, 1986). Affiliative communication between males often involves hip-clasping, mounting, and genital manipulation (i.e., one individual reaches out and fondles the other's genitalia). In contrast, females often embrace each other (Dixson, 1977; Maestripieri, 1996a; Thierry, 1984). Mounting, clasping, and embracing may be accompanied by lip-smacking or teeth-chattering by one or both partners (Chevalier-Skolnikoff, 1974b; Maestripieri, 1996a). These signals may express excitement and/or function to minimize risk of aggression and promote bonding. Particular affiliative interactions between adults known as "greetings" probably serve to negotiate dominance relationships, alliance formation, and decision-making processes relative to the direction of travel (also in baboons; see Colmenares, 1991; Smuts & Watanabe, 1990; Figure 12.5).

Play is frequently characterized by the occurrence of a distinctive facial expression known as the "play face" (van Hooff, 1962, 1967). This expression

Figure 12.5 Greeting interaction between two adult Guinea baboons. (Photo: D. Maestripieri)

consists of a wide opening of the mouth, as if attempting to bite, but without clenching the teeth. The top teeth are typically covered, but the bottom teeth may be visible. Play faces are usually displayed simultaneously by two play partners and may be associated with soft vocalizations (Symons, 1978). In macaques, the play face does not occur in contexts other than play, is not used to initiate play from a distance, and occurs most often during wrestling play (Preuschoft, 1992). Traditional explanations of play signals in monkeys and other animals maintain that they are a form of "meta-communication," that is, they would communicate to the partner "I want you to know that this is only play" (Altmann, 1962; Bateson, 1953). This explanation, however, implies quite sophisticated cognitive processes, notably the ability to attribute knowledge to others. Some authors have hypothesized that the play face may simply be a form of play rather than a signal with a complex meaning (Maestripieri, 1997; Pellis & Pellis, 1996; Tanner & Byrne, 1999). Other explanations concerning the meaning of the play face are also plausible.

One final context in which nonvocal signals play an important role involves interactions with infants. Although macaque infants communicate their basic needs (food, transport, and protection) to their mothers mainly with vocalizations (Maestripieri & Call, 1996), mothers communicate with their infants primarily with facial expressions and body postures. As in other contexts, the most likely function of these signals is to allow for reduction of distance between mother and infant. Macaque mothers retrieve their infants from a distance by using the pucker, the bared-teeth, the lip-smack, or the presentation, depending on the species or the circumstance (Jensen & Gordon, 1970; Maestripieri, 1996d; Maestripieri & Wallen, 1997). Some of these signals are used interchangeably or occur in rapid succession; for example, a mother will first lip-smack or bare-teeth to her infant and then turn around and raise her tail (Hinde & Simpson, 1975). These interactions are particularly frequent in the first weeks of an infant's life when mothers display these expressions to encourage their infants' independent locomotion (Maestripieri, 1995a, 1996d).

The Interpretation of Macaque Nonvocal Signals

From the perspective of proximate causation, it is likely that some facial expressions, gestures, or body postures reflect underlying emotional states. From a functional point of view, these and other signals may communicate information about the signaler's impending behavior, requests to approach and engage in affiliation, mating, or play, or requests to inhibit behaviors such as aggression or fleeing. Many nonvocal signals in the macaque repertoire can be used in different contexts with different meanings, and the same communicative function can often be served by different signals. Thus, the relation between the structure of signals and their meaning seems to be probabilistic rather than fixed, with important information being provided by the social context. The appropriate use of signals in relation to context probably requires social learning during development (see above). Thus, if flexibility in the use of signals and social learning are among the hallmarks of intentional communication, then most, if not all, macaque nonvocal communication can be interpreted as intentional. This, however, does not necessarily imply that macaques can take another individual's perspective or understand how another individual may perceive the meaning of their expressions.

To a human observer, macaque signals appear to be graded rather than discrete. Facial expressions seem to merge into one another in both their morphology and their meaning. For example, the defensive threat merges with the bared-teeth, the bared-teeth merges with teeth-chattering, teeth-chattering merges with lip-smacking, and lip-smacking merges with the pucker (Shirek-Ellefson, 1972). Thus, Burling's (1993) characterization of primate nonvocal signals as graded appears to be correct. It cannot be ruled out, however, that what appear as graded signals to humans are in fact discrete signals to macaques. Macaques may be better than humans not only at discriminating their signals from one another, but also at identifying subtle differences in the signal structure associated with different contexts or the identities of different senders (Zeller, 1980). To understand fully the nature of macaque nonvocal signals, we need better techniques for recording and analyzing signals and a better understanding of macaque perceptual abilities relative to our own.

The view of macaque facial expressions and gestures as graded signals merging into one another is in part the result of the notion that nonvocal signals only reflect emotional and motivational states. In this view, the flexibility in the combination of elements in each signal would be an adaptation to reflect the intensity of the emotion or motivation underlying the signal. Similarly, the structural similarities among signals would allow the expression of rapid transitions in emotional or motivational states (Shirek-Ellefson, 1972). Although macaque nonvocal signals do not appear to be used to communicate about aspects of the external environment, such as the presence of food or predators, they convey information about social activities occurring within the group, and often also contain an indication of the location where the activity will take place (e.g., grooming or direction of travel). Thus, although macaque nonvocal signals are not used symbolically, some of them do have iconic or indexical properties (Emory & Harris, 1978; Maestripieri, 1996b). These signals convey information about external events in addition to the emotion and motivation of individuals, but the external events that are often most relevant to macaques are those occurring in their social group and involve the activity of other individuals.

Nonvocal Communication in the Great Apes

The facial expressions and gestures of the great apes, particularly chimpanzees, have been described by many authors. Köhler (1925) argued that chimpanzees communicate about not only "subjective moods and emotional states" but also "definite desires and urges" (p. 273). According to Köhler (1925):

> [a] considerable proportion of all desires is naturally expressed by slight initiation of the actions which are desired. Thus, one chimpanzee, who wishes to be accompanied by another, gives the latter a nudge, or pulls his hand, looking at him and making the movement of "walking" in the direction desired. One who wishes to receive bananas from another initiates the movement of snatching or grasping, accompanied by intensely pleading glances and pouts. The summoning of another animal from a considerable distance is often accompanied by a beckoning gesture very human in character . . . In all cases their mimetic actions are characteristic enough to be distinctly understood by their comrades. (p. 274)

More than four decades after Köhler, other researchers such as Goodall (1968, 1986), van Hooff (1973), and Berdecio and Nash (1981) described many facial expressions and gestures used by wild and captive chimpanzees in a variety of social contexts. According to van Hooff (1973), most chimpanzee nonvocal displays occurred in relation to play, aggression and submission, affiliation, and excitement. De Waal (1988) noted that, similar to macaques, chimpanzees and bonobos use bared-teeth displays for submission and appeasement purposes, the pout face to request contact, nursing, and food, and various types of play faces during play. In addition to the typical play face observed in macaques, chimpanzees, bonobos, and orangutans also exhibit "smile" and "laughter" (Chevalier-Skolnikoff, 1982; van Hooff, 1967). Unlike monkeys, all great apes are reported to exhibit novel facial expressions, particularly in the context of play, in which the facial muscles are contorted in highly variable shapes and combinations (Chevalier-Skolnikoff, 1982; de Waal, 1988). Hand-begging gestures are rare or nonexistent among Old World monkeys but have been reported in all four species of great apes. De Waal (1988) believes that bonobos use this gesture as an overture for reconciliation after a fight more than chimpanzees do, whereas chimpanzees use this gesture to request food or agonistic support more than bonobos do. Other gestures and postures observed among chimpanzees and bonobos include wrist-shaking, arm-waving, arm-up, stretch-over, hunch-over, hand- and foot-clapping, chest-beating, various types of rhythmic movements involving the hands and feet, and embraces (see Goodall, 1968). Some of these signals are presumably used as attention-getters (e.g., arm-waving), while others are more explicit requests for sex or grooming. McGrew and Tutin (1978) reported a "hand-clasp" posture exhibited by two individuals engaged in allogrooming among wild chimpanzees (see also de Waal & Seres, 1997). It is unclear, however, if the posture has any communicative significance to the individuals engaged in this behavior or to other group members. A form of attention-getting behavior, "leaf-clipping," has been reported among chimpanzees in the Mahale Mountains, mostly in the context of requesting sex or food (Nishida, 1980).

The distinction between intentional and nonintentional gestures in chimpanzee nonvocal communication was first explicitly made by Plooij (1978, 1984). Plooij described several gestures used by infants during interactions with their mothers or with their peers: a "hands around the head" gesture to request tickling, an "arm-high" gesture to initiate grooming; a "food-beg" gesture to request food; a "leaf-grooming" and "running away with an object" gestures to encourage social play. Plooij identified gestures as being intentional when they were used flexibly and/or were accompanied by gaze alternation. Flexibility meant that the same signal could be used to achieve different goals, and different signals could be used for the same goal. For example, Plooij observed a juvenile who, in some cases, used an "arm-high" gesture to invite grooming under its arm, and in other cases used the same gesture in an appeasement context. Gaze alternation involved monitoring the response of another individual to the signal and suggested that the sender had some understanding of the effect of the signal on the recipient. For example, Plooij observed that when begging for food, infants alternated their gaze between their mother's face and their hand (see also Bard, 1992, for similar interactions in orangutans). Plooij (1984) argued that some gestures develop ontogenetically from goal-directed actions but then become true signals in part owing to the influence of social learning and shaping. He suggested that, at some point during development, the infant understands that the mother is an independent agent with her own communicative ability, and at this point, most gestures begin to be used intentionally.

Building on Plooij's work, Tomasello and colleagues (1985, 1989) focused on intentional gestures used by juvenile chimpanzees during interactions with their mothers or other group members. In addition to flexibility in the use of signals and gaze-alternation, intentional gestures were also identified through response-waiting and audience effects. Response-waiting meant that the individual waited for a response from another individual after sending the signal, thus suggesting that the goal of the gesture was to communicate. Finally, audience effects occurred when an individual used a signal differently depending on the identity or attentional state of the recipient. Some of the intentional gestures studied by Tomasello and collaborators were used to get the mother's attention and initiate nursing (e.g., touching her body), solicit carrying (placing one arm on the back of another individual or pulling the other individual along), request grooming (exposing the body part to be groomed or placing the other individual's hand on this part), request food sharing (placing the hand under the adult's mouth in a begging gesture), or invite play (arm-raising, ground-slapping, head-bobbing, hand-clapping, foot-stomping, running away and looking back). Tomasello et al. (1985) reported that some gestures were used quite flexibly in different contexts and that the older juveniles used some novel behaviors not observed among other individuals. Tomasello et al. (1989) also reported the creation of new gestures when new materials were introduced to the group (e.g., some chimpanzees used newly introduced wood chips to initiate play by throwing them at others).

In follow-up studies, Tomasello and colleagues (1994; Tomasello, Call, Warren, et al., 1997) reported that juveniles could initiate play with one of eight different gestures and exhibited three different gestures to solicit nursing, three to request carrying, and two to beg for food. Visual signals were used only if the recipient was looking and tactile signals only if the recipient was attending to the behavior of the signaler. Many communicative or social interactions (e.g., play) were initiated with an "attention-getter" gesture such as "throwing chips," "poking at," or "ground slapping." Play gestures represented 60–70 percent of gestures across all ages, and play was also the most flexible context in which gestures occurred. The comparison of gestures across time periods and generations showed that there was little overlap among gestures either within or between groups. Specifically, (1) some juveniles used gestures that no other group member used; (2) some juveniles used gestures that had not been directed to them and that they had little opportunity to observe; (3) juveniles raised only with peers developed many of the same gestures as those raised with adults; (4) within-group variability in the use of gestures was very high. Two individuals were taught new food-begging gestures by human experimenters and then reintroduced into the group, but other individuals kept using their own gestures and did not adopt the new ones during the course of the study.

On the basis of these findings, Tomasello and Camaioni (1997) argued that chimpanzees use two basic types of intentional gestures with their conspecifics: "attractors" and "incipient actions." Attractors are imperative gestures aimed at getting other individuals' attention, while incipient actions also include imperative gestures, but they are used to communicate information about impending behavior or to request specific activities. According to Tomasello and Camaioni (1997), both attractors and incipient actions are mostly used in dyadic contexts and are never used for declarative purposes, such as to share interest in or comment on something or someone. Furthermore, many gestures rely on physical contact between signaler and recipient, or else are incipient movements anticipating contact. Thus, in Tomasello and Camaioni's (1997) view, gestures are more closely related to the mechanical manipulation of another's body than to the psychological manipulation of another's mind. Finally, intentional gestures are probably learned by a process of ontogenetic ritualization (or conventionalization) and not by observational learning. Ontogenetic ritualization is similar to a social shaping process in which each individual learns the effects of its behavior on the other's behavior. Tomasello and Camaioni (1997) emphasized the differences between the intentional gestures of chimpanzees and those of human children and adults. In their view, the latter are often used triadically and for declarative purposes, are often indexical or symbolic, are meant to influence others psychologically, not mechanically, and are learned through social rather than individual learning.

One may argue that the above characterization of chimpanzee gestures is overly strict and that the differences between ape and human gestures are overstated. For example, chimpanzees and other primates can use both attention-getters and requests for action in triadic ways (e.g., as alarm calls, food begs, or recruitment solicitations). Furthermore, among both human and nonhuman primates, many gestures are used to manipulate behavior rather than either the body or the mind. It is also likely that the gestural repertoires of both nonhuman primates and humans are the result of a combination of genetic expression and individual and social learning processes. Finally, some of the distal and declarative uses of human gestures are strictly related to language, and there is evidence that when great apes learn the rudiments of human language, the use of indexical and symbolic gestures follows closely (see below).

The use of iconic or indexical gestures in the context of communicative interactions with conspecifics has also been reported for great apes that had not been language trained, particularly bonobos and gorillas. Savage-Rumbaugh and colleagues (Savage et al., 1977; Savage, Rumbaugh & Wilkerson, 1978) reported high variability in both bonobo copulation positions and the facial expressions and gestures that accompanied them, including prolonged mutual gaze, and a number of different gestural and postural signals. Savage et al. (1977) argued that some gestures were iconically related to the desired change in the partner's behavior. For example, the male would often physically push the female's body into a desired copulatory position or move his hand across the female's body, a gesture the investigators interpreted as an iconic indication of what he wanted the female to do (but see Tomasello & Call, 1997, for a different interpretation).

Tanner and Byrne (1993, 1996, 1999) have argued that some captive lowland gorillas use both iconic and indexical gestures similar to those observed by Savage et al. (1977) among bonobos. Most of the observed gestures occurred during play, a few in agonistic contexts, and none in feeding situations. In the context of play, an adult male appeared to use his arms iconically to indicate to another individual the direction in which he wanted her to move, or the action he wanted her to perform. Many of these gestures appeared interchangeable in function. There were marked individual differences in the use of these gestures, including an increase in their expression during development, as well as changes in the preferred types of gestures as individuals matured. Although some of their developmental data were consistent with the conventionalization hypothesis, Tanner and Byrne (1999) also argued that some aspects of gestural communication, notably the comprehension of gestures, are not learned but somehow "biologically encoded." Furthermore, they argued that gorillas are anatomically and cognitively preadapted to use iconic and indexical gestures and have the potential for symbolic communication as well. For example, they noted some similarities between the iconic gestures used by zoo gorillas and the signs used by language-trained gorillas, who often elaborated on species-typical gestures in their symbolic communication. In this view, gestural communication in humans and in the great apes shares some of its cognitive underpinnings, and the sharp distinction made by some authors between the symbolic use of gestures in humans and the nonsymbolic nature of primate gestures is not entirely warranted.

Nonvocal Communication between Apes and Humans

Human-reared apes, and particularly those that have been language-trained, differ from other apes in many aspects of their nonvocal communicative skills, most notably in their understanding and use of gaze, pointing, and symbolic gestures (see also Chapter 8). In fact, the communicative competence of human-reared apes can approximate that of human children (e.g., Tomasello & Camaioni, 1997). In children, gaze following and joint attention are important for the development of language and the association of words and objects. Monkeys and apes (as well as other animals including domestic cats and dogs) routinely follow the gaze of other individuals as a way to acquire information about their environment (Tomasello et al., 1998). They also often engage in gaze alternation to direct another individual's attention to something or someone. For example, monkeys use gaze alternation to recruit support from conspecifics when challenged (de Waal et al., 1976). Gaze alternation in conjunction with gestures has also been observed among captive chimpanzees (Plooij, 1979; Russell et al., 1997; Tomasello et al., 1985), and it is relatively common in the context of interactions with humans, particularly in association with requests for food and pointing (Leavens & Hopkins, 1998). Human-reared apes are more likely to use gaze alternation while gesturing than mother-reared apes, and language-trained chimpanzees are the most likely to engage in this behavior (Hopkins & Leavens, 1998; Leavens & Hopkins, 1998; Miles, 1990). Tomasello and Camaioni (1997), however, have argued that gaze alternation or gaze following in chimpanzees does not necessarily imply an understanding of other individuals' intentionality. In other words, although gaze cues are interpreted to predict another individual's behavior, it is unlikely that they are used to interpret their mental states or knowledge (but see Emery, 2000; Emery et al., 1997). For example, when chimpanzees look into the face of a conspecific, they may simply be checking its physical status, gauging its orientation toward a target, or predicting something about that individual's future action in that they understand that the eyes have something to do with action. Such explanations are both plausible and difficult to disprove, but they can be easily applied to many gaze-related interactions in humans as well.

Along with joint attention, pointing plays an important role in language acquisition. Children begin pointing at about eight months of age, and the frequency of this gesture increases dramatically in the following year (Butterworth, 1991). Human children initially point without regard for the attentional status of an observer, but gradually integrate their pointing behavior with visual interactions, notably joint attention or gaze alternation. In children, pointing is often accompanied by vocalizations, especially naming behavior (Butterworth, 1991). Thus, pointing serves an important declarative function. Pointing has been observed in all four species of great apes, and in some species of monkeys as well (Leavens & Hopkins, 1999). Nevertheless, pointing is generally exhibited by primates in captivity who have been exposed to a great deal of contact with humans (but see Vea & Sabater-Pi, 1998). In other words, pointing is not part of the natural communicative repertoire of monkeys and apes, but rather represents an adaptation by nonhuman primates to facilitate interactions with their caregivers. Although Leavens and Hopkins (1999) argued that most pointing observed in apes develops in the absence of specific human training, it occurs most often in the context of interactions with humans, in which some reinforcement processes are likely to occur, even if these were not deliberate attempts to shape behavior. Pointing is most common in language-trained apes, but this may reflect the fact that these animals have more experience in face-to-face interactions with humans rather than reflecting their language training per se (Leavens & Hopkins, 1999).

Captive chimpanzees typically point by extending their arm and stretching all fingers (whole-hand point) rather than by using their index finger only. This may be due, at least in part, to differences in hand anatomy and the resting state of the index finger (Povinelli & Davis, 1994). Pointing with the index finger is most common in captive chimpanzees that have been raised by humans, and in particular those that have been language-trained, again suggesting that contact with humans is the critical factor. In the great apes, pointing is often accompanied by gaze monitoring, or gaze alternation between the human and an object or location in the environment. Pointing is rarely accompanied by vocalizations. Pointing in the apes is mostly observed in the context of requests for food, but it is also used with a declarative function by language-trained apes. Leavens and colleagues (1996; Leavens and Hopkins, 1998, 1999) argued that chimpanzee pointing is a clear example of intentional communication because it is accompanied by attention-getting behaviors, gaze alternation, and audience effects: apes are less likely to point when the observer is absent or is looking away (see also Call & Tomasello, 1994; Leavens et al., 1996; Krause & Fouts, 1997).

It is possible that although captive apes may learn how to point in the context of interactions with humans, they do not fully understand pointing behavior because they do not understand that other individuals are intentional beings with their own subjectivity (Call & Tomasello, 1994; Tomasello & Camaioni, 1997; see also Chapter 8). Therefore, although captive chimpanzees may learn how to use pointing to demand objects, or more generally to affect the behavior of human beings, they may not understand the communicative function of this gesture (see Povinelli, Bering, & Giambrone, in press, and Chapter 8 for further discussion of chimpanzees' comprehension of pointing and other referential gestures). Thus, in this view, the humanlike behavior of captive apes is not fundamentally different from that of other domesticated animals such as cats and dogs, which, after extended contact with humans, begin to show humanlike emotional and communicative responses. Many researchers, however, and particularly those involved with ape language research, disagree with this interpretation (see Chapter 14).

Overall, the studies of nonvocal communication in monkeys and apes reviewed in this chapter suggest that the distinction between innate and learned facial expressions and gestures is probably not a useful one. Most nonvocal communication in human and nonhuman primates probably results from an interaction between species-typical genetic predispositions and processes of individual and social learning. There is evidence, however, that the production of facial expressions in primates is less dependent on environmental influence than the comprehension and use of these signals. The notion that all primate nonvocal signals reflect emotional states or impending behavior in a reflex-like manner is also unsupported. Rather, there is little disagreement that most of these signals can be used flexibly in different communicative contexts and with different meanings.

The nonvocal communicative abilities of nonhuman primates, particularly the great apes, have attracted the attention of both linguists and cognitive psychologists. The former have made comparisons between the linguistic properties of primate and human nonvocal signals, whereas the latter have focused on the cognitive processes underlying the acquisition and use of facial expressions and gestures. Both linguists and cognitive psychologists have emphasized the differences, rather than the similarities, between primate and human nonvocal signals. Using concepts and operational definitions imported from human research to investigate the communicative processes of nonhuman primates is certainly appropriate. Ultimately the value of any research paradigm will depend on the amount of new knowledge that the paradigm is able to produce, regardless of where it comes from. But whenever monkeys and apes are compared to humans according to criteria such as their ability to use human cognitive skills or acquire human language, it is important to keep in mind that nonhuman primates are not human, and therefore there is no reason to expect that they should be able to think or communicate in a humanlike fashion.

LAP wrote the section of this chapter on face recognition and DM the section on nonvocal signals.

13

Nonlinguistic Vocal Communication

Michael J. Owren, Drew Rendall,

and Jo-Anne Bachorowski

Over the past several decades, researchers interested in comparing vocal communication in nonhuman primates and humans have tended to adopt one of
two quite different positions. On the one hand, primatologists working both
in laboratory and in field settings have argued that vocalizations in nonhuman primates (hereafter "primates") can exhibit a number of important
commonalities with speech behavior in humans (e.g., Ghazanfar & Hauser,
1999; Hauser, 1996; Seyfarth, 1987; Snowdon et al., 1997). Parallels have, for
example, been proposed at neurobiological, perceptual, and cognitive levels,
with particular attention given to the possibility that humanlike mental representations are involved in mediating linguistic-like, referential vocalizing. Others have emphasized that nonhuman primate vocalizations are fundamentally
different from human speech (Lancaster, 1975; Lieberman, 1984), suggesting
instead that the most fruitful comparison is to emotion-related sounds like
laughter and sobbing (Deacon, 1989, 1997). All roads seem nonetheless to lead
back to language, in that researchers studying these nonlinguistic sounds typically also adopt a linguistic-like, meaning-based approach in proposing that
these signals convey representational messages.

We suggest that a greater effort should be made to understand nonlinguistic signaling in its own right. Specifically, we argue that approaching communication with an emphasis on linguistically inspired constructs such as representation may be useful in certain cases but is a poor fit for some of the most ubiquitous vocalizations in both primates and humans. Accordingly, we outline an "affect-induction" account of signaling that eschews these sorts of concepts. This approach specifically concerns vocal behavior, although it may also be relevant to other signal modalities. The approach is grounded in the observation that the medium of sound presents particular opportunities for influencing others by means other than conveying encoded, representational messages. We argue that a simpler and likely earlier-emerging function of vocal production was specifically to influence the affective states of listeners, thereby also modulating their behavior. Two particular strategies we propose are producing vocalizations whose acoustic features have an immediate impact on listener attention, arousal, and emotion, as well as pairing individually distinctive sounds with important affective changes so as to elicit learned emotional responses later with these sounds. Both strategies are routinely available to all primates including humans, and we therefore suggest this overall approach as one that can potentially be applied to nonlinguistic vocal behavior throughout the Primate order.

In making these arguments, we first describe two approaches in which vocal behavior is proposed to have linguistic-like meaning, representing either the vocalizer's internal motivational state or its external circumstances. While often discussed as alternative characterizations, these interpretations are fundamentally similar in that both rest on demonstrating a correlation between the occurrence of particular acoustic features and a proposed referent. We then sketch out the affect-induction framework, using several well-studied primate vocalizations to exemplify the approach. Finally, the same framework is applied to human laughter as we argue that it accounts for detailed aspects of the acoustics and rates of laughter shown by human participants in two laboratory experiments. The goal throughout is to examine why—counter to representational explanations—both primates and humans can routinely produce sounds that are not specific to particular motivational states or behavioral circumstances. While seldom viewed that way, this problem may be the key to eventually understanding the widespread occurrence of nonlinguistic signaling.

Representational Accounts of Nonlinguistic Vocalizations

Representational interpretations have long been a part of understanding communication, as illustrated by the widely held view that the function of many animal signals is to convey encoded information concerning the signaler's motivational state or behavioral tendencies. While therefore often referred to as the "motivational" approach, it is nonetheless fundamentally a representational one. Particular signal features are claimed to be correlated with and hence representative of critical internal states in the signaler, providing information that reduces recipient uncertainty about intentions or behavior (Smith, 1977, 1997). As the internal states in question are very difficult to measure (but see Bayart et al., 1990; Fichtel et al., 2001; Friedman et al., 1995; Jürgens, 1979), researchers have typically tested for associations between the occurrence of a signal and some observable aspect of subsequent signaler actions.

Overall, the results of such work have been mixed, with some researchers arguing that signals are only modestly related to upcoming behavior (e.g., Caryl, 1979; Paton & Caryl, 1986) and others proposing much stronger associations (e.g., Dabelsteen & Pedersen, 1990; Hansen, 1986). Nelson's (1984) analysis of attack and flight displays in pigeon guillemots showed both outcomes occurring within a single circumstance: intruders attempting to displace territory owners from their nesting sites. While male territory owners in this shorebird species sometimes signaled when attacking intruders and sometimes did not, displays by these interloper opponents were strongly correlated with departure.

It has also become increasingly common to argue for "reference" rather than motivational representational value in animal signals, meaning that the signals act as symbols of objects or events external to the signaler. A key event in prompting this change was the finding that vervet monkeys produce acoustically distinct alarm vocalizations to predators such as snakes, eagles, and leopards, and that the calls can by themselves precipitate functionally differentiated escape responses in listeners (Seyfarth et al., 1980a, b; Struhsaker, 1967). These sounds have therefore been proposed to convey information about the three predator types with language-like semanticity, having representational value that is emancipated from signaler motivational state. An important component in the interpretation is that there is no particular relationship between the acoustic form of each signal and its referent, meaning that these calls show the same arbitrariness that typifies the relationship between a human word and its semantic content. Thus there is no obvious relationship between the snake alarm call and an actual snake, any more than there is between the word "snake" and the animal itself.

Research in both acoustic primatology and other areas blossomed following this seminal demonstration, often aiming to uncover further evidence of external reference or to demonstrate other language-like phenomena. The strongest subsequent examples of semantic-like function have tended to be predator-specific alarm calls, although referential signaling has also been claimed in other contexts (e.g., Cheney & Seyfarth, 1982, 1988; Gouzoules et al., 1984; Hauser, 1998a). To help sort out when it is legitimate to invoke referentiality, Macedonia and Evans (1993; Evans, 1997) argued that two criteria should be met. First, they suggested that signaling should demonstrate "production specificity," occurring almost exclusively in response to a common, specifiable category of eliciting events or objects. Second, responses by others should show "perceptual specificity," with occurrence of the signal alone being sufficient for receivers to select an appropriate response. While the degree of specificity needed to meet these two requirements was not precisely specified, the authors suggested "near-exclusivity" as a guideline.

Expanding the Boundaries of Signaling

While presented as alternative views, the motivational and referential approaches are both inherently representational; the function of communication is assumed to be conveying encoded information from one individual to another. Each therefore rests on demonstrating corresponding correlations between particular features of signals and one or more observable aspects of the signaling circumstances. In the motivational view that relationship concerns signaler state or upcoming behavior, while in the referential approach the connection is to external objects or events. Yet even when empirical study has failed to demonstrate specificity in either domain, the representational approach itself has not been drawn into question. Instead, such outcomes have been taken to indicate either that the information encoded in the signal is requisitely vague (e.g., Smith, 1977, 1997) or explicitly false (e.g., Dawkins & Krebs, 1978; Krebs & Dawkins, 1984).

The other possibility, of course, is that signals are not conveying representational information, and that any inference made by recipients concerning signaler circumstances is a secondary outcome. The function of signaling must ultimately always be to influence the current or future behavior of the signal recipient, and there is no a priori reason to believe that sending encoded information is the only way in which to do so. Nonetheless, at least for studies involving animals such as birds and mammals, few if any such alternatives have been suggested. The absence of such alternatives puts the representational approach itself at risk of being circular and unfalsifiable, as happens when the representational value of purportedly motivational communication is tested using the signals themselves as evidence of underlying states (e.g., Dabelsteen & Pedersen, 1990). A common experimental design used to test the referential value of vocal signals is similarly problematic. In a number of studies, the referential similarity of two different call types has been operationally defined as the degree to which habituation occurring to the one generalizes to the other. The procedure involves first presenting a series of exemplars drawn from one of the sound categories, each time measuring how quickly and how long the target listener looks in the direction of the sound source. When this call type no longer elicits a looking response, an exemplar from the other category is played. If the looking response is reinstated, the two types are deemed to have dissimilar meanings. If, however, habituation generalizes to the novel call type, researchers infer that the two are referentially similar (e.g., Cheney & Seyfarth, 1988; Hauser, 1998a; Zuberbühler et al., 1999).

Experiments using this methodology have not tested the role that psychological processes other than reference-related decoding might be playing in guiding looking responses. In effect, the design contrasts a psychologically based argument (that the two sounds have similar referents) with an acoustically based alternative (that their physical dissimilarity should prevent generalization). The crucial issue, however, is not whether the two sound types are physically different but how they affect the psychological mechanisms that influence listener attention and orienting responses. In effect, this generalization-of-habituation design assumes that referential value is the only significant factor in motivating an animal to look in the direction of a conspecific's call. Auditory and visual attention is of course influenced by a variety of factors, including the abruptness of a sound and whether it shows frequency or amplitude modulation. Later in this chapter we describe evidence showing that these and other features can routinely influence listener attention and arousal at the lowest levels of brain organization. While having no necessary connection to representation or the higher-level cognitive processes implicated by typical referentiality arguments, these attributes do contribute to the overall salience of a sound. The nonrepresentational alternative that has not been addressed when primate calls are played back to test referentiality is that generalization occurs from one call type to another based on relative salience, or because the sounds share attention-getting auditory features like abrupt onsets that are nowhere near so obvious in visual representations such as spectrograms that experimenters rely on.

Our argument is that vocalizations likely first evolved precisely because callers could affect the behavior of others by taking advantage of both the salience of acoustic events and the listener's own capacity for affective learning. In this view, any linguistic-like signaling that occurs in an otherwise nonlinguistic system has probably become evolutionarily possible because nonrepresentational vocalizations were already routinely occurring. The larger point at this juncture, however, is that testing representational explanations of signaling requires that alternative, nonrepresentational accounts of the same behavior also be explicitly included. In other words, the motivational and referential interpretations must be made subject to possible disproof, for instance, when results show that a given signal does not meet the specificity criteria that would make a signal representational in the first place. A priori exclusion of nonrepresentational interpretations creates circularity and tautology, significantly detracting from the inferential power afforded by otherwise well executed studies.

Lack of Specificity in Primate Calls and Human Laughter

In this section we present three examples of nonlinguistic vocal signals in nonhuman primates and humans that available evidence indicates do not meet either of the specificity criteria outlined earlier. The two primate examples are macaque coos and screams, illustrating both that acoustically similar sounds can occur in diverse circumstances and that acoustically diverse sounds can occur in similar circumstances. The third example is human laughter, which, while less well studied than macaque calls, shows much the same pattern of diversity in form and context. We focus on these instances in particular because each is one of the most commonly used vocal types in their respective species' repertoires. Rather than being an exception to the rule, then, these and other sounds that fail to show the specificity associated with representational signaling may instead be the more general case.

Macaque coos. Coo calls are vowel-like vocalizations produced by a variety of macaque species. These have arguably received more research attention over the decades than have any other primate call. They are tonal calls characterized by stable, quasi-periodic vocal-fold vibration occurring at a relatively low fundamental frequency (F_0) of, for example, 400–500 Hz in adult females (illustrated in Figure 13.1). Significant energy is also present at integer multiples of this base rate of vibration, and coos therefore often show a rich spectrum of these harmonics. Caller age, sex, individual identity, and species can all be associated with variation in F_0, the number of higher harmonics present, details of amplitude patterning over the frequency spectrum, and the relative tonality of the call. This call type has been documented as a common vocalization in the repertoires of stumptail (Lillehei & Snowdon, 1978), long-tailed (Goustard, 1963; Palombit, 1992), Japanese (Green, 1975; Itani, 1963), 1975), rhesus (Rowell & Hinde, 1962), pigtailed (Grimm, 1967), bonnet (Hohmann, 1989), lion-tailed (Hohmann & Herzog, 1985; Sugiyama, 1968),

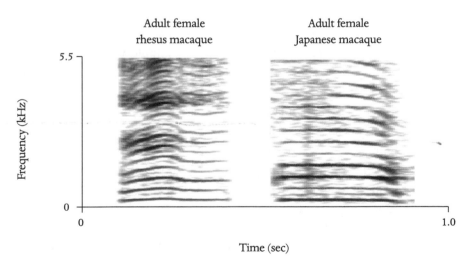

Figure 13.1 Narrowband spectrograms of coo calls by adult female rhesus and Japanese macaques (45 Hz bandwidth, Gaussian analysis window, 11,025 Hz sampling rate).

Tonkean (Masataka & Thierry, 1993), and Barbary macaques (Todt, 1988). While early evidence from some species suggested that context-specific acoustic variation might be present in coos, the larger picture does not suggest specificity in either acoustics or use.

For example, Green (1975) characterized Japanese macaque coos as showing continuous acoustic gradation, while also arguing that the sounds consist of seven separable acoustic types differentially associated with ten different circumstances. Though demonstrating statistical heterogeneity across these contexts, Green's analysis did not take into account variation due to either the age, sex, or individual identity of the caller. Thus, seven of the categories were age- or sex-specific, including "separated male," "female minus infant," "nonconsorting female," "female at young," "young alone," "young to mother," and "estrous female." Furthermore, each of these categories could be associated with multiple coo variants. For instance, separated males produced sounds from five different purported subtypes, and offspring used four different types when calling to their mothers. The categories "dominant at subordinate," "subordinate to dominant," and "dispersal" were not confounded by age-sex class, but were also associated with multiple acoustic variants. In addition to the contexts included in Green's analysis, coos have been found to be produced by Japanese macaques of all ages as they wait for daily food provisions provided by caretakers (Green, 1975; Masataka & Fujita, 1989; Owren et al., 1992), by youngsters either playing among themselves or vocalizing to an adult male (Owren et al., 1993), and in response to a coo from another animal (Itani, 1963; Mitani, 1986). Here again, multiple versions have been documented in each circumstance (Masataka, 1992; Owren & Casale, 1994; Sugiura, 1998).

Coos have also been found to occur in a variety of situations and acoustic forms in other species, specifically including Tonkean (Masataka & Thierry, 1993) and lion-tailed macaques (described as "whoo" calls by Hohmann & Herzog, 1985). Examining the distribution of ten proposed acoustic variants of these sounds in lion-tailed macaques, Hohmann (1989) looked for differential associations with the ten contexts Green (1975) had used earlier. Here, each situational category was associated with 2.9 of these purported subtypes, with each subtype in turn being associated with 2.9 different circumstances. In rhesus monkeys, coos are produced during individual or group movement (Hauser, 1991), foraging (Hauser & Marler, 1993), anticipation of provisioning

(Masataka & Fujita, 1989; Owren et al., 1992), weaning (Owren et al., 1993), play (Owren et al., 1993), by youngsters calling to an adult male (Kalin et al., 1992), and by infants separated from their mothers (e.g., Kalin et al., 1992; Newman & Symmes, 1974) or experiencing other kinds of distress (Maestripieri et al., 2000). In Hauser and Marler's (1993) study of free-ranging rhesus, coos were by far the most common call type occurring in food-related circumstances and could not be distinguished acoustically from coos recorded in other situations.

Rhesus monkey screams.   Screaming is another sort of sound that has been documented in a variety of species, including both primates and humans. Such sounds have also been suggested to be representational in some cases, with the strongest claims being made by Gouzoules et al. (1984) for rhesus monkey screams. These researchers proposed that juvenile rhesus monkeys produce five different scream types in agonistic situations, and that the sounds help recruit aid from mothers and other biological kin by conveying specific information about the fight. In this interpretation, screams can inform listeners about the opponent's relative rank and genetic relationship to the victim, and whether severe physical contact has occurred during the attack. Like coos, rhesus monkey screams show acoustic gradation (illustrated in Figure 13.2), which Gouzoules et al. nonetheless argued fell into discrete categories based on visual inspection of spectrograms and careful listening.

The data that Gouzoules et al. presented, however, provide little support for this claim. As in the case of Japanese macaque coos, the researchers were

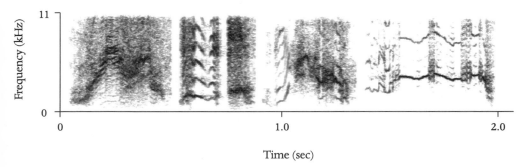

Figure 13.2   Wideband spectrograms of four screams from a single adult female rhesus macaque (300 Hz bandwidth, Gaussian analysis window, 22,050 Hz sampling rate).

able to demonstrate statistical heterogeneity in the distribution of purported subtypes across calling circumstances. Their data also clearly showed, however, that "noisy" screams were consistently predominant. These were the most frequent sounds in three of the five contexts examined, were tied for most frequent in the fourth, and were the second most frequent in the fifth (see Table 5 in Gouzoules et al., 1984). Even in this last case, noisy screams far outnumbered the three scream types that occurred at lower rates. In other words, the result was not that each circumstance was differentially associated with a particular scream type, but rather that there was statistically significant variation in the relative rates of producing the various calls when viewed across all types and contexts taken together.

The authors buttressed their representational interpretation using the latencies and durations of looking responses by free-ranging animals hearing screams played back from a hidden speaker. Here, playback of screams proposed to convey information indicative of increasing agonistic severity was found to be correlated with shorter latencies and longer looking times. Whereas responses to noisy screams were clear, however, target animals were very inconsistent in looking toward the speaker when the three other types were tested. For noisy screams, response latencies averaged approximately 2.1 seconds and mean looking time was about 3.3 seconds, while those for "arched," "tonal," and "pulsed" versions were both highly variable and very slow. Here, mean latencies were approximately 5.3, 7.2, and 10.8 seconds, respectively, while corresponding mean looking times were 3.5, 1.8, and 0.9 seconds (after Gouzoules et al., 1984). Hence, it is difficult to argue that the target animals were routinely responding to the sounds at all. Our interpretation is that while outcomes for noisy screams provide good evidence that listeners were responding to the sounds, in the other three cases it is difficult to conclude that the subjects were even interested.

Subsequent attempts to classify rhesus monkey screams based on the criteria described by Gouzoules et al. have failed, as these sounds show pronounced acoustic variability (Kitko et al., 1999). In fact, these sounds exhibit the hallmarks of a nonlinear vocal-production system (Wilden et al., 1998), including virtually instantaneous shifts from tonality to noise, segments showing simultaneous periodicity and noise, multiple apparent energy sources, and frequency-domain sidebands that result from amplitude modulation in the temporal domain (Owren & Nederhouser, 2000). Because mammalian sound production is based on vibrations occurring in coupled vocal folds, interactions between the two folds readily produce highly unstable behavior that leads to rapid shifts among multiple possible vibration modes. Wilden et al. (1998) therefore argue that the noise routinely observed in mammalian sounds can be chaotic rather than stochastic, with important implications for the predictability of their acoustic features and the degree to which vocalizers can control them (see also Tokuda et al., 2002). Specifically, minute changes in factors such as air pressure and muscle tension will cause the vocal folds to flip almost instantaneously from one vibration mode to another, unpredictably producing dramatic qualitative changes in acoustic output that are unlikely to be under close control by the vocalizer. The apparent chaos visible in the frequency spectra of screams in rhesus monkeys and other primates has been empirically confirmed in analogous components of human infant crying (Mende et al., 1990). The upshot is that if the underlying vocal production system behaves in a chaotic fashion, it becomes difficult to argue for stable, nuanced variation in the acoustic domain. Because such chaotic phenomena are evident in the acoustic features of sounds such as screams and other calls of distressed animals (e.g., Riede et al., 1997), it constitutes evidence against the possibility of discrete, representational vocal categories.

Human laughter.   The last example we will consider is the case of human laughter. Like coos in macaques, laughter in humans is a routine accompaniment to interindividual interaction, occurring roughly thirty times more often in social than in solitary situations (Provine & Fischer, 1989). Despite the prominent role laughter plays in human sociality, however, it is difficult to link these vocalizations to either particular motivational states or external circumstances. For example, many have suggested that laughter signals the occurrence of positive emotions such as happiness and joy (e.g., Darwin, 1872; McComas, 1923; Nwokah et al., 1993; van Hooff, 1972). It has also been pointed out, however, that laughter can be correlated with negative states such as anger, shame, and nervousness (Darwin, 1872; McComas, 1923), self-deprecation (Glenn, 1991/1992), and appeasement or submission (Adams & Kirkevold, 1978; Deacon, 1997; Dovidio et al., 1988; Grammer & Eibl-Eibesfeldt, 1990). Other arguments have connected laughter to cognitive incongruity and other forms of humor (Apte, 1985; Black, 1984; Deacon, 1989; Edmonson, 1987; Milford, 1980; Weisfeld, 1993), tension-release (Hayworth, 1928; Milford,

1980; Sroufe & Waters, 1976), cueing individuality (Edmonson, 1987), expression of sexual interest (Grammer, 1990; Grammer & Eibl-Eibesfeldt, 1990), and "mobbing" of individuals being excluded from a group (Eibl-Eibesfeldt, 1989).

A remarkable aspect of this list is that all of these suggestions seem plausible, at least based on everyday experience. But while each seems at least partly right, the various proposals are not uniformly compatible from a representational point of view. Instead, the list includes claims that laughter can represent both positive and negative emotional states, and can have meanings that range from very broad to quite specific. Adding to the difficulty, it has recently been found that laugh acoustics are much more variable than previously thought (Bachorowski et al., 2001), and that the rate and acoustics of laughter shown under experimental conditions can vary significantly even when participants almost uniformly report experiencing positive emotions (Bachorowski et al., under review). Although Provine and Yong (1991) have argued that laughter is a stereotyped vocalization consisting of a series of vowel-like bursts produced using regular vocal-fold vibration ("voicing"), Bachorowski et al. (2001) instead found that only 30 percent of the 1,024 laughter bouts produced by ninety-seven participants of both sexes were predominantly voiced, whereas 48 percent were mainly unvoiced, and 22 percent were a mix of the two (Figure 13.3). These outcomes are particularly important because a series of perceptual experiments then showed that human listeners were very consistent in rating voiced laughs more positively than unvoiced versions (Bachorowski & Owren, 2001). In other words, although produced in uniformly positive circumstances, the laughs themselves were acoustically variable, and that variability was in turn functionally significant with regard to how positive later ratings were when listeners scored either the sounds or their own affect, or inferred vocalizer characteristics. Laugh bouts were also highly variable in overall duration and the number of calls involved. Truly remarkable variation was found in F_0 values for voiced versions, both within and among laughers and laugh bouts. While mean F_0 was found to be approximately twice that of normative speech for both males and females, F_0 excursions could also exceed the highest notes of tenor and soprano singers, respectively. With this more detailed information becoming available, the relationship between laugh acoustics, laugher states, and the contexts of laughing has become even more muddled than before.

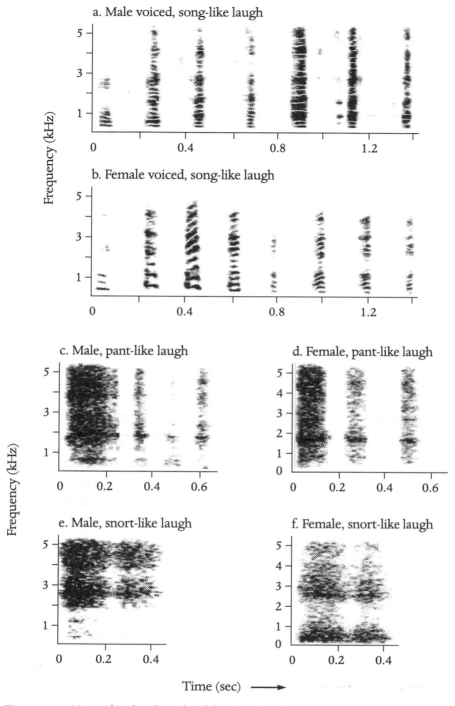

Figure 13.3 Narrowband (a, b) and wideband (c, d, e, f) spectrograms of representative laughs by human adult males and females, illustrating three of the multiple laugh types identified by Bachorowski & Owren (2001). (Redrawn from Bachorowski & Owren, 2001)

An Affect-Induction Account of Nonlinguistic Vocalizations

The problem for understanding nonlinguistic vocalizations such as primate calls and human laughter is thus that a linguistic-like, representational perspective appears a poor fit, whether the signaling is proposed to involve either internal or external reference. These sounds illustrate that in both primates and humans, particular nonlinguistic vocalizations need not be uniquely associated with particular contexts, nor are particular contexts necessarily associated with particular sounds or acoustic features. A further difficulty is that there have been few if any alternatives to these kinds of representational, "meaning"-based interpretations. We suggest that there is value in rethinking the problem from the beginning, for instance, by assuming only that the function of signaling is to influence the behavior of others—no matter how that influence is achieved. If so, there are some very simple ways that vocalizers can change listener behavior, specifically by producing sounds that have an impact on that individual's affective state. Here, the term "affect" is being broadly construed to include processes such as arousal, attention, motivation, and emotion, with an important difference from previous approaches in that it is the listener's state that is of primary concern, not the vocalizer's. This affect-induction approach (also referred to as "affect-conditioning"; see Owren & Rendall, 1997, 2001; Rendall & Owren, 2002) to vocal signaling proposes that the medium of sound affords at least two ways to change the affective state of listeners without needing to bring linguistic-like meaning into the picture. We refer to these strategies as exerting "direct" and "indirect" effects, respectively.

Direct Effects of Vocalizations

The most basic means by which one individual can influence another's use of sounds is to produce vocalizations whose features in and of themselves elicit affective responses. Sound is known to have exactly that kind of power, as illustrated by the acoustic-startle response (e.g., Davis, 1984). This reflexive interruption of ongoing behavior is elicited by acoustic features such as abrupt onsets and high amplitudes and is believed to be present in every hearing species (e.g., Eaton, 1984). While vocalizations are unlikely to have evolved specifically to elicit startle, the phenomenon demonstrates that sound itself can affect a listener's nervous system at the most basic levels—those that control arousal and attention. It is therefore not surprising that abrupt onsets and rapid upward frequency sweeps have been linked to increased arousal and activity in listeners across a diversity of mammalian species, whereas slower-onset sounds that gradually fall in frequency have been to found to decrease arousal and concomitant activity in listeners (reviewed by Owren & Rendall, 2001). In the affect-induction view, such links show that vocalizers can directly impact listeners by incorporating these sorts of features in calls. In other words, regardless of other aspects of the vocalizations or social circumstances, a caller can potentially modulate a listener's nervous system activity simply by producing sounds with particular kinds of acoustic properties.

From this perspective, it seems highly unlikely that even the most linguistic-like primate vocalizations are truly arbitrary in acoustic structure. In the case of vervet monkey alarm calls, for example, each of the sounds investigated by Seyfarth and colleagues (1980a, b) is well designed to elicit listener attention. Leopard alarm calls are explosive, high-amplitude sounds (Seyfarth et al., 1980b; Struhsaker, 1967), and the quieter snake and eagle alarm calls consist of a series of abrupt-onset, broadband energy pulses (Owren & Bernacki, 1988). While neither high amplitude nor pulsed structure appears to be inherently related to any referential value these calls may have, they point to an underlying evolutionary process in which vocal acoustics were shaped by attention-getting functions as well as predator-specific differentiation.

Those principles are likely still at work in the click-like "gecker" calls that young macaques use to solicit attention from their mothers in "fussing" or "protesting" contexts (e.g., Maestripieri et al., 2000; Owren et al., 1993). Adopting a representational perspective, one would say that the distressed youngster's calls convey information about its emotional state or need for help. Because mothers are in many cases strongly predisposed to come to an infant's aid, however, invoking metaphorical constructs such as information encoding and transmission appears superfluous. It is likely sufficient that the acoustics of these vocalizations are well suited to drawing the mother's attention and increasing her arousal level, thereby helping trigger a response.

In fact, relying on direct effects of acoustics may be one of the few ways in which relatively impotent vocalizers can gain some leverage over the behavior of more powerful individuals. Infants and other dependent offspring routinely face this situation with their caretakers, but the same holds for any subordinate animal vis-à-vis more dominant conspecifics. The caretaker or dominant individual has the upper hand, having by far the greater control over whether the outcome of the interaction is positive or negative for the other. The youngster or subordinate can nonetheless make use of the direct impact that sounds can have. Though a crude and likely energetically expensive strategy, vocalizations can be used as a sort of acoustic bludgeon in situations in which the interests of callers and listeners are in conflict. In this case, the calls should have salient acoustic features such as abrupt onset, dramatic frequency and amplitude modulations, high amplitudes, and any available annoying or aversive qualities. Callers are therefore expected to vocalize again and again, mounting an aural assault of intrusive sounds that vary over time so as to preclude habituation by the listener. Calling should stop only when vocalizers get what they want or when the costs of vocal production exceed the benefit to be gained.

This is what occurs when infants experience rough handling or aggression, are separated from or rejected by their mothers, or are being weaned. Indeed, in these instances youngsters are known to produce streams of acoustically salient sounds. For example, Grimm (1967) described infant pig-tailed macaques separated from their mothers as producing long series of intense sounds that included both noisy and tonal versions (see also Simons & Bielert, 1973). Newman and Symmes's (1974) study of isolate-reared rhesus infants included three comparably aged control offspring raised with their mothers. When the latter were placed in an isolation cage, they also produced a variety of calls, described as coos, pant-threats, geckers, squeaks, and screeches. Variety has also been the hallmark of the calls produced by young rhesus monkeys experimentally separated from their mothers. Newman (1995) described such sounds as including four coo variants and a total of seventeen qualitatively different vocalizations, even though the infants being recorded were separated from their mothers for only five minutes at a time. Coos were the most common and became increasingly prevalent as the infants matured, but other sounds included all five scream types described by Gouzoules et al. (1984), as well as geckers and squeaks (see also Bayart et al., 1990; Kalin et al., 1992; Levine et al., 1987; Maestripieri et al., 2000). Variability and acoustic salience have also been documented in the calls of distressed infants in Japanese macaques, in the form of coos that escalate to screeches and screams (Green, 1981), multi-segment "rise" calls that begin as a coo and end as a scream (Green, 1975; Owren, unpublished data), and a variety of nonlinear phenomena occurring in calls from ten different infants recorded by Riede et al. (1997).

During weaning, infant macaques can no longer elicit maternal responses and are commonly observed to produce hundreds of diverse and ongoing sounds even when ignored for only a few minutes. From a representational perspective, there is no obvious function for either repeating or varying the calls in these situations. The youngster is unlikely to be experiencing corresponding changes in its needs, nor is there any evident role for representational information transmission. If information content were the critical factor, mothers should respond as soon as they hear and "decode" the calls. From the affect-induction point of view, however, the whole point is that mothers are not always motivated to respond. By combining reiterative sound production with a diversity of acoustic features, an infant can use the energy of the signals themselves as a barrage of aural pinpricks that wear down the mother's resistance and help induce that motivation.

This interpretation is also consistent with the finding of Hammerschmidt et al. (1994) that six- to eighteen-month-old Barbary macaque youngsters produce long call sequences when unable to settle among the clusters of groupmates in sleeping trees. The authors noted that if the vocalizers are "announcing" their desire for a spot, it is puzzling that the sounds should be so conspicuous and should continue at such length. Under natural circumstances, this is risky behavior that can attract predators while making it more difficult to hear them approaching. Consistent with an affect-induction interpretation, however, the vocalizations were found to include calls with a variety of acoustic characteristics. Hammerschmidt et al. (1994) examined six different types whose spectrograms include coo-like, scream-like, and gecker-like sounds. Cluster analysis suggested a total of fourteen call types when the features of about 1,800 vocalizations were examined. These calls were produced either in this dusk-calling context or when infants were attempting to escape the clutches of a male alloparent (Hammerschmidt & Todt, 1995).

The affect-induction framework arguably also makes much better sense of the remarkable variability and evident nonlinearities in rhesus monkey screams. Gouzoules et al. (1984) argue that these sounds are simply too loud not to be designed for distance transmission and hence for recruiting potential allies from afar. An alternative interpretation is that a primary function of the calls is precisely to be loud and jarring to the caller's opponent (Rendall, under review). In this view, subordinate vocalizers are attempting to make themselves less appealing as objects of aggression, with variable high-amplitude sounds being exactly what is needed for such a strategy. Whereas the representational account of Gouzoules et al. implies that scream types should fall into discrete call categories with close, mutually exclusive relationships to the particular circumstances involved, the affect-induction view argues instead that acoustic variability is crucial. That is not to say that call features will be random, because key production factors such as tracheal air pressures and laryngeal muscle tension will certainly be affected by vocalizer arousal level and emotional state. Rather, the argument is that any attendant regularities between vocalizer state and acoustic output will be a secondary outcome, with the primary function of calling being to have a direct auditory impact on the opponent.

Finally, in the case of coos, many investigators have noted that pitch contour is a salient component of these sounds. Green (1975) proposed that variations in pitch patterning over time could be used to differentiate some coo types in Japanese macaques, which seemed to be confirmed by a long series of subsequent laboratory tests in which subjects were specifically trained to categorize coos on this basis (reviewed by Stebbins & Moody, 1994). This interpretation notwithstanding, what was more likely being confirmed is that pitch contour differences were very evident to the macaque subjects, who then learned to make the contour-based distinction that the experimenters required. In the first study in which subjects were tested for categorical responding in the absence of such training, the effect disappeared (Le Prell & Moody, 2000). This result may thus make better sense of two otherwise puzzling results, namely, Owren and Casale's (1994) finding that captive Japanese macaques produce highly variable coos in food-anticipation contexts, and Hauser's (1991) observation that coos given by a single adult female rhesus monkey encouraging presumably food-related group movement over a single eighty-seven-minute period included seventy-two different pitch contours.

Here the interpretation is that by repeatedly producing variable but generally upward-sweeping pitch contours in their coos, hungry animals can exert an arousing influence on others. Under natural circumstances such calling could be used to facilitate group foraging behavior after periods of rest. Nonetheless, affect alone probably does not account for the range of circumstances in which coos are used. This sound type is also particularly well suited to being individually distinctive (Rendall, 1996; Rendall et al., 1996), a factor that plays a central role in the other general function we hypothesize to be important in nonlinguistic vocalizations.

Indirect Effects of Vocalizations

A second simple strategy that vocalizers can use to influence the behavior of others is to pair their calls with events that induce arousal or emotion, thereafter being able to use the sounds to induce learned affect. In other words, if a vocalizer can associate its sounds with salient, affective responses occurring in listeners for other reasons, the resulting conditioning gives them leverage over an individual's state and concomitant behavior. We consider this strategy to be indirect in the sense that it is not the acoustic energy itself that exerts an effect. Rather, the critical element is how vocalizers and listeners have interacted in the past, with learned affect in the latter accruing to individually distinctive features of particular call types emitted by the former.

"Affiliative" and "threat" calls are two examples. In the first case, dominant primates in a number of species have been found to vocalize when amenable to a grooming interaction with a subordinate. From a representational point of view, the call is said to encode the caller's desire for grooming, peaceful intent, low likelihood of being aggressive, or some similar message. The recipient decodes the signal content, decides whether it is reliable, and behaves accordingly. From the affect-induction point of view, the dominant animal's call influences the listener through learned affect. If the call has preceded peaceful grooming episodes in the past, it will have been associated with a positive outcome, producing requisitely positive affect or at least a decrease in the anxiety and wariness the subordinate likely feels when interacting with a dominant animal. The call thus become a means whereby the dominant animal can decrease the subordinate listener's negative emotions upon approaching or being approached, thereby facilitating the interaction. Quite different associations will accrue to threat vocalizations produced by a dominant animal, however, just before or as it aggressively hits, bites, or chases the subordinate. Laboratory studies of learned fear suggest that as little as a single pairing of a dominant animal's threat call with violent behavior should suffice to induce strong learned affect (LeDoux, 2000; Rescorla & Wagner, 1972). When the former victim hears that sound in the future, the vocalization alone will elicit arousal and learned fear, thereby encouraging it to decide to move away and avoid confrontation.

In both cases the principles being invoked are those of Pavlovian conditioning, a ubiquitous form of learning documented in animals ranging from the simplest metazoa to humans (e.g., Dimberg & Öhman, 1996; Domjan, 1998; Turkhan, 1989). In each of the species tested, it was found that if an individual experiences a salient but otherwise innocuous stimulus that precedes the occurrence of significant positively or negatively toned events, the stimulus alone can come to elicit learned responses. We suggest that rather than conveying encoded information about their intentions in vocalizations such as grooming or threat calls, socially potent animals can use their vocalizations as conditioned stimuli, thereby modulating the behavior of others by inducing learned affective responses. While the caller's motivational state and the predictive value of the call are likely involved here as well, there are several important distinctions between this account and the motivational approach to signaling.

First, the affect-induction explanation ascribes the effectiveness of the vocalization to the recipient's history of experience with such sounds, most significantly with the particular vocalizer involved in a given interaction. But whereas the motivational explanation links particular motivational states to particular call types, no such specificity is implied by the affect-induction view. Caller motivational state may very well be involved as a mechanism triggering the vocalizations, but the key point for the call's function is that it is paired with an event whose affective consequences on the listener help the vocalizer achieve the desired effect. In other words, vocalizers in diverse motivational states (e.g., wanting physical proximity, grooming, or sexual contact, wishing to approach and handle an unrelated infant) might in each case benefit from producing a call that has in the past been associated with positive affect in the listener. Conversely, an animal experiencing a particular motivational state could potentially benefit from producing more than one call type, so long as each helps facilitate an affective response that makes the listener more likely to behave as the caller desires.

A second important difference is that because the affect-induction view specifically invokes basic learning principles, vocal behavior in primates should be found to reveal the same phenomena uncovered in the conditioning laboratory. For example, learning theory predicts that the degree of associative strength accruing to a conditioned stimulus followed by some positive or negative event will in many instances mirror the degree to which this stimulus is predictive of that outcome. This is the expected outcome for the case of affiliative calls and grooming, in which the effectiveness of a given individual's call should reflect the relative number of times the listener has experienced this call in conjunction with a positive outcome. Laboratory studies have also shown that there are conditions under which learned affect does not simply reflect the number or proportion of times two stimuli are paired. For example, a host of experiments with subjects such as monkeys, rats, and dogs have examined the learned fear that results when the subject is exposed to electric shock. If the shock is preceded by an otherwise innocuous stimulus like a tone or a light, subjects in these preparations quickly learn to move to a different compartment in the experimental apparatus before this aversive stimulus arrives. The result is that even though the animal may no longer experience the shock, it continues to perform the avoidance response. In fact, the shock may be turned off completely for quite some time before that learned avoidance behavior begins to decline (e.g., Domjan, 1998). Traditionally, at least part of the explanation is that the avoidance behavior is maintained because it is being reinforced on every trial by decreasing the learned fear elicited by the conditioned stimulus. By implication, a dominant primate's threat calls can be effective even if they are only occasionally paired with actual agonism. A subordinate's learned fear response will then be maintained by the animal's own avoidance responses upon hearing it. Note that a motivational interpretation has much greater difficulty accounting for what threat calls encode if, like the guillemot's threat displays discussed earlier, they are sometimes followed by attack but often are not.

Finally, two critical differences should be pointed out between the two affect-induction strategies we have outlined. First, while the strategy of producing vocalizations with direct impact on listeners is likely of most value to socially impotent vocalizers, the opposite is generally the case for inducing learned affect. This strategy is of greatest benefit to those whose power in a given relationship allows them to induce positive or negative affect reliably in the other. Examples include dominant animals interacting with subordinates and caregivers interacting with dependent offspring. In both cases there is a decided asymmetry in the ability of each party to shape the course of the interchange and its ultimate outcome. It might also be noted, however, that

even relatively impotent individuals sometimes have the opportunity to associate their calls with affect-inducing events in others, for instance, when they
groom higher-ranking animals, provide sexual favors, or discover high-quality
food. Second, whereas relying on the direct effects of vocalizations implies
using a larger number of acoustically variable, high-impact sounds, calls used
to elicit learned affective responses should be fewer in number and well suited
to individual distinctiveness. This expectation is based on the assumption that
vocalizers will be more effective in inducing learned affect in a particular listener if their sounds are both salient auditory events and readily discriminable
from similar calls produced by others. A dominant caller will most successfully
elicit learned fear using a threat call, for instance, if its vocalization is distinct
from those a prospective victim routinely hears from animals that do not in
fact hit, bite, or chase it afterwards.

Individual Distinctiveness in Close-Range Calls

While there are a variety of possible ways in which that distinctiveness can be
achieved, the nature of mammalian vocal production again likely has a significant impact on the form of vocalizations used for indirect affect induction.
Various issues involved are reviewed at some length by Owren and Rendall
(1997, 2001; see also Bachorowski & Owren, 1999; Rendall, 1996), where the
overall conclusion drawn is that individual distinctiveness is most reliable,
even obligatory, in sounds that reveal the resonance (or formant) effects of the
supralaryngeal airways. The resulting pattern of peaks and troughs in the frequency spectrum of a sound is in essence a filtering imprint created by the
individual's vocal-tract cavities and tissues. Because no two individuals have
exactly the same anatomy, sounds that reveal variation in the fine features of
the supralaryngeal filter are inherently well suited for use as conditioned stimuli. In nonhuman primates and humans alike, two kinds of source energy are
known to be good media for the filter to act on. One is stable, relatively low
frequency vocal-fold vibration that produces a harmonically rich frequency
spectrum—in other words, the characteristic structure of many coo calls. The
prominent role of vocal-tract filtering in individual distinctiveness in rhesus
monkey coos has been demonstrated both in acoustic analysis (Rendall et al.,
1998) and through playback experiments with free-ranging monkeys (Rendall
et al., 1996).

The other source energy that is particularly well suited to revealing supralaryngeal filtering effects is broadband random noise created through turbulent airflow in the vocal tract. This outcome has been demonstrated in rhesus macaque "pant-threats" (Fitch, 1997), which are comparatively low amplitude, noise-based calls that dominant animals use to intimidate subordinates and sometimes before attacking them. Although pant-threats often occur when animals are virtually face-to-face and there is no mystery as to who the caller is, these sounds carry more prominent acoustic cues to individual identity (Owren, unpublished data) than do rhesus monkey screams (Rendall et al., 1998). The latter would of course require substantial individual- or kin-based distinctiveness in order to function as recruitment signals to distant allies. Gouzoules and colleagues (1986) report playback experiment results suggesting that they do, whereas Rendall and colleagues conclude from both acoustic analyses and playback experiments that they do not (Rendall, under review; Rendall et al., 1996, 1998). Another call of interest is the tonal, strongly vowel-like grunt produced by baboons. This sound specifically resembles human vowels in being rich in formant-based cues to individual identity (Owren et al., 1997). Grunts are extremely common in chacma baboons, who produce them in several different contexts. This call can for instance play a pivotal role when an adult female approaches another after an altercation has occurred between them (Cheney et al., 1995; Cheney & Seyfarth, 1997; Silk et al., 1996; see also Palombit et al., 1999). Data from both observational studies and playback experiments show that subordinate victims are more likely to act affiliatively toward dominant animals after hearing those individuals' grunt calls. While grunting thus plays a "signaling" function, the crucial question concerns the mechanisms involved. A motivational interpretation would be that grunts encode information about the vocalizer's benign intent, with subordinates behaving affiliatively after decoding that message. In contrast, we suggest that affect learning explains the result much more parsimoniously. In this view, the "signal" value involved is no more than a conditioned association between the dominant individual's grunts and positive emotions that resulted from a subsequent peaceful interaction on some previous occasion.

Based on the ubiquity of Pavlovian conditioning effects, an impact on listener affect appears inevitable when distinctive calls are differentially paired with positive or negative outcomes. As argued by Cheney et al. (1995), grunts function to mollify former victims and facilitate friendly interactions. We suggest that the mechanism serving this "appeasing" function involves a conditioned decrease in the victim's negative affect. We further propose that this kind of interpretation is potentially applicable to many cases of purportedly representational signaling in animal communication, including the macaque coos considered earlier. These sounds have also been found to play a role in facilitating affiliative interactions among adult females (Bauers & de Waal, 1991), to exhibit stable acoustic variation among individual callers (e.g., Hauser, 1991; Masataka, 1992; Owren et al., 1993; Rendall, 1996; Rendall et al., 1998), and to demonstrate requisite distinctiveness in playback testing (Hansen, 1976; Masataka, 1985; Pereira, 1986; Rendall et al., 1996).

In the next section, we argue that the approach exemplified here with various primate calls can also be applied to human laughter. This is not to argue that laughter bears any particular analogous or homologous relationship to nonhuman sounds, but rather to illustrate potentially that a framework based on affect induction can apply to a range of nonlinguistic vocal signals. In the face of evidence that nonlinguistic vocalizations in both primates and humans with a variety of acoustic features can occur in diverse contexts, we ask whether laughers may be playing out the expected strategies of direct and indirect affect induction in accordance with the differential opportunities afforded them in various situations.

Applying the Affect-Induction Perspective to Human Laughter

As discussed earlier, human laughter is observed in a variety of different social circumstances and proposed to have a requisite diversity of "meanings." Empirical studies are now demonstrating, however, that laughter induces both arousal and positive emotion in human listeners (Bachorowski & Owren, 2001; Bradley & Lang, 2000). Furthermore, laughter has been found to show significant acoustic variability in the absence of corresponding differences in production circumstances (Bachorowski et al., 2001, under review), with significant functional implications for listener responses. For example, Grammer and Eibl-Eibesfeldt (1990) found that when males and females met for the first time and interacted for ten minutes, a male's interest in meeting that female again was significantly correlated with the amount of voiced laughter she had produced. In Bachorowski and colleagues' experiment (under review), participants rated their emotional responses after watching each of a series of video clips, two of which were specifically intended to elicit laughter. In addition to providing laugh sounds for acoustic analysis (Bachorowski et al., 2001), the study tested whether participants watching in dyads would produce different amounts or kinds of laughter depending on who they were paired with. Comparisons involved males and females in same- and different-sex pairs, with dyad members being either friends or strangers. Across these conditions, both males and females laughed copiously and produced a similar number of vocalizations. Laugh rates and acoustics varied significantly, however, depending on dyad composition.

Results from Laughter Experiment

In accounting for the differences, Bachorowski et al. (under review) looked for evidence of vocalizers producing laughter with either presumed direct effects, presumed indirect effects, or both. For voiced laughter, for instance, vocalizers were considered to be relying on a strategy of direct effects when their laughs showed "acoustically extreme" values on measures such as call and bout duration, maximum F_0, mean F_0, and F_0-range. As would be the case for virtually any sort of sound, laugh acoustics themselves likely have a higher direct impact on listeners when unusually long, very high pitched, or highly frequency modulated. Operationally, a sound was considered acoustically extreme on one of these measures when the value fell at least one standard deviation above the grand mean for the overall laugh sample. Because virtually any laughter is likely to have some direct impact on the listener's state, this approach was a requisitely conservative way of examining whether reliance on such effects would be particularly prominent in some pairings but not others. Vocalizer reliance on indirect effects was gauged by the amount of laughter produced relative to a control condition in which participants were alone while watching the video clips. This approach was grounded in the finding of Bachorowski et al. (2001) that both voiced and unvoiced laughter includes prominent formant-related acoustic cues to individual identity. The reasoning therefore was that increasing the amount of laughter produced relative to the "alone" condition would be consistent with a strategy of both associating individually distinctive sounds with positive affect in the listener and taking advantage of conditioned affective responses that would be present if the parties had interacted positively in the past.

To summarize the results, variation in male laughter was found to be largely mediated by the presence or absence of preexisting friendships, whereas female laughter was more consistently associated with whether the testing partner was male or female. For example, males produced both the most laughter and the most acoustically extreme sounds when tested with a friend—especially a male friend. When paired with a stranger of either sex, however, they neither produced more laughter than did control males who watched the clips alone nor showed acoustically extreme features. In contrast, females produced the most acoustically extreme laughs when paired with a male stranger, showing unusually high mean-F_0 and large F_0-range. Females tested with a male friend also exhibited a larger number of especially long bouts and calls, as well as laughing significantly more than in other conditions. Females tested with another female laughed no more than when alone, regardless of whether this partner was a friend or a stranger. While difficult to explain if one views laughter as a representational signal of internal state, outcomes from at least six of the eight pairing conditions can be explained by arguing that vocalizers are using their sounds to induce or accentuate positive affect in listeners through indirect and direct effects of laugh acoustics (Owren & Bachorowski, 2001). From this perspective, situational differences are specifically expected because those effects are necessarily dependent on the relationship between vocalizer and listener.

When paired with an unfamiliar individual, for example, a vocalizer can elicit arousal and perhaps positive affect by producing laughter, but in the absence of any history of interaction cannot induce partner-specific learned emotional responses. As a result, the affect-induction approach predicts that vocalizer behavior should primarily reflect the potential benefit of accentuating the listener's preexisting internal state using direct effects of laughter. This approach predicts a net benefit in the case of females paired with an unfamiliar male, but the opposite for males paired with unfamiliar females or males. The rationale is that males can overall be expected to show a positive affective predisposition toward stranger females, based on both human sociobiology (e.g., Daly & Wilson, 1983; Geary, 1998) and a wealth of empirical evidence showing that males routinely overestimate the degree of sexual interest that unfamiliar females have in them (e.g., Abbey, 1982; Koeppel et al., 1993; Muehlenhard et al., 1986; Saal et al., 1989; Shotland & Craig, 1988).

Producing arousal-inducing sounds can therefore benefit a female by playing on this positively biased affective stance in the male and encouraging a strategy of friendly pursuit. Females have been found to use similar sexually tinged strategies in other circumstances regardless of any actual interest they may or may not have, thereby rendering male behavior more positive and malleable (Dunn & Cowan, 1993; Johnson, 1976; Singer, 1964; see also Grammer, 1990). Females, however, do not have the opportunity to induce conditioned positive responses to laughter in males with whom they have not interacted, and in the experiment were requisitely found not to laugh at higher overall rates with unfamiliar males than in the control condition. Nonetheless, the laughter they did produce showed significantly more acoustically extreme features than in other conditions, consistent with the view that they were exerting direct effects on male arousal states.

The same reasoning leads to markedly different expectations for females paired with unfamiliar males. While being alone with an unfamiliar female can be considered an inherently positive experience for a male, from the female's point of view it is a mixed blessing. A stranger male can represent a significant threat to a female, whose affective stance should therefore be one of vigilance and wariness even when she is potentially interested in that individual. In the absence of any previous interactions, the male's best strategy is therefore generally to avoid vocal acoustics whose direct impact on the female would be to accentuate her negatively tinged predisposition, and instead to produce laughs modest in both number and acoustics that are paired as well as possible with positive states occurring in the female. Thus, even as females experimentally paired with unfamiliar males were producing acoustically extreme laughter, these same males were giving acoustically innocuous laughs that occurred at the same rate as when they were watching alone.

The same logic applies to dyads involving unfamiliar males, in light of the undercurrent of competition that can be expected for same-sex relationships between stranger males—whether in humans or in primates (e.g., Boyd & Silk, 1997; Geary, 1998; Smuts et al., 1987). As a result, arousal-inducing sounds produced by either male are expected to elicit negatively rather than positively toned responses in his partner. Consistent with this reasoning, males paired with unfamiliar male partners laughed no more than those in the control condition, and did not produce acoustically extreme sounds. The affect-induction approach does not produce similarly clear-cut expectations in the case of stranger female pairs, because human females across cultures can exhibit significant levels of same-sex competition (reviewed by Björkqvist, 1994; Burbank, 1987) while also showing higher levels of trust, cooperation, and situationally dependent accommodation behavior than males (Garza & Borchert, 1990; Moely et al., 1979). One could therefore argue that female strangers should either show the same competitiveness as males or be more positively predisposed. The outcome observed was that these females produced neither a large number of laughs nor acoustically extreme sounds, a result that favors the former.

The overall rationale changes significantly when individuals are already friends, as did Bachorowski and colleagues' empirical outcomes. In this case a vocalizer can still use the acoustics of laughter to accentuate the listener's friendly affective stance through direct effects. Here, however, positive conditioned responses resulting from past interactions become a salient feature. When two parties are friends, the laugher's sounds will have been repeatedly paired with positive affect occurring in the listener. There is no inherent sex-based asymmetry involved so long as the individuals are genuinely positively predisposed toward each other as a result of these past experiences. Using this rationale, we therefore expected participants to laugh copiously while watching the video clips, regardless of dyad composition. That clearly occurred for males, who when paired with a friend both produced more acoustically extreme sounds and laughed at higher rates than did males tested either alone or with a stranger. Females tested with a male friend showed similar though less dramatic effects, while females paired with female friends were found not to produce either especially large numbers of laughs or a high proportion of acoustically extreme versions.

This last result was thus the only outcome among the various testing conditions that was at odds with the affect-induction rationale. One interpretation is that females are less apt than males to rely on laughter as a strategy of emotional influence with same-sex friends. Laughter may in fact be of particular importance to male friends, acting as a countermeasure to the ever-present pull of intrasex competition. Though similarly susceptible to intrasex conflict, females are also reported to form deeper and more complex same-sex friendships than males (e.g., Barth & Kinder, 1988; Block, 1980; Weiss & Lowenthal, 1975; Wright, 1982), for example, showing significantly more intimacy through conversational self-disclosure (reviewed by Hill & Stull, 1987; McKinney & Donaghy, 1993). Females may thus be able to draw on a greater range of affective mechanisms in their same-sex relationships than males, with strategic use of laugh sounds playing a requisitely smaller role. Nevertheless, this account does not directly follow from the logic of affect induction, and outcomes for females with female friends are thus less well explained than the other findings.

Taken together then, the affect-induction view could specifically account for experimental outcomes in six testing conditions, was agnostic but compatible with the results of another, and was inconsistent in the last case. In contrast, few if any of the differences among the dyads would be anticipated from viewing laughter as a representational signal of motivational state, because participants in every condition reported experiencing positive affect while viewing the film clips. In other words, while there was dramatic variation in the acoustics of laughter recorded in this experiment (Bachorowski et al., 2001), as well as systematic differences in the rate and acoustic salience of the laughter produced among the various conditions, there was little evidence of covariation in associated affective states.

Other Experimental Findings

The affect-induction perspective can also account for important aspects of data reported in Grammer and Eibl-Eibesfeldt's (1990) investigation of laugh production and function. These investigators used a "strangers-meet" paradigm in which participants came to a laboratory and waited briefly together while being surreptitiously recorded. The researchers subsequently asked participants to evaluate privately their interest in the person they had waited with and tested for covariation between those ratings and the amount and kind of laughter that had been produced during the interaction. Most important for differentiating representational and affect-induction accounts, Grammer and Eibl-Eibesfeldt found no relationship between the amount of laughter produced by participants and subsequent reported interest in their partners. Contrary to the basic expectation that vocal output should reflect important aspects of internal state, the laughter produced by each individual as he or she interacted with the other showed no correlation with how positive their stance toward that individual was ultimately reported to be. The researchers did find the converse, however, namely that *exposure* to laughter was an important predictor of listener interest in several analyses. Furthermore, it was specifically female laughter that was a statistically significant mediator of male interest, whereas male laughter bore no relationship to the level of interest reported by their female dyad partners.

Grammer and Eibl-Eibesfeldt (1990; see also Grammer, 1990) went on to suggest that the participants were using their laughter to communicate "interest," "readiness," and "playfulness" to their dyad partners, and that the females in particular were also signaling "submissiveness." These suggestions are difficult to accept, however, in the absence of corresponding correlations between laugh production and vocalizer interest in their partners. In contrast, the affect-induction perspective does predict the outcomes observed, based on the same reasoning used in accounting for the results of Bachorowski et al. (under review). As before, the most basic principle is that vocalizations work by influencing listener state, not by conveying representational information about vocalizer state. Therefore, the overall expectation in Grammer and Eibl-Eibesfeldt's situation is specifically that reported interest should be correlated with the laughter being heard rather than the laughter being produced. The effect is necessarily mediated, however, by the opportunities and constraints of the circumstances of each pairing condition. The dyads were always strangers in this experiment, corresponding to the stranger conditions in the Bachorowski et al. paradigm. If we assume as before that females have significantly more opportunity to use their laughter to encourage and accentuate positive affect in males than vice versa, the sex difference observed by Grammer and Eibl-Eibesfeldt is also expected.

Viewing laughter as a representational expression of motivational state, one can argue that listeners should become more interested in partners who laugh more rather than less in their presence. If so, however, it is still not clear why laugh production was not correlated with vocalizer interest in either males or females. One might propose that laughter does in fact represent positive emotion but was being used dishonestly. There is no evidence for that suggestion, but it can be pushed to account for the sex difference in reported interest by inferring that female listeners understood that their partners' laughter was deceptive while the males did not. Grammer and Eibl-Eibesfeldt's explanation is instead that males are interested in females who laugh as a signal of submissiveness. We find it much more compelling to suggest that these females were using their sounds to induce arousal and concomitant positive affect in the males, who became more interested as a result (see Owren & Bachorowski, 2001, for further discussion).

In the early part of this chapter, we raised several basic points about understanding nonlinguistic communication. To start, we noted that although the traditional motivational and referential interpretations have been presented as alternative interpretations, both are grounded in viewing signaling as encoding, transmission, and subsequent decoding of information. They are therefore both representational formulations, each of which ultimately requires that observable relationships be demonstrated between signals and some aspect of either signaler state, signaler behavior, or the external environment. Whatever the representation is proposed to be, both approaches stress that specific information is being conveyed. While representational accounts may be tenable for some vocal signals, however, it is also clear that some of the best-studied and most ubiquitous primate calls do not show the requisite links between call acoustics and either vocalizer state or circumstance. The macaque coos and screams reviewed here are neither unique to particular situations nor the only type of call produced in a given situation. Although less is known concerning human laughter, the evidence that is available points in the same direction. Here again, it is difficult to link particular sounds to any given vocalizer state or social situation, and the acoustics of the vocalizations used can be extremely variable even within a particular circumstance.

Given such outcomes, investigators must be willing to question the utility of representational accounts. Unfortunately, to date there have been few other possibilities to consider. To help fill the need for other perspectives, we have proposed a nonrepresentational approach that arguably better fits the calls reviewed here and that may be applicable in other cases as well. Starting from the broader view that the ultimate function of signaling is to influence the behavior of others rather than to convey specific representational information, we have outlined two ways in which sound can be used as a means of influencing the affective systems that help guide the behavior of listeners. One is to produce sounds whose physical properties have a direct impact on listener arousal, attention, or emotion. Another is for the individual to pair sounds that have conspicuous cues to vocalizer identity with behavior that has salient affective consequences for listeners, thereafter being able to use these calls to induce learned affect in those particular listeners.

Both strategies thus emphasize inherent links between signal structure and function, which is markedly different than in either representational explanation. Arguments for external reference in primate vocalizations, for example, typically stress the absence of such relationships. Further, neither the motivational nor the referential perspectives specifically address the question of individual distinctiveness in vocalizations, even though such differences readily emerge when we compare different call types produced by a given species (e.g., Mitani et al., 1996; Rendall et al., 1998). Both representational views also put listener evaluation of signal content in a primary role. If that content is motivational, the recipient must be assessing the implications for upcoming signaler behavior—implying evaluative and hence cognitive processing. In the referential approach, semantic calls are explicitly proposed to activate cognitive representations of signal meaning, in that case involving psychological processes that directly parallel those associated with language comprehension in humans.

The last and most general difference we point out between the representational and affect-induction views is in the degree to which mechanism and function are given their due as separate levels of analysis. The two are unquestionably conflated in the motivational view, where signaler state comes into play both as a mechanism (signaler state plays a key role in triggering vocal production) and as a function (the purpose of signaling is to convey that the state is occurring). The problem is of course that there is no necessary connection between the mechanisms involved in producing a behavior and the adaptive value of performing it. While it seems quite likely that internal state plays an important role both when primates call and when humans laugh, that does not explain why vocalizers benefit from producing the sounds. Rather, functional value can only derive from effects that sound production exerts on listeners. The affect-induction stance explicitly disentangles mechanism and function by treating vocalizers and listeners separately throughout. In this view, for instance, caller motivational states have no necessary relationship to why listeners are affected by associated vocalizations. There is no claim that signaler states must be uniquely linked to specific sounds in order to be effective, nor the converse. Instead, the relationship between vocal acoustics and listener states and behavior becomes primary.

Mechanism and function are also easily conflated in the referential perspective, depending on how one approaches the concept of information. Trouble is particularly apt to arise when metaphorical notions such as meaning, reference, and semanticity are taken literally. Each construct relies on treating information *as if* it were a concrete entity or material thing that lies encoded in a physical signal and passes from one individual to another. Nevertheless, although a listener may certainly respond to a sound *as if* encoded information has been received, it is not in fact literally true. Instead, this abstracted notion of information transmission is conceptual shorthand that draws from multiple elements of the communication event. Most important, the "information" in question reflects a combination of the physical signal and the psychological processing that occur in perceivers upon hearing the sound. In other words, viewing signals as having encoded "content" blurs the respective roles played by signalers, the signal itself, and perceivers that respond to the event. As a result, information transmission becomes both a mechanism and a function of the interaction, and the distinction between signaler and recipient fitness interests is lost (for detailed further discussion, see Owings & Morton; 1997, Owren & Rendall, 1997; Thompson, 1997).

Taken together, these and other differences between the representational and affect-induction views lead to a variety of divergent expectations about naturally occurring vocal communication. To illustrate, three kinds of examples will be considered, with the goal in each case being to deduce testable, alternative predictions.

The first example involves *correlations between signals and behavior.* Whereas the motivational perspective argues that signal characteristics related to vocalizer states are designed to convey information about those states to others, the affect-induction view proposes that the function is to influence listener behavior by eliciting arousal, attention, and emotional responses. Therefore, while the former predicts correlations between vocalizations and signaler states and behaviors, the latter suggests that the stronger and more consistent relationship will be between acoustic signaling and *listener* states and behaviors. Given that nonspecific calls can be among the most commonly used in a given species' vocal repertoire, it must be concluded that vocalizers are on average deriving a net benefit from producing them in spite of the apparent absence of representational information involved. The affect-induction view ascribes that benefit to the impact that these sounds have on listener behavior.

The second example concerns *correlations between signal structure and function*. Other predictions derive from the structure-function relationships posited between acoustics and vocal usage. Specifically, individuals with less social potency in a given interaction with another are expected to produce both a larger number of calls and a larger proportion of sounds with high-impact acoustic features such as abrupt onsets and dramatic frequency modulations. These features are in turn predicted to be demonstrably more salient to listeners than equivalent control sounds that do not have them. The proposed noxiousness of screams should be revealed by testing the willingness of experimental subjects to listen to them, in circumstances in which they can either act to terminate the sounds or simply move away. The comparison here would again be to equivalent control sounds that do not have the chaotic features that are deemed critical to this effect. Transmission effects on screams should also be revealing. On the one hand, if these sounds are in fact representational and are used to recruit distant allies by transmitting information to them, they should be demonstrably less subject to broadcast distortion than are other calls. If, on the other hand, screams instead function by influencing affect and motivation in a nearby opponent, the fidelity of the sounds is secondary, and their features should be better suited for direct impact on that listener's nervous system. Predictions are quite different for animals with high social potency. When these individuals vocalize at close quarters when interacting with others, they are expected to produce a smaller number of vocalizations and to rely much more on sounds that are rich in cues to caller identity.

Finally, there is the role of *normative learning effects*. It is important to note the implications of the central place that learning is afforded in the affect-induction account. If conditioning is in fact a critical component of vocal behavior, it follows that learning phenomena documented over decades of laboratory study should be found in natural communication settings as well. An example discussed earlier is the role that active avoidance responses can have in maintaining learned fear even when the conditioned stimulus is only sometimes paired with the aversive outcome. The stimulus "preexposure" effect is another classic phenomenon predicted to be applicable to vocal behavior. In a preexposure experiment, subjects first experience a stimulus presented several times or more on its own, that is, without being paired with any affect-inducing event. When that stimulus subsequently is paired with an important outcome such as electric shock or food presentation, conditioning is significantly retarded relative to a control condition in which pairing occurs from the beginning. This situation can readily be generalized to both group-living primates and human social interactions. For example, the preexposure effect predicts that fear-learning in a primate should proceed more quickly to the individually distinctive threat call of an unfamiliar animal that has recently immigrated from another group than to the same vocalization of a longtime resident trying to rise in rank by threatening and attacking others. Conditioned affect in response to laughter should similarly develop faster in response to sounds from an unfamiliar person who becomes part of a social group than to those from an acquaintance whose distinctive laughter is familiar but has not been consistently paired with the occurrence of positive affect in the listener. The larger expectation is thus that if affective learning is indeed an important mechanism underlying the effectiveness of nonlinguistic vocalizations, these and other previously documented conditioning phenomena should demonstrably play out in normal usage. Failure to show such effects would constitute evidence against the importance of learned affect induction.

In conclusion, if language is ultimately the inspiration for understanding both nonhuman and human nonlinguistic signaling systems, it seems paradoxical that in spoken language, so-called paralinguistic and paralanguage features of an utterance that are themselves not linguistic nonetheless play a critical role in shaping its linguistic function. While thus literally labeled as peripheral phenomena, attributes such as the overall pitch contour of a sentence and details of voice quality can clearly modify the meaning of speech without having representational value. As a result, these may be the very features that researchers interested in nonlinguistic vocal communication should be working hardest to understand. Better still, one can reverse the field entirely and work toward an explicitly nonlinguistic understanding of sounds such as coos, screams, and laughter, and thereby potentially shed light on both "para-" and linguistic aspects of speech. Rather than making linguistically grounded concepts both the beginning and end of the inquiry, we suggest casting the theoretical net both more widely and more imaginatively in order to discover the common principles that can unite primate and human signalers alike in a common framework for vocal communication.

Thanks to Dario Maestripieri and an anonymous reviewer for helpful comments on a previous version of this chapter. Correspondence concerning this article can be addressed to Michael J. Owren, Department of Psychology, 224 Uris Hall, Cornell University, Ithaca, NY, 14853. E-mail correspondence should go to *mjo9@cornell.edu*.

14

Language

Duane M. Rumbaugh, Michael J. Beran,

and E. Sue Savage-Rumbaugh

The twentieth century witnessed many serious efforts to teach animals to speak and to comprehend language. Interest in animals' (and especially apes') abilities to learn language has been driven by a number of scientific questions, with some of the more focused ones being:

1. What is the essence of language?
2. Do animals talk in their own languages—languages that we don't comprehend?
3. Can animals learn to speak as we do?
4. Can the essence of "natural" human language become an integral component of artificial languages, ones that have been constructed/ written for projects with animals?
5. Is it possible to have a conversation with an animal through an acquired language?
6. Must there be clear evidence of syntax, both in use and in comprehension, for it to be concluded that an animal has learned language?
7. If animals learn to use symbols, are the symbols semantically meaningful to them in the same way that words are meaningful to us? Do the symbols represent things and events?

8. If animals learn to use symbols, do they comprehend them when they are used by others who wish to communicate with them?
9. If animals can learn languages that we challenge them with, does that mean that the animals are smart?
10. Are they smarter than animals that don't learn the languages?
11. Do learned languages uniquely help animals solve problems?
12. Can animals literally think in terms of learned languages?
13. Do animals learn language spontaneously—without formal training? If yes, does their acquisition of language parallel the normal course of language acquisition by children?
14. If animals learn language, do their brains mediate it in the same manner that our brains mediate language?
15. Will animals teach others, and notably their young, the language that they have been taught or have learned in the research programs in which they have participated?
16. Should animals be treated with greater respect and be granted rights because they have learned language?

We hold that there are good data from the great apes (the gorilla, the orangutan, the chimpanzee, and the bonobo), marine mammals, and parrots with which to address each of the above questions, and we hope the reader will retain these questions in order to recognize the answers as they are presented. Although there has been and still is controversy regarding the meaning and meaningfulness of language research with animals, systematic and long-term efforts have now been in place for over fifty years. Those who have stayed the course are confident that theirs is a rich field for learning and is here to stay. History will tell.

Regardless of the question driving a particular researcher or reader, it seems inevitable that the data obtained from the field of animal language research will serve to define language and to define the relationship between human and ape cognitive foundations for language acquisition.

What Is Language?

The first and third authors of this chapter have proposed the following as a working definition of language:

> Language is a neurobehavioral, multidimensional system that provides for the construction and use of symbols in a manner that enables the conveyance and receipt of information and novel ideas between individuals. The meanings of symbols in this system are basically defined and modulated through social interactions. (Rumbaugh & Savage-Rumbaugh, 1994, p. 309)

To these researchers, language is a system constructed through the use of symbols in meaningful social interactions with others. From this definition, there is no a priori reason to exclude the possibility that nonhuman animals also could possess such a communicative system. For many years there has been a sense that some animal species "must be talking to one another." The whistles of dolphins and the songs of whales in the oceans, the chirps of birds, the cries of monkeys in the trees, and even the howls of wolves all could possess meaning. Language, however, requires a specific quality in addition to providing meaning to others: it must be intentional. The communicator must intend for the recipient of the communication to understand its meaning, and the communicator must understand something about the mental state of the receiver. Therefore, the vocalizations and other communicative utterances of animals must be closely examined to determine the intentionality of the individual producing the communication, and if intentionality of this form is evident, then such utterances constitute language. Language is also a constructed system, and it must be open to the addition of new symbols, there must be new ways for symbols to be used and combined, and there must be novel ways in which meaning is created through interaction with others. Therefore, "language" as a label for the communicative capacities of nonhuman animals is applicable in only a few experimental studies conducted primarily with apes (which are the focus of this chapter), dolphins, and parrots. The questions that interest us about nonhuman animals and language are best addressed through a summary of research conducted over the past several decades with these species.

The great apes, with their large brains, dextrous hands, and close genetic relatedness to humans, were considered the best candidates for language acquisition research. These animals have been the most actively involved in language acquisition projects. Darwin (1859) suggested that in addition to biological continuity among all animals, there may also be psychological continuity. If there is such psychological continuity, and if human language is viewed as a psychological mechanism, precursors of such a mechanism should be evident in our closest living relatives, the great apes. And it is here that animal language studies began.

Ape Language Acquisition

Early Attempts to Teach Language to Apes

Initially, interest in ape language centered on teaching great apes to produce human speech. At that time, speech was considered the epitome of human language, and efforts to "talk" to the animals involved having the animals "talk" back. At the beginning of the 1900s, Furness (1916) undertook the first extensive project to teach an ape to speak. Working with an orangutan, Furness reported that the animal learned a number of different speech sounds which it voluntarily produced in settings that were deemed appropriate. Whatever promise the project showed, however, was terminated with the orangutan's death. Kellogg and Kellogg (1933) followed the attempt of Furness by rearing a chimpanzee with their own child. The interest of these two scientists was the role of early environment and its influence on the cognitive development of the ape (an issue which received much attention throughout the history of ape language research). Although Kellogg and Kellogg proposed the theoretical possibility of raising a human child in "wild" conditions devoid of human culture and interactions with conspecifics, they believed instead that the reverse condition, of taking a "wild" animal and providing it with an enculturated environment, would also readily (and ethically) address the issue of nature versus nurture. This female chimpanzee, Gua, was seven and a half months old when it was included in the daily life of its nine-and-a-half-month-old human counterpart, Donald. Very quickly, differences between the two infants were evident. Gua proved to be superior to Donald in all motor aspects, moving easily throughout the house. Gua was also the more advanced tool user, learning to use many of the items that were available in her environment. Gua proved to be sensitive to human speech, and responded appropriately to some simple statements and requests from her human companions. By fourteen and a half months of age, however, Donald's comprehension of language exceeded that of Gua, and by the end of the nine-month experiment, Gua trailed Donald in speech comprehension. Additionally, Gua's comprehension (and also Donald's at that age) was often tied to contextual cues, a claim that would be made against other apes reported to comprehend speech.

The chimpanzee continued to be the great ape species of choice for most subsequent ape language experiments. Hayes and Hayes (1951) undertook the next major ape language project with the female chimpanzee Viki. Viki entered the Hayes' home at the age of three days, and for the next seven years she was the object of all their attention. Viki participated in numerous psychological experiments, and an extensive database of chimpanzee cognition was collected. Viki showed that the chimpanzee was capable of far more human-like intelligence than had previously been thought possible. She could imitate movements, sort pictures into categories, and even produce vocal sounds on command. Viki was taught to vocalize with four distinct sounds. These were "ma ma," "pa pa," "cup," and "up." She could not vary these sounds to any greater degree, however. Viki also came to understand the spoken English of her caregivers. But, as with Gua, this comprehension was limited to specific contexts. Without the necessary contextual information, Viki could not respond appropriately to the requests and statements of the humans around her.

Sign Language Studies and Sarah's Plastic Tokens

Based on the difficulties encountered in trying to teach apes to "speak," and with greater understanding of the differences in the ape vocal tract compared to that of the human (differences that restrict the range of sounds these animals can make), Allen and Beatrix Gardner embarked on an ape language project in which the medium of communication was the gestural language American Sign Language. Working with another female chimpanzee, Washoe, Gardner and Gardner (1969, 1971) began teaching the chimpanzee to make signs when she was one year of age. There was no English spoken around Washoe, only the use of gestural signs. Caregivers were able to shape Washoe's hands into appropriate signs, and by three years of age Washoe was able to make eighty-five different signs. Washoe's signing was limited, however, and rarely did she produce multi-sign utterances that adhered to subject-verb or verb-object structure. This approach of using sign language with apes was adopted by other researchers, including Roger Fouts with chimpanzees, Lyn Miles with an orangutan, and Penny Patterson with gorillas (Fouts, 1973; Fouts & Fouts, 1989; Miles, 1983, 1990; Patterson, 1978; Patterson & Linden,

1981). In all of these studies, the ape subjects readily learned to make and use signs. Many of these apes continue to use sign language to this day. For example, sign language–competent chimpanzees raised around both humans and other chimpanzees use signs in a conversational manner as they adjust and reiterate statements made to them so as to continue linguistic interactions with their human companions (Jensvold & Gardner, 2000).

With the emphasis in ape language research now firmly separated from the realm of vocal communication, other researchers began implementing nonverbal linguistic systems into ape language studies. David Premack worked with the female chimpanzee Sarah, using a system consisting of plastic tokens that could be arranged on a magnetic board. Sarah proved a capable subject in many psychological experiments, and she readily learned to answer "questions" posed to her through use of her symbol system (Premack, 1971, 1986; Premack & Premack, 1983). Although Premack attributed conceptual processes to Sarah's symbol use, other researchers believed that the problems faced by Sarah could as easily be solved through recourse to conditional match-to-sample strategies (Savage-Rumbaugh & Rumbaugh, 1979; Savage-Rumbaugh, Rumbaugh, & Boysen, 1980). Additionally, Sarah's use of the plastic tokens was usually noncommunicative. It was, rather, a system for solving abstract tasks. Sarah could not use her system to provide information to the people around her, and there was no "conversational" aspect to her token use.

Project Lana

Duane Rumbaugh also began working with a female chimpanzee, Lana. Rumbaugh (1977) used geometric symbols, called lexigrams. These were embossed on keys that were lighted when they were depressed by Lana, and the entire system was mediated through computerized automation. Lana was taught to string lexigrams together from her earliest training. The central element in this study was the design of the language system. The lexigrams designed for Lana to use were strikingly artificial. Rumbaugh and Ernst von Glasersfeld, a linguist associated with the project, named Lana's language "Yerkish" in honor of the Yerkes primate laboratory. Each design element corresponded to some grammatical category. The artificial language was strikingly similar to "natural" human language systems. Colors of the lexigrams corresponded to rough semantic classes; for example, any violet lexigram corresponded to an autonomous actor, and any red lexigram represented something edible. Nine different shapes were presented in various combinations to make discriminable lexigrams. This allowed the lexigrams chosen by Lana to be presented in a row of symbols projected on a display above the computer so that Lana could see what she had said (Figure 14.1).

Lana expressed her representations of situations through the choice and ordering of her lexigrams. She was reinforced for grammatical strings but not for nongrammatical ones. Rumbaugh and Gill (1976) described the language skills that Lana acquired via these training techniques. Lana was very cooperative, and within two weeks learned stock sentences of the form "Please machine give juice" and "Please machine give piece-of chow." After extensive use of lexigrams in requests, Lana was trained to name objects. It took 1,600 trials for her to learn to name M&Ms as opposed to banana slices, but later she transferred the naming notion to other incentives and objects in a relatively

Figure 14.1   Lana at her keyboard. On each key was a lexigram (a geometric symbol representing a word or phrase). When Lana pressed a key, it was illuminated, and a facsimile of the lexigram was produced in an array of small projectors above her keyboard. The projectors allowed her to check her accuracy and make statements of request to humans, and allowed humans to question her.

small number of trials. She apparently "got" the idea of naming at some point, and named slides with no specific training. Lana also generalized the meaning of "no" to other than its original context, using it to express disapproval as well as negation in many contexts.

After learning to produce stock sentences, Lana was presented with certain word strings, some of which could correctly be used to begin a sentence (e.g., "Please machine give"), while others were grammatically incorrect (e.g., "Please give machine"). With about 90 percent accuracy, Lana erased the incorrect word strings by use of the "period" key rather than trying to complete the strings (the "period" key was used as a termination of a string of lexigrams, and thus Lana learned to terminate incorrect strings). Lana also correctly completed the strings that were grammatically valid about 90 percent of the time. Rumbaugh and his associates (Rumbaugh, 1977; Rumbaugh et al., 1973) concluded from these studies that Lana had mastered the basics of reading as required by the computerized apparatus while also showing a skill analogous to writing in her sentence constructions.

In addition, Lana used some stock sentences in new contexts. Her first conversation with the humans occurred on March 6, 1974. Rumbaugh (1977) recounted the scene:

> Tim was drinking a Coke about 4:00 P.M., standing just outside of Lana's room and in her full view. Lana knew the word for Coke and could ask for it when it was in one of her liquid dispensers. Lana looked at Tim and, by either a highly improbable sequence of chance events or with comprehension, asked, *? Lana drink this out-of room period.* She had come to use the word *this* primarily in reference to things for which she had no formal name, an unplanned development which grew out of the name-learning studies discussed above. Tim responded *Yes,* opened the door, and shared his "this" with Lana "out-of room." The scenario was repeated twice; if chance was originally the causal agent, it had been supplanted by learning. We talked about the significance of the exchange and decided that we should conduct continued observations to determine whether Lana knew that "this" was really "Coke." Another 15 cents was invested in the Coca-Cola Company, and at 4:41 Tim presented himself outside Lana's room with another Coke. Lana's first response was the stock sentence *Please machine give Coke period.,* which was correct but not appropriate since the machine had no Coke to vend. Next she said *Please Lana drink Coke this room period.* Perhaps she

intended to say *out-of room* instead of *this room*, but she did not. Tim said *No*. Lana came back with the original composition, *? Lana drink this out-of room period* to which Tim responded with a question for clarification, *? Drink what period*. Lana answered, *? Drink Coke out-of room period*. Tim said *Yes*, the door was opened, the Coke was shared; and Lana's first conversation, one she had both initiated and successfully negotiated, had been recorded. (Rumbaugh, 1977, pp. 173–174; italics added for indication of lexigram use by Lana and Tim)

In one series of tests (Essock et al., 1977), any of six different items (a ball, a bowl, a box, a can, a cup, and a shoe) could be presented to Lana in each of six different colors (black, blue, orange, red, violet, and yellow), and she could give either the lexigram name of the object or its color depending on the question presented to her. If the question was of the kind "What color this?" Lana would give the color name, whereas the question "What name-of this?" was answered with the object name. Additionally, Lana answered questions pertaining to sets of multiple items, each in a different color. She would answer questions that took the form "What name-of this which-is red?" when both a red shoe and a blue cup were present, or questions of the form "What color this ball?" when both a ball and a box were present.

When Lana was four and a half years old, she was tested for her conversational abilities by having technicians tell her lies or present problems to her (Gill, 1977). She was asked for morning decisions in the afternoon (What want drink?) and vice versa, or shown desirable but unavailable foods or drinks. She nearly always engaged in "conversation" until she got something acceptable, and the conversations were, as expected, longer when her expectations were contradicted.

At the close of the Lana Project, criticisms of ape language projects emerged (Petitto & Seidenberg, 1979; Seidenberg & Petitto, 1979; Terrace et al., 1979). One concern was that the apes trained with various language systems were not, in fact, building and modulating word meanings but rather were simply learning to mimic the utterances of the human caretakers (Terrace et al., 1979). For instance, Herbert Terrace, who worked with the chimpanzee Nim (Terrace, 1979), and his associates initially concluded that Nim appeared to be competent in using sign language. In reality, Nim was producing long utterances by delayed imitation of signing by his caretakers. The long sequences of signs that Nim produced did not add any additional meaning to the statements made by the humans. In addition, Nim did not show evidence of turn-taking in his "conversations" with his caretakers, and there was no orderly relationship between the growth of his vocabulary and the length of his utterances (Terrace et al., 1979). This critique of sign language–competent apes also was leveled by Terrace et al. (1979) against the Gardners' work and against the work of Patterson (1978) by Petitto and Seidenberg (1979). Later, more controlled studies of sign language use and vocabulary size were conducted with other apes by both the Gardners (Gardner & Gardner 1975; Gardner & Gardner, 1978, 1984; Gardner et al., 1989) and Fouts (Fouts et al., 1976, 1989), but criticisms of this method of language research remained.

Accusations also were leveled at the work with Lana (Terrace et al., 1977). Her sentences were portrayed as involving not a syntactic system but rather learned chained associations no more complex than those performed by pigeons (Thompson & Church, 1980). Despite these claims, the work with Lana was defended on numerous fronts. When Lana was five and a half years old, her computer system was expanded to allow her to construct sentences of up to ten lexigrams as opposed to the earlier limit of seven lexigrams. Betz (1981) examined all of the utterances that exceeded seven lexigrams in length during the first twenty-four days after the new length modification was added to the computer system. During that time, Lana made thirty-six novel "sentences" that contained at least eight lexigrams. An analysis of these sentences showed that of all the lexigrams used in these sentences, 92 percent were first used by Lana in a given conversation. Here, analyses embraced the time span of fifteen minutes prior to Lana's social utterance. Lana was therefore not imitating the utterances of her caretakers. Her sentences were unique to that time frame, and they were in response to humans communicating with her. Additionally, it was found that Lana, unlike Nim (Terrace et al., 1979), seldom repeated words within an utterance (Betz, 1981).

The Lana Project established a number of important developments and findings (Rumbaugh & Savage-Rumbaugh, 1994). First, it showed that computerized keyboards facilitated research on apes' language skills. This system was later used by researchers in Japan working with the chimpanzee Ai. Ai also learned to classify colors and objects and developed spontaneous word-order preferences (Matsuzawa, 1989) as well as numerical competence in labeling sets of items (Matsuzawa, 1985). Second, the Lana Project showed that

Lana was capable of learning and using numerous lexigrams. Third, it provided evidence that Lana was innovative in her communicative statements and requests, despite never having been trained to demonstrate such innovation. Fourth, it demonstrated that Lana was capable of cross-modal naming of objects with only a few exposures to such a task (e.g., Lana could name objects that she could not see but that she could feel with her hand). Fifth, it demonstrated that Lana's color vision was practically indistinguishable from that of humans. Finally, one of the most practical benefits of this research was a conversation board that proved to be fully functional and useful in promoting cross-species communication.

Lana was able, from the outset and without major training, to respond to the statements made by the humans around her by "reading" the projectors above her keyboard. This system allowed her to understand when questions were addressed to her, and to understand what responses and utterances she herself had made. In other words, it facilitated conversations between Lana and the humans working with her. This lexigram system later was used with handicapped children (Parkel et al., 1977). A number of children with no productive language and difficulties with vocal production learned several words when they were given a lexigram board and taught to use it (Romski et al., 1984, 1985; Sevcik et al., 1991).

Additionally, the lexigrams Lana learned became deeply embedded in her memory. After more than twenty years, during which time Lana had neither seen nor used many of the lexigrams that were a part of her original keyboard system, she was presented again with these lexigrams and was asked to label various foods, colors, and objects. Showing remarkable memory for those lexigram names, Lana correctly identified a majority of the items with the correct lexigram on the very first trial (Beran et al., 2000). In the food category, Lana retained the meaning of the lexigram *apple*. In the color category, she retained the meanings of the lexigrams *black, blue, orange,* and *yellow.* In the object category, Lana remembered the meanings of the lexigrams for *ball, box, can, cup,* and *shoe.* Overall, she selected the correct lexigram for five of the seven lexigrams she had not seen for more than twenty years, and she had "forgotten" only two of them (the name of the color violet and the name for bowl; Beran et al., 2000). In addition to her involvement in language acquisition research, Lana has also manifested symbolic comprehension of another set of items, Arabic numerals, which she uses as indicators of the number of items she is to select on a computer monitor (Beran & Rumbaugh, 2001; Rumbaugh et al., 1989; Rumbaugh & Washburn, 1993).

Project Sherman and Austin

The next major step in ape language research involved the co-rearing of two male chimpanzees, Sherman and Austin, by Sue Savage-Rumbaugh (Savage-Rumbaugh, 1986). These chimpanzees remained socially housed, and they interacted with human caretakers on a daily basis. Lexigrams again were the medium of communication between chimpanzees and humans, but the environment for use of those symbols was drastically different than that employed with Lana.

Sherman and Austin participated in training and test sessions together, and their use of lexigrams took place in a myriad of situations. Initial training with these symbols, however, still took the form of associative learning. When Sherman or Austin used a lexigram, he received the item represented by that symbol. The chimpanzees quickly learned to select the appropriate lexigrams based on the foods that were available.

Savage-Rumbaugh interacted with the apes as though they "knew" that their lexigrams represented the things they had used for making their requests known. Later, however, this "knowledge" was tested by asking the animals to give the name of each specific item requested by them earlier (Savage-Rumbaugh, 1986). When Sherman and Austin initially failed, specific training ensued. During this training, Sherman and Austin were presented with the items that they had successfully requested and were reinforced if they responded by pressing the correct name-lexigram for each one. A highly preferred food was used on all correct trials as the reinforcement, regardless of the object used for naming on a given trial. In other words, the reinforcing food was different from the item referred to by the lexigrams used by the chimpanzees. Initially the chimpanzees appeared to be confused by always getting the same consequence regardless of their selections. It was as if they expected to get each item named, just as they had in the request task.

Because of their difficulties, the procedures were modified. Instead of being given the same reinforcer for correct naming, Sherman and Austin first received a small portion of the item correctly named before receiving the larger reward used from the beginning of this training. Sherman and Austin learned naming in this context rather quickly, and soon it was no longer necessary to give them even a taste of the item named. They learned that naming an item was quite a different task from requesting it.

Demonstrating that apes knew the difference between naming and requesting items was the first major contribution of this project. Sherman and Austin demonstrated comprehension of their lexigram system, and their lexigram use was referential. That is, the lexigrams had acquired symbolic meaning for the chimpanzees and were an integral part of their daily interactions with each other and with the humans around them. From this referential base, Sherman and Austin demonstrated numerous competencies in language use. The chimpanzees asked for things not physically present (Savage-Rumbaugh et al., 1978a, b). Sherman and Austin responded to the requests of others using the keyboard (Savage-Rumbaugh et al., 1978a, b), and they demonstrated symbol-based, cross-modal matching (Savage-Rumbaugh et al., 1988). In the cross-modal tests, Sherman and Austin first saw an object and then found its identical match in an opaque box. Next the chimpanzees were presented with a lexigram representing an item, and they had to find that item in the box. Then the chimpanzees were allowed to hold and feel and object while they were blindfolded, and then they were to indicate the lexigram name of that object. Finally, the chimpanzees labeled olfactory stimuli with the lexigram name of that stimulus. They performed all of these tasks at very high levels (Savage-Rumbaugh et al., 1988). Therefore, language competence manifested itself in human-like intelligent behaviors rarely seen in other animals. Symbolic representation allowed for greater cognitive functioning in these animals (a phenomenon repeatedly seen with these animals and other language-competent animals when compared with conspecifics without language competence).

The symbol system allowed Sherman and Austin to accomplish tasks that otherwise would have been impossible to complete. The chimpanzees learned to request food items from each other using the lexigram system (Figure 14.2). Additionally, the chimpanzees would enter a room full of items, return to the test area, announce which item they would retrieve, and then go and retrieve it (Savage-Rumbaugh et al., 1983). The chimpanzees also would request tools from each other so as to access baited food sites (Savage-Rumbaugh et al., 1978a). One chimpanzee would know which tool was needed, but only the other chimpanzee had access to that tool. Through

Figure 14.2 Sherman and Austin at the keyboard. The apparatus with compartments between the two chimpanzees was used to hold foods and other items. Here, Sherman is using the keyboard as Austin watches. (a) Sherman surveys the tray of foods. (b) Sherman indicates a food name as Austin watches. (c) Austin selects the named food and hands a piece to Sherman. (d) Both chimpanzees eat a piece of the named food. Both chimpanzees learned to make requests and respond to the requests of the other on the keyboard using the appropriate lexigram names for items in the compartments.

lexigram-mediated communication, the appropriate tool was obtained for accessing the food, the task was accomplished, and the food reward was shared by the two chimpanzees. Therefore, the learned lexigram language uniquely allowed these animals to solve the problem. Sherman and Austin spontaneously came to use labels for items even when their keyboard system was intentionally made unavailable, demonstrating that their referential abilities transferred to novel symbols (the labels for the items; Savage-Rumbaugh, 1986).

Sherman and Austin most clearly demonstrated the referential nature of their lexigram use in a study on categorization (Savage-Rumbaugh, Rumbaugh, Smith, & Lawson, 1980). The chimpanzees first learned to sort six individual items into one of two bins, each representing a different category of object. In the food category were the following items: beancake, bread, and orange. In the tool category were these items: key, money, and stick. After they learned to sort these items into the correct bins (and categories), the bins were removed. Now only the lexigrams representing "tool" and "food" were presented, and the chimpanzees had to select one of these lexigrams when presented with an item. Next the chimpanzees were presented with photographs of those items, and they had to learn to label these photographs as either tools or foods. Finally, the chimpanzees were presented with lexigrams for all six items; again the lexigrams were labeled categorically. In the test (transfer) phase, novel items, photographs of those items, and lexigrams of those items were presented. These items (and their corresponding photographs and lexigrams) were foods and tools not used during the training phases (i.e., they did not include the original six items). The chimpanzees correctly labeled the items and the photographs of those items with the categorical lexigram for either "food" or "tool." Additionally, of the eighteen different lexigrams that were presented, only one categorization error was made, and it involved the labeling of the lexigram *sponge* as a food rather than a tool by Sherman (who, incidentally, often ate the sponges that were used to provide the apes with liquids for consumption). This study clearly demonstrated that the lexigrams represented real-world items for these chimpanzees, and that the lexigrams themselves had "meaning."

As with previous ape language studies, opponents attempted to reduce the behavior of Sherman and Austin by applying the behavioristic principles of shaping and chaining. Epstein and colleagues (1980) conditioned pigeons to chain a sequence of key presses and argued that such learned behavior was equivalent to the behavior of Sherman and Austin with the lexigram keyboard. As already noted, however, Sherman and Austin were capable of using alternate symbol systems such as labels, they could categorize their symbols, and they could announce their intentions. These skills were not simulated by opponents of the project, and thus the simulations with pigeons were not equivalent. The project with Sherman and Austin demonstrated the following important aspects of language and symbol use: (1) apes can comprehend symbols, but production of symbols does not necessarily lead to comprehension; (2) to function "referentially," symbols that are used by apes must be freed from contextual constraints and be used in novel situations; (3) apes will use symbols to communicate with one another; (4) apes will make statements about intended future actions; and (5) referential comprehension and usage are prerequisites to syntactic competence.

Kanzi

Another species of great ape, the bonobo, next became involved in ape language research. Sue Savage-Rumbaugh began work with a mature female bonobo, Matata, who had been caught wild in Zaire and had lived until she was a young adult at another primate facility before coming to the Language Research Center at Georgia State University. Using the same training techniques that had been successful with Sherman and Austin, researchers found that Matata showed little promise. When Matata was absent from the laboratory for breeding purposes, however, her young son, Kanzi, at the age of two and a half years, began to evidence remarkable abilities with both the lexigram keyboard and, soon after, the comprehension of human speech (Savage-Rumbaugh & Lewin, 1994). Kanzi, who was believed to be too young to learn the lexigram system, had merely been "around" the testing sessions employed with his mother. There was no overt indication that Kanzi was actively attending to these sessions or that he was engaged in any form of learning regarding these lexigrams. That is to say, Kanzi was not observed to be engaged in the experimental sessions in the same way his mother was engaged. He was not presented discrete training trials with exemplars. With his mother absent, however, and with more attention focused on him, Kanzi began to show that he had learned much of what his mother had failed to absorb. Kanzi responded appropriately to the spoken requests of his caregivers and used his lexigrams appropriately in a variety of contents (Savage-Rumbaugh et al., 1986).

Kanzi then was immersed in environmental situations that were believed to be important to his initial learning of language (Savage-Rumbaugh, 1991). Rather than receiving formal discrete trial training and associative learning, Kanzi "experienced" language in the same way that human children do. He traveled to locations in the forest with caretakers, assisted in preparing meals, and was otherwise immersed in structured, consistent daily routines which provided him with the framework for gleaning the meaning of both spoken words and the corresponding lexigrams. Although other apes had also been raised in such an environment (Hayes & Hayes, 1951; Kellogg & Kellogg,

1933) without learning to comprehend human speech, Kanzi also was exposed to a system whereby he could participate in conversations. Kanzi was immersed in conversations about past, present, and future events, and within these conversations the caretakers around him not only spoke aloud but also used the visual keyboard system. This allowed Kanzi to participate directly in the conversations, and when he used lexigrams, his utterances were treated by the caretakers as being meaningful. Thus, Kanzi not only observed the concurrence of lexigram use and spoken English by humans in conjunction with the occurrence of events but also experienced the outcomes of using those lexigrams himself. Kanzi's inclusion in daily activities was mediated in part by his communicative utterances to the caretakers around him.

Kanzi's competence with the lexigram system exceeded all success enjoyed by Lana, Sherman, and Austin (Savage-Rumbaugh et al., 1990). In addition, Kanzi became the first ape with a documented comprehension of a large vocabulary of spoken English words. And not only did Kanzi understand individual words, but also he was capable of deciphering short sentences involving objects, actions, and locations (Figure 14.3). In an extensive battery of tests (Table 14.1), both Kanzi, aged seven years, and Alia, a two-and-a-half-year-old child, were presented with 660 novel sentences (e.g. neither familiar, trained, nor modeled by others) of request (Savage-Rumbaugh et al., 1993). For example, Kanzi and Alia were asked to take an object to a stated person or location ("Take the shoe to the bedroom," "Give Karen an apple," "Take the vacuum cleaner outdoors"), to do something to an object ("Put the pine needles in the refrigerator"), or to go somewhere and retrieve an object ("Get the lettuce that's in the microwave oven"). Sentence types also included several types of reversals such as "Make the doggie bite the snake" and "Make the snake bite the doggie." Both Alia and Kanzi were correct on about 70 percent of the sentences. Kanzi was correct on 81 percent of the "reversed" sentences in which the key words were presented in both orders, while Alia was correct on only 64 percent of the same questions (Table 14.2).

Observational Language Acquisition in Bonobo and Chimpanzee

Kanzi's success led to a new method of instantiating language skills in great apes (Savage-Rumbaugh et al., 1989). In addition, the finding that a bonobo evidenced skills never seen in chimpanzees also made it necessary to look

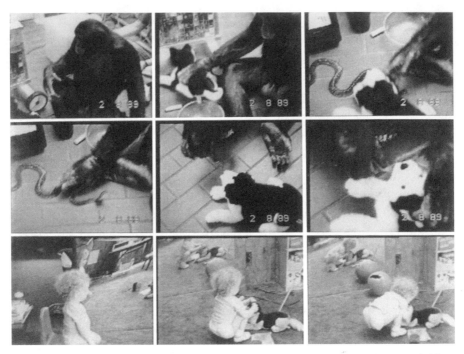

Figure 14.3 An example of English comprehension by a bonobo and a human child. In the first six panels, Kanzi was asked to "make the snake bite the doggie." After examining the area (top row, left), Kanzi first selected (top row, center) and moved the doggie to the snake (top row, right). Kanzi then picked up the snake (middle row, left), moved it toward the doggie (middle row, center), and pressed the mouth against the doggie (middle row, right). When asked the same question, the human child, Alia, (bottom row) instead went to the doggie and bit it herself.

at both rearing differences (strict training measures versus observational learning through immersion in language-rich environments) and species differences (chimpanzees versus bonobos). To this end, a female bonobo, Panbanisha, and a female chimpanzee, PanPanzee (referred to as Panzee), were co-reared from their first few weeks of life until they were a few years of age. Of interest were the developmental differences exhibited by the two species in their language acquisition. Panzee and Panbanisha, like Kanzi, experienced a highly social environment that included activities such as traveling to different locations and experiencing daily routines which human caretakers and apes could then talk about. These routines were sets of interindividual interactions that could occur on different occasions but in a similar manner, and these

Table 14.1 The performance of Kanzi *(Pan paniscus)* and Alia *(Homo sapiens)* in response to spoken English sentences

Sentence type	Kanzi % correct	Alia % correct
1	64	72
2	74	72
3	81	81
4	33	57
5	89	87
6	71	60
7	70	65
8	63	85
9	75	69
10	85	51
11	77	52
12	70	60
13	64	30
Total	72	66

Note: Data are from Savage-Rumbaugh et al., 1993. Numbers in col. 1 refer to sentences of the following construction:

1. Put object X in/on transportable object Y—"Put the ball on the pine needles."
2. Put object X in/on non-transportable object Y—"Put the ice water in the potty."
3. Give (or show) object X to animate A—"Give the lighter to Rose."
4. Give object X and object Y to animate A—"Give the peas and the sweet potatoes to Kelly."
5. Do action A on animate A—"Give Rose a hug."
6. Do action A on animate A with object X—"Get Rose with the snake."
7. Do action A on object X (with object Y)—"Knife the sweet potato."
8. Announce information—"The surprise is hiding in the dishwasher."
9. Take object X to location Y—"Take the snake outdoors."
10. Go to location X and get object Y—"Go to the refrigerator and get a banana."
11. Go get object X that is in location Y—"Go get the carrot that is in the microwave."
12. Make pretend-animate A do action A on recipient Y—"Make the doggie bite the snake."
13. All other sentence types (11 sentence types that did not form a distinct category).

interactions were embedded within a larger context. For example, cooking meals and receiving blankets could be part of the larger context, "evening," when preparations for ending the day occurred (Brakke & Savage-Rumbaugh, 1995, 1996; Savage-Rumbaugh et al., 1993). Panzee and Panbanisha came to demonstrate language abilities, including comprehension of human speech, similar to those of Kanzi, although the rate of language acquisition for the chimpanzees was somewhat slower than for the bonobo (Brakke & Savage-Rumbaugh, 1995, 1996; Rumbaugh & Savage-Rumbaugh, 1994). During a

Table 14.2 The performance of Kanzi (*Pan paniscus*) and Alia (*Homo sapiens*) in response to spoken English sentences presented with multiple word orders creating three distinct sentence subtypes

Sentence subtype	Kanzi % correct	Alia % correct
1		
Overall	83	59
Both sentences correct	74	38
2		
Overall	79	67
Both sentences correct	57	38
3		
Overall	79	69
Both sentences correct	57	39
Total	72	66

Note: Data are from Savage-Rumbaugh et al., 1993. Subtype 1 refers to verb plus word-order changes, and the appropriate response must differ between these sentences ("Could you take the pine needles outdoors?"—"Go outdoors and get the pine needles"). Subtype 2 refers to constant word order, but the appropriate response must differ ("Take the lighter outdoors"—"Go get the lighter that is outdoors"). Subtype 3 refers to changed word order, and the appropriate response must differ ("Pour the Coke in the lemonade"—"Pour the lemonade in the Coke"). "Overall" refers to total performance with all sentences of that type. "Both sentences correct" refers to the percentage of correct pairs of reversed sentences of each type.

three-year period in which utterances were recorded for Panzee and Panbanisha (when they were both between one and a half and four years of age), the bonobo Panbanisha responded correctly to 92 percent of the utterances directed to her while the chimpanzee, Panzee, responded correctly to 81 percent of them.

These apes' accuracy rates remained consistent over this time period, and they were able to respond equally well to utterances consisting only of spoken English as well as utterances containing spoken English and lexigrams (Brakke & Savage-Rumbaugh, 1995). Panzee, therefore, was the first chimpanzee reared in this environment to show comprehension of a large vocabulary of spoken English words, and her comprehension of spoken English remained high across the years (Beran et al., 2000). The results of this co-rearing study (Brakke & Savage-Rumbaugh, 1995, 1996) demonstrated that, although the bonobo may show a slightly greater affinity for language learning, early environment and contextual surroundings have a huge impact on language learning by apes. This finding has important implications for language development in human children as well as in animals. An elaborated and linguistically rich environment produces competence with symbols and understanding of speech exceeding that from more restricted surroundings. Such a finding should not be surprising. Studies of nonhuman primates have suggested that aspects of the early rearing environment have a direct impact on the cognitive development of apes. Denied appropriate and timely stimulation by peers and adults for the first two years of life, apes will suffer essentially irreparable deficits in the development of their cognition, intelligence, social communication skills, and breeding and parenting skills (Bard & Gardner, 1996; Davenport et al., 1973; Menzel et al., 1970).

Rumbaugh and colleagues (1996) stated that Kanzi's (and also Panzee's and Panbanisha's) extensive opportunities to observe reliable, predictable, meaningful, consistent, and communicative patterns of language coupled with salient events helped make possible his spontaneous language acquisition. With everything else that occurred around him, "it was through Kanzi's reliable access to the patterned experiences afforded by the *logic structure* of his environment (e.g., the speech of the experimenters and their use of word-lexigrams on a keyboard that structured his mother's instructional sessions) that he perceptually *discerned* and *learned* the relationships between symbols and events that provided for him the basic processes and competencies with language" (Rumbaugh et al., 1996, p. 119).

Savage-Rumbaugh (1991) stated that daily routines are important to the emergence of language in apes. Lexical markers come to signal various parts of the routine. Through this interaction of the routines and the markers referring to those routines, the referents of the symbolic markers become clarified. Later, when comprehension of these markers is separated from the routine, the ape can then deal with the details of specific word associations.

Language Acquisition in Dolphins and a Parrot

Although the focus of this chapter (and this volume) is on primate behavior, we would be remiss in not presenting the important research conducted with other species and their language capacities. These studies rival many of the ape projects in terms of longevity and the magnitude of data that have been collected and applied to the question of animal language. These studies provide a comparative framework for many of the issues of syntax, semantics, and rearing variables that influence the emergence of language skills.

Because of their complex use of vocal communication (in the form of whistles) as well as their large brain size, dolphins also made good candidates for language acquisition research. Louis Herman taught each of two dolphins an artificial language system (Herman et al., 1984, 1993). One dolphin, Akeakamai (Ake), was taught to respond to the hand and arm gestures of humans, whereas the other dolphin, Phoenix, was taught computer-generated vocal sounds. The words in each of these artificial language systems referred to objects, actions, properties, and relationships. These words could be combined into meaningful sentences that the dolphins could interpret. Some sentences were relatively simple and required an action be done to an object. Other sentences, however, were more complex and involved relations between words within the sentence as well as multiple relevant item words. By using the same words, multiple sentences with multiple meanings could be constructed, and this required that the dolphins learn not only the semantic meaning of the symbols but also the syntactic rules governing the combinations of those symbols (Herman, 1988; Herman et al., 1993).

Initially the question was whether the dolphins could respond appropriately to requests asking either for an action to be done to a single object or that the dolphin perform an action with two objects. Later, semantic categories could be combined according to syntactic rules in such a way that three-, four-, and five-word relational sequences could be directed to the dolphin (Herman et al., 1993). The dolphins organized their responses in such way as to take into account both the syntactical aspect of the sentence and the meanings of all referents within that sentence. For example, the animals demonstrated syntactic understanding through responding correctly to sentences containing the same words but in a different word order (such as LEFT HOOP PIPE FETCH, which asked the dolphin to take the pipe to the hoop on her left, versus HOOP LEFT PIPE FETCH, which asked the dolphin to take the pipe on her left to the hoop). Semantic understanding was demonstrated through selection of the correct items. Syntactic understanding also was shown through responses to syntactically anomalous sentences, to structurally novel sentences, to sentences with word modification within the sentence (i.e., in which one word modified the meaning of another), to contrasting interrogative and imperative sentence forms, and to sentences that contained variations in the placement of modifiers (Herman et al., 1993).

The research with dolphins as well as research by Ronald Schusterman with sea lions (Schusterman & Gisiner, 1988; Schusterman & Krieger, 1984, 1986) has focused heavily on syntactic aspects of language comprehension as well as semantic understanding of word meaning. Herman (1988) believed that comprehension of this language system was more important to investigate than production (as in the early ape language studies), because investigations of productive use of language in apes had revealed no convincing evidence for the understanding or use of syntax—although, as already noted, Rumbaugh (1977) believed that Lana did show evidence of grammar use and understanding. Additionally, Herman (1988), as well as Savage-Rumbaugh (1986), noted that language comprehension precedes language production in human children, and thus is a more likely place to start an investigation of language in nonhuman animals. Therefore, Herman was interested in the ways in which his dolphins responded to different types of linguistic statements by humans. In addition to responding appropriately to commands in this linguistic situation, the dolphins also answered questions pertaining to the status of named objects (such as HOOP QUESTION to ask whether a hoop was present in the tank).

In addition to responding to the requests of humans who were present, the dolphins were also capable of responding appropriately to degraded video displays of this gestural language and to anomalous gestural signals in which the semantic rules and syntactic constraints of the language were violated (Herman et al., 1990). The abilities of the dolphins indicated that they had a referential understanding of the signs used in their language. As with apes, the dolphins utilized abstract mental representations when responding to language-mediated tasks. In both symbol systems, the dolphins demonstrated an understanding of both the syntactic rules and the semantic content of the symbols in their system. The dolphins responded appropriately not only to the meaning of individual gestures and signals, but also to the relation of those gestures and signals to each other within a sentence.

Despite the data indicating language comprehension by these dolphins, there is a lack of any productive language data. Unlike the work with lexigram-competent apes in which both comprehension of lexigram use and spoken English (including for syntactic structure) was investigated and demonstrated, the work with dolphins has remained tied to the realm of comprehension. Therefore, little is known at the present time about what the dolphins would "say" if they had the means to produce linguistic output that was comprehensible to humans (dolphins use a seemingly complex communicative system of whistles among conspecifics, and this points even more strongly to a need to look to productive language use in the dolphin).

At nearly the same time that Savage-Rumbaugh was investigating Kanzi's language competence, Irene Pepperberg was training an African grey parrot, Alex, using what she called the "model-rival" approach (Pepperberg, 1999). To teach Alex to communicate, human experimenters would engage in conversations with one another centering on things of interest to Alex. Alex observed these interactions, and the humans served alternatively as "models" and "rivals" for Alex in that they received items that Alex would want to receive. When Alex entered the conversation, he was rewarded by having the humans respond as if he "meant" what he had said. That is, the requests made by Alex were honored as reflecting his communicative intentions. Pepperberg (1999) believed that it was because Alex could observe and enter the conversation at will that he eventually attained the ability to ask for, choose, and describe the color, shape, and materials of the objects with which he was familiar. This finding is very important because it closely reflects the views of Savage-Rumbaugh (1991) regarding the ideal situation in which apes attain language skills, and it also resembles the contextual way in which human infants learn language both by interacting with parents and by observing parents and others interacting with one another. In these contexts Alex responded spontaneously, like human children and like the apes raised in a language environment at the Language Research Center. A broad range of responses is available to all of these language learners, rather than the single responses that have for so long been the standard in the typical classical or operant conditioning experiment.

From this approach, Pepperberg (1990a) found that Alex learned many names for items, and he used those names appropriately in a variety of contexts. Alex also learned to comprehend spoken English as he responded appropriately to the spoken requests of his caretakers. He could be queried on the names of items, the quantity of items in an array, the color of items, and even the material from which an item was made. Alex could decompose longer requests into relevant subcomponents so that the correct category and answer within that category could be determined. To do this, Alex not only had to understand word meanings but also had to perform tasks involving recursive processing in many instances (Pepperberg, 1990b).

Research into the cognitive capabilities of nonhuman primates, and especially the great apes, has contributed much to our understanding of the evolution of language. Large and complex brains presumably have been selected for their enhanced processing capacities and their increased learning and memory capacities (Rumbaugh & Savage-Rumbaugh, 1994). Early rearing in a complex and logically structured environment, together with the intelligence afforded by brains larger than would be expected for primates of their size (Jerison, 1973, 1985; Rumbaugh & Pate, 1984), makes the great apes facile learners of both spoken words and visual symbols (Rumbaugh, Savage-Rumbaugh, & Washburn, 1996; Rumbaugh, Washburn, & Hillix, 1996). This is not to assert that the ape learns language to the extent that the human child learns language or that any nonhuman animal has the full capacity for language that is enjoyed by humans. Rather, the assertion is that chimpanzees have some of the competencies for language skills that were once considered unique to humans. These language skills are perhaps the best example of what Rumbaugh, Washburn, and Hillix (1996) have termed emergents. Emergents (in contrast to basic operant and respondent conditioning) occur as the result of integrative processes of complex brains. Emergents are new capabilities and behavior patterns noted for being synergistic, clever, and appropriate. Such behaviors have no specific reinforcement history; they emphasize classes of experience; they often occur "silently" or covertly; and they are noted for their immediate adaptiveness to specific and general situations (Rumbaugh, Washburn, & Hillix, 1996). As we have noted throughout this chapter, language acquisition by apes occurs in just such a manner and is clearly an example of an emergent.

The ape, like the human child, is equipped to learn from early experience and to organize its world in a meaningful way. Language research with apes has revealed that if there are "basics" in language acquisition, they surely involve the ability to symbolize, or represent things not present in time or space; the ability to use symbols to communicate intentions, information, and perhaps even emotions; and the ability to respond to such communicative interactions in social situations (Rumbaugh & Savage-Rumbaugh, 1999). In

Table 14.3   Answers to the questions posed in the chapter as they pertain to apes

Question	Answer	Relevant citations
Do animals talk in their own languages?	Perhaps, but some evidence indicates that apes may use symbols in the wild.	Savage-Rumbaugh et al., 1996
Can animals learn to speak as we do?	Some birds can, but no clear data for apes. The more important question, however, is whether or not apes can use language through another medium.	Kellogg & Kellogg, 1933; Gardner & Gardner, 1969, 1971; Pepperberg, 1999; Rumbaugh & Savage-Rumbaugh, 1994
Can the essence of human language become an integral component of artificial languages?	Yes. Syntactic and semantic requirements can be learned by apes.	Herman, 1988; Herman et al., 1984, 1993; Rumbaugh, 1977; Savage-Rumbaugh, 1986; Savage-Rumbaugh et al., 1993
Is it possible to have a conversation with an animal through an acquired language?	Yes. The use of nonverbal symbolic systems allows for two-way communication.	Gill, 1977; Pepperberg, 1999; Rumbaugh, 1977; Savage-Rumbaugh, 1986
Must there be clear evidence of syntax, both in use and in comprehension, for it to be concluded that an animal has language?	No. Semantic understanding is the essential aspect of language as it provides for meaning.	Rumbaugh & Savage-Rumbaugh, 1994; Savage-Rumbaugh, 1986; Savage-Rumbaugh & Brakke, 1996; Savage-Rumbaugh & Lewin, 1994
When animals learn to use symbols, are the symbols meaningful to them in the same way that words are meaningful to us?	Yes. These symbols are representations of real-world items, activities, and locations.	Brakke & Savage-Rumbaugh, 1995, 1996; Herman, 1988; Herman et al., 1984; Miles, 1990; Pepperberg, 1990a; Rumbaugh, 1977; Savage-Rumbaugh, 1986; Savage-Rumbaugh & Lewin, 1994; Savage-Rumbaugh et al., 1986, 1989
If animals learn to use symbols, do they comprehend them when they are used by others who wish to communicate with them?	Yes. This comprehension occurs when both other animals and humans use symbols.	Herman, 1988; Herman et al., 1984, 1993; Rumbaugh, 1977; Savage-Rumbaugh 1986, 1991; Savage-Rumbaugh et al., 1978a, b, 1983, 1986, 1989

Question	Answer	References
If animals can learn languages, does that mean that the animals are intelligent?	Yes, but this does not mean that non–language-competent animals are not intelligent.	Rumbaugh et al., 1996; Savage-Rumbaugh & Lewin, 1994
Are language-competent animals different from animals that don't learn language?	Language seemingly facilitates some abilities that are not present otherwise in animals.	Pepperberg, 1999; Rumbaugh & Savage-Rumbaugh, 1994; Rumbaugh et al., 1996; Savage-Rumbaugh & Lewin, 1994
Do learned languages uniquely help animals solve problems?	Yes. Symbolic communication facilitates the ability to communicate specific information.	Savage-Rumbaugh et al., 1978a, b, 1988; Savage-Rumbaugh, Rumbaugh, & Boysen, 1980
Can animals think in terms of language?	Yes. Apes categorize symbols and perform cross-modal matching of items to symbols.	Savage-Rumbaugh et al., 1988; Savage-Rumbaugh, Rumbaugh, & Boysen, 1980
Do animals learn language spontaneously?	Yes. With rearing similar to that of human children, apes spontaneously learn language. Parrots also learn through observation.	Brakke & Savage-Rumbaugh, 1995, 1996; Pepperberg, 1999; Savage-Rumbaugh, 1991; Savage-Rumbaugh & Lewin, 1994; Savage-Rumbaugh et al., 1986, 1989, 1993
Does the acquisition of language in animals parallel the normal course of language acquisition by children?	Yes. Although human children eventually exceed the language skills of apes, early development is very similar.	Savage-Rumbaugh et al., 1993; Savage-Rumbaugh & Lewin, 1994
Do ape brains mediate language in the same manner that human brains mediate language?	Not clear. The chimpanzee brain, however, shows many similarities to the human brain in areas related to language.	Cantalupo & Hopkins, 2001; Gannon et al., 1998; Hopkins et al., 1998
Do animals teach others, notably their young, the language that they have been taught?	Not clear. There are some indications, however, from signing chimps.	Fouts et al., 1989
Should animals be treated with greater respect and granted rights because they have learned language?	Yes; but all nonhuman animals should be treated with respect.	

addition, given the close genetic relationship between humans and members of the genus *Pan* (Andrews & Martin, 1987; Sibley & Ahlquist, 1987), comparative studies of ape cognition should be expected to demonstrate aspects of shared psychological competencies such as language acquisition.

Having established in our laboratory that great apes learn and use language in much the same way that young human children do, and having determined that this visual language system has referential meaning for these animals, we propose a question for the twenty-first century: To what extent does the ape brain works like the human brain? Studies have demonstrated that chimpanzee and other great ape brains are much more closely related in morphology to the human brain than previously thought for specific "language" areas such as the planum temporale (Gannon et al., 1998; Hopkins et al., 1998) and Brodmann's area 44 (Cantalupo & Hopkins, 2001; see also Chapter 15). These studies all provide evidence that neuroanatomical substrates for left-hemisphere dominance in speech production areas are not unique to humans or hominid evolution but may have been evident at least 5 million years ago.

With the advent of brain imaging technologies such as positron emission tomography (PET) and functional magnetic resonance imaging (fMRI), it will be important to determine what role the great ape brain plays in both comprehension of speech and symbol use as well as in the production of symbol statements and even species-specific vocal communicative utterances. We have undertaken one such study in which Panzee (a chimpanzee) engaged in computerized speech comprehension. After the task, Panzee was given a PET scan. Although homologous areas (to human activation) were not activated in Panzee, she did activate bilateral dorsomedial frontal cortex, cerebellum, and thalamus. This suggests that Panzee processed spoken English using different brain regions than human subjects, but this study was the first ever conducted with a language-competent animal, and many questions remain to be addressed.

In addition, we believe that it will be important to define further the role of the environment in language acquisition by nonhuman animals. Language does not occur in a vacuum; it is necessary only within the social and cultural context in which it is learned and employed. Therefore, cultural traditions and transmission are key aspects of the evolution of language and cognition in general in humans (Tomasello, Kruger, & Ratner, 1993; Whiten, 1999a, 2000a). With an emerging picture suggesting that cultural learning occurs in wild chimpanzees (Boesch, 1996; Boesch et al., 1994; McGrew, 1992, 1998; Sugiyama, 1997; Whiten, 2000a; Whiten et al., 1999; Wrangham et al., 1994), it is even more apparent that the chimpanzee is equipped to learn some of its behaviors within a social context. One interesting aspect of this is to examine the extent to which ape mothers teach their offspring about the lexigram symbol system. With some (although controversial) evidence that such "teaching" by nonhuman animals, and especially chimpanzees, may occur both in the wild and in captive populations (Boesch, 1991; Fouts et al., 1989; Goodall, 1986; Whiten, 1999b; see also Caro & Hauser, 1992; Maestripieri, 1995b), it is important to establish the capacity, the extent, and the course of such teaching. For that, time is the important variable. Having had the opportunity to work with chimpanzees and bonobos for more than thirty years using the lexigram symbol system, we are only now beginning to understand the truly important questions. These questions can be answered only through continuing to provide these animals with an environment in which language competence can flourish.

As we conclude this chapter, we turn our attention again to the questions posed at the beginning. Our answers are summarized in Table 14.3. We hope that the reader has reached the same conclusions based on the data presented throughout the chapter. The reader will note that the last question is one of the most difficult, and perhaps there is no answer. Rights imply responsibilities, and perhaps for that reason animals should have no rights. It was not so long ago, however, that the same claims were made for people of different races, genders, religions, and sexual persuasions from those of the "majority." What is clear is that the way in which humans are monopolizing this planet is not a good thing. We must learn to live with the animals that inhabit this planet as our neighbors. Our claim for the great apes is that both biological and psychological continuity between us and them is established. They are our relatives, and if rights should be granted to any animal, perhaps they have the greatest claim.

Preparation of this chapter was supported by NICHD Grants HD-38051 and HD-06016 to the Language Research Center of Georgia State University and by the College of Arts and Sciences, Georgia State University. The authors thank all the people who have cared for the apes at the Language Research Center throughout the years.

15

Brain Substrates for Communication, Cognition, and Handedness

William D. Hopkins, Dawn L. Pilcher,

and Claudio Cantalupo

Numerous comparative studies have documented overall changes in the organization and structure of the central nervous systems of primates. In primate evolution, the size of the brain has grown exponentially in comparison to the size of the body. As a result, the human brain is three times larger than would be predicted for a nonhuman primate of the same body size (Jerison, 1973; Stephan et al., 1981). Rather than focusing solely on brain size, more recent studies have examined the relative change in specific structures or regions of the brain relative to overall brain and body size (Deacon, 1997). For example, after adjusting for differences in overall brain size, it has been estimated that the frontal lobe is roughly 202 percent larger in humans than in an average nonhuman primate of the same body size. This is in contrast to other lobes such as the occipital lobe, which is only 60 percent larger in humans than in primates of the same body size (Deacon, 1990). These findings have been interpreted as suggesting that the expansion of specific areas of the brain accounts for variations in behavioral, perceptual, and cognitive differences between primate species.

The focus of this chapter is on comparative aspects of cerebral dominance or hemispheric specialization in human and nonhuman primates. Hemispheric specialization (HS) refers to the degree to which the left or right cerebral hemisphere exhibits control or dominance for a specific behavioral, perceptual, or cognitive ability. At the neuroanatomical level, hemispheric specialization refers to morphological or cellular differences between the left and right hemisphere in homologous regions of the brain.

In humans, the most pronounced behavioral manifestations of hemispheric specialization are handedness and lateralization of function for communication, notably speech production and perception (Hellige, 1993). In terms of handedness, approximately 85–90 percent of the population is right-handed (Annett, 1985; Porac & Coren, 1981). The prevalence of right-handedness is fairly consistent across cultures, although some variation does exist; it should be noted, however, that there are no reports of population-level left-handedness in either modern human studies or in the archaeological record. In terms of language, both speech perception and production appear to be lateralized to the left hemisphere, particularly among right-handed subjects (Beaton, 1997; Foundas et al., 1995; Moffat et al., 1998). Visual half-field, dichotic listening and functional imaging studies have fairly consistently reported the left hemisphere asymmetry for language processing. In addition, left hemisphere dominance for language is not modality specific; studies in deaf populations have shown a left hemisphere superiority in processing sign language (Hickok et al., 1996).

The neurobiological basis for language lateralization appears to be associated with distinct neuroanatomical regions in the left hemisphere. Specifically, lesions in the posterior portion of the left but not the right temporal lobe induce deficits in the perception of language. Lesions in the ventral posterior portion of the left but not the right frontal lobe induce deficits in speech production. These deficits are referred to as Wernicke's and Broca's aphasias, which were named after the scientists who initially reported these observations. More recent morphological studies in neurologically intact subjects have reported differences in the size of left and right regions of the hemispheres corresponding to Broca's and Wernicke's area (Foundas et al., 1995). For example, the planum temporale (PT), a region of the posterior temporal lobe that encompasses Wernicke's area, has been reported to be larger in the left hemisphere, particularly among right-handed subjects. The pars triangularis and pars opercularis, regions of the ventral frontal lobe that correspond to Broca's area, have also been found to be larger in the left than in the right hemisphere (Amunts et al., 1999; Foundas et al., 1998).

From the earliest reports of behavioral and neuroanatomical asymmetries in humans, there has been a theoretical and empirical interest in whether similar asymmetries are present in nonhuman animals, notably primates (see Harris, 1993, for historical review). It was not until relatively recently that researchers have accumulated enough data to assess truly whether behavioral or brain asymmetries in nonhuman primates in any way resemble human patterns of asymmetry. Before these data emerged, the relatively sparse data, particularly from great apes, led to numerous theories regarding the origin of hemispheric specialization in humans. For example, some posited that HS was unique to humans and was associated with the evolution of language abilities (Ettlinger, 1988; Warren, 1980). Others have posited that handedness and other behavioral asymmetries evolved in the context of increasing bipedalism in hominoid evolution and the emergence of motor skill associated with tool-using abilities (see Bradshaw & Rogers, 1993, for review). In these evolutionary scenarios, HS was viewed as uniquely human and functionally distinct from asymmetries in other nonhuman animals. Recent findings clearly challenge this long-held belief. Evidence of population-level asymmetries have now been reported in fish, amphibians, reptiles (Bisazza et al., 1998), birds, rats, and nonhuman primates (Bradshaw & Rogers, 1993), clearly suggesting that language, bipedalism, and tool use are not necessary conditions for the expression of HS. In fact, it can be argued on the basis of these findings that HS is a fundamental attribute of the vertebrate central nervous system. What is unclear from the existing research is whether there is phylogenetic continuity in the distribution, structure, and function of HS in nonhumans compared to humans, or even primates compared to non-primates.

In this chapter we first summarize the literature on neuroanatomical asymmetries in nonhuman primates. All aspects of neuroanatomical asymmetries are reviewed as we compare findings among different primate species. We then discuss the neuroanatomical findings in relation to behavioral data on hand preferences and other aspects of functional lateralization in different primate species. We focus on studies of hand preference rather than other behavioral manifestations of hemispheric specialization because the overwhelming comparative data are on hand preference (see Hopkins & Fernandez-Carriba, 2002, for review of recent perceptual and cognitive studies in nonhuman primates).

Neuroanatomical Asymmetries in Nonhuman Primates

Methodologies Used in Measuring
Neuroanatomical Morphological Asymmetries

A variety of methodologies have been used to measure brain asymmetries in nonhuman primates. Each method has both strengths and weaknesses, and it is important to understand some of them in the context of interpreting the findings from different studies. Thus, before reviewing the findings, we present some evaluation of the different methods. Investigators have taken three basic approaches in the assessment of neuroanatomical asymmetries in human and nonhuman primates. They include (1) direct measurement of the brain, (2) measurements of the shape of the skull (petalias) and measurement of sulci developed from endocasts, and more recently (3) use of brain imaging technologies, notably magnetic resonance imaging (MRI).

Direct measures of brain asymmetries from the surface of the brain are ideal but require a postmortem specimen. Postmortem brain studies are difficult in primates, however, because of their long life spans and ethical concerns about euthanasia. Furthermore, postmortem brains are often fixed in a chemical preservative, such as formalin, that can cause some shrinkage in brain tissue, which may or may not affect one hemisphere more than another. The use of endocasts to measure the shape of the brain and/or lengths of various sulci has been by far the most widely used procedure. To form an endocast, the inside of the skull is filled with latex, which creates a fossil-like representation of any impressions left by the brain, such as the brain's general shape, any protrusions, and some sulci. Petalias, which are anterior and posterior protrusions of one hemisphere of the brain in comparison to the other, are posited to be the result of the brain's torquing nature during development (Best, 1988). The most common pattern of petalia asymmetry found in humans is a larger left occipital region and a greater right frontal pole. Petalias have been studied in a host of species. Endocast methodology is a relatively easy procedure to use, and many specimens can be obtained in this manner compared to other techniques. The most serious limitation of this technique is that the measurement of asymmetry is qualitative, with the observers simply judging whether there is a left or right protrusion of the frontal or posterior region of the brain. In terms of sulci, the principal problem with endocasts is that posterior portions of the brain do not cast very well. Moreover, lengths of some sulci can be easily quantified, but only those that project to the surface of the brain. Many sulci or some portions of a specific sulcus do not project to the surface, and these would not be detected by endocasts, leading to some potential error in measurement of the sulci.

Finally, studies have begun to use in vivo brain imaging techniques, such as MRI, to assess brain asymmetries. MRI offers several advantages over traditional measurement techniques. First, since excellent representation of brain areas can be obtained with this technique, euthanasia is not necessary to obtain an objective measure of asymmetry for specific regions of interest. Second, researchers can obtain multiple planes of assessment for a specific sulcus or brain region rather than having to rely on surface measurements alone. In some cases 3-D renderings or reconstructions can be obtained for regions of interest. Concerns about the use of MRI technology include its cost and pragmatic limitations such as the size of the head coils relative to the size of the primate. That is, most MRI head coils are designed for human skulls. They work effectively for most primates, but adult male gorillas and orangutans have skulls larger than the human head coil. Thus, different scanning procedures need to be used with these individuals, which may introduce some error when their data are compared to data from other individuals of the same or different species.

In conclusion, some of the limitations of endocasts and cadaver specimens, such as the difficulty of locating the median border of smaller brains, poor casting of the occipital region in endocasts (Falk, 1987; LeMay et al., 1982), and difficulty identifying sulcal endpoints owing to shrinkage caused by chemical preservation (Falk, 1986; Frontera, 1958; Heilbronner & Holloway, 1988), may be influencing measurement of asymmetries in some species but not others. In addition, although 3-D coordinates can be taken from the surface of the cadaver or endocast, many of the structures being investigated for asymmetries do not exist solely on the lateral surface of the brain. MRI, by contrast, allows for alignment and reformatting or slicing into different planes to measure both surface and internal anatomy. Specifically, structural MRI is a noninvasive technique that can provide a detailed representation of neural structures in vivo. These features, which make MRI so versatile, are not possible with either conventional postmortem procedures or endocasts.

Skulls and Petalia

Groves and Humphrey (1973) were among the first to observe a difference between the right and left hemisphere of the great ape brain. This difference was observed as a leftward asymmetry in the length of the skull for a significant proportion of 38 mountain gorillas. In studies that followed, the inside of the skull was used to determine hemispheric differences. Holloway and De La Coste-Lareymondie (1982) qualitatively scored the frontal and occipital petalia from the endocasts of gorillas, chimpanzees, bonobos, and orangutans as either leftward, rightward, or no asymmetry. Petalia asymmetries in the occipital pole were reported for 91 of 135 great apes, with 65 having a left hemisphere bias (Holloway & De La Coste-Lareymondie, 1982). Asymmetries in the frontal pole were reported for 65 of 135 great apes, with 53 exhibiting a right hemisphere bias. The occurrence of a combined right frontal and left occipital petalia asymmetry was reported in only 34 of 135 great apes.

Rather than directly measuring petalias from endocasts, LeMay (1976) measured petalia width from photographs of seventeen chimpanzee cadaver brain specimens and reported that eleven of those subjects had a wider left than right occipital petalia, and six had a wider right compared to left occipital region, a proportion consistent with that reported by Holloway and De La Coste-Lareymondie (1982). In another study that used photographs of cadaver specimens, Cain and Wada (1979) observed the frontal lobe petalia from photographs of seven baboon cadaver brains. Of the seven specimens, six had a significantly longer right than left frontal pole, which is also consistent with reports of human and great ape frontal lobe patterns.

In an attempt to quantify the previously reported endocranial petalia asymmetries of monkeys and apes, one of the first MRI studies in nonhuman primates investigated the differences in the width of the cerebral hemispheres. Specifically, Hopkins and Marino (2000) identified the first axial slice above the third ventricle and determined the length of each hemisphere by measuring from the frontal to the occipital pole, 5 mm lateral to the midline. The midline was defined by placing a point in the middle of the interhemispheric cleft at the frontal and occipital poles and connecting a straight line between these points. Four regional width measurements were taken corresponding to 10 percent and 30 percent of the length from the frontal and occipital pole within each hemisphere. These were collectively referred to as the anterior-frontal

(10 percent from frontal pole), posterior-frontal (30 percent from the frontal pole), parietal lobe (30 percent from occipital pole), and occipital lobe (10 percent from occipital pole). Widths were measured from the midline to the lateral surface of the brain (in tenths of a millimeter). Measurements were collected from nineteen great apes (nine chimpanzees, four orangutans, two gorillas, and four bonobos), fifteen Old World monkeys (nine rhesus monkeys, two baboons, and four sooty mangabeys), and eight New World monkeys (four capuchin monkeys and four squirrel monkeys). The results indicated that the ape sample showed a right frontal and left occipital asymmetry in cerebral width. In contrast, neither the Old nor the New World monkey samples exhibited population-level biases in hemispheric width.

A related study by Pilcher and colleagues (2001) investigated cerebral volumetric asymmetries in a sample of twenty-three lesser and great apes (nine chimpanzees, four orangutans, two gorillas, four bonobos, and four gibbons), fourteen Old World monkeys (eight rhesus monkeys, two baboons, and four mangabeys), and eight New World monkeys (four capuchin and four squirrel monkeys). Pilcher and colleagues (2001) calculated asymmetry quotients from volume measures of the anterior and posterior regions of the frontal and occipital poles in the left and right hemisphere. The landmarks used to identify the posterior border of the frontal region were the lateral commissures. For the anterior border of the occipital region, the scan numerically halfway between the posterior commissural fibers and the lateral surface of the occipital lobe was identified. In the sample of Old and New World monkeys, there were no population-level asymmetries in any of the volumetric measures. In the ape sample, the left volumetric occipital region was larger than the right, but the frontal areas were not asymmetric at the population level. In addition, within the ape sample, these occipital asymmetries as well as a rightward posterior frontal asymmetry predicted the leftward direction of the planum temporale (PT). (The PT was also measured with MRI and will be discussed later on.)

In another study that considered the volume of petalia asymmetries in chimpanzees, differences in the "patchy or confluent local protrusions" were measured between the cerebral hemispheres using 3-D reconstruction of nine cadaver specimen MR images (Zilles et al., 1996). Zilles and colleagues (1996) reported that when a 3-D image of the right hemisphere was superimposed on the left hemisphere and the surface areas were subtracted, there were no significant differences between the petalia for the hemispheres.

Sulci Length and Variation

In a series of studies using endocasts of Old World monkey skulls, Falk and colleagues measured the length of a number of different fissures from both hemispheres. In one of their initial studies, photographs were taken from the right and left lateral surface of the endocast and then scanned for digitized length analysis (Falk, 1978). Falk reported a rightward asymmetry in length for the sylvian fissure (SF) of patas monkeys, colobus monkeys, and pig-tailed langurs, as well as a rightward superior temporal sulcus (STS) in colobus monkeys and pig-tailed langurs. In the frontal region, she reported a rightward asymmetry in the orbital sulcus for the pig-tailed langurs and guenons.

In a later study, Falk and colleagues (1986) used a slightly different method of length analysis by recording the 3-D coordinates of SF length from the endocasts of ten rhesus monkeys. From these coordinates, computer software was then used to calculate SF length. Contrary to Falk's (1978) previous study in other Old World monkeys, Falk et al. (1986) reported a left rather than right hemisphere asymmetry in SF length for this sample of rhesus monkeys. Falk continued her analyses of sulcal patterns by investigating additional areas of the Old World monkey brain. A subsequent study by Falk et al. (1990) investigated sulcal asymmetries specific to the frontal lobes from 335 rhesus monkey endocasts. Again, the 3-D coordinates of the imprinted sulcal patterns were collected from the endocasts and analyzed using a computer aided design system. It was reported that the left central sulcus, right rectus principal, and right lateral orbital sulcus were significantly longer than those in the left hemisphere (Falk et al., 1990).

Rather than using endocasts, LeMay and Geschwind (1975) were among the first to measure asymmetries directly in the nonhuman primate cadaver brain. As in Falk's (1978) methodology, photographs were taken of the left and right hemispheres in nine New World monkeys, twenty-one Old World monkeys, and thirty-nine lesser and great ape cadaver specimens. Tracings on transparent paper placed over these photographs were used to observe the frequency of differences in the height of the terminating point of the left and right SF. Subjects with differences greater than 3 mm in the height of the SF were classified as left or right dominant (depending on which hemisphere was larger). All other subjects were classified as having no hemispheric bias. From the monkey sample, LeMay and Geschwind (1975) reported a right hemisphere bias in one spider monkey, one crested black macaque, and one leaf monkey. It was further reported by LeMay and Geschwind (1975) that none of the lesser apes had an asymmetry greater than 3 mm, whereas fifteen of the twenty-eight great apes had a higher right than left SF, and one had a higher left than right SF terminating point. Twelve great ape subjects had no hemispheric bias.

Yeni-Komshian and Benson (1976) were the first to examine the direct length of the SF in postmortem brains of twenty-five humans, twenty-five chimpanzees, and twenty-five rhesus monkeys. Methodologically, a thread was aligned in the sulcus and cut at the length of the SF from the anterior to the posterior border of the left and right hemispheres. This method revealed that the SF was significantly longer in the left hemisphere than in the right hemisphere in both the human and chimpanzee brains, but no significant asymmetry was found in the rhesus monkey brains. More recently, Heilbronner and Holloway (1988) also investigated SF length in a sample of Old and New World monkey cadaver brains with the use of a very similar methodology. Their sample included thirty rhesus monkeys, thirty long-tailed macaques, twenty-seven cotton-top tamarins, thirty-one common squirrel monkeys, and twenty-six common marmosets. The anterior and posterior borders of the SF were anchored with pins on the cadaver surface, and the length was measured from each hemisphere with a flexible measuring tape. In contrast to the findings in the monkey sample of Yeni-Komshian and Benson (1976) as well as the endocast studies by Falk et al. (1990), population-level left hemisphere asymmetries were found for rhesus monkeys, long-tailed macaques, common marmosets, and cotton-top tamarins.

In another study by Heilbronner and Holloway (1989), sulcal asymmetries were measured in the frontal, temporoparietal, and limbic areas of sixty rhesus and long-tailed macaque cadaver specimens. As in previous investigations of sulcal patterns in cadaver brains, a flexible measuring tape was used to record fissure length directly on the surface of the specimen. The authors reported, however, that the only pronounced surface asymmetry was "a small parietal lobe sulcus," which appeared to be more developed in the left than the right hemisphere. There were no significant asymmetries in the superior temporal sulcus, horizontal or vertical limb of the arcuate, interparietal sulcus, cingulate gyrus, or principal sulcus.

Most recently, Hopkins, Pilcher, and MacGregor (2000) investigated the length of the SF and superior temporal sulcus (STS) in nonhuman primates including eight New World primates, fifteen Old World monkeys, and twenty-six great apes using MRI. The method used to measure these fissures within the undulating cortex was as follows. Three regions from the sagittal plane of reformatted MRI images were chosen based on their consecutive position before the insula (lateral, medial, and insular). The lengths of the fissures were then traced using a mouse-driven free-hand line tool and quantified using NIH Image (see Figure 15.1). Depicted in Table 15.1 are the number of subjects classified as left or right hemisphere dominant for each taxonomic family and region of the SF. For the SF, the ape sample was not significantly lateralized on the surface measures of the brain, but strong leftward asymmetries were

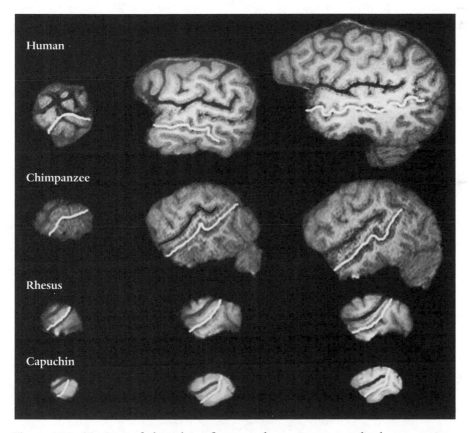

Figure 15.1 Tracings of the sylvian fissure and superior temporal sulcus at various sagittal depths in different primate species. (Redrawn from Hopkins, Pilcher, & MacGregor, 2000)

Table 15.1 The frequency of leftward, rightward, or absence of asymmetry according to the asymmetry quotient values for each region of the sylvian fissure and superior temporale sulcus across taxonomic family; the binomial z score is also presented based on the probability of a leftward asymmetry

	Lateral				Medial				Insular			
	L > R	R > L	R = L	z	L > R	R > L	R = L	z	L > R	R > L	R = L	z
Sylvian fissure												
Great apes	11	16	1	-1.13	28	0	0	**5.29**	22	4	2	**3.02**
Old World monkeys	4	12	0	**-2.00**	16	0	0	**4.00**	9	5	2	0.50
New World monkeys[a]	4	4	0	0.00	6	0	0	**2.45**	4	4	0	0.00
Superior temporal sulcus												
Great apes	12	13	3	-0.76	11	12	5	-1.13	12	12	4	-0.76
Old World monkeys	2	13	1	**-3.00**	1	9	6	**-3.50**	4	8	4	**-2.00**
New World monkeys[a]	4	4	0	0.00	2	3	1	-0.82	3	3	2	-0.71

a. Two New World monkeys had only one scan before the opening of the insula; therefore only a lateral and insular measure were possible. Numbers highlighted in bold indicate significant z-scores.

found in the medial and insular regions. The STS was not significantly lateralized for any of the three planes in the great ape sample. In the sample of Old and New World monkeys, a leftward medial SF was present at the population level, but there were no significant asymmetries at the insular region. On the lateral surface, both the SF and STS of the Old World monkey sample were significantly rightward, yet the New World monkeys exhibited no lateral asymmetries in either fissure.

Planum Temporale

In regards to HS, one of the most striking cerebral differences in humans is the planum temporale (PT). Because of methodological limitations, surface measures of the SF and STS have been used to speculate the directional asymmetries in the PT, specifically in nonhuman primates (Yeni-Komshian & Benson, 1976). In fact, the greater length of the SF in the left hemisphere is commonly thought to be a direct consequence of the larger expansion of the PT in the same hemisphere. Studies in both great apes (Cantalupo et al., in press) and humans (Foundas et al., 1999), however, have failed to reveal any evidence of a direct relationship between PT and SF asymmetries. This converging evidence clearly raises questions about the common practice of considering SF asymmetry as a predictor of PT asymmetry. Therefore, direct measures of the PT in nonhuman primates are necessary before speculations about the potential similarities with the human PT are possible. The assumption that the PT's leftward asymmetry is unique to humans has been challenged in two studies. Gannon et al. (1998) assessed PT asymmetries in a sample of eighteen chimpanzee cadaver brain specimens. These authors localized the borders of the PT (applying the same borders used in humans: posterior margin of Heschl's sulcus, superolateral margin of the superior temporal gyrus, and the horizontal limb of the SF) from the surface of the cadaver brain, and a thin piece of black plastic was shaped to fit precisely inside the PT area. The plastic shape was then attached to a white card and scanned to measure what was considered the surface area of the PT. Gannon et al. (1998) reported that in seventeen of the eighteen cadaver specimens, the PT was larger in the left than in the right hemisphere. Although this leftward asymmetry in the PT of this sample of great apes is similar to that in humans, the proportion is much higher than that found in humans (see Geschwind & Levitsky, 1968).

Hopkins and colleagues (1998) measured the PT in a sample of twenty-one great apes (twelve chimpanzees, four orangutans, two gorillas, and three bonobos) from MRI. The PT was identified and measured from both the sagittal and coronal planes following procedures not unlike those used in humans (see Figure 15.2). From the sagittal plane of the PT, sixteen had a significantly leftward bias, two had a rightward bias (two orangutans), and three subjects (two chimpanzees and one bonobo) had no bias. In the coronal plane, the PT was significantly leftward for eleven apes, rightward for two (one chimpanzee and one orangutan), and there was no significant bias for seven apes (two chimpanzee, two bonobos, one gorilla, and two orangutans). Overall, the statistical majority of these apes had a larger planum temporale area in the left

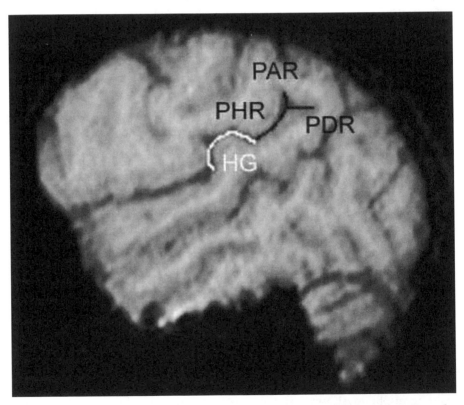

Figure 15.2 Outline of landmarks used to quantify the planum temporale (PT) in the sagittal plane. HG = Heschl's gyrus, PHR = posterior horizontal ramus, PAR = posterior ascending ramus, PDR = posterior descending ramus.

than in the right hemisphere. Attempts to localize the PT in Old World monkeys, New World monkeys, and lesser apes were unsuccessful because Heschl's gyrus, the principal neuroanatomical landmark used to define the PT, could not be identified on the MR images. Gilissen (2001) also investigated PT asymmetries in common chimpanzees with the use of MRI. In this study, he discussed collecting MRI scans from ten chimpanzee cadaver specimens for subsequent measurement of the PT surface area from the sagittal plane using similar landmarks to those employed by Hopkins et al. (1998). Unlike Gannon et al. (1998) and Hopkins et al. (1998), Gilissen reported that although the mean asymmetry value reflects a leftward asymmetry ($M = -.206$), the PT was not significantly asymmetrical at the population level for this sample of chimpanzees $t(9) = -1.69$, $p < .06$. It should be noted, however, that seven of the ten chimpanzees had a left hemisphere asymmetry, a percentage not unlike that reported by Hopkins et al. (1998) in a larger sample of apes.

Hand/Motor Area (Knob)

In the first study of its kind in nonhuman primates, Hopkins and Pilcher (2001) examined asymmetries in the precentral sulcus in a sample of great apes. Studies in humans have identified a region of the precentral gyrus referred to as the "knob" (Yousry et al., 1997). Functional imaging studies (e.g., PET and fMRI) in human subjects have shown that the knob is where the hand and fingers are represented in the motor cortex (Pizzella et al., 1999). Moreover, when this structure is stimulated with transcranial magnetic stimulation, there are significant differences in the threshold of the right and left hemisphere (Boroojerdi et al., 1999; Triggs et al., 1997).

Following procedures employed with humans, Hopkins and Pilcher (2001) localized the knob on successive axial slices through the dorsal and medial region of the precentral sulcus in a sample of twenty great apes (two gorillas, four orangutans, ten chimpanzees, and four bonobos; see Figure 15.3). Attempts to localize this region in Old and New World monkeys failed owing to the lack of significant gyrification in this region of the motor strip to clearly define anatomical regions of interest. Hopkins and Pilcher (2001) measured both the width (which replicated the method used by Yousry et al., 1997) and volume of this knob structure on consecutive axial slices and found that the knob was significantly larger and wider in the left than in the

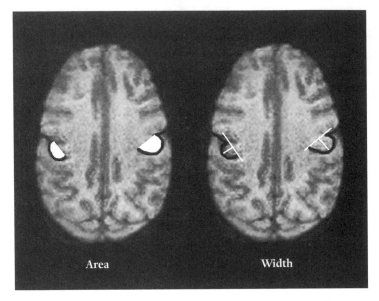

Figure 15.3 Axial view of a chimpanzee's brain showing the "knob."

right hemisphere. Of the twenty great ape subjects, fifteen had a left hemi-sphere asymmetry, two had a right hemisphere asymmetry, and three had no bias.

Corpus Callosum

The corpus callosum (CC) is the tract of fibers that connects cortical areas between the left and right cerebral hemispheres. In humans, variation in CC morphology has been implicated in a host of behavioral phenomena. In particular, a larger CC has been associated with left-handedness in humans (Driesen & Raz, 1995; Witelson, 1985, 1989). Moreover, larger CC size has been associated with anomalous or reversed asymmetry in the planum temporale (Aboitiz & Ide, 1998). It has also been reported that individuals born without a CC (agenesis of the CC) have unusually large neuroanatomical asymmetries (Jancke & Steinmetz, 1994). In short, to some extent, discussions of brain asymmetry need to be interpreted within the context of CC morphology.

Very few studies have examined CC morphology in nonhuman primates, but the results that do exist are intriguing from the standpoint of the evolution of brain asymmetries. In one of the first studies to compare CC morphology systematically in nonhuman primates, De Lacoste and Woodward (1988) measured total callosal area and splenial width in a sample of twenty strepsirhines, seven ceboids, fourteen cercopithecoids, and fifteen pongids. Both total CC area and splenial widths were divided by total brain weight and compared among families. Pongids had the lowest ratio scores (smaller corpus callosum size than expected for brain size), followed by cercopithecoids, ceboids, and strepsirhines. These data suggest that as the brain has grown larger, the CC has not grown allometrically at the same rate.

More recently, Rilling and Insel (1999) examined the relationship between brain volume, neocortical surface area, and CC size using MRI in a sample of nonhuman primates representing four different taxonomic families. Rilling and Insel (1999) divided the size of the corpus callosum by both the overall brain volume (CC:VOL) and total neocortical surface area (CC:NEO) for each subject. They found that the species more distantly related to humans (i.e., New World monkeys) have larger ratios than more closely related species (i.e., great apes). Overall, humans have the lowest ratio, followed by great apes, lesser apes, Old World monkeys, and New World monkeys. In addition, Rilling and Insel (1999) have reported that the slope of the regression line between brain volume and CC size in a sample of forty-one nonhuman primates was .70. They interpret the .70 slope (given that it is less than 1.0) to mean that the CC has not evolved at the same pace as brain volume in primate evolution, and therefore there has been a reduced need for interhemispheric communication and a greater need for intrahemispheric specialization.

The results reported by Rilling and Insel (1999) in primates are consistent with a report on the size of the CC relative to brain size in a host of mammals including rats, rabbits, cats, dogs, horses, cows, and humans. Olivares and colleagues (2000) reported a negative association between CC size and brain weight. These effects were particularly evident in the posterior portion of the CC or splenium. Olivares and colleagues (2000) traced these findings to inherent differences in the degree of bilateral overlap of the visual system. One interpretation of the species differences in ratio measures is that as the brain increased in size during evolution, each hemisphere became increasingly independent, and this led to greater specialization within a hemisphere. In other words, "fewer" callosal fibers were necessary because of reduced communication between the hemispheres. According to Rilling and Insel (1999), the increased behavioral and neuroanatomic lateralization in humans is reflected in greater intrahemispheric compared to interhemispheric connectivity.

To test this hypothesis, Hopkins and Rilling (2000) regressed the brain asymmetry measures of cerebral width previously reported by Hopkins and Marino (2000) on the CC:VOL and CC:NEO ratio measures. A significant negative correlation was found, particularly for the asymmetries in the posterior occipital lobe. Subjects with larger left hemisphere occipital asymmetries had smaller CC:VOL and CC:NEO ratio scores, a finding that supports the theory proposed by Rilling and Insel (1999) and others (Ringo et al., 1994).

The evidence of neuroanatomical asymmetries in nonhuman primates allows for several general conclusions to be drawn. First, as the brain has increased in size, each hemisphere has become increasingly independent and isolated as reflected in the fact that the ratio in the size of the CC is smaller in primates more closely related to humans. Second, great apes appear to exhibit several neuroanatomical asymmetries that are not morphologically present in nonhuman primate species more distantly related to humans, notably the PT and "knob." The most likely explanation for these species differences is the greater gyrification seen in the great ape brain compared to other nonhuman primates (see Zilles et al., 1989). Third, where common measures of asymmetry have been employed, monkeys do show some similarities to apes and humans, including SF length and sulci found primarily in the frontal lobe.

The findings in great apes resemble in many ways those reported in humans. In particular, the three most common measures of brain asymmetry reported in humans are a left hemisphere asymmetry in the PT, left hemisphere asymmetry in length of the SF, and a left occipital, right frontal petalia asymmetry. In great apes, the exact patterns of results are found as in humans. The findings in monkeys are less clear, but similar effects are found for SF length.

Implications of the Evolution of Neuroanatomical Asymmetries for the Emergence of Handedness, Cognition, and Communication

We began this chapter by discussing the robust nature of behavioral and neuroanatomical asymmetries in humans, particularly in the realm of handedness and language. The neuroanatomical data clearly point to the presence of landmarks in the great ape brain (and to a lesser extent the monkey brain) that have been implicated in the neurobiology of language in humans. Thus, one can ask whether there are neurobiological substrates underlying the expression of lateralized communicative behaviors or hand use in monkeys and apes which are potentially homologous to those found in humans. In the remaining portion of this chapter, we discuss whether there are homologs to handedness and lateralization in communicative behavior between monkeys, apes, and humans.

Handedness

The issue of whether nonhuman primates exhibit population-level handedness remains a topic of considerable debate. A full review of all the relevant studies is beyond the scope of this paper (see Fagot & Vauclair, 1991; Hopkins, 1996; MacNeilage et al., 1987; McGrew & Marchant, 1997; Ward & Hopkins, 1993, for reviews), but there are clear differences in perspectives that center on both methodological and theoretical grounds (see Hopkins, 1999; McGrew & Marchant, 1997).

In our laboratory, we have examined laterality in hand use of chimpanzees for a host of behavioral measures including simple reaching, bipedal reaching, throwing, bimanual feeding, and coordinated bimanual actions and have consistently found evidence of population-level right-handedness. The extent to which these results generalize to other great apes and monkeys is not exactly clear because there are very few studies that have examined hand preference using the same measures in different species. When considering hand preferences averaged across multiple measure of hand use (at least two) from a number of different studies, Hopkins and Pearson (2000) have reported that chimpanzees, bonobos, and gorillas are all right-handed at the population level while orangutans are not. If subjects are classified as left- or right-handed according to several different criteria, one finds about a 2:1 ratio of right- to left-handed individuals, a proportion that is significantly different from chance but lower than those reported in Western human cultures. The cause of this difference is not clear but warrants some discussion (see below).

With respect to comparative analyses of primates using the same measures, Westergaard and colleagues (1998) reviewed all the findings on posture and reaching in nonhuman primates. Primarily, hand preferences were assessed for tripedal and bipedal reaching. Tripedal reaching involves reaching for food while one hand and both feet are maintained on the ground and the axis of the body is parallel to the ground surface. Bipedal reaching involves reaching for food while in an erect posture with two feet on the ground and the axis of the body perpendicular to the ground surface. Westergaard and colleagues (1998) reported the mean handedness indices for each posture for a number of primate species. Handedness indices, which reflect the directional bias in hand use, are computed by subtracting the number of left-hand responses from the number of right-hand responses and dividing by the total number of response [HI = (#R − #L)/(#R + #L)] (see Hopkins, 1999, for discussion). Westergaard and colleagues (1998) found a gradual increase in the degree of right-handedness in species more closely related to humans than those more distantly related.

For other measures of hand use, the findings are less consistent than the posture and reaching findings. For example, coordinated bimanual actions have been studied as means of examining hand preferences in the complementary use of the hands. Chimpanzees, gorillas, orangutans, rhesus monkeys, baboons, and capuchin monkeys have, to some extent, been tested on comparable tasks (Hopkins, 1995a; Spinozzi et al., 1998; Westergaard & Suomi, 1996; Westergaard et el., 1998; One bimanual coordination task, referred to as the tube task, has been used to assess hand preferences in capuchin monkeys, rhesus monkeys, orangutans, and chimpanzees (Hopkins, 1995a, 1999; Spinozzi et al., 1998; Westergaard & Suomi, 1996). In the tube task, peanut butter is placed on the inside edges of PVC pipe and handed to the subjects. The subjects are required to extract the peanut butter by holding the PVC pipe with one hand and using the opposite hand to remove it. The hand used to extract the peanut butter was recorded over a series of test sessions. As with the studies on reaching, HI values were used to calculate directional biases in hand use. Table 15.2 depicts the findings from the various tube task studies. In capuchin monkeys, Westergaard and Suomi (1996) did not find a population bias in hand use for the tube task, while Spinozzi and colleagues (1998) did report a hand bias. Indeed, the capuchin monkey data from Spinozzi and colleagues (1998) are the strongest evidence of right-hand use for any species. In rhesus monkeys, Westergaard and Suomi (1996) reported a right hand bias for the tube task in one study, but a left hand bias was reported in a second study (Westergaard et al., 1997). Westergaard et al. (1997) attributed the contradictory findings to age, but the left-handed sample comprised all nursery-reared

Table 15.2 Handedness studies using the tube task in nonhuman primates

Author(s)	Species	N	Mean HI	Population effect
Westergaard & Suomi, 1997	Capuchin	24	.04	No bias
Spinozzi et al., 1998	Capuchin	26	.44	Right-handed
Westergaard & Suomi, 1996	Rhesus	55	.14	Right-handed
Westergaard et al., 1997	Rhesus	18	−.32	Left-handed
Hopkins, 1995a	Chimpanzee	110	.14	Right-handed
Hopkins, unpublished	Orangutan	6	.15	No bias

subjects, while the rearing history of Westergaard and Suomi's (1997) sample was not reported. In great apes (all the tube task data have come from the Yerkes colony), the chimpanzees were right-handed, but there was no bias for orangutans.

Vocal and Gestural Communication

There have been three central areas of investigation on behavioral asymmetries in communication, including (1) perception of vocalizations, (2) asymmetries in the perception and production of facial expressions, and (3) lateralized hand use in gestural communication (great apes only).

Perception of species-specific vocalizations.   In one of the most frequently cited papers in this literature, Petersen and colleagues (1978) reported a left hemisphere advantage in the discrimination of species-specific calls by Japanese macaques. In this study, ten monkeys, including five Japanese macaques, two bonnet macaques, two pigtail macaques, and one vervet monkey, were trained to discriminate two types of "coo" vocalizations produced by Japanese macaques that differ in the temporal location of the peak frequency. During training, the stimuli were presented to either the left or right ear of the subjects, and the number of correct trials per ear was tallied for each training session and subject. For each training session, the subjects could be characterized as performing better with their left ear, right ear, or neither ear. Petersen et al. (1978) then determined how many subjects performed significantly better with the right or left ear. They found that all five Japanese macaques reached criterion on the task faster when the stimuli were presented to the right ear, whereas only one of the remaining five monkeys showed the same right ear advantage. Note that none of subjects showed a significant left ear advantage.

Capitalizing on the study by Petersen and colleagues, Heffner and Heffner (1984) trained Japanese macaques similarly to perform auditory discriminations of the "coo" vocalizations, then performed unilateral lesions to either the left or right posterior temporal lobes in subsamples of individuals. Monkeys with a left hemisphere lesion showed a greater decrement in postoperative performance and took longer to relearn the discrimination than monkeys with a right hemisphere lesion. Heffner and Heffner (1984) concluded that the left hemisphere was the dominant half of the brain in performing this discrimination task, but emphasized that to some extent the asymmetry was transient, as all monkeys subsequently reacquired the discrimination.

More recently, rather than utilizing lesion approaches or more traditional laboratory approaches to the assessment of auditory laterality, Hauser and Andersson (1994) examined orienting asymmetries to different auditory stimuli in rhesus monkeys living on Cayo Santiago. To summarize briefly, an audio speaker was positioned in a field setting, and the monkeys were lured to the feeding site by the experimenters. Once positioned centrally behind the speakers (which varied from 5 to 10 meters from the subject), an acoustic stimulus was presented to the subject, and the experimenters recorded which way the monkeys turned (left or right) to orient toward the sound. There were four types of acoustic stimuli, including three types of conspecific calls (rhesus monkey vocalizations) and one type of heterospecific call (a local bird call). A total of eighty monkeys were tested, with one trial presented to each subject. Each of the four types of calls was played to twenty different monkeys, and the distribution of left- as opposed to right-orienting monkeys was compared across the different stimulus types. Hauser and Andersson (1994) reported that significantly more monkeys oriented to the right than to the left for the three conspecific calls but not for the heterospecific call. The authors interpreted these findings as evidence that the left hemisphere is dominant in processing species-specific calls by rhesus monkeys. Subsequently, Hauser and colleagues (1998) tested for orienting asymmetries in the same format described above but manipulated the interpulse interval for three different types of rhesus monkey vocalizations including grunts, shrills, and copulation screams. Variations in the interpulse intervals were either longer or shorter than population mean pulse interval for each of the three call types. For two types of calls (grunts and shrills), extreme variation in the interpulse interval resulted in a lack of any observed orienting asymmetry, while presentation with the "normal" range resulted in a right-sided orienting bias. For the copulation scream, changes in the interpulse interval did not significantly influence the overall right-sided orienting asymmetry.

Laterality and manual gestures. Several studies have reported data on hand use in gestural communication in gorillas, bonobos, and chimpanzees that are not language-trained. In an observational study, Shafer (1993) reported that gorillas gesture more with the right than with the left hand. Specifically, hand use in gestures was noted in thirty-two gorillas living in zoos. It was reported that twenty-two preferred the right and ten preferred the left hand. Using the same type of ethogram, Shafer (1997) studied gestures in a sample of twelve captive bonobos. All twelve bonobos showed a bias in hand use, nine preferring the right and three preferring the left hand. The data from Shafer (1997) are consistent with previous findings by Hopkins and de Waal (1995), who studied laterality in hand use for gestures in a sample of twenty captive bonobos. Of the eighteen subjects that showed a bias in hand use for gestures, sixteen preferred the right hand and two preferred the left. Krause and Fouts (1997) reported hand preference for non–American Sign Language (ASL) gestures in two chimpanzees trained in ASL. Overall, one subject was left-handed and the other showed no bias for all types of gestures. Both chimpanzees, however, gestured significantly more with their right hand for indexical pointing compared to whole-hand pointing. The results reported by Krause and Fouts (1997) are nearly identical to those reported by Leavens and colleagues (1996) in a sample of three non–language-trained chimpanzees. Leavens et al. (1996) reported that the subjects used the left hand more often for whole-hand pointing, whereas right hand use was more prevalent for indexical pointing.

More recently, Hopkins and Leavens (1998) have designed a series of studies explicitly to elicit manual gestures in a sample of 115 chimpanzees. With specific reference to laterality, Hopkins and Leavens (1998) reported a population-level right hand bias in manual gestures, particularly for food begs, contrasted with whole-hand pointing (see Leavens & Hopkins, 1999, for description of this distinction). Hopkins and Leavens (1998) also found that right hand use for manual gestures was observed significantly more often when gestures were accompanied by a vocalization than when they were not.

Interestingly, hand use for gestures did not significantly correlate with hand preferences for bimanual feeding or a coordinated bimanual task (see Hopkins, 1994, 1995b). In a follow-up study, Hopkins and Wesley (2002) examined the effect of situational factors on hand use in gestures. In this study, an experimenter elicited gestures from a sample of 113 chimpanzees when positioned either to the left, the right, or directly in front of the subject. Hand use and gesture type were recorded for each subject. Subjects were classified with regard to experimenter position as strongly left-handed (all three gestures with the left hand), mildly left-handed (use of the left hand on two of the three trials), mildly right-handed (use of the right hand on two of the three trials), or strongly right-handed (all three gestures with the right hand). There were significantly more mildly right-handed than mildly left-handed subjects. In addition, there were significantly more strongly right-handed subjects than there were strongly left-handed subjects.

Perception and production of facial expressions. With respect to behavioral studies, Hamilton (1977a, b) was the first to investigate extensively asymmetries in the discrimination of facial expressions by monkeys. The results were equivocal. In a later study, Overman and Doty (1982) similarly reported no evidence of asymmetries in processing facial stimuli. A different picture has emerged as more data on discrimination of facial expressions have been collected. Hamilton and Vermeire (1983, 1988) found a right hemisphere superiority when split-brain rhesus monkeys discriminate the facial expressions of conspecific individuals. In nineteen out of twenty-seven split-brain monkeys (70 percent), the right hemisphere was more adept at discriminating two photographs of two different individuals with the same facial expression and two photographs of the same individual with two different facial expressions. In a retest carried out six months later, the right hemisphere continued to have advantage over the left in the previously learned facial discriminations. Hamilton and Vermeire (1988) replicated these findings using new photographs of the same subjects, demonstrating that these lateralized processes implicated the use of facial attributes rather than incidental details in the photographs (see also Vermeire & Hamilton, 1998a).

One study has aimed to explore hemispheric asymmetries in the perception of facial stimuli in apes. Morris and Hopkins (1993) found a left visual field advantage in three chimpanzees discriminating human chimeric stimuli.

Using Levy's free visual discrimination paradigm (Levy et al., 1983), these authors taught the subjects to select the photograph with the happy face as opposed to the same individual posing with a neutral facial expression. Stimuli were made of a neutral half and a smiling half. Each half could be the left or the right half of a whole expression and could be placed in its original position or on the opposite side (in the latter case by using the mirror-reversed duplicate of the original). In 62 percent of the trials, the chimpanzees selected the stimuli with the smiling half on the left side and, hence, in their left visual field.

Much less is known about the relative contribution of the right and left hemisphere in the production of facial expressions in nonhuman primates. Ifune and colleagues (1984) presented video sequences of human and nonhuman primates to each visual field of split-brain rhesus monkeys, along with segments that contained other animals and scenes in an attempt to elicit emotional responses in these subjects. A higher number of submissive and aggressive facial expressions were elicited during stimulation of the right as opposed to the left hemisphere.

Hauser (1993) reported that during the production of facial expressions by neurologically intact rhesus monkeys, the left side of the face began to move first and was more expressive than the right side (as reflected in the number of skin folds and height of the corner of the mouth). This facial asymmetry was tested for four different facial expressions, including the fear grimace, copulation grimace, open mouth threat, and ear flap threat, which took place spontaneously while subjects were engaged in social interactions.

Hook-Costigan and Rogers (1998) examined facial asymmetries in marmosets but found slightly different results from those reported by Hauser (1993). The marmosets were videotaped while producing three facial expressions. Two expressions were accompanied by vocalization and were referred to as the "tsik" (characterized as fearful) and the "twitter" (defined as a social contact call). The third expression was simply referred to as the silent fear expression. For each call, the experimenters recorded areas left and right of midline of the mouth to quantify laterality in the intensity of the expression. They also recorded the distance from midline to the side of the mouth as an indicator of asymmetry. For the area measure, a left side asymmetry was found for the fear and tsik expressions, while a right side asymmetry was found for the twitter expression. For the distance to midline measure, a left side bias was found for the fear and tsik expressions, but no effect was found for the twitter expression.

This brief review of findings on handedness and laterality in communicative behavior clearly shows that monkeys and apes exhibit population-level asymmetries as pronounced as those reported at the neuroanatomical level. What is unclear from the existing data on brain asymmetries in nonhuman primates is how they relate to the evolution of handedness and other higher cognitive functions in primates. The principal problem is the lack of studies attempting to correlate behavioral and neuroanatomical measures of asymmetry. Moreover, it is unclear whether certain asymmetrical brain areas that have known functions in humans have comparable functions in apes and monkeys. For example, the great apes clearly show a left hemisphere asymmetry in the PT, a region that in the human brain is involved in language and speech comprehension (and other possible functions). Studies indicate that bonobos and chimpanzees can comprehend human speech (Savage-Rumbaugh, 1987), but whether the PT area is involved in this perceptual process in apes remains unknown. This is an intriguing question because it has direct bearing on theoretical issues concerning whether apes have "language" as well as whether speech perception or speech production is the driving force for both the evolution and ontogeny of language (e.g., MacNeilage, 1998). Of course, it is also possible that the PT in great apes does not have the same function as in the human brain. Then the question becomes: What is the function of the PT in apes, and how is this related to the evolution of its role in speech perception in humans?

A related matter involves the findings of asymmetries in the production of facial expressions. Some facial expressions are accompanied by vocalizations and some are not, but little attention has been paid to how this variable may or may not influence facial asymmetries. Moreover, the issue of semanticity contrasted with affective valence in vocalizations is fundamentally related to questions of asymmetries in facial expressions. For example, in rhesus monkeys, Gouzoules and colleagues (1984) have reported that certain calls have a semantic function in that they signal the social status of the individual monkey. Many of the vocalizations studied by Gouzoules et al. (1984) were also studied by Hauser (1993) in his research on asymmetries in facial expressions. Hauser described these calls as having an affective function, which explained the right hemisphere dominance for production. This perspective is inconsistent with the functional description of these calls by Gouzoules et al. (1984). These two areas of investigation clearly complement each other, but consistent definitions are needed if we are to comprehend the collective results more fully.

Another theoretical issue that emerges from the studies on communication and brain asymmetries is the matter of subcortical and cortical representation of vocalizations. There is little evidence from brain stimulation studies that nonhuman primate vocalizations are controlled by cortical areas. Rather, the data point to structures within the limbic systems including the cingulate gyrus, amygdala, and periaquiductal gray matter (Jürgens, 1995). Yet nearly all studies of brain asymmetries in nonhuman primates have focused on cortical areas (but see Heilbronner & Holloway, 1989). What is needed are studies that focus on subcortical areas or structures within the limbic system that have been implicated in the control of vocalizations and facial expressions.

Last, generally speaking, there has been very little research on the development of behavioral and neuroanatomical asymmetries. Warren (1980) posited that different mechanisms governed the development of laterality in humans as opposed to nonhumans. Some studies have questioned Warren's premise (see Hopkins, Dahl, & Pilcher, 2000, 2001), and behavioral-genetic and longitudinal studies are absolutely critical for discerning whether the experience of the developing individual influences the development of brain asymmetries or vice versa. Heritability studies and data from primate species that frequently have twin offspring would also provide valuable information on the interaction between genetic and non-genetic factors in the expression of behavioral and brain asymmetries.

In conclusion, studies collected in our laboratory have focused on behavioral and structural neuroanatomical asymmetries as measured from cadaver specimens and MRI scans. Clearly, MRI will allow investigators to establish structure-function relationships between brain and behavior. But these approaches will not address other aspects of asymmetries, notably electrophysiological or metabolic. These approaches are now being widely used in studies with human subjects, and these techniques need to be developed in nonhuman primates. Some techniques, such as PET and fMRI, have been successfully used with monkeys (Logothetis et al., 1999; Rilling et al., 2001; Stefanacci et al., 1998) to evaluate visual processing of various stimuli and affective valence. Despite the pragmatic constraints of functional imaging techniques, the development of these techniques in great apes would be worthwhile because it would allow investigators to bridge the gap between the typical invasive techniques used with monkeys and the imaging procedures used with humans. Functional neuroimaging would also allow for great apes to be (re)considered as valuable subjects in the growing and broadening field of neuroscience without compromising the ethical constraints placed on the use of these magnificent animals.

This work was supported in part by NIH grants RR-00165, NS-29574, NS-36605, and HD-38051. Significant proportions of the MRI scans were collected during many hours of dedicated work by Jim Rilling and Tom Insel. Special thanks are also directed to Brent Swenson and the rest of the veterinary staff for assisting in the care of the animals during scanning. We recognize the important contributions of Leslie MacGregor to this research.

16

Psychopathology

Alfonso Troisi

The study of psychopathology in nonhuman primates is an important topic for both practical and theoretical reasons. Investigators continue to use large numbers of nonhuman primates for psychobiological research. The use of the primates as "models" to obtain a better understanding of human physiology and behavior requires continuous empirical verification that the experimental subjects are not only physically but also psychologically healthy. Data deriving from individuals suffering from behavioral pathologies can lead to wrong conclusions and are likely to lack general validity. Another practical reason for the of study of primate psychopathology is that cultural attitudes toward the welfare of captive animals have changed dramatically over the last few decades, and there has been an increasing concern for the effects of captivity on nonhuman primates' psychological well-being. Since many institutions continue to house nonhuman primates in captivity for educational and research purposes, ethical considerations dictate that any kind of suffering, including that related to psychopathology, should be rapidly diagnosed and its causes removed. In addition, since nonhuman primates suffering from behavioral pathologies may convey misleading information on the behavioral repertoire of the species, institutions that pursue educational goals (e.g., zoos) cannot accomplish their task in the absence of effective strategies for diagnosing and preventing psychopathology.

Apart from these practical reasons, the study of psychopathology in non-human primates is extremely interesting for theoretical reasons. The study of mental illness in species other than *Homo sapiens* inevitably leads us to consider a variety of questions that are crucial for clinical psychiatry. Does the expression of psychopathology require cognitive and emotional capacities that are unique to human beings? Is human psychopathology an evolutionary novelty, or can similar disorders occur in other species as well? Did mental illness afflict human beings in their natural environment, or is it instead an unfortunate byproduct of civilization?

To answer these questions, human psychopathology should be a science with solid foundations in the field of evolutionary biology. Unfortunately, this is not the case. Both research and clinical psychiatrists rarely take into consideration the issues of adaptation and phylogeny (McGuire et al., 1992). In particular, those researchers who use animal models (and should therefore be aware of the importance of an evolutionary perspective) often limit their approach to the investigation of either neurobiological mechanisms that are common to human and nonhuman primates or behavioral symptoms that seem to approximate the clinical manifestations of the human syndromes (Troisi, 1994).

In this chapter I take a different approach. I believe that the study of psychopathology in nonhuman primates is a complex task which requires the integration of concepts and methods deriving from distinct disciplines, including ethology, behavioral ecology, comparative psychology, and clinical psychiatry. Such an integration is possible only when empirical data and theoretical hypotheses are framed within an evolutionary perspective. Therefore, the approach I take to analyze and discuss some of the issues related to the study of psychopathology in nonhuman primates is that of Darwinian psychiatry, defined as the study of mental illness from an evolutionary perspective (McGuire & Troisi, 1998). Inevitably, taking such an approach will orient the discussion toward certain issues to the detriment of others. I do not address the variety of experimental models that have been developed to study the pathogenic effects of psychotropic drugs on primate behavior (e.g., the amphetamine-induced model of psychosis), nor do I examine the experimental paradigms that have been developed to induce abnormal behaviors in nonhuman primates through exposure to extremely severe and highly artificial stressors (e.g., newborn social isolation). Excellent reviews of these studies have already been published (McKinney, 2000; Miczek, 1983), and I refer the interested reader to these sources of information. Rather, I focus on studies suggesting that, under specific circumstances, nonhuman primates living in natural or seminatural settings can manifest spontaneous psychopathology.

The chapter is divided into two sections. First, I discuss the problem of defining psychopathology in nonhuman primates. I argue that current definitions of mental illness as conceptualized in clinical psychiatry are not directly applicable to nonhuman subjects and that an alternative, evolutionary-based definition is required to conduct a comparative analysis of psychopathology. I then examine the phenomenology of major disorders in nonhuman primates, including depression and anxiety. I conclude with a brief discussion of possible ways to improve the methodology of studies of psychopathology in nonhuman primates.

Defining Psychopathology in Nonhuman Primates

The fourth edition of the *Diagnostic and Statistical Manual of Mental Disorders (DSM-IV)* of the American Psychiatric Association defines a mental disorder as "a clinically significant behavioral or psychological syndrome or pattern that occurs in an individual and that is associated with present distress (e.g., a painful symptom) or disability (i.e., impairment in one or more important areas of functioning) or with a significantly increased risk of suffering death, pain, disability, or an important loss of freedom" (American Psychiatric Association, 1994, p. xxi).

Clearly, such a definition of psychopathology is not directly applicable to nonhuman subjects, including primates, because of its emphasis on subjective suffering. Not knowing how a subject feels or what are its mental experiences restricts the researcher interested in studying psychopathology in nonhuman primates to inferences based on observable behaviors. Conversely, in clinical psychiatry very little attention is paid to the description and measurement of behavioral changes. The diagnostic process is based almost exclusively on the evaluation of psychic symptoms as voiced by the patient, and among these symptoms, mental distress in its various forms (e.g., depression, anxiety, irritability) plays a central role. Because of this major methodological difference, the definition of mental illness currently used in clinical psychiatry is difficult to apply to nonhuman primates.

An alternative approach to conceptualizing mental illness in nonhuman primates is to use a definition that focuses exclusively on behavioral aspects. Erwin and Deni (1979) suggested that psychopathology in nonhuman primates consists of behaviors that deviate in terms of quality or quantity from species-typical behaviors in the natural setting.

The methodological merit of this definition is that assessment of mental health in nonhuman primates is based on quantitative comparisons with behavioral data collected under ecologically valid conditions. Field studies of a variety of nonhuman primate species have provided a large database on the distribution in wild populations of those behaviors which form the normal (in the statistical sense) repertoire of the species. These data make it possible to implement a "diagnostic evaluation" of the behavioral profile of a single individual by comparing the form and frequency of its behaviors with those typical of individuals of the same age and sex living in the natural environment.

There is, however, a conceptual problem with a definition based on the normative criterion: it equates the statistical norm with the biological norm and, therefore, fails to distinguish deviations from the norm that are harmful from those that are neutral and those that are beneficial in terms of adaptive significance. Such a distinction may be critical for judging whether or not an abnormal (in the statistical sense) behavior performed by a primate is pathological. Let us consider two examples taken from the primatological literature that demonstrate how the normative criterion of psychopathology can lead to wrong conclusions when applied to statistically deviant behaviors displayed by nonhuman primates either in the natural environment or in captivity.

In the mid-1960s, reports of infant killing in the common langur first appeared in the primatological literature. Field researchers observed that unweaned infants were attacked and killed by individuals from all-male bands that had invaded the one-male harem groups and evicted the resident male. Following the male takeover, the females became sexually receptive and mated with the new resident male (Mohnot, 1971; Sugiyama, 1965).

Infanticide was initially regarded as an aberrant behavior and explained as a pathological response to recent habitat disturbance and human interference (e.g., Dolhinow, 1977). The normative criterion played a relevant role in suggesting such an explanation: the phenomenon was rarely seen, and langurs were not observed to commit infanticide over wide parts of their geographical range. After decades of controversy, most primatologists now agree that infanticide among langurs is best interpreted as a male reproductive strategy that curtails the female's investment in the offspring of another male and thereby increases the infanticidal male's chances of siring offspring of his own (Hiraiwa-Hasegawa & Hasegawa, 1994; van Schaik & Janson, 2000). This shift in thinking about infanticide was in large part due to the capacity of the sexual selection model to predict the specific circumstances of infanticidal events (Hrdy, 1984). This shift would not have been possible, however, if primatologists had continued to apply the normative criterion. In statistical terms, infanticide is not a species-typical behavior: its global frequency is extremely low, and a relevant percentage of immigrant males are not infanticidal (Sommer, 1994). Yet the demonstration that the behavior may have adaptive consequences for the perpetrators makes these observations irrelevant for judging whether or not infanticidal males are affected by psychopathology.

Compared with infanticide, the case of "chimpanzee art" is much less controversial but just as instructive about the weakness of the normative criterion. When offered the opportunity, many captive chimpanzees appear to enjoy drawing and painting, as discussed fully in Morris (1962). For no material reward, an individual may work for minutes on end with intense concentration. No one would dare to suggest that drawing and painting are manifestations of psychopathology, even though these activities have never been observed in natural settings and are not species-typical behaviors. Certainly this is not due simply to the fact that chimpanzees seem to experience pleasurable feelings while engaging in these "unnatural" behaviors. In addition, chimpanzees that, for experimental purposes, are allowed to consume drugs of abuse experience pleasure when intoxicated. As is the case for their human counterparts, however, these individuals are rightly considered to be suffering from a behavioral disorder. The reason why drawing and painting are not considered abnormal behaviors is that they do not compromise biological adaptation. These unusual and statistically deviant activities reflect the high degree of plasticity and versatility of primate behavior and are best conceived of as a product of chimpanzees' cognitive potential that become actual under the enriched conditions of captivity.

The preceding discussion underscores one important point: a valid definition of psychopathology in nonhuman primates cannot be based solely on the criteria of subjective suffering and/or statistical deviance. According to Darwinian psychiatry, the single attribute that best characterizes psychopathology is its maladaptive consequences (Troisi & McGuire, 1998). If a behavioral profile is associated with an impairment of one or more of those functional capacities that are essential for biological adaptation, then such a condition can be defined as psychopathology.

Surely, mental suffering and statistical deviance are frequent correlates of psychopathology in both human and nonhuman subjects. From a Darwinian perspective, mental pain has evolved as an emotional indicator that biological goals have not been or are not being achieved, that is, that one's fitness has been or is being compromised. Therefore, it is plausible to assume that maladaptive behavioral disorders are associated with negative emotions in nonhuman primates as well. As for statistical deviance, natural selection tends to eliminate maladaptive behavior patterns, and thus their frequency of occurrence in natural populations should be quite low. Nevertheless, we should not overlook the possibility that atypical behaviors reflect behavioral plasticity or alternative strategies. Assessment of functional capacities and adaptive consequences of behavior remains a crucial step in the process of diagnosing psychopathology in nonhuman primates.

Spontaneous Psychopathology in Nonhuman Primates

According to an epidemiological study conducted in the United States, the cumulative lifetime prevalence of all psychiatric disorders is 48 percent (Kessler et al., 1994). In an article that addressed the question whether primate models are useful for studying psychiatric disorders, McGuire and colleagues (1983) wrote: "[We] have probably spent over 20,000 hours observing vervet monkeys in quasi-natural environments and we have never seen behavior suggestive of schizophrenia, mania, agitated depression, or involutional depression." (p. 324). Taken together, these findings lead to an indisputable conclusion: spontaneous psychopathology is much more frequent in human beings than in nonhuman primates.

Such a discrepancy in prevalence rates can be explained in part by hypothesizing that the etiology of human psychiatric disorders involves genetic and environmental contributions for which there are no well-established counterparts among feral nonhuman primates. As discussed in the previous section, however, the role of diagnostic criteria should not be overlooked. An experienced clinical psychiatrist who spent twenty thousand hours observing the nonverbal behavior of human subjects interacting in their natural environments (e.g., at home, at work) would be able to identify only a small minority of those individuals who are affected by psychiatric disorders as defined by current diagnostic criteria. If the diagnostic criteria for different types of psychiatric disorders could be reformulated in ethological terms, the same or similar criteria could be applied in naturalistic primate studies and the detection of psychopathology in nonhuman primates might then become more feasible (Troisi, 1999).

These considerations explain why available data on spontaneous psychopathology in nonhuman primates are often limited to case studies and anecdotal reports. In the rest of this section I describe the phenomenology of different types of spontaneous psychopathology that have been observed in nonhuman primates living in natural or quasi-natural environments.

Depression

As described in human subjects, the core symptoms of clinical depression are sad mood and the loss of interest or pleasure in nearly all activities. Associated behavioral changes include reduced appetite, insomnia, psychomotor retardation, and characteristic postures and facial expressions (e.g., sad frown).

Darwinian psychiatry suggests that depression can be an adaptive strategy for coping with adverse external conditions (McGuire et al., 1997). This hypothesis builds on the idea that depression is a psychological and behavioral response to an actual or potential reduction in goal achievement that has evolved to promote escape from and avoidance of situations that decrease fitness. The negative emotional feelings associated with depression provide information about one's negative cost-benefit state, and physiological slowing further constrains costly behavior. Co-evolution has favored the capacity to recognize and respond to the affective manifestations of depressed individuals; thus depression can elicit social support from and care by others. In mild forms, depression often correlates with the implementation of alternative strategies. If the strategies are effective, a negative cost-benefit situation may be offset, and the negative mood may dissipate. If, however, the negative situation cannot be overcome or the individual is not able to disengage from an unreachable goal, serious pathology may arise (Nesse, 2000).

Primate studies have documented the occurrence of depression in nonhuman primates living in natural environments. In line with the predictions of Darwinian psychiatry, these cases of naturally occurring depression range in severity from transient adaptive responses to serious pathology and are triggered by stressful events associated with reduced fitness payoffs.

Traumatic disruption of attachment bonds invariably leads to the development of grief reactions and can precipitate severe depression in some vulnerable individuals. Goodall (1986) described in detail the severe effects of death of their mother on wild chimpanzees of the Gombe Stream Reserve. Being nutritionally independent, juvenile chimpanzees (aged five to seven years) generally survive the death of their mother. Yet, in response to the traumatic event, they show evident symptoms of emotional distress that, in some cases, can leave lasting scars. All orphaned chimpanzees initially become listless and show the typical signs of depression, including huddled posture, sad facial expression, withdrawal from social activities, and a marked reduction of social play. In most cases these orphaned youngsters are adopted by elder siblings or adult females, and their depression gradually decreases in the following four to eight months. Recovery from loss may be more problematic for some individuals, however. Goodall observed a variety of long-lasting disorders following the death of the mother including development of abnormal behaviors such as rocking and plucking out one's own hair, growth retardation, and increased vulnerability to infectious diseases. Two of the six young chimpanzees at Gombe who lost their mothers when they were between four and six years of age died within eighteen months. In this regard, the case of Flint was exceptional because of his older age and premorbid personality traits. He was nine years old when his very old mother, Flo, died. Flint was unusually dependent on his mother. After Flo's death he was unable to cope; he showed gradually increasing signs of lethargy, loss of appetite, and a dramatic reduction of social play. His physical condition deteriorated markedly throughout the following three weeks, and eventually he died of gastroenteritis and peritonitis.

In both nonhuman primates and human beings, not only traumatic disruption of attachment bonds but also social defeat can trigger a depressive reaction. Clinical studies have demonstrated that social defeat is a powerful stressor and may result in a variety of psychiatric and psychosomatic symptoms (Björkqvist, 2001; Rohde, 2001). There are interesting analogies between these clinical data and those that have emerged from studies of competitive loss and social defeat in nonhuman primates.

Studying a population of wild baboons in East Africa, Sapolsky and colleagues (1997) found that social subordinance was associated with hyperactivity of the hypothalamic-pituitary-adrenal (HPA) axis. In humans, elevated glucocorticoid activity is present in 20 to 40 percent of depressed outpatients and 40 to 60 percent of depressed inpatients. In particular, hypercortisolism is one of the most common biologic correlates of recurrent and severe depression (Thase, 2000). Among vervet monkeys, high-status males have peripheral serotonin (5-HT) levels averaging between 1.5 and 2.0 times the levels of low-status males. In addition, dominant males show proportionally greater behavioral responses to substances such as tryptophan and fluoxetine that influence central nervous system (CNS) serotonin concentrations (Raleigh et al., 1985). When a dominant male becomes subordinate, his peripheral 5-HT levels and CNS 5-HT responsivity change to those characteristic of his new low-ranking status. Human findings are consistent with those in nonhuman primates. Serotonergic dysfunction and low cerebrospinal fluid (CSF) concentrations of 5-hydroxyindoleacetic acid (5-HIAA, the major metabolite of 5-HT) have been found in at least a significant subgroup of depressed patients and individuals at genetic risk for this disorder (Maes & Meltzer, 1995).

It is likely that a traumatic change in social status, rather than subordination per se, is associated with an increased risk of depression. Shively et al. (1997) manipulated the social status of adult female long-tailed macaques by reorganizing social groups. These authors observed that, following social status manipulation, a subset of subordinate females developed a behavioral depression response characterized by an extreme form of social withdrawal. Interestingly, these subordinate females showed reduced brain dopaminergic activity in comparison to socially dominant females (Shively, 1998), a biological finding consistent with the observation that depression in human beings may be accompanied by reduced CSF concentrations of the dopamine metabolite homovanillic acid (Mann & Kapur, 1995). Kummer (1995) described in detail the morphological and behavioral changes associated with loss of dominant status in male baboons (i.e., weight and hair loss, psychomotor retardation, and suppression of sexual activity). In a study of a semifree-ranging mandrill group, Setchell and Dixson (2001a) showed that loss of alpha status in males is consistently linked with morphological, endocrine, and behavioral changes. Males that fell in rank declined in group association and began spending some time either peripheral to the group or solitary. These behavioral changes were associated with a decrease in testicular volume, a significant loss of body mass, and a reduction in the extension of the red sexual skin and in the activity of the sternal gland. Competitive loss may also result in impaired immunological function as indicated by the finding that socially defeated male macaques show an increased susceptibility to respiratory infection (Cohen et al., 1997).

Anxiety

In human beings, anxiety is an aversive emotional state characterized by apprehensive worry, increased vigilance, tension, and somatic changes (e.g., dyspnea, palpitations, and dizziness). Darwinian psychiatry views anxiety as an evolved emotional response warning that high-priority biological goals may be jeopardized (McGuire & Troisi, 1998). Time-limited, threat-initiated anxiety is adaptive because it contributes to avoidance of dangerous situations and the overriding of other potentially competing goal priorities. Pathological anxiety is an exaggeration of normal anxiety and is likely to arise from dysregulation of the same mechanisms that modulate adaptive responses to impending dangers. The *DSM-IV* distinguishes among several anxiety disorders. From an evolutionary perspective, anxiety subtypes (e.g., agoraphobia, social anxiety, separation anxiety, etc.) probably exist because of the benefits of having specialized responses to deal with different types of threats (Marks & Nesse, 1994).

The existence in nonhuman primates of behavioral manifestations of anxiety is well documented (Castles et al., 1999; Maestripieri, 1993a) and validated by pharmacological studies (Schino et al., 1996). In most cases, nonhuman primates show behavioral symptoms of anxiety under circumstances that involve potential threats to their physical safety or social relationships. This form of anxiety is adaptive and does not reflect the presence of psychopathology. Yet there is evidence that, like human beings, nonhuman primates can suffer from genuine anxiety disorders. In her monograph on wild baboons, Strum (1987) describes the case of David, an extremely shy male baboon who was abnormally sensitive to social rebuff and whose dependence on his natal group hindered his emigration. Researchers studying rhesus macaques in both laboratory and field settings have identified a subgroup of "high-reactive" monkeys who tend to be shy and who consistently respond to mildly stressful stimuli with behavioral expressions of fear and anxiety, significant and prolonged cortisol elevations, unusually high heart rates, and dramatic increases in CNS metabolism of norepinephrine (see Suomi, 1999, for a review).

The most convincing data on the spontaneous occurrence of anxiety disorders in nonhuman primates are probably those concerning macaque mothers who physically abuse their infants because of pathological levels of separation anxiety. Abnormal levels of maternal anxiety associated with physical abuse of infants were first observed in free-ranging Japanese macaques. Hiraiwa (1981) reported that orphaned primiparous mothers often maltreated their infants by alternating physical abuse with extremely protective behavior. These observations were in line with the large body of experimental data showing that, in nonhuman primate females, separation from the mother during infancy can have deleterious, long-lasting effects on the development of maternal behavior (Suomi & Ripp, 1983). Subsequent studies of captive Japanese monkeys suggested that the pathogenetic mechanism underlying abusive behavior was dysfunctional separation anxiety (Troisi & D'Amato, 1994). The monkey mothers studied by Troisi and co-workers alternated violent physical abuse with attentive maternal care, but, unlike normal mothers, they also had extremely possessive relationships with their infants, scoring highest on maternal protectiveness and lowest on maternal rejection (Troisi et al., 1989). These abusive mothers seemed to be highly sensitive to stressful situations and unable to cope with stress with appropriate parental behavior typical of normal monkey mothers. They seemed to be in constant anxiety lest they lose their infants and reacted with physical abuse to minor stressors such as attempts by their infants to leave them, infants' failure to respond to their retrieval signals, and infants' screaming and convulsive jerks (Troisi & D'Amato, 1984). Interestingly, physical abuse did not include aggressive behaviors such as biting or striking but consisted of behavior patterns that macaques normally display while manipulating inanimate objects. Abusive mothers were observed to sit and step on their infants, to crush them against the ground, or to drag them by one limb or the tail across the ground (see Figures 16.1 and 16.2). Treatment of one these mothers with diazepam, an anxyolitic drug,

Figure 16.1 Okame, an abusive mother of the Rome colony of Japanese macaques, drags her infant by the tail across the ground. (Photo: A. Troisi)

Figure 16.2 Chiocciola, an abusive mother of the Rome colony of Japanese macaques, steps on her infant. (Photo: A. Troisi)

during the first eight weeks postpartum caused a complete suppression of physical abuse and a dramatic reduction of parental behaviors reflective of maternal anxiety (Troisi & D'Amato, 1991).

More recently, the parenting style of abusive mothers has been systematically investigated in rhesus macaques living in stable social groups (Maestripieri, 1998b). This study has confirmed the tendency of abusive mothers to exhibit a protective and controlling style of mothering. These rhesus mothers spent a higher percentage of time in contact with their infants, initiated more contacts with them, and restrained them more often than control mothers. Unlike the Japanese monkey mothers studied by Troisi and co-workers, however, the abusive rhesus mothers studied by Maestripieri also rejected their infants much more than controls.

The existence of dysfunctional attachment patterns and abnormal levels of separation anxiety is well documented in human mothers who physically abuse their children. Attachment theorists explain the apparent paradox of the association between physical abuse and possessive mothering by emphasizing the interrelationships between separation anxiety, helplessness, and anger, observing that "abusing mothers, who exhibit a high degree of sensitivity to separation from significant others, interpret the normal behavior of their children as if it were actual or threatened rejection. If so, these mothers would be likely to respond to such a misperception with dysfunctional levels of anxiety and anger" (DeLozier, 1982, p. 114). It is possible that the same pathogenetic mechanisms are at work in some cases of primate infant abuse. Even though such a hypothesis is difficult to accept because it implies the existence of rather sophisticated cognitive capabilities in monkey mothers, some analogies between abusive monkey and human mothers are impressive.

In a study of the relation between attachment patterns and parental behavior, McKinsey Crittenden and colleagues (2000) compared normal and abusing mothers. The mothers in this study were videotaped for three minutes playing with their children as naturally as possible. The videotapes of the play interaction were coded using an instrument that yields scores for maternal sensitivity, control, and unresponsiveness. Mothers with high scores on sensitivity are those who attend to the child's signals and respond in ways that increase dyadic synchrony and shared affect. Mothers who score high on control attend to their children's behavior but use this information to increase their interference with the children's activity. Mothers who score high on unresponsiveness are inattentive to their children and uninvolved in their play. The authors found that, whereas the normal mothers were the most sensitive, the abusing mothers scored lowest on unresponsiveness and had relatively high scores on both the sensitivity and control scales. The other major finding was that the attachment patterns of the mothers were significantly associated with maltreatment status. The majority of abusing mothers were classified as having a dysfunctional attachment pattern, suggesting substantial risk that these mothers would respond to their child's distress with inconsistent and hostile behavior.

Other Disorders

Current psychiatric nosography includes a variety of disorders characterized by impaired impulse control. Patients diagnosed with these disorders exhibit anger attacks, impulsive aggression, risk-taking behavior, and poor social affiliation. Many studies have demonstrated that low CNS serotonergic activity is a consistent neurobiological correlate of impulse control deficit, irrespective of the specific psychiatric diagnosis. Low CSF 5-HIAA concentrations have been found among depressed patients who had used violent means to attempt suicide, violent criminal offenders, women with bulimia, and individuals who engage in impulsive fire setting (Coccaro & Kavoussi, 1996).

Paralleling these clinical findings, studies of free-ranging rhesus monkeys have found that interindividual differences in impulse control are correlated with brain serotonergic function (see Chapter 2 on aggression). Subjects with low CSF 5-HIAA concentrations are more likely to exhibit behaviors characteristic of impaired impulse control, such as spontaneous long leaps at dangerous heights (Mehlman et al., 1994). Direct observation of aggressive behavior showed high rates of impulsive and unrestrained aggression among these subjects, who would eventually be killed during violent encounters. Of forty-nine two-year-old males followed longitudinally until they became adults, 91 percent of the subjects who died from violence came from the two lowest quartiles of CSF 5-HIAA concentrations (Higley, Mehlman, Higley, et al., 1996). Male rhesus monkeys with low CSF 5-HIAA concentrations also show deficits in social functioning as evidenced by a variety of different behavioral measures. They have few social partners and spend a lot of time alone, and they are unlikely to obtain high social dominance ranking. During the breeding season, they are also less likely to be sought as consorts by breeding females. Finally, males with lower CSF 5-HIAA levels emigrate from their natal groups at a younger age than those with higher CSF 5-HIAA concentrations (Mehlman et al., 1995). Earlier emigration is likely to be the combined result of both social ostracism by members of the natal group and a greater propensity toward risk-taking behavior. In fact, the process of emigration increases mortality and morbidity for the age class of young, sexually mature males. Overall, the overlap between the behavioral profile of rhesus monkeys with reduced CNS 5-HT function and the clinical profile of psychiatric patients with disorders of impulse control is remarkable.

Another class of psychiatric disorders that may occur spontaneously in nonhuman primates is the somatizing disorders. The essential feature of these disorders is the conscious or unconscious production of physical symptoms that lack an organic basis. Patients diagnosed with these disorders present to their doctors with somatic complaints or objective signs of disease for which there are no demonstrable organic findings (Ford, 1984). Current psychiatric nosology classifies somatizing disorders under distinct categories that include somatoform disorders (hypochondriasis and hysterical disorders), factitious disorders, and malingering.

Caine and Reite (1983) reported that one macaque female in their colony "showed signs of what could possibly be called 'hysterical paralysis' or malingering. Whenever she was placed in her social group she limped badly, although, upon examination, no evidence of injury or disease was found. Furthermore, the limping disappeared when the animal was housed alone" (p. 25). De Waal (1982) described a case of disease simulation by an adult male chimpanzee: "Yeroen hurts his hand during a fight with Nikkie . . . Yeroen walks past the sitting Nikkie from a point in front of him to a point behind him and the whole time Yeroen is in Nikkie's field of vision he hobbles pitifully, but once he has passed Nikkie his behavior changes and he walks normally again. For nearly a week, Yeroen's movement is affected in this way whenever he knows Nikkie can see him" (p. 47).

Troisi and McGuire (1991) have advanced the hypothesis that patients with somatizing disorders use disease simulation to manipulate their social environment because they lack more adaptive strategies for social interaction. The cases of somatization in nonhuman primates reported above are compatible with such a hypothesis and suggest that, under specific circumstances, nonhuman primates can simulate a disease to minimize the competitive disadvantages that accrue from limited capacities to interact socially.

In humans, manipulation of others through deception is one of the most relevant features of another psychiatric disorder, namely psychopathy or antisocial personality disorder. As defined in the *DSM-IV*, (American Psychiatric Association, 1994), antisocial personality disorder is a pervasive pattern of disregard for the rights of others associated with distinctive emotional and behavioral features. Individuals with this personality disorder are frequently deceitful and manipulative in order to gain personal profit. Even though they may display a glib, superficial charm, these subjects lack empathy and tend to be callous, cynical, and contemptuous of the feelings, rights, and suffering of others. Finally, they display a reckless disregard for their personal safety and are free of symptoms of anxiety.

The ethological literature on chimpanzee behavior provides some support for the contention that certain features of human psychopathy may be applicable to great apes. De Waal (1982) referred to one female chimpanzee in his Arnhem colony, named Puist, as "deceitful or mendacious . . . She may invite her opponent to reconciliation in the customary way. She holds out her hand and when the other hesitantly puts her hands in Puist's, she suddenly grabs hold of her" (p. 55). Goodall (1986) described in detail the unusual personality features of Passion and of her daughter Pom, two chimpanzee females of the Gombe population. Acting together, Passion and Pom brutally killed and cannibalized at least eight infants over a period of four years. Like human subjects with antisocial personality disorder, these females appeared to be emotionally cold and fearless. They did not form attachment bonds with other group members and did not display any signs of anxiety or fear in dangerous situations. Using a standardized observational measure known as the Emotional Profile Index, Buirski and Plutchik (1991) found that Passion exhibited markedly higher levels of aggressiveness and distrust and markedly lower levels of timidity compared to the normative values derived from ten female chimpanzees. Lilienfeld et al. (1999) have developed a measure of psychopathy for use in chimpanzees, the Chimpanzee Psychopathy Measure (CPM). In a sample of thirty-four captive chimpanzees, the scores on the CPM (based on the assessment of six raters familiar with the chimpanzees) correlated positively with ethogram measures of aggression, daring behaviors, and gentle teasing, and correlated negatively with ethogram measures of generosity and anxiety. Even though these findings do not imply the existence of the full syndrome of psychopathy in the chimpanzees with the highest scores, the observed individual differences are likely to reflect a continuous distribution along the dimension of antisocial personality.

Deviant Behavior: Psychopathology or Adaptation?

As mentioned above, a Darwinian approach to the study of psychopathology emphasizes the importance of distinguishing between deviant or atypical behaviors that are harmful (true psychopathologies) and those that are neutral or beneficial in terms of adaptive significance (pseudo-psychopathologies). Such a distinction is not always simple to make and often requires a continuous reevaluation based on the progressive accumulation of new empirical data. Some of the behavioral syndromes described in this chapter illustrate this issue well.

The analogies between the clinical symptoms caused by social defeat in humans and the behavioral and physiological consequences of competitive loss in nonhuman primates are so impressive that a diagnosis of "depression with psychosomatic symptoms" seems warranted for the cases observed among male baboons and mandrills (see the studies by Kummer, 1995, and Setchell and Dixson, 2001a, quoted above). Data on the evolution of alternative mating strategies in male primates suggest, however, that the reproductive consequences of social defeat could be adaptive responses rather than pathological symptoms.

In mandrills, there are two morphological and behavioral variants of adult males that differ in terms of secondary sexual adornments and reproductive strategies (Setchell & Dixson, 2001b). "Fatted" males have highly developed sex skin coloration, large testes, high plasma testosterone levels, and fat rumps, whereas "nonfatted" males have paler sex skin, smaller testes, lower plasma testosterone, and slimmer rumps. While "fatted" males mate-guard fertile females, less developed males remain in the periphery of the group and mate sneakily with females. Similar inter-male differences have been observed in orangutans (Maggioncalda et al., 2000). In the presence of many dominant males, adolescent male orangutans undergo a developmental arrest: they become fertile but do not develop fully adult secondary sexual features, such as cheek flanges, laryngeal sac, beard and mustaches, large body size, and a musky odor. Developmental arrest is associated with a distinct hormonal profile (Maggioncalda et al., 1999). Arrested males lack levels of luteinizing hormone (LH), testosterone, and dihydrotestosterone (DHT) necessary for development of secondary sexual traits. They do, however, have sufficient testicular steroids, LH, and follicle-stimulating hormone (FSH) to fully develop primary sexual function and fertility. As in mandrills, the two morphological variants of male orangutans use different mating strategies (Galdikas, 1985). Developed males are frequently involved in male-male aggression, are attractive to females and typically consort with them, and may sire many offspring over a relatively short period of time. In contrast, being inconspicuous and less attractive to females, arrested males adopt a low-cost, low-benefit reproductive strategy based on sneaky matings and forced copulations.

In the short run, the reproductive success of "nonfatted" male mandrills and arrested male orangutans is lower than that of fully developed males. There are advantages associated with the use of the "sneak and rape" mating strategy, however. While the "combat and consort" strategy imposes costs on dominant males in terms of metabolic energy and exposure to inter-male aggression, the suppression of secondary sexual traits allows subordinate males to minimize aggression and injury from dominant, fully mature males, while still being able to sire. Both the strategies are maintained by natural selection because the disadvantages associated with each strategy are balanced by advantages in a different context. In both mandrills and orangutans, subordinate males can rapidly switch over to the "combat and consort" strategy if the density of dominant males decreases. Arrested male orangutans develop into flanged males if a more favorable reproductive situation occurs, and subordinate male mandrills develop secondary sexual traits when they become dominant. Such a flexibility is a further indication that the "sneak and rape" strategy is an adaptive alternative strategy based on a continuous assessment of reproductive opportunity and risk of intermale aggression.

In light of these data, one could speculate that the depressive and psychosomatic symptoms following competitive loss in nonhuman primates are also an adaptive response that has evolved to facilitate the adjustment to the new subordinate role and to minimize the risk of further aggression. Similar adaptationist explanations can be advanced for other deviant behaviors discussed in this chapter. Transient depression in orphaned chimpanzees could be a means to elicit help and support from other group members. Disease simulation could be a form of social deception that would exploit the mechanisms for the care of the sick that have evolved in many animal species, including non-human primates. In rhesus macaques, an anxious temperament could be a behavioral polymorphism associated with optimal developmental outcomes when combined with specific early experiences. In line with this hypothesis, Suomi (1991) found that "high-reactive" rhesus macaques reared by nurturant mothers became especially adept at recruiting and retaining other group members as allies during agonistic encounters, and that these individuals often rose to dominant positions in the social hierarchy.

It would be erroneous to conclude from the preceding discussion that the existence of true psychiatric disorders is limited to humans because all deviant behavioral profiles observed in monkeys and apes are in reality alternative strategies maintained by natural selection. On the one hand, the dysfunctional nature of some primate psychopathologies cannot be called into question (for example, maternal abuse of infants in macaques and fatal depression following death of the mother in adolescent chimpanzees). In addition, most of the adaptationist explanations of deviant behaviors in nonhuman primates are still based on suggestive speculations rather than on proven evidence. On the other hand, the adaptationist approach is gaining in popularity among clinical psychiatrists, and the hypothesis that several conditions currently classified as psychiatric disorders may in fact be adaptive responses is now viewed as legitimate (McGuire & Troisi, 1998). For example, the hypothesis that depression following competitive loss may be adaptive has first been advanced by clinical psychiatrists, not primatologists. The social competition hypothesis of depression (Price et al., 1994) explains depressive signs and symptoms linked to hierachical defeat as an unconscious, involuntary losing strategy that enables the individual to accept defeat and to accommodate to what would otherwise be unacceptably low social rank.

The traditional approach to the study of psychopathology in nonhuman primates has taken for granted the validity of the definition and classification of mental disorders as they are formulated in clinical psychiatry. Since such an approach largely neglects behavioral descriptions of disorders and pays scarce attention to their functional consequences, knowledge of the occurrence and phenomenology of spontaneous psychopathology in nonhuman primates has progressed slowly. Yet the data reviewed in this chapter suggest that monkeys and apes can suffer from symptoms similar to those that afflict human subjects with psychiatric disorders. To emerge, these symptoms do not require the exposure to the extremely severe and highly artificial stressors that have been commonly employed in experimental studies.

The integration of two key concepts of Darwinian psychiatry into studies of psychopathology in nonhuman primates can increase the validity and usefulness of this area of research: individual differences and functional assessment. On the one hand, the search for spontaneous psychopathology in nonhuman primates living in natural or quasi-natural environments should not be limited to the individuation of gross behavioral abnormalities. Subtle individual differences at the physiological and behavioral level may reflect an increased vulnerability to psychiatric disorders that becomes evident only under stressful circumstances. The data of Suomi and co-workers on the "high-reactive" subgroup of rhesus macaques and those of Troisi and co-workers on the macaque mothers who maltreated their infants but exhibited normal social behavior are informative in this regard (see the section on anxiety disorders). On the other hand, we should not overlook the possibility that what are currently interpreted as deviant behavior patterns in fact reflect the normal operation of adaptive mechanisms. Formerly, if an animal was seen behaving in a different way from the majority of the population, it was thought to be abnormal. At present, whenever behavioral ecologists see an animal engaging in atypical behavior, they first explore the possibility that it is employing an alternative strategy to compete successfully with its rivals. Considering the behavioral plasticity and flexibility of nonhuman primates, such a theoretical shift is particularly relevant for primatology. Here the importance of functional assessment clearly emerges. The evaluation of the impact on individual adaptation is crucial for deciding if a behavioral profile that is atypical in statistical terms and that resembles a human psychiatric disorder can be correctly classified as a real case of spontaneous psychopathology in a nonhuman primate.

References

Contributors

Index

References

Abbey, A. (1982). Sex differences in attributions for friendly behavior: Do males misperceive females' friendliness? *Journal of Personality and Social Psychology, 42,* 830–838.

Abbott, D. H. (1984). Behavioral and physiological suppression of fertility in subordinate marmoset monkeys. *American Journal of Primatology, 6,* 169–186.

Aboitiz, F., & Ide, A. (1998). Anatomical asymmetries in language-related cortex and their relation to callosal function. In B. Stemmer & H. A. Whitaker (Eds.), *Handbook of neurolinguistics* (pp. 393–404). San Diego: Academic Press.

Adams, R., & Laursen, B. (2001). The organization and dynamics of adolescent conflict with parents and friends. *Journal of Marriage and Family, 63,* 97–110.

Adams, R. M., & Kirkevold, B. (1978). Looking, smiling, laughing, and moving in restaurants: Sex and age differences. *Environmental Psychology and Nonverbal Behavior, 3,* 117–121.

Adams-Curtis, L., & Fragaszy, D. M. (1995). Influence of a skilled model on the behavior of conspecific observers in tufted capuchin monkeys *(Cebus apella). American Journal of Primatology, 37,* 65–71.

Adolphs, R., Damasio, H., Tranel, D., & Damasio, A. R. (1996). Cortical systems for the recognition of emotion in facial expressions. *Journal of Neuroscience, 16,* 7678–87.

Aggleton, J. P., & Young, A. W. (2000). The enigma of the amygdala: On its contribution to human emotion. In R. D. Lane & L. Nadel (Eds.), *Cognitive neuroscience of emotion* (pp. 106–128). New York: Oxford University Press.

Ainsworth, M. D. (1979). Attachment as related to mother-infant interaction. *Advances in the Study of Behavior, 9,* 1–51.

Ainsworth, M. D., Blehar, M. C., Waters, E., & Wall, S. (1978). *Patterns of attachment: A psychological study of the strange situation.* Hillsdale, NJ: Erlbaum.

Alberts, S. C. (1999). Paternal kin discrimination in wild baboons. *Proceedings of the Royal Society of London, B, 266,* 1501–6.

Alberts, S. C., & Altmann, J. (1995). Preparation and activation: Determinants of age at reproductive maturity in male baboons. *Behavioral Ecology and Sociobiology, 36,* 397–406.

Alexander, R. D. (1979). *Darwinism and human affairs.* Seattle: University of Washington Press.

Alexander, R. D. (1987). *The biology of moral systems.* New York: Aldine.

Allison, T., Puce, A., Spencer, D. D., & McCarthy, G. (1999). Electrophysiological studies of human face perception. I: Potentials generated in occipitotemporal cortex by face and non-face stimuli. *Cerebral Cortex, 9,* 415–430.

Allport, G. W., & Odbert, H. S. (1936). Trait-names: A psycho-lexical study. *Psychological Monographs, 47,* no. 211.

Almagor, M., Tellegen, A., & Waller, N. G. (1995). The Big Seven model: A cross-cultural replication and further exploration of the basic dimensions of natural language trait descriptors. *Journal of Personality and Social Psychology, 69,* 300–307.

Altmann, J. (1980). *Baboon mothers and infants.* Cambridge, MA: Harvard University Press.

Altmann, J., Hausfater, G., & Altmann, S. A. (1988). Determinants of reproductive success in savannah baboons, *Papio cynocephalus.* In T. H. Clutton-Brock (Ed.), *Reproductive success: Studies of individual variation in contrasting breeding systems* (pp. 403–418). Chicago: University of Chicago Press.

Altmann, J., & Samuels, A. (1992). Costs of maternal care: Infant-carrying in baboons. *Behavioral Ecology and Sociobiology, 29,* 391–398.

Altmann, S. A. (1962). A field study of the sociobiology of rhesus monkeys, *Macaca mulatta. Annals of the New York Academy of Sciences, 102,* 338–345.

Ambert, A. M. (1992). *The effect of children on parents.* New York: Haworth Press.

American Psychiatric Association. (1994). *Diagnostic and statistical manual of mental disorders* (2nd ed.) *(DSM-IV).* Washington, DC: American Psychiatric Association.

Amunts, K., Schleicher, A., Bürgel, U., Mohlberg, H., Uylings, H. B., & Zilles, K.

(1999). Broca's region revisited: Cytoarchitecture and intersubject variability. *Journal of Comparative Neurology, 412,* 319–341.

Anderson, B., Erwin, N., Flynn, D., Lewis, L., & Erwin, J. (1977). Factors influencing aggressive behavior and risk of trauma in the pigtail macaque *(Macaca nemestrina). Aggressive Behavior, 3,* 33–46.

Anderson, C. M., & Bielert, C. F. (1994). Adolescent exaggeration in female catarrhine primates. *Primates, 35,* 283–300.

Anderson, J. R., & Mitchell, R. W. (1999). Macaques but not lemurs co-orient visually with humans. *Folia Primatologica, 70,* 17–22.

Anderson, J. R., Montant, M., & Schmitt, D. (1996). Rhesus monkeys fail to use gaze direction as an experimenter-given cue in an object-choice task. *Behavioural Processes, 37,* 47–55.

Anderson, J. R., Sallaberry, P., & Barbier, H. (1995). Use of experimenter-given cues during object-choice tasks by capuchin monkeys. *Animal Behaviour, 49,* 201–208.

Anderson, K., Kaplan, H., & Lancaster, J. (1999). Paternal care by genetic fathers and stepfathers. I: Reports from Albuquerque men. *Evolution and Human Behavior, 20,* 405–431.

Andrew, R. J. (1963a). Evolution of facial expressions. *Science, 142,* 1034–1041.

Andrew, R. J. (1963b). The origin and evolution of the calls and facial expressions of the primates. *Behaviour, 20,* 1–109.

Andrews, M. W., & Rosenblum, L. A. (1991). Attachment in monkey infants raised in variable- and low-demand environments. *Child Development, 62,* 686–693.

Andrews, M. W., & Rosenblum, L. A. (1993). Assessment of attachment in differentially reared infant monkeys *(Macaca radiata):* Response to separation and a novel environment. *Journal of Comparative Psychology, 107,* 84–90.

Andrews, P., & Martin, L. (1987). Cladistic relationships of extant and fossil hominids. *Journal of Human Evolution, 16,* 101–108.

Angleitner, A., Ostendorf, F., & John, O. P. (1990). Towards a taxonomy of personality descriptors in German: A psycho-lexical study. *European Journal of Personality, 4,* 89–118.

Annett, M. (1985). *Left, right, hand, and brain: The right-shift theory.* London: Erlbaum.

Antinucci, F. (Ed.). (1989). *Cognitive structure and development in nonhuman primates.* Hillsdale, NJ: Erlbaum.

Apte, M. L. (1985). *Humor and laughter: An anthropological approach.* Ithaca, NY: Cornell University.

Apter, A., van Praag, H. M., Plutchik, R., Sevy, S., Korn, M., & Brown, S. L. (1990). Interrelationships among anxiety, aggression, impulsivity, and mood: A serotonergically linked cluster? *Psychiatry Research, 32,* 191–199.

Archer, J. (1988). *The behavioral biology of aggression.* Cambridge: Cambridge University Press.

Archer, J. (1991). The influence of testosterone on human aggression. *British Journal of Psychology, 82,* 1–28.

Ardrey, R. (1966). *The territorial imperative.* New York: Atheneum.

Argyle, M., & Cook, M. (1976). *Gaze and mutual gaze.* Cambridge: Cambridge University Press.

Arruda, M. F., Yamamoto, M. E., & Bueno, O. F. A. (1986). Interactions between parents and infants and infant-father separation in the common marmoset *(Callithrix jacchus). Primates, 27,* 215–228.

Atz, J. W. (1970). The application of the idea of homology to behavior. In L. R. Aronson, E. Tobach, D. S. Lehrman, & J. S. Rosenblatt (Eds.), *Development and evolution of behavior* (pp. 53–74). San Francisco: W. H. Freeman.

Aureli, F. (1992). Post-conflict behaviour among wild long-tailed macaques *(Macaca fascicularis). Behavioral Ecology and Sociobiology, 31,* 329–337.

Aureli, F. (1997). Post-conflict anxiety in nonhuman primates: The mediating role of emotion in conflict resolution. *Aggressive Behavior, 23,* 315–328.

Aureli, F., Cords, M., & van Schaik, C. P. (2002). Conflict resolution following aggression in gregarious animals: A predictive framework. *Animal Behavior, 64,* 325–343.

Aureli, F., Cozzolino, R., Cordischi, C., & Scucchi, S. (1992). Kin-oriented redirection among Japanese macaques: An expression of a revenge system? *Animal Behaviour, 44,* 283–291.

Aureli, F., Das, M., & Veenema, H. C. (1997). Differential kinship effect on reconciliation in three species of macaques *(Macaca fascicularis, M. fuscata, and M. sylvanus). Journal of Comparative Psychology, 111,* 91–99.

Aureli, F., & de Waal, F. B. M. (1997). Inhibition of social behavior in chimpanzees under high-density conditions. *American Journal of Primatology, 41,* 213–228.

Aureli, F., & de Waal F. B. M. (2000a). *Natural conflict resolution.* Berkeley: University of California Press.

Aureli, F., & de Waal, F. B. M. (2000b). Why natural conflict resolution? In F. Aureli & F. B. M. de Waal (Eds.), *Natural conflict resolution* (pp. 3–9). Berkeley: University of California Press.

Aureli, F., Preston, S. D., & de Waal, F. B. M. (1999). Heart rate responses to social interactions in free-moving rhesus macaques *(Macaca mulatta):* A pilot study. *Journal of Comparative Psychology, 113,* 59–65.

Aureli, F., & Schaffner, C. M. (2002). Relationship assessment through emotional mediation. *Behaviour, 139,* 393–420.

Aureli, F., & Smucny, D. A. (2000). The role of emotion in conflict and conflict resolution. In F. Aureli & F. B. M. de Waal (Eds.), *Natural conflict resolution* (pp. 199–224). Berkeley: University of California Press.

Aureli, F., van Panthaleon van Eck, C. J., & Veenema, H. C. (1995). Long-tailed macaques avoid conflicts during short-term crowding. *Aggressive Behavior, 21,* 113–122.

Aureli, F., & van Schaik, C. P. (1991a). Post-conflict behaviour in long-tailed macaques *(Macaca fascicularis)*. I: The social events. *Ethology, 89,* 89–100.

Aureli, F., & van Schaik, C. P. (1991b). Post-conflict behaviour in long-tailed macaques *(Macaca fascicularis)*. II: Coping with the uncertainty. *Ethology, 89,* 101–114.

Aureli, F., van Schaik, C. P., & van Hooff, J. A. R. A. M. (1989). Functional aspects of reconciliation among captive long-tailed macaques *(Macaca fascicularis)*. *American Journal of Primatology, 19,* 39–51.

Avis, J., & Harris, P. L. (1991). Belief-desire reasoning among Baka children: Evidence for a universal conception of mind. *Child Development, 62,* 460–467.

Axelrod, R. (1984). *The evolution of cooperation.* New York: Basic Books.

Bachmann, G. A., & Leiblum, S. R. (1991). Sexuality in sexagenarian women. *Maturitas, 13,* 43–50.

Bachmann, G. A., Leiblum, S. R., Sandler, B., Ainsley, W., Narcessian, R., Shelden, R., & Hymans, H. N. (1985). Correlates of sexual desire in post-menopausal women. *Maturitas, 7,* 211–216.

Bachorowski, J.-A., & Owren, M. J. (1995). Vocal expression of emotion: Acoustic properties of speech are associated with emotional intensity and context. *Psychological Science, 6,* 219–224.

Bachorowski, J.-A., & Owren, M. J. (1999) Acoustic correlates of talker sex and individual talker identity are present in a short vowel segment produced in running speech. *Journal of the Acoustical Society of America, 106,* 1054–63.

Bachorowski, J.-A., & Owren, M. J. (2001). Not all laughs are alike: Voiced but not unvoiced laughter elicits positive affect in listeners. *Psychological Science, 12,* 252–257.

Bachorowski, J.-A., Smoski, M., Tomarken, A. J., & Owren, M. J. (under review). Laugh rate and acoustics are associated with social context.

Bachorowski, J.-A., Smoski, M., & Owren, M. J. (2001). Acoustic features of laughter. *Journal of the Acoustical Society of America, 110,* 1581–97.

Bagatell, C. J., Heiman, J. R., Rivier, J. E., & Bremner, W. J. (1994). Effects of endogenous testosterone and estradiol on sexual behavior in normal young men. *Journal of Clinical Endocrinology and Metabolism, 78,* 711–716.

Baker, K. C., & Aureli, F. (1996). The neighbor effect: Other groups influence intragroup agonistic behavior in captive chimpanzees. *American Journal of Primatology, 40,* 283–291.

Baker, K. C., & Aureli, F. (1997). Behavioural indicators of anxiety: An empirical test in chimpanzees. *Behaviour, 134,* 1031–50.

Baker, K. & C., Smuts, B. (1994). Social relationships of female chimpanzees: Diversity between captive social groups. In R. W. Wrangham, W. C. McGrew, F. B. M. de Waal, & P. Heltne (Eds.), *Chimpanzee cultures* (pp. 227–242). Cambridge, MA: Harvard University Press.

Baldwin, D. A. (1993). Early referential understanding: Infants' ability to recognize referential acts for what they are. *Developmental Psychology, 29,* 832–843.

Baldwin, D. A., & Baird. J. A. (1999). Action analysis: A gateway to intentional inference. In P. Rochat (Ed.), *Early social cognition.* (pp. 215–240). Hillsdale, NJ: Erlbaum.

Bancroft, J., & Wu, F. C. (1983). Changes in erectile responsiveness during androgen replacement therapy. *Archives of Sexual Behavior, 12,* 59–66.

Bandura, A. (1973). *Aggression: A social learning analysis.* Englewood Cliffs, NJ: Prentice-Hall.

Bandura, A. (1989). Social cognitive theory. *Annals of Child Development, 6,* 1–60.

Banse, R., & Scherer, K. R. (1996). Acoustic profiles in vocal emotional expression. *Journal of Personality and Social Psychology, 70,* 614–636.

Bard, K. A. (1991). Distribution of attachment classifications in nursery chimpanzees. *American Journal of Primatology, 24,* 88.

Bard, K. A. (1992). Intentional behavior and intentional communication in young free-ranging orangutans. *Child Development, 63,* 1186–97.

Bard, K. A. (1998). Social-experiential contributions to imitation and emotion in chimpanzees. In S. Braten (Ed.), *Intersubjective communication and emotion in early ontogeny* (pp. 208–227). Cambridge: Cambridge University Press.

Bard, K. A. (2000). Crying in infant primates: Insights into the development of crying in chimpanzees. In R. G. Barr, B. Hopkins, & J. A. Green (Eds.), *Crying as a sign, a symptom, and a signal* (pp. 157–175). London: Mac Keith Press.

Bard, K. A. & Gardner, K. H. (1996). Influences on development in infant chimpanzees: Enculturation, temperament, and cognition. In A. E. Russon, K. A. Bard, & S. T. Parker (Eds.), *Reaching into thought: The minds of the great apes* (pp. 235–256). New York: Cambridge University Press.

Barkow, J. H., Cosmides, L., & Tooby, J. (1992). *The adapted mind: Evolutionary psychology and the generation of culture.* Oxford: Oxford University Press.

Barnland, D. C., & Yoshioko, M. (1990). Apologies: Japanese and American styles. *International Journal of Intercultural Relations, 14,* 193–206.

Barr, C., Becker, M., & Higley, J. D. (under review). Early life events as predictors of aggression and violence in adult.

Barros, M., Boere, V., Huston, J. P., & Tomaz, C. (2000). Measuring fear and anxiety in the marmoset *(Callithrix penicillata)* with a novel predator confrontation model: Effects of diazepam. *Behavioural Brain Research, 108,* 205–211.

Barth, R. J., & Kinder, B. N. (1988). A theoretical analysis of sex differences in same-sex friendships. *Sex Roles, 19,* 349–363.

Barton, R. A. (1987). Allogrooming and mutualism in diurnal lemurs. *Primates, 28,* 539–542.

Bastian, M. L., Sponberg, A. C., Suomi, S. J., & Higley, J. D. (2003). Long-term effects of infant rearing condition on the acquisition of dominance rank in juvenile and adult rhesus macaques *(Macaca mulatta). Developmental Psychobiology, 42,* 44–51.

Bates, E. (1979). *The emergence of symbols: Cognition and communication in infancy.* New York: Academic Press.

Bates, E., Camaioni, L., & Volterra, V. (1975). The acquisition of performatives prior to speech. *Merrill-Palmer Quarterly, 21,* 205–226.

Bateson, G. (1953). Metalogue: About games and being serious. *Review of General Semantics, 10,* 213–217.

Batson, C. D. (1998). Altruism and prosocial behavior. In D. T. Gilbert, S. T. Fiske, & G. Lindzey (Eds.), *The handbook of social psychology* (pp. 282–316). New York: McGraw-Hill.

Batson, C. D., & Ahmad, N. (2001). Empathy-induced altruism in a prisoner's dilemma. II: What if the target of empathy has defected? *European Journal of Social Psychology, 31,* 25–36.

Batson, C. D., & Moran, T. (1999). Empathy-induced altruism in a prisoner's dilemma. *European Journal of Social Psychology, 29,* 909–924.

Bauer, H. R., & Philip, M. (1983). Facial and vocal individual recognition in the common chimpanzee. *Psychological Record, 33,* 161–170.

Bauer, R. M. (1998). Physiological measures of emotion. *Journal of Clinical Neurophysiology, 15,* 388–396.

Bauers, K., & de Waal, F. B. M. (1991). "Coo" vocalizations in stumptailed macaques: a controlled analysis. *Behaviour, 119,* 143–160.

Baum, M. J., Everitt, B. J., Herbert, J., & Keverne, E. B. (1977). Hormonal basis of proceptivity and receptivity in female primates. *Archives of Sexual Behavior, 6,* 173–192.

Baum, M. J., Slob, A. K., de Jong, F. H., & Westbroek, D. L. (1978). Persistence of sexual behavior in ovariectomized stumptail macaques following dexamethasone treatment or adrenalectomy. *Hormones and Behavior, 11,* 323–347.

Baumrind, D. (1971). Current patterns of parental authority. *Developmental Psychology Monographs, 4,* 1–103.

Bayart, R., Hayashi, K. T., Faull, K. F., Barchas, J. D., & Levine, S. (1990). Influence of maternal proximity on behavioral and physiological responses to separation in infant rhesus monkeys (Macaca mulatta). *Behavioral Neuroscience, 104,* 98–107.

Beach, F., & Levinson, G. (1950). Effects of androgen on the glans penis and mating behavior of castrated rats. *Journal of Experimental Zoology, 114,* 159–171.

Beaton, A. A. (1997). The relation of planum temporale asymmetry and morphology of the corpus callosum to handedness, gender, and dyslexia: A review of the evidence. *Brain and Language, 15,* 255–322.

Bell, D. C., & Richard, A. J. (2000). Caregiving: The forgotten element in attachment. *Psychological Inquiry, 11,* 69–83.

Belsky, J. (1993). Etiology of child maltreatment: A developmental-ecological analysis. *Psychological Bulletin, 114,* 413–434.

Belsky, J. (1999). Modern evolutionary theory and patterns of attachment. In J. Cassidy & P. R. Shaver (Eds.), *Handbook of attachment* (pp. 141–161). New York: Guilford Press.

Belsky, J., Steinberg, L., & Draper, P. (1991). Childhood experience, interpersonal development, and reproductive strategy: An evolutionary theory of socialization. *Child Development, 62,* 647–670.

Bennett, A. J., Tsai, T., Pierre, P. J., Suomi, S. J., Shoaf, S. E., Linnoila, M., & Higley, J. D. (1998). Behavioral response to novel objects varies with CSF monoamine concentrations in rhesus monkeys. *Society for Neuroscience Abstracts, 24,* 954.

Benton, A. L. (1980). The neuropsychology of faces. *American Psychologist, 35,* 176–186.

Beran, M. J., Pate, J. L., Richardson, W. K., & Rumbaugh, D. M. (2000). A chimpanzee's (Pan troglodytes) long-term retention of lexigrams. *Animal Learning and Behavior, 28,* 201–207.

Beran, M. J., & Rumbaugh, D. M. (2001). "Constructive" enumeration by chimpanzees (Pan troglodytes) on a computerized task. *Animal Cognition, 4,* 81–89.

Beran, M. J., Savage-Rumbaugh, E. S., Brakke, K. E., Kelley, J. W., & Rumbaugh, D. M. (1998). Symbol comprehension and learning: A "vocabulary" test of three chimpanzees (Pan troglodytes). *Evolution of Communication, 2,* 171–188.

Berard, J. D. (1989). Life histories of male Cayo Santiago macaques. *Puerto Rico Health Sciences Journal, 8,* 61–64.

Bercovitch, F. B. (1993). Dominance rank and reproductive maturation in male rhesus macaques (Macaca mulatta). *Journal of Reproduction and Fertility, 99,* 113–120.

Bercovitch, F. B., & Berard, J. D. (1993). Life-history costs and consequences of rapid reproductive maturation in female rhesus macaques. *Behavioral Ecology and Sociobiology, 32,* 103–109.

Bercovitch, F. B., & Goy, R. W. (1990). The socioendocrinology of reproductive development and reproductive success in macaques. In T. E. Ziegler & F. B. Bercovitch (Eds.), *Socioendocrinology of primate reproduction* (pp. 59–93). New York: Wiley Liss.

Bercovitch, F. B., Roy, M. M., Sladky, K. K., & Goy, R. W. (1988). The effects of isosexual rearing on adult sexual behavior in captive male rhesus macaques. *Archives of Sexual Behavior, 17*, 381–388.

Bercovitch, F. B., & Strum, S. C. (1993). Dominance rank, resource availability, and reproductive maturation in female savanna baboons. *Behavioral Ecology and Sociobiology, 33*, 313–318.

Berdecio, S., & Nash, L. T. (1981). Chimpanzee visual communication. *Anthropology Papers, no. 26*, Arizona State University.

Bereczkei, T. (2001). Maternal trade-off in treating high-risk children. *Evolution and Human Behavior, 22*, 197–212.

Berenbaum, S. A., & Hines, M. (1992). Early androgens are related to childhood sex-typed toy preferences. *Psychological Science, 3*, 203–206.

Berenbaum, S. A., & Snyder, E. (1995). Early hormonal influences on childhood sex-typed activity and playmate preferences: Implications for the development of sexual orientation. *Developmental Psychology, 31*, 31–42.

Bering, J. M., Bjorklund, D. F., & Ragan, P. (2000). Deferred imitation of object-related actions in human-reared juvenile chimpanzees and orangutans. *Developmental Psychobiology, 36*, 218–232.

Bering, J. M. (2001). Theistic percepts in other species: Can chimpanzees represent the minds of non-natural agents? *Journal of Cognition and Culture, 1*, 107–137.

Berman, C. M. (1982a). The ontogeny of social relationships with group companions among free-ranging infant rhesus monkeys. I: Social networks and differentiation. *Animal Behaviour, 30*, 149–162.

Berman, C. M. (1982b). The ontogeny of social relationships with group companions among free-ranging infant rhesus monkeys. II: Differentiation and attractiveness. *Animal Behaviour, 30*, 163–170.

Berman, C. M. (1984). Variation in mother-infant relationships: Traditional and non-traditional factors. In M. F. Small (Ed.), *Female primates* (pp. 17–36). New York: Alan R. Liss.

Berman, C. M. (1990a). Consistency in maternal behavior within families of free-ranging rhesus monkeys: An extension of the concept of maternal style. *American Journal of Primatology, 22*, 159–169.

Berman, C. M. (1990b). Intergenerational transmission of maternal rejection rates among free-ranging rhesus monkeys. *Animal Behaviour, 39*, 329–337.

Berman, C. M., & Kapsalis, E. (1999). Development of kin bias among rhesus monkeys: Maternal transmission or individual learning? *Animal Behaviour, 58*, 883–894.

Berman, C. M., Rasmussen, K. L. R., & Suomi, S. J. (1993). Reproductive consequences of maternal care patterns during estrus among free-ranging rhesus monkeys. *Behavioral Ecology and Sociobiology, 32*, 391–399.

Berman, C. M., Rasmussen, K. L. R., & Suomi, S. J. (1997). Group size, infant development and social networks in free-ranging rhesus monkeys. *Animal Behaviour, 53,* 405–421.

Bernstein, I. S. (1964). Role of the dominant male rhesus monkey in response to external challenges to the group. *Journal of Comparative and Physiological Psychology, 57,* 404–406.

Bernstein, I. S. (1970). Primate status hierarchies. In L. A. Rosenblum (Ed.), *Primate behavior: Developments in field and laboratory research,* Vol. 1 (pp. 71–109). New York: Academic Press.

Bernstein, I. S. (1981). Dominance: The baby and the bathwater. *Behavioral and Brain Sciences, 4,* 419–457.

Bernstein, I. S., & Ehardt, C. L. (1985a). Agonistic aiding: Kinship, rank, age, and sex influences. *American Journal of Primatology, 8,* 37–52.

Bernstein, I. S., & Ehardt, C. L. (1985b). Intragroup agonistic behavior in rhesus monkeys *Macaca mulatta. International Journal of Primatology, 6,* 209–226.

Bernstein, I. S., & Ehardt, C. L. (1985c). Age-sex differences in the expression of agonistic behavior in rhesus monkey *(Macaca mulatta)* groups. *Journal of Comparative Psychology, 99,* 115–132.

Bernstein, I. S., & Ehardt, C. L. (1986a). Modification of aggression through socialization and the special case of adult and adolescent male rhesus monkeys *(Macaca mulatta). American Journal of Primatology, 10,* 213–227.

Bernstein, I. S., & Ehardt, C. L. (1986b). Selective interference in rhesus monkey *(Macaca mulatta)* intragroup agonistic episodes by age-sex class. *Journal of Comparative Psychology, 100,* 380–384.

Bernstein, I. S., & Ehardt, C. (1986c). The influence of kinship and socialization on aggressive behaviour in rhesus monkeys *(Macaca mulatta). Animal Behavior, 34,* 739–747.

Bernstein, I. S., Gordon, T. P., & Rose, R. M. (1974a). Aggression and social controls in rhesus monkey *(Macaca mulatta)* groups revealed in group formation studies. *Folia Primatologica, 21,* 81–107.

Bernstein, I. S., Gordon, T. P., & Rose, R. M. (1974b). Factors influencing the expression of aggression during introductions to rhesus monkey groups. In R. L. Holloway (Ed.), *Primate aggression, territoriality, and xenophobia: A comparative perspective* (pp. 211–240). New York: Academic Press.

Bernstein, I. S., Judge, P. G., & Ruehlmann, T. E. (1993a). Kinship, association, and social relationships in rhesus monkeys *(Macaca mulatta). American Journal of Primatology, 31,* 41–53.

Bernstein, I. S., Judge, P. G., & Ruehlmann, T. E. (1993b). Sex differences in adolescent rhesus monkey *(Macaca mulatta)* behavior. *American Journal of Primatology, 31,* 197–210.

Bernstein, I. S., Rose, R. M., & Gordon, T. P. (1977). Behavioural and hormonal responses of male rhesus monkeys introduced to females in the breeding and nonbreeding seasons. *Animal Behaviour, 25,* 609–614.

Bernstein, I. S., Rose, R. M., Gordon, T. P., & Grady, C. L. (1979). Agonistic rank, aggression, social context, and testosterone in male pigtailed monkeys. *Aggressive Behavior, 5,* 329–339.

Bernstein, I. S., Ruehlmann, T. E., Judge, P. G., Lindquist, T., & Weed, J. L. (1991). Testosterone changes during the period of adolescence in male rhesus monkeys *(Macaca mulatta)*. *American Journal of Primatology, 24,* 29–38.

Bernstein, I. S., & Sharpe, L. G. (1966). Social roles in a rhesus monkey group. *Behaviour, 26,* 91–104.

Berntson, G. G., & Boysen, S. T. (1989). Specificity of the cardiac response to conspecific vocalizations in chimpanzees. *Behavioral Neuroscience, 103,* 235–245.

Berntson, G. G., Boysen, S. T., Bauer, H. R., & Torello, M. S. (1989). Conspecific screams and laughter: Cardiac and behavioral reactions of infant chimpanzees. *Developmental Psychobiology, 22,* 771–787.

Berntson, G. G., Sarter, M., & Cacioppo, J. T. (1998). Anxiety and cardiovascular reactivity: The basal forebrain cholinergic link. *Behavioural Brain Research, 94,* 225–248.

Best, C. (1988). The emergence of cerebral asymmetries in early human development: A literature review and a neuroembryological model. In D. L. Molfese & S. J. Segalowitz (Eds.), *Brain lateralization in children: Developmental implications* (pp. 5–34). New York: Guilford Press.

Betz, S. K. (1981). Sentence expansion by Lana chimpanzee. Master's thesis, Georgia State University.

Bielert, C., Girolami, L., & Anderson, C. (1986). Male chacma baboon *(Papio ursinus)* sexual arousal: Studies with adolescent and adult females as visual stimuli. *Developmental Psychobiology, 19,* 369–383.

Bingham, H. C. (1932). Gorillas in a native habitat. *Publications of the Carnegie Institute, 426,* 65.

Bisazza, A., Rogers, L. J., & Vallortigara, G. (1998). The origins of cerebral asymmetry: A review of evidence of behavioural and brain lateralization in fishes, reptiles, and amphibians. *Neuroscience and Biobehavioral Reviews, 22,* 411–426.

Bjorklund, D. F., & Pellegrini, A. D. (2002). *The origins of human nature: Evolutionary developmental psychology.* Washington, DC: American Psychological Association Press.

Bjorklund, D. F., Yunger, J. L., Bering, J. M., & Ragan, P. (2002). The generalization of deferred imitation in enculturated chimpanzees. *Animal Cognition, 5,* 49–58.

Björkqvist, K. (1994). Sex differences in physical, verbal, and indirect aggression: A review of recent research. *Sex Roles, 30,* 177–188.

Björkqvist, K. (2001). Social defeat as a stressor in humans. *Physiology and Behavior, 73,* 435–442.

Black, D. W. (1984). Laughter. *Journal of the American Medical Association, 252,* 2995–98.

Blanchard, D. C., Hebert, M., & Blanchard, R. J. (1999). Continuity vs. (political) correctness: Animal models and human aggression. *HFG Review, 3,* 6–10.

Blest, A. D. (1957). The function of eyespot patterns in the Lepidoptera. *Behaviour, 11,* 209–255.

Block, J. (1961). *The Q-sort method in personality assessment and psychiatric research.* Springfield, IL.: Thomas.

Block, J. (1977). Advancing the psychology of personality: Paradigmatic shift or improving the quality of research? In D. Magnusson & N. S. Endler (Eds.), *Personality at the crossroads: Current issues in interactional psychology* (pp. 37–63). New York: John Wiley & Sons.

Block, J. (1995). A contrarian view of the five-factor approach to personality description. *Psychological Bulletin, 117,* 187–215.

Block, J., Weiss, D. S., & Thorne, A. (1979). How relevant is a semantic similarity interpretation of personality ratings? *Journal of Personality and Social Psychology, 37,* 1055–74.

Block, J. D. (1980). *Friendship: How to give it, how to get it.* New York: Macmillan.

Blurton Jones, N. G. (1967). An ethological study of some aspects of social behaviour of children in nursery school. In D. Morris (Ed.), *Primate ethology* (pp. 347–368). London: Weidenfeld.

Blurton Jones, N. G. (1993). The lives of hunter-gatherer children: Effects of parental behavior and parental reproductive strategy. In M. E. Pereira & L. A. Fairbanks (Eds.), *Juvenile primates* (pp. 309–326). New York: Oxford University Press.

Blurton Jones, N. G., Hawkes, K., & O'Connell, J. F. (1989). Modeling and measuring costs of children in two foraging societies. In V. Standen & R. A. Foley (Eds.), *Comparative socioecology* (pp. 367–390). Oxford: Blackwell Scientific Publications.

Boccia, M. L. (1986). Grooming site preferences as a form of tactile communication and their role in the social relations of rhesus monkeys. In D. M. Taub & F. A. King (Eds.), *Current perspectives in primate social dynamics* (pp. 505–518). New York: Van Nostrand Reinhold.

Boccia, M. L., Reite, M., & Laudenslager, M. (1989). On the physiology of grooming in a pigtail macaque. *Physiology and Behavior, 45,* 667–670.

Boehm, C. (1992). Segmentary "warfare" and the management of conflicts: Comparison of East African chimpanzees and patrilineal-patrilocal humans. In A. H. Harcourt & F. B. M. de Waal (Eds.), *Coalitions and alliances in humans and other animals* (pp. 137–173). Oxford: Oxford University Press.

Boesch, C. (1991). Teaching among wild chimpanzees. *Animal Behaviour, 41,* 530–532.

Boesch, C. (1996). Three approaches for assessing culture among wild chimpanzees. In A. E. Russon, K. A. Bard, & S. T. Parker (Eds.), *Reaching into thought: The minds of the great apes* (pp. 404–429). Cambridge, MA: Harvard University Press.

Boesch, C., & Boesch, H. (1990). Tool use and tool making in wild chimpanzees. *Folia Primatologica, 54,* 86–99.

Boesch, C., & Boesch-Achermann, H. (2000). *The chimpanzees of the Taï Forest: Behavioural ecology and evolution.* Oxford: Oxford University Press.

Boesch, C., Marchesi, P., Marchesi, N., Fruth, B., & Joulian, F. (1994). Is nut cracking in wild chimpanzees cultural behavior? *Journal of Human Evolution, 4,* 325–338.

Bolig, R., Price, C. S., O'Neill, P. L., & Suomi, S. J. (1992). Subjective assessment of reactivity level and personality traits of rhesus monkeys. *International Journal of Primatology, 13,* 287–306.

Booth, A., Shelley, G., Mazur, A., Tharp, G., & Kittok, R. (1989). Testosterone, and winning and losing in human competition. *Hormones and Behavior, 23,* 556–571.

Boothe, R., & Sackett, G. P. (1975). Perception and learning in infant rhesus monkeys. In G. H. Bourne (Ed.), *The rhesus monkey* (pp. 343–363). New York: Academic Press.

Borkenau, P. (1992). Implicit personality theory and the Five-Factor Model. *Journal of Personality, 60,* 295–327.

Bornstein, M. H. (Ed.) (1995). *Handbook of parenting,* Vols. 1–4. Hillsdale, NJ: Erlbaum.

Boroojerdi, B., Foltys, H., Krings, T., Spetzger, U., Thron, A., & Topper, R. (1999). Localization of the motor hand area using transcranial magnetic stimulation and functional magnetic resonance imaging. *Clinical Neurophysiology, 110,* 699–704.

Borries, C., Sommer, V., & Srivastava, A. (1994). Weaving a tight social net: Allogrooming in free-ranging female langurs *(Presbytis entellus). International Journal of Primatology, 15,* 421–443.

Botchin, M. B., Kaplan, J. R., Manuck, S. B., & Mann, J. J. (1993). Low versus high prolactin responders to fenfluramine challenge: Marker of behavioral differences in adult male cynomolgus macaques. *Neuropsychopharmacology, 9,* 93–99.

Boulton, M. J. (1993). Proximate causes of aggressive fighting in middle school children. *British Journal of Educational Psychology, 63,* 231–244.

Bourne, G. H. (1971). *The ape people.* New York: G. P. Putnam's Sons.

Bowlby, J. (1952). *Maternal care and mental health.* Monograph of the World Health Organization, Geneva.

Bowlby, J. (1969). *Attachment and loss: Attachment,* Vol. 1. New York: Basic Books.

Bowlby, J. (1973). *Attachment and loss: Separation: Anxiety and anger,* Vol. 2. New York: Basic Books.

Bowlby, J. (1980). *Attachment and loss: Loss: Sadness and depression,* Vol. 3. New York, Basic Books.

Bowlby, J. (1984). Caring for the young: Influences on Development. In R. S. Cohen, B. J. Cohler, & S. H. Weissman (Eds.), *Parenthood: A psychodynamic perspective* (pp. 269–284). New York: Guilford Press.

Bowlby, J. (1988). *A secure base.* New York: Basic Books.

Boyce, W. T., O'Neill-Wagner, P., Price, C. S., Haines, M., & Suomi, S. J. (1998). Crowding, stress, and violent injuries among behaviorally inhibited rhesus macaques. *Health Psychology, 17,* 285–289.

Boyd, R., & Silk, J. B. (1997). *How humans evolved.* New York: Norton.

Boysen, S. T., & Berntson, G. G. (1989). Conspecific recognition in the chimpanzee *(Pan troglodytes):* Cardiac responses to significant others. *Journal of Comparative Psychology, 103,* 215–220.

Bradley, M. M., & Lang, P. J. (2000). Affective reactions to acoustic stimuli. *Psychophysiology, 37,* 204–215.

Bradshaw, J., & Rogers, L. J. (1993). *The evolution of lateral asymmetries, language, tool use, and intellect.* San Diego: Academic Press.

Bradshaw, R. H. (1993). Displacement activities as potential covert signals in primates. *Folia Primatologica, 61,* 174–176.

Brakke, K. E., & Savage-Rumbaugh, E. S. (1995). The development of language skills in bonobo and chimpanzee. I: Comprehension. *Language and Communication, 15,* 121–148.

Brakke, K. E., & Savage-Rumbaugh, E. S. (1996). The development of language skills in Pan. II: Production. *Language and Communication, 16,* 361–380.

Bretherton, I. (1985). Attachment theory: Retrospect and prospect. In I. Bretherton & E. Waters (Eds.), Growing points of attachment theory and research. *Monographs of the Society for Research in Child Development, 50,* 3–35.

Brooks-Gunn, J., & Furstenberg, F. F. (1986). Antecedents and consequences of parenting: The case of adolescent motherhood. In A. Fogel & G. F. Melson (Eds.), *Origins of nurturance* (pp. 233–258). Hillsdale, NJ: Erlbaum.

Brothers, L. (1990). The neural basis of primate social communication. *Motivation and Emotion, 14,* 81–91.

Brown, G. L., Ebert, M. H., Goyer, P. F., Jimerson, D. C., Klein, W. J., Bunney, W. E. J., & Goodwin, F. K. (1982). Aggression, suicide, and serotonin: Relationships to CSF amine metabolites. *American Journal of Psychiatry, 139,* 741–746.

Brown, G. L., Goodwin, F. K., Ballenger, J. C., Goyer, P. F., & Major, L. F. (1979).

Aggression in humans correlates with cerebrospinal fluid amine metabolites. *Psychiatry Research, 1,* 131–139.

Brown, G. R., & Dixson, A. F. (2000). The development of behavioural sex differences in infant rhesus macaques *(Macaca mulatta). Primates, 41,* 63–77.

Brown, J. V., & Pieper, W. A. (1973). Non-nutritive sucking in great ape and human newborns. *American Journal of Physical Anthropology 38,* 549–554.

Bruce, C. (1982). Face recognition by monkeys: Absence of an inversion effect. *Neuropsychologia, 20,* 515–521.

Bruce, V., Burton, M., & Dench, N. (1994). What's distinctive about a distinctive face? *Quarterly Journal of Experimental Psychology, 47A,* 119–141.

Bruner, J. (1972). The nature and uses of immaturity. *American Psychologist, 27,* 687–708.

Bruner, J. (1990). *Acts of meaning.* Cambridge, MA: Harvard University Press.

Bryant, B. K. (1992). Conflict resolution strategies in relation to children's peer relations. *Journal of Applied Developmental Psychology, 13,* 35–50.

Buchanan, C. M., Eccles, J. S., & Becker, J. B. (1992). Are adolescents the victims of raging hormones? Evidence for activational effects of hormones on moods and behavior at adolescence. *Psychological Bulletin, 111,* 62–107.

Buchanan-Smith, H. M. (1999). Exploration of unfamiliar areas and detection of potentially threatening objects in single- and mixed-species groups of tamarins. *International Journal of Comparative Psychology, 12,* 2–20.

Buck, R. (1988). The perception of facial expression: Individual regulation and social coordination. In Alley, T. R. (Ed.), *Social and applied aspects of perceiving faces* (pp. 141–165). Hillsdale, NJ: Erlbaum.

Buck, R. (1994). Social and emotional functions in facial expression and communication: The readout hypothesis. *Biological Psychology, 38,* 95–115.

Buirski, P., Kellerman, H., Plutchik, R., Weninger, R., & Buirski, N. (1973). A field study of emotions, dominance, and social behavior in a group of baboons *(Papio anubis). Primates, 14,* 67–78.

Buirski, P., & Plutchick, R. (1991). Measurement of deviant behavior in a Gombe chimpanzee: Relation to later behavior. *Primates, 32,* 207–211.

Buirski, P., Plutchik, R., & Kellerman, H. (1978). Sex differences, dominance, and personality in the chimpanzee. *Animal Behaviour, 26,* 123–129.

Bunk, S. (2001). Mind-body research matures. *Scientist, 15,* 8.

Burbank, V. (1987). Female aggression in cross-cultural perspective. *Behavior Science Research, 21,* 70–100.

Burger, J., Gochfeld, M., & Murray, B. G. Jr. (1991). Role of a predator's eye size in risk perception by basking black iguana *(Ctenosaura similes). Animal Behaviour, 42,* 471–476.

Burghardt, G. M., & H. W. Greene. (1988). Predator stimulation and duration of death feigning in neonate hognose snakes. *Animal Behaviour, 36,* 1842–44.

Burke, C. W., & Anderson, D. C. (1972). Sex-hormone binding globulin is an oestrogen amplifier. *Nature, 240,* 38–40.

Burling, R. (1993). Primate calls, human language, and nonverbal communication. *Current Anthropology, 34,* 25–37.

Burnstein, E., Crandall, C., & Kitayama, S. (1994). Some neo-Darwinian decision rules for altruism: Weighing cues for inclusive fitness as a function of the biological importance of the decision. *Journal of Personality and Social Psychology, 67,* 773–789.

Burton, F. D. (1972). The integration of biology and behavior in the socialization of *Macaca sylvana* of Gibraltar. In F. E. Poirier (Ed.), *Primate socialization* (pp. 29–62). New York: Random House.

Burton, L. M., & Dilworth-Anderson, P. (1991). The intergenerational family roles of aged black Americans. *In Families: Intergenerational and generational connections* (pp. 311–330). Binghamton, NY: Haworth.

Buss, A. H. (1995). *Personality: Temperament, social behavior, and the self.* Needham Heights, MA: Allyn & Bacon.

Buss, D. M. (1984). Evolutionary biology and personality psychology: Toward a conception of human nature and individual differences. *American Psychologist, 39,* 1135–47.

Buss, D. M. (1991). Evolutionary personality psychology. *Annual Review of Psychology, 42,* 459–491.

Butovskaya, M. L., & Kozintsev, A. G. (1994). Affiliative behaviour in an all-male group of stumptail macaques. In J. J. Roeder, B. Thierry, J. R. Anderson, & N. Herrenschmidt (Eds.), *Current primatology,* Vol. 2. (pp. 157–164). Strasbourg: Université Louis Pasteur.

Butovskaya, M., & Kozintsev, A. (1999). Aggression, friendship, and reconciliation in Russian primary schoolchildren. *Aggressive Behavior, 25,* 125–139.

Butovskaya, M., Verbeek, P., Ljungberg, T., & Lunardini, A. (2000). A multicultural view of peacemaking among young children. In F. Aureli & F. B. M. de Waal (Eds.), *Natural conflict resolution* (pp. 243–258). Berkeley: University of California Press.

Butterworth, G. (1991). The ontogeny and phylogeny of joint visual attention. In A. Whiten (Ed.), *Natural theories of mind* (pp. 223–232). Oxford: Basil Blackwell,

Butterworth, G., & Jarrett, N. (1991). What minds have in common is space: Spatial mechanisms serving joint visual attention in infancy. *British Journal of Developmental Psychology, 9,* 55–72.

Buydens-Branchey, L., Branchey, M., Hudson, J., & Fergeson, P. (2000). Low HDL cholesterol, aggression, and altered central serotonergic activity. *Psychiatry Research, 93,* 93–102.

Byrne, G., & Suomi, S. J. (1995). Development of activity patterns, social interactions, and exploratory behavior in infant tufted capuchins *(Cebus apella). American Journal of Primatology, 35,* 255–270.

Byrne, R. W. (1997). Machiavellian intelligence. *Evolutionary Anthropology, 5,* 172–180.

Cacioppo, J. T., & Berntson, G. G. (1994). Relationship between attitudes and evaluative space: A critical review, with emphasis on the separability of positive and negative substrates. *Psychological Bulletin, 115,* 401–423.

Cacioppo, J. T., Berntson, G. G., & Klein, D. J. (1992). What is an emotion? The role of somatovisceral afference, with special emphasis on somatovisceral "illusions". In M. S. Clark (Ed.), *Emotion and social behavior* (pp. 63–98). Newbury Park, CA: Sage Publications.

Cacioppo, J. T., Berntson, G. G., Sheridan, J. F., & McClintock, M. K. (2000). Multilevel integrative analyses of human behavior: Social neuroscience and the complementing nature of social and biological approaches. *Psychological Bulletin, 126,* 829–843.

Cacioppo, J. T., & Gardner, W. L. (1999). Emotion. *Annual Review of Psychology, 50,* 191–214.

Cacioppo, J. T., Gardner, W. L., & Berntson, G. G. (1999). The affect system has parallel and integrative processing components: Form follows function. *Journal of Personality and Social Psychology, 76,* 839–855.

Cacioppo, J. T., Klein, D. J., Berntson, G. G., & Hartfield, E. (1993). The psychophysiology of emotion. In M. Lewis & J. M. Haviland (Eds.), *The handbook of emotions* (pp. 119–142). New York: Guilford Press.

Cacioppo, J. T., & Tassinary, L. G. (1990). *Principles of psychophysiology: Physical, social, and inferential elements.* Cambridge: Cambridge University Press.

Cacioppo, J. T., Tassinary, L. G., & Berntson, G. G. (2000). *Handbook of psychophysiology* (2nd ed.). New York: Cambridge University Press.

Cacioppo, J. T., Tassinary, L. G., & Fridlund, A. J. (1990). The skeletomotor system. In J. T. Cacioppo & L. G. Tassinary (Eds.), *Principles of psychophysiology: Physical, social, and inferential elements* (pp. 325–384). Cambridge: Cambridge University Press.

Cain, D. P., & Wada, J. A. (1979). An anatomical asymmetry in the baboon brain. *Brain, Behavior, and Evolution, 16,* 222–226.

Caine, N. G. (1986). Behavior during puberty and adolescence. *Comparative Primate Biology, 2A,* 327–361.

Caine, N. G., Earle, H., & Reite, M. (1983). Personality traits of adolescent pig-tailed monkeys *(Macaca nemestrina):* An analysis of social rank and early separation experience. *American Journal of Primatology, 4,* 253–260.

Caine, N. G., & Mitchell, G. D. (1979). The relationship between maternal rank and companion choice in immature macaques *(Macaca mulatta* and *M. radiata). Primates, 20,* 583–590.

Caine, N. G., & Reite, M. (1983). Infant abuse in captive pig-tailed macaques: Relevance to human child abuse. In M. Reite & N. G. Caine (Eds.), *Child abuse: The nonhuman primate data* (pp. 19–27). New York: Alan Liss.

Call, J. (in press). Social knowledge and social manipulation in monkeys and apes. In C. Harcourt & B. Sherwood (Eds.), *New perspectives in primate evolution and behaviour.* Otley, UK: Westbury Publishing.

Call, J., Agnetta, B., & Tomasello, M. (2000). Social cues that chimpanzees do and do not use to find hidden objects. *Animal Cognition, 3,* 23–34.

Call, J., & Carpenter, M. (2001). Do chimpanzees and children know what they have seen? *Animal Cognition, 4,* 207–220.

Call, J., Hare, B., & Tomasello, M. (1998). Chimpanzee gaze following in an object-choice task. *Animal Cognition, 1,* 89–99.

Call, J., & Tomasello, M. (1994). Production and comprehension of referential pointing by orangutans *(Pongo pygmaeus). Journal of Comparative Psychology 108,* 307–317.

Call, J., & Tomasello, M. (1996). The effect of humans on the cognitive development of apes. In A. E. Russon, K. A. Bard, & S. T. Parker (Eds.), *Reaching into thought* (pp. 371–403). Cambridge: Cambridge University Press.

Call, J., & Tomasello, M. (1998). Distinguishing intentional from accidental actions in orangutans *(Pongo pygmaeus),* chimpanzees *(Pan troglodytes),* and human children *(Homo sapiens). Journal of Comparative Psychology, 112,* 192–206.

Call, J., & Tomasello, M. (1999). A nonverbal false belief task: The performance of children and great apes. *Child Development, 70,* 381–395.

Call, J., & Tomasello, M. (in press). What chimpanzees know about seeing revisited: An explanation of the third kind. In N. Eilan, C. Hoerl, T. McCormack, & J. Roessler (Eds.), *Joint attention.* Oxford: Oxford University Press.

Camaioni, L. (1991). Mind knowledge in infancy: The emergence of intentional communication. *Early Development and Parenting, 1,* 15–22.

Campbell, S., & Whitehead, M. (1977). Oestrogen therapy and the menopausal syndrome. *Clinical Obstetrics and Gynecology, 4,* 31–47.

Canli, T., Desmond, J. E., Zhao, Z., Glover, G., & Gabrieli, J. D. E. (1998). Hemispheric asymmetry for emotional stimuli detected with fMRI. *NeuroReport, 9,* 3233–39.

Cannon, W. B. (1929). *Bodily changes in pain, hunger, fear, and rage,* Vol. 2. New York: Appleton.

Cantalupo, C., & Hopkins, W. D. (2001). Asymmetric Broca's area in great apes. *Nature, 414,* 505.

Cantalupo, C., Pilcher, D., & Hopkins, W. D. (in press). Are planum temporale and sylvian fissure asymmetries directly related? An MRI study in great apes. *Neuropsychologia.*

Capitanio, J. P. (1984) Early experience and social processes in rhesus macaques *(Macaca mulatta).* I: Dyadic social interaction. *Journal of Comparative Psychology, 98,* 35–44.

Capitanio, J. P. (1986). Behavioral pathology. In G. Mitchell & J. Erwin (Eds.), *Comparative primate biology: Behavior, conservation, and ecology,* Vol. 2 (pp. 411–454). New York: Alan R. Liss.

Capitanio, J. P. (1999). Personality dimensions in adult male rhesus macaques: Prediction of behaviors across time and situation. *American Journal of Primatology, 47,* 299–320.

Capitanio, J. P., Mendoza, S. P., & Baroncelli, S. (1999). The relationship of personality dimensions in adult male rhesus macaques to progression of Simian Immunodeficiency Virus disease. *Brain, Behavior, and Immunity, 13,* 138–154.

Caporael, L. R., Dawes, R. M., Orbell, J. M., & van de Kragt, A. J. C. (1989). Selfishness examined: Cooperation in the absence of egoistic incentives. *Behavioral and Brain Sciences, 12,* 683–739.

Carey, S., & Diamond, R. (1977). From piecemeal to configurational representation of faces. *Science, 195,* 312–314.

Carey, S. & Spelke, E. S. (1994). Domain-specific knowledge and conceptual change. In L. Hirschfeld & S. Gelman (Eds.), *Mapping the mind: Domain-specificity in cognition and culture* (pp. 169–200). Cambridge: Cambridge University Press.

Caro, T. M., & Hauser, M. D. (1992). Is there teaching in nonhuman animals? *Quarterly Review of Biology, 67,* 151–171.

Carpenter, C. R. (1934). A field study of the behavior and social relations of howling monkeys *(Alouatta palliata). Comparative Psychology Monographs, 10,* 1–168.

Carpenter, C. R. (1942). Sexual behavior in free ranging rhesus monkeys *(Macaca mulatta).* I: Specimens, procedures, and behavioral characteristics of estrus. II: Periodicity of estrus, homosexual, autoerotic, and non-conformist behavior. *Journal of Comparative Psychology, 33,* 113–162.

Carpenter, C. R. (1974). Aggressive behavioral systems. In R. L. Holloway (Ed.), *Primate aggression, territoriality, and xenophobia: A comparative perspective* (pp. 459–496). New York: Academic Press.

Carpenter, M., Akhtar, N., & Tomasello, M. (1998). Fourteen through 18-month-old infants differentially imitate intentional and accidental actions. *Infant Behavior and Development, 21,* 315–330.

Carpenter, M., Tomasello, M., & Savage-Rumbaugh, E. S. (1995). Joint attention and imitative learning in children, chimpanzees, and enculturated chimpanzees. *Social Development, 4,* 217–237.

Caryl, P. G. (1979). Communication by agonistic displays: What can game theory contribute to ethology? *Behaviour, 68,* 136–169.

Castles, D. L., Aureli, F., & de Waal, F. B. M. (1996). Variation in conciliatory tendency and relationship quality across groups of pigtail macaques. *Animal Behaviour, 52,* 389–403.

Castles, D. L., & Whiten, A. (1998a). Post-conflict behaviour of wild olive baboons. I: Reconciliation, redirection, and consolation. *Ethology, 104,* 126–147.

Castles, D. L., & Whiten, A. (1998b). Post-conflict behaviour of wild olive baboons. II: Stress and self-directed behaviour. *Ethology, 104,* 148–160.

Castles, D. L., Whiten, A., & Aureli, F. (1999). Social anxiety, relationships, and self-directed behaviour among wild female olive baboons. *Animal Behaviour, 58,* 1207–15.

Chadwick-Jones, J. K. (1989). Presenting and mounting in non-human primates: Theoretical developments. *Journal of Social and Biological Structures, 12,* 319–333.

Chagnon, N. A. (1977). *Yanomamo: The fierce people.* New York: Holt, Rinehart, & Winston.

Chagnon, N. A. (1979). Is reproductive success equal in egalitarian societies? In N. A. Chagnon & W. Irons (Eds.), *Evolutionary biology and human social behavior: An anthropological perspective* (pp. 374–401). North Scituate, MA: Duxbury.

Chagnon, N. A. (1982). Sociodemographic attributes of nepotism in tribal populations: Man the rule-breaker. In King's College Sociobiology Group (Eds.), *Current problems in sociobiology* (pp. 291–318). Cambridge: Cambridge University Press.

Chamberlain, B., Ervin, F. R., Pihl, R. O., & Young, S. N. (1987). The effect of raising or lowering tryptophan levels on aggression in vervet monkeys. *Pharmacology Biochemistry and Behavior, 28,* 503–510.

Chamove, A. S. (1974) A new primate behavior category system. *Primates, 15,* 85–99.

Chamove, A. S., Eysenck, H. J., & Harlow, H. F. (1972). Personality in monkeys: Factor analyses of rhesus social behavior. *Quarterly Journal of Experimental Psychology, 24,* 496–504.

Champoux, M., Higley, J. D., & Suomi, S. J. (1997). Behavioral and physiological characteristics of Indian and Chinese-Indian hybrid rhesus macaque infants. *Developmental Psychobiology, 31,* 49–63.

Champoux, M., Hwang, L., Lang, O., & Levine, S. (2001). Feeding demand conditions and plasma cortisol in socially housed squirrel monkey mother-infant dyads. *Psychoneuroendocrinology, 26,* 461–477.

Champoux, M., & Suomi, S. J. (1994). Behavioral and adrenocortical responses of rhesus macaque mothers to infant separation in an unfamiliar environment. *Primates, 35,* 191–202.

Champoux, M., Suomi, S. J., & Schneider, M. L. (1994). Temperament differences between captive Indian and Chinese-Indian hybrid rhesus macaque neonates. *Laboratory Animal Science, 44,* 351–357.

Chapais, B. (1983). Matriline membership and male rhesus reaching high ranks in the natal troop. In R. A. Hinde (Ed.), *Primate social relationships: An integrated approach* (pp. 171–175). Sunderland, MA: Sinauer Associates.

Chapais, B. (1986). Why do adult male and female rhesus monkeys affiliate during the birth season? In R. G. Rawlins & M. J. Kessler (Eds.), *The Cayo Santiago macaques* (pp. 173–200). Albany: SUNY Press.

Chapais, B. (1988). Rank maintenance in female Japanese macaques: Experimental evidence for social dependency. *Behaviour, 102,* 41–59.

Chapais, B. (1992). The role of alliances in social inheritance of rank among female primates. In A. Harcourt & F. B. M. de Waal (Eds.), *Coalitions and alliances in humans and other animals* (pp. 29–59). New York: Oxford University Press.

Chapais, B. (1995). Alliances as a means of competition in primates: Evolutionary, developmental, and cognitive aspects. *Yearbook of Physical Anthropology, 38,* 115–136.

Chapais, B. (2001). Primate nepotism: What is the explanatory value of kin selection? *International Journal of Primatology, 22,* 203–229.

Chapais, B., Gauthier, C., & Prud'homme, J. (1995). Dominance competition through affiliation and support in Japanese macaques: An experimental study. *International Journal of Primatology, 16,* 521–536.

Chapais, B., Savard, L., & Gauthier, C. (2001). Kin selection and the distribution of altruism in relation to degree of kinship in Japanese macaques. *Behavioral Ecology and Sociobiology, 49,* 493–502.

Chapais, B., & Schulman, S. (1980). An evolutionary model of female dominance relations in primates. *Journal of Theoretical Biology, 82,* 47–89.

Cheney, D. L. (1977). The acquisition of rank and the development of reciprocal alliances among free-ranging immature baboons. *Behavioral Ecology and Sociobiology, 2,* 303–318.

Cheney, D. L. (1981). Intergroup encounters among free-ranging vervet monkeys. *Folia Primatologica, 35,* 124–146.

Cheney, D. L. (1987). Interactions and relationships between groups. In B. B. Smuts, D. L. Cheney, R. M. Seyfarth, R. W. Wrangham, & T. T. Struhsaker (Eds.), *Primate societies* (pp. 267–281). Chicago: University of Chicago Press.

Cheney, D. L. (1992). Intragroup cohesion and intergroup hostility: The relation between grooming distributions and intergroup competition among female primates. *Behavioral Ecology, 3,* 334–345.

Cheney, D. L., & Seyfarth, R. M. (1982). How vervet monkeys perceive their grunts. *Animal Behaviour, 30,* 739–751.

Cheney, D. L., & Seyfarth, R. M. (1988). Assessment of meaning and the detection of unreliable signals by vervet monkeys. *Animal Behaviour, 36,* 477–486.

Cheney, D. L., & Seyfarth, R. M. (1990a). Attending to behaviour versus attending to knowledge: Examining monkeys' attribution of mental states. *Animal Behaviour, 40,* 742–753.

Cheney, D. L., & Seyfarth, R. M. (1990b). *How monkeys see the world: Inside the mind of another species.* Chicago: University of Chicago Press.

Cheney, D. L., & Seyfarth, R. M. (1997). Reconciliatory grunts by dominant female baboons influence victim's behavior. *Animal Behaviour, 54,* 409–418.

Cheney, D. L., Seyfarth, R. M., & Silk, J. B. (1995) The role of grunts in reconciling opponents and facilitating interactions among adult female baboons. *Animal Behaviour, 50,* 249–257.

Chepko-Sade, B. D., & Olivier, T. J. (1979). Coefficient of relatedness and the probability of intragenealogical fission in *Macaca mulatta. Behavioral Ecology and Sociobiology, 5,* 263–278.

Chevalier-Skolnikoff, S. (1973). Facial expression of emotion in nonhuman primates. In P. Ekman (Ed.), *Darwin and facial expression* (pp. 11–89). New York: Academic Press.

Chevalier-Skolnikoff, S. (1974a). *The ontogeny of communication in the stumptail macaque (Macaca arctoides).* Basel: Karger.

Chevalier-Skolnikoff, S. (1974b). Male-female, female-female, and male-male sexual behavior in the stumptail monkey, with special attention to the female orgasm. *Archives of Sexual Behavior, 3,* 95–116.

Chevalier-Skolnikoff, S. (1982). A cognitive analysis of facial behavior in Old World monkeys, apes, and humans. In C. Snowdon, C. H. Brown, & M. Petersen (Eds.), *Primate communication* (pp. 303–368). Cambridge: Cambridge University Press.

Cheverud, J. M., Buettner-Janusch, J., & Sade, D. (1978). Social group fission and the origin of intergroup genetic differentiation among the rhesus monkeys of Cayo Santiago. *American Journal of Physical Anthropology, 49,* 449–456.

Chikazawa, D., Gordon, T. P., Bean, C. A., & Bernstein, I. S. (1979). Mother-daughter dominance reversals in rhesus monkeys *(Macaca mulatta). Primates, 20,* 301–305.

Chisholm, J. (1993). Death, hope, and sex: Life history theory and the development of reproductive strategies. *Current Anthropology, 34,* 1–24.

Chivers, (1974). The siamang in Malaya. *Contributions to primatology,* Vol. 4. Basel: Karger.

Chongthammakun, S., & Terasawa, E. (1993). Negative feedback effects of estrogen on luteinizing-hormone-releasing hormone release occur in pubertal, but not prepubertal, ovariectomized female rhesus monkeys. *Endocrinology, 132,* 735–743.

Christiansen, K., & Knussmann, R. (1987). Androgen levels and components of aggressive behavior in men. *Hormones and Behavior, 21,* 170–180.

Christopher, S. B., & Gelini, H. R. (1977). Sex differences in use of a species-typical facial gesture by pigtail monkeys *(Macaca nemestrina). Primates, 18,* 565–577.

Cialdini, R. B., & Kenrick, D. T. (1976). Altruism as hedonism: A social developmental perspective on the relationship of negative mood state and helping. *Journal of Personality and Social Psychology, 34,* 907–914.

Cilia, P., & Piper D. C. (1997). Marmoset conspecific confrontation: An ethologically based model of anxiety. *Pharmacology Biochemistry and Behavior, 58,* 85–91.

Clark, L. A., & Watson, D. (1995). Constructing validity: Basic issues in objective scale development. *Psychological Assessment, 7,* 309–319.

Clarke, A. S., & Boinski, S. (1995). Temperament in nonhuman primates. *American Journal of Primatology, 37,* 103–125.

Clarke, A. S., & Lindburg, D. G. (1993). Behavioral contrasts between male cynomolgus and lion-tailed macaques. *American Journal of Primatology, 29,* 49–59.

Clarke, A. S., & Mason, W. A. (1988). Differences among three macaque species in responsiveness to an observer. *International Journal of Primatology, 9,* 347–364.

Clarke, A. S., Mason, W. A., & Moberg, G. P. (1988). Differential behavioral and adrenocortical responses to stress among three macaque species. *American Journal of Primatology, 14,* 37–52.

Clarke, A. S., & Snipes, M. (1998). Early behavioral development and temperamental traits in mother- vs. peer-reared rhesus monkeys. *Primates, 39,* 433–448.

Clarke, M. R., Kaplan, J. R., Bumsted, P. T., & Koritnik, D. R. (1986). Social dominance and serum testosterone concentration in dyads of male *Macaca fascicularis. Journal of Medical Primatology, 15,* 419–432.

Cleckley, H. (1941/1988). *The mask of sanity.* St. Louis: Mosby.

Clements, W. A., & Perner, J. (1994). Implicit understanding of belief. *Cognitive Development, 9,* 377–395.

Cleveland, J., & Snowdon, C. T. (1984). Social development during the first 20 weeks in the cotton-top tamarin *(Saguinus o. oedipus). Animal Behaviour, 32,* 432–444.

Cloninger, C. R. (1987). A systematic method for clinical description and classification of personality variants: A proposal. *Archives of General Psychiatry, 44,* 573–588.

Cloninger, C. R. (1988). A unified biosocial theory of personality and its role in the development of anxiety states: A reply to commentaries. *Psychiatric Developments, 6,* 83–120.

Clutton-Brock, T. H. (1991). *The evolution of parental care.* Princeton: Princeton University Press.

Coccaro, E. F., & Kavoussi, R. J. (1996). Neurotransmitter correlates of impulsive aggression. In D. M. Stoff & R. B. Cairns (Eds.), *Aggression and violence: Genetic, neurobiological, and biosocial perspectives* (pp. 67–85). Hillsdale, NJ: Erlbaum.

Cochran, C. G. (1979). Proceptive patterns of behavior throughout the menstrual cycle in female rhesus monkeys. *Behavioral and Neural Biology, 27,* 342–353.

Coe, C. L. (1990). Psychobiology of maternal behavior in nonhuman primates. In N. A. Krasnegor & R. S. Bridges (Eds.), *Mammalian parenting: Biochemical, neurobiological, and behavioral determinants* (pp. 157–183). New York: Oxford University Press.

Coe, C. L., & Levine, S. (1981). Normal responses to mother-infant separation in nonhuman primates. In D. F. Klein & J. Rabkin (Eds.), *Anxiety: New research and changing concepts* (pp. 155–177). New York: Raven Press.

Coelho, A. M. J., & Bramblett, C. A. (1981). Sexual dimorphism in the activity of olive baboons *(Papio cynocephalus anubis)* housed in monosexual groups. *Archives of Sexual Behavior, 10,* 79–91.

Cohen, S., Line, S., Manuck, S. B., Rabin, B. S., Heise, E. R., & Kaplan, J. R. (1997). Chronic stress, social status, and susceptibility to upper respiratory infections in nonhuman primates. *Psychosomatic Medicine, 59,* 213–221.

Collaer, M. L., & Hines, M. (1995). Human behavioral sex differences: A role for gonadal hormones during early development? *Psychological Bulletin, 118,* 55–107.

Collet, C., Vernet-Maury, E., Delhomme, G., & Dittmar, A. (1997). Autonomic nervous system response patterns specificity to basic emotions. *Journal of the Autonomic Nervous System, 62,* 45–57.

Collins, R. (1981). On the microfoundation of macrosociology. *American Journal of Sociology, 86,* 984–1014.

Collins, R. (1993). Emotional energy as the common denominator of rational action. *Rationality and Society, 5,* 203–230.

Collins, W. A., Maccoby, E. E., Steinberg, L., Hetherington, E. M., & Bornstein, M. H. (2000). Contemporary research on parenting: The case for nature and nurture. *American Psychologist, 55,* 218–232.

Colmenares, F. (1991). Greeting behaviour between male baboons: Oestrus females, rivalry, and negotiation. *Animal Behaviour, 41,* 49–60.

Colvin, J. (1982). Social integration and emigration of immature male rhesus macaques. Ph.D. dissertation, University of Cambridge.

Colvin, J. (1983). Familiarity, rank, and the structure of rhesus male peer network. In R. A. Hinde (Ed.), *Primate social relationships: An integrated approach* (pp. 190–200). Oxford: Blackwell.

Colvin, J. D. (1985). Breeding-season relationships of immature male rhesus monkeys with females. I: Individual differences and constraints on partner choice. *International Journal of Primatology, 6,* 261–287.

Colvin, J. D. (1986). Proximate causes of male emigration at puberty in rhesus monkeys. In R. G. Rawlins & M. Kessler (Eds.), *The Cayo Santiago macaques* (pp. 131–157). Albany: SUNY Press.

Conaway, C. H., & Koford, C. B. (1964). Estrous cycles and mating behavior in a free-ranging band of rhesus monkeys. *Journal of Mammalogy, 45,* 577–588.

Conrade, G. & Ho, R. (2001). Differential parenting styles for fathers and mothers: Differential treatment for sons and daughters. *Australian Journal of Psychology, 53,* 29–35.

Coope, J. (1976). Double-blind crossover study of estrogen replacement. In S. Campbell (Ed.), *The management of menopausal and post-menopausal years* (pp. 159–168). Baltimore: University Park Press.

Coplan, J. D., Andrews, M. W., Rosenblum, L. A., Owens, M. J., Friedman, S., Gorman, J. M., & Nemeroff, C. B. (1996). Persistent elevations of cerebrospinal fluid concentrations of corticotropin-releasing factor in adult nonhuman primates exposed to early life stressors: Implications for the pathophysiology of mood and anxiety disorders. *Proceedings of the National Academy of Sciences USA, 93,* 1619–23.

Cords, M. (1988). Resolution of aggressive conflicts by immature long-tailed macaques, *Macaca fascicularis. Animal Behaviour, 36,* 1124–35.

Cords, M. (1992). Post-conflict reunions and reconciliation in long-tailed macaques. *Animal Behaviour, 44,* 57–61.

Cords, M., & Aureli, F. (1993). Patterns of reconciliation among juvenile long-tailed macaques. In M. E. Pereira & L. A. Fairbanks (Eds.), *Juvenile primates: Life history, development, and behavior* (pp. 271–284). New York: Oxford University Press.

Cords, M., & Aureli, F. (2000). Reconciliation and relationship qualities. In F. Aureli & F. B. M. de Waal (Eds.), *Natural conflict resolution* (pp. 177–198). Berkeley: University of California Press.

Cords, M., & Killen, M. (1998). Conflict resolution in human and nonhuman primates. In J. Langer & M. Killen (Eds.), *Piaget, evolution, and development* (pp. 193–218). Mahwah, NJ: Erlbaum.

Cords, M., & Thurnheer, S. (1993). Reconciliation with valuable partners by long-tailed macaques. *Ethology, 93,* 315–325.

Cosmides, L. (1989). The logic of social exchange: Has natural selection shaped how humans reason? Studies with the Wason selection task. *Cognition, 31,* 187–276.

Cosmides, L., & Tooby, J. (1992). Cognitive adaptations for social exchange. In J. Barkow, L. Cosmides, & J. Tooby (Eds.), *The adapted mind: Evolutionary psychology and the generation of culture* (pp. 163–228). New York: Oxford University Press.

Crawford, M. P. (1938). A behavior rating scale for young chimpanzees. *Journal of Comparative Psychology, 26*, 79–91.

Crook, J. H. (1989). Introduction: Socioecological paradigms, evolution, and history: Perspectives for the 1990s. In V. Standen & R. Foley (Eds.), *Comparative socioecology: The behavioural ecology of humans and other mammals* (pp. 1–36). Oxford: Blackwell Scientific Publications.

Crockett, C. M., & Pope, T. (1988). Inferring patterns of aggression from red howler monkey injuries. *American Journal of Primatology, 15*, 289–308.

Cronbach, L. J., & Meehl, P. E. (1955). Construct validity in psychological tests. *Psychological Bulletin, 52*, 281–302.

Cummings, E. M., Iannotti, R. J., & Zahn-Waxler, C. (1989). Aggression between peers in early childhood: Individual continuity and developmental change. *Child Development, 60*, 887–895.

Czaja, J. A., & Bielert, C. (1975). Female rhesus sexual behavior and distance to a male partner: Relation to stage of the menstrual cycle. *Archives of Sexual Behavior, 4*, 583–597.

Dabbs, J. M. J., Frady, R. L., Carr, T. S., & Besch, N. F. (1987). Saliva testosterone and criminal violence in young adult prison inmates. *Psychosomatic Medicine, 49*, 174–182.

Dabelsteen, T., & Pedersen, S. M. (1990). Song and information about aggressive responses of blackbirds, *Turdus merula*: Evidence from interactive playback experiments with territory owners. *Animal Behaviour, 40*, 1158–68.

Daly, M., & Wilson, M. (1982). Who are newborn babies said to resemble? *Ethology and Sociobiology, 3*, 69–78.

Daly, M., & Wilson, M. (1983). *Sex, evolution, and behavior* (2nd ed.). Boston: Willard Grant.

Daly, M., & Wilson, M. (1995). Discriminative parental solicitude and the relevance of evolutionary models to the analysis of motivational systems. In M. Gazzaniga (Ed.), *The cognitive neurosciences* (pp. 1269–86). Cambridge, MA: MIT Press.

Daly, M., & Wilson, M. (1996). Violence against stepchildren. *Current Directions in Psychological Science, 5*, 77–81.

Daly, M., & Wilson, M. (1997). Evolutionary psychology: Adaptationist, selectionist, and comparative. *Psychological Inquiry, 8*, 34–38.

Daly, M., & Wilson, M. (1999). Human evolutionary psychology and animal behaviour. *Animal Behaviour, 57*, 509–519.

Damasio, A. R. (1994). *Descartes' error: Emotion, reason, and the human brain.* New York: Grosset/Putnam.

Damasio, A. R. (1996). The somatic marker hypothesis and the possible functions of the prefrontal cortex. *Philosophical Transactions of the Royal Society of London, B, 351,* 1413–20.

Damasio, A. R. (1998). Emotion in the perspective of an integrated nervous system. *Brain Research Reviews, 26,* 83–86.

Damasio, A. R. (2000). *The feeling of what happens: Body, emotion, and the making of consciousness.* London: Vintage.

Damasio, A. R., & van Hoesen, G. W. (1983). Emotional disturbances associated with focal lesions of the limbic frontal lobe. In K. H. Heilman & P. Satz (Eds.), *Neuropsychology of human emotion* (pp. 85–110). New York: Guilford Press.

D'Amato, F. R., Troisi, A., Scucchi, S., & Fuccillo, R. (1982). Mating season influence on allogrooming in a confined group of Japanese macaques: A quantitative analysis. *Primates, 23,* 220–232.

Darling, N., & Steinberg, L. (1993). Parenting style as context: An integrative model. *Psychological Bulletin, 113,* 487–496.

Darwin, C. (1859). *On the origin of species.* London: Murray.

Darwin, C. (1872). *The expression of the emotions in man and animals.* London: Murray.

Das, M., Penke, Z., & van Hooff, J. A. R. A. M. (1997). Affiliation between aggressors and third parties following conflicts in long-tailed macaques *(Macaca fascicularis). International Journal of Primatology, 18,* 159–181.

Das, M., Penke, Z., & van Hooff, J. A. R. A. M. (1998). Post-conflict affiliation and stress-related behavior of long-tailed macaque aggressors. *International Journal of Primatology, 19,* 53–71.

Dasser, V. (1987). Slides of group members as representations of the real animals *(Macaca fascicularis). Ethology, 76,* 65–75.

Dasser, V. (1988). A social concept in Java monkeys. *Animal Behaviour, 36,* 225–230.

Datta, S. B., & Beauchamp, G. (1991). Effects of group demography on dominance relationships among female primates .1. Mother-daughter and sister-sister relations. *American Naturalist, 138,* 201–226.

Davenport, R. K., Rogers, C. W., & Rumbaugh, D. M. (1973). Long-term cognitive deficits in chimpanzees associated with early impoverished rearing. *Developmental Psychology, 9,* 343–347.

Davidson, R. J. (1995). Cerebral asymmetry, emotion, and affective style. In R. J. Davidson & K. Hugdahl (Eds.), *Brain asymmetry* (pp. 361–387). Cambridge, MA: MIT Press.

Davidson, R. J. (2000). The neuroscience of affective style. In M. S. Gazzaniga (Ed.), *The new cognitive neurosciences* (2nd ed.) (pp. 1149–66). Cambridge, MA: The MIT Press.

Davidson, R. J., & Cacioppo, J. T. (1992). New developments in the scientific study of emotion: An introduction to the special section. *Psychological Science, 3,* 21–22.

Davidson, R. J., Ekman, P., Saron, C. D., Senulis, J. A., & Friesen, W. V. (1990). Approach-withdrawal and cerebral asymmetry: Emotional expression and brain physiology I. *Journal of Personality and Social Psychology, 58,* 330–341.

Davidson, R. J., & Hugdahl, K. (1995). *Brain asymmetry.* Cambridge, MA: MIT Press.

Davidson, R. J., Putnam, K. M., & Larson, C. L. (2000). Dysfunction in the neural circuitry of emotion regulation—a possible prelude to violence. *Science, 289,* 591–594.

Davidson, R. J., & Sutton, S. K. (1995). Affective neuroscience: The emergence of a discipline. *Current Opinion in Neurobiology, 5,* 217–224.

Davis, M. (1984). The mammalian startle reflex. In R. C. Eaton (Ed.), *Neural mechanisms of startle behavior* (pp. 287–351). New York: Plenum Press.

Davis, M., & Whalen, P. J. (2001). The amygdala: Vigilance and emotion. *Molecular Psychiatry, 6,* 13–34.

Dawkins, R., & Krebs, J. R. (1978). Animal signals: Information or manipulation? In J. R. Krebs & N. B. Davies (Eds.), *Behavioural ecology* (pp. 282–309). Oxford: Blackwell Scientific.

Deacon, T. W. (1989). The neural circuitry underlying primate calls and human language. *Human Evolution, 4,* 367–401.

Deacon, T. W. (1990). Problems of ontogeny and phylogeny in brain-size evolution. *International Journal of Primatology, 11,* 237–282.

Deacon, T. W. (1997). *The symbolic species: The co-evolution of language and the brain.* New York: W. W. Norton.

Deag, J. M., & Crook, J. H. (1971). Social behaviour and "agonistic buffering" in the wild Barbary macaque, *Macaca sylvana L. Folia Primatologica, 15,* 183–200.

De Lacoste, M. C., & Woodward, D. J. (1988). The corpus callosum in non-human primates: Determinants of size. *Brain, Behavior, and Evolution, 31,* 318–323.

DeLozier, P. P. (1982). Attachment theory and child abuse. In C. M. Parkes & J. Stevenson-Hinde (Eds.), *The place of attachment in human behavior* (pp. 95–117). New York: Basic Books.

Dennerstein, L., Burrows, G. D., Wood, C., & Hyman, G. (1980). Hormones and sexuality: Effect of estrogen and progestogen. *Obstetrics and Gynecology, 56,* 316–322.

Dennerstein, L., Gotts, G., Brown, J. B., Morse, C. A., Farley, T. M. M., & Pinol, A. (1994). The relationship between the menstrual cycle and female sexual interest in women with PMS complaints and volunteers. *Psychoneuroendocrinology, 19,* 293–304.

Dennerstein, L., Wood, C., & Burrows, G. D. (1977). Sexual response following hysterectomy and oophorectomy. *Obstetrics and Gynecology, 49,* 92–96.

Desimone, R. (1991). Face-selective cells in the temporal cortex of monkeys. *Journal of Cognitive Neuroscience, 3,* 1–8.

Desmond, A., & Moore, J. (1991). *Darwin.* London: Michael Joseph.

Desrochers, S., Morisette, P., & Ricard, M. (1995). Two perspectives on pointing in infancy. In C. Moore & P. J. Dunham (Eds.), *Joint attention: Its origins and role in development* (pp. 85–101). Hillsdale, NJ: Erlbaum.

Deutsch, J., & Larsson, K. (1974). Model-oriented sexual behavior in surrogate-reared rhesus monkeys. *Brain, Behavior, and Evolution, 9,* 157–164.

DeVinney, B. J., Berman, C. M., & Rasmussen, K. L. R. (2001). Changes in yearling rhesus monkey's relationships with their mothers after sibling birth. *American Journal of Primatology, 54,* 193–210.

DeVore, I. (1963). Mother-infant relations in free-ranging baboons. In H. L. Rheingold (Ed.), *Maternal behavior in mammals* (pp. 305–335). New York: John Wiley & Sons.

de Waal, F. B. M. (1982). *Chimpanzee politics.* New York: Harper & Row.

de Waal, F. B. M. (1984). Coping with social tension: Sex differences in the effect of food provision to small rhesus monkey groups. *Animal Behaviour, 32,* 765–773.

de Waal, F. B. M. (1986a). The integration of dominance and social bonding in primates. *Quarterly Review of Biology, 61,* 459–479.

de Waal, F. B. M. (1986b). Class structure in a rhesus monkey group: The interplay between dominance and tolerance. *Animal Behaviour, 34,* 1033–40.

de Waal, F. B. M. (1987). Tension regulation and nonreproductive functions of sex in captive bonobos *(Pan paniscus). National Geographic Research, 3,* 318–335.

de Waal, F. B. M. (1988). The communicative repertoire of captive bonobos *(Pan paniscus)* compared to that of chimpanzees. *Behaviour, 106,* 183–251.

de Waal, F. B. M. (1989a). *Peacemaking among primates.* Cambridge, MA: Harvard University Press.

de Waal, F. B. M. (1989b). Dominance "style" and primate social organization. In V. Standen & R. A. Foley (Eds.), *Comparative socioecology: The behavioural ecology of humans and other mammals* (pp. 243–263). Oxford: Blackwell Scientific.

de Waal, F. B. M. (1989c). The myth of a simple relation between space and aggression in captive primates. *Zoo Biology Supplement, 1,* 141–148.

de Waal, F. B. M. (1990). Do rhesus mothers suggest friends to their offspring? *Primates, 31,* 597–600.

de Waal, F. B. M. (1992). Appeasement, celebration, and food sharing in the two *Pan* species. In T. Nishida, W. C. McGrew, P. Marler, M. Pickford, & F. B. M. de Waal (Eds.), *Topics in primatology.* Vol. 1, *Human origins* (pp. 37–50). Tokyo: University of Tokyo Press.

de Waal, F. B. M. (1996a). Conflict as negotiation. In W. C. McGrew, L. F. Marchant, & T. Nishida (Eds.), *Great ape societies* (pp. 159–172). Cambridge: Cambridge University Press.

de Waal, F. B. M. (1996b). *Good natured: The origins of right and wrong in humans and other animals.* Cambridge, MA: Harvard University Press.

de Waal, F. B. M. (1996c). Macaque social culture: Development and perpetuation of affiliative networks. *Journal of Comparative Psychology, 110,* 147–154.

de Waal, F. B. M. (1997). The chimpanzee's service economy: Food for grooming. *Evolution and Human Behavior, 18,* 375–386.

de Waal, F. B. M., & Aureli, F. (1996). Consolation, reconciliation, and a possible cognitive difference between macaques and chimpanzees. In A. E. Russon, K. A. Bard, & S. T. Parker (Eds.), *Reaching into thought: The minds of the great apes* (pp. 80–110). Cambridge: Cambridge University Press.

de Waal, F. B. M., & Johanowicz, D. L. (1993). Modification of reconciliation behavior through social experience: An experiment with two macaque species. *Child Development, 64,* 897–908.

de Waal, F. B. M. & Luttrell, L. M. (1985). The formal hierarchy of rhesus monkeys: An investigation of the bared-teeth display. *American Journal of Primatology, 9,* 73–85.

de Waal, F. B. M., & Luttrell, L. M. (1986). The similarity principle underlying social bonding among female rhesus monkeys. *Folia Primatologica, 46,* 215–234.

de Waal, F. B. M., & Luttrell, L. (1988). Mechanisms of social reciprocity in three primate species: Symmetrical relationship characteristics or cognition? *Ethology and Sociobiology, 9,* 101–118.

de Waal, F. B. M., & Ren, R. (1988). Comparison of the reconciliation behavior of stumptail and rhesus macaques. *Ethology, 78,* 129–142.

de Waal, F. B. M., & Seres, M. (1997). Propagation of handclasp grooming among captive chimpanzees. *American Journal of Primatology, 43,* 339–346.

de Waal, F. B. M., & van Hooff, J. A. R. A. M. (1981). Side-directed communication and agonistic interactions in chimpanzees. *Behaviour, 77,* 164–198.

de Waal, F. B. M., van Hooff, J. A. R. A. M., & Netto, W. J. (1976). An ethological analysis of types of agonistic interaction in a captive group of Java monkeys *(Macaca fascicularis). Primates, 17,* 257–290.

de Waal, F. B. M., & van Roosmalen, A. (1979). Reconciliation and consolation among chimpanzees. *Behavioral Ecology and Socioecology, 5,* 55–66.

Diamond, M., & Sigmundson, H. K. (1997). Sex reassignment at birth: Long-term review and clinical implications. *Archives of Pediatrics and Adolescent Medicine, 151,* 298–304.

Diamond, R., & Carey, S. (1986). Why faces are and are not special: An effect of expertise. *Journal of Experimental Psychology, 115,* 107–117.

Di Bitetti, M. S. (2000). The distribution of grooming among female primates: Testing hypotheses with the Shannon-Wiener diversity index. *Behaviour, 137,* 1517–40.

Diezinger, F., & Anderson, J. R. (1986). Starting from scratch: A first look at a "displacement activity" in group-living rhesus monkeys. *American Journal of Primatology, 11,* 117–124.

Di Fiore, A., & Rendall, D. (1994). Evolution of social organization: A reappraisal for primates by using phylogenetic methods. *Proceedings of the National Academy of Science, USA, 91,* 9941–45.

Dimberg, U., & Öhman, A. (1996). Behold the wrath: Psychophysiological responses to facial stimuli. *Motivation and Emotion, 20,* 149–182.

Dittus, W. P. J. (1979). The evolution of behaviors regulating density and age-specific sex ratios in primate population. *Behaviour, 69,* 265–302.

Dittus, W. P. J., & Ratnayeke, S. M. (1989). Individual and social behavioral responses to injury in wild toque macaques *(Macaca sinica). International Journal of Primatology, 10,* 215–234.

Dixson, A. F. (1977). Observations on the displays, menstrual cycles, and sexual behaviour of the "Black Ape" of Celebes *(Macaca nigra). Journal of Zoology, 182,* 63–84.

Dixson, A. F. (1986). Plasma testosterone concentrations during postnatal development in the male common marmoset. *Folia Primatologica, 47,* 166–170.

Dixson, A. F. (1987). Effects of adrenalectomy upon proceptivity, receptivity, and sexual attractiveness in ovariectomized marmosets *(Callithrix jacchus). Physiology and Behavior, 39,* 495–499.

Dixson, A. F., & Nevison, C. M. (1997). The socioendocrinology of adolescent development in male rhesus monkeys *(Macaca mulatta). Hormones and Behavior, 31,* 126–135.

Dolan, R. J., & Morris, J. S. (2000). The functional anatomy of innate and acquired fear: Perspectives from neuroimaging. In R. D. Lane & L. Nadel (Eds.), *Cognitive neuroscience of emotion* (pp. 225–241). New York: Oxford University Press.

Dolhinow, P. (1977). Normal monkeys? *American Scientist, 65,* 266.

Dolhinow, P., & Fuentes, A. (Eds.). (1999). *The nonhuman primates.* Mountain View, CA: Mayfield.

Domjan, M. (1998). *Principles of learning and behavior* (4th ed.). New York: Brooks-Cole.

Donald, M. (1991). *Origins of the modern mind.* Cambridge, MA: Harvard University Press.

Donald, M. (2000). The central role of culture in cognitive evolution: A reflection on the myth of the "Isolated Mind." In L. P. Nucci, G. B. Saxe, & E. Turiel (Eds.), *Culture, thought, and development* (pp. 19–38). Mahwah, NJ: Erlbaum.

Doudet, D., Hommer, D., Higley, J. D., Andreason, P. J., Moneman, R., Suomi, S. J., & Linnoila, M. (1995). Cerebral glucose metabolism, CSF 5-HIAA levels, and aggressive behavior in rhesus monkeys. *American Journal of Psychiatry, 152,* 1782–87.

Dovidio, J. F., Brown, C. E., Heltman, K., Ellyson, S. L., & Keating, C. F. (1988). Power displays between women and men in discussions of gender-linked tasks: A multi-channel study. *Journal of Personality and Social Psychology, 55,* 580–587.

Dow, M. G. T., Hart, D. M., & Forrest, C. A. (1983). Hormonal treatments of sexual unresponsiveness in postmenopausal women: A comparative study. *British Journal of Obstetrics and Gynaecology, 90,* 361–366.

Drescher, V. M., Gantt, W. H., & Whitehead, W. E. (1980). Heart rate response to touch. *Psychosomatic Medicine, 42,* 559–565.

Drickamer, L. C. (1974). A ten-year summary of reproductive data for free-ranging *Macaca mulatta. Folia Primatologica, 21,* 61–80.

Drickamer, L. C. (1976). Quantitative observations of grooming behavior in free-ranging *Macaca mulatta. Primates, 17,* 323–335.

Driesen, N. R., & Raz, N. (1995). The influence of sex, age, and handedness on corpus callosum morphology: A meta-analysis. *Psychobiology, 23,* 240–247.

Driver, J., Davis, G., Ricciardelli, P., Kidd, P., Maxwell, E., & Baron-Cohen, S. (1999). Gaze perception triggers reflexive visuospatial orienting. *Visual Cognition, 6,* 509–540.

Dugatkin, L. A. (1997). *Cooperation among animals: An evolutionary perspective.* Oxford: Oxford University Press.

Dum, J., & Herz, A. (1987). Opioids and motivation. *Interdisciplinary Science Reviews, 12,* 180–190.

Dunbar, R. I. M. (1984). *Reproductive decisions. An economic analysis of gelada baboon social strategies.* Princeton: Princeton University Press.

Dunbar, R. I. M. (1991). Functional significance of social grooming in primates. *Folia Primatologica, 57,* 121–131.

Dunbar, R. I. M. (1993). Coevolution of neocortical size, group size, and language in humans. *Behavioral and Brain Sciences, 16,* 681–735.

Dunbar, R. I. M. (2000). Causal reasoning, mental rehearsal, and the evolution of primate cognition. In C. Heyes & L. Huber (Eds.), *The evolution of cognition* (pp. 205–219). Cambridge, MA: MIT Press.

Dunbar, R. I. M., & Dunbar, P. (1988). Maternal time budgets of gelada baboons. *Animal Behaviour, 36,* 970–980.

Dunbar, R. I. M., Marriott, A., & Duncan, N. D. C. (1997). Human conversational behavior. *Human Nature, 8,* 231–246.

Dunbar, R. I. M., & Sharman, M. (1984). Is social grooming altruistic? *Zeitschrift für Tierpsychologie, 64,* 163–173.

Dunn, J., & Herrera, C. (1997). Conflict resolution with friends, siblings, and mothers: A developmental perspective. *Aggressive Behavior, 23,* 343–357.

Dunn, K. F., & Cowan, G. (1993). Social influence strategies among Japanese and American college women. *Psychology of Women Quarterly, 17,* 39–52.

Dutton, D. M., Clark, R. A., & Dickins, D. W. (1997). Personality in captive chimpanzees: Use of a novel rating procedure. *International Journal of Primatology, 18,* 539–552.

Eaton, G. G. (1976). The social order of Japanese macaques. *Scientific American, 235,* 96–105.

Eaton, G. G., Johnson, D. F., Glick, B. B., & Worlein, J. M. (1985). Development in Japanese macaques *(Macaca fuscata):* Sexually dimorphic behavior during the first year of life. *Primates, 26,* 238–248.

Eaton, G. G., Johnson, D. F., Glick, B. B., & Worlein, J. M. (1986). Japanese macaques *(Macaca fuscata)* social development: Sex differences in juvenile behavior. *Primates, 27,* 141–150.

Eaton, G. G., & Resko, J. A. (1974). Plasma testosterone and male dominance in a Japanese macaque *(Macaca fuscata)* troop compared with repeated measures of testosterone in laboratory males. *Hormones and Behavior, 5,* 251–259.

Eaton, G. G., Worlein, J. M., & Glick, B. B. (1990). Sex differences in Japanese macaques *(Macaca fuscata):* Effects of prenatal testosterone on juvenile social behavior. *Hormones and Behavior, 24,* 270–283.

Eaton, R. C. (Ed.). (1984). *Neural mechanisms of startle behavior.* New York: Plenum Press.

Eberhart, J. A., Yodyingyuad, U., & Keverne, E. B. (1985). Subordination in male talapoin monkeys lowers sexual behaviour in the absence of dominants. *Physiology and Behavior, 35,* 673–677.

Eddy, T. J., Gallup, G. G. Jr., & Povinelli, D. J. (1993). Attribution of cognitive states to animals: Anthropomorphism in comparative perspective. *Journal of Social Issues, 49,* 87–101.

Edmonson, M. S. (1987). Notes on laughter. *Anthropological Linguistics, 29,* 23–34.

Ehardt, C. L., & Bernstein, I. S. (1987). Patterns of affiliation among immature rhesus monkeys *(Macaca mulatta). American Journal of Primatology, 13,* 255–269.

Ehardt, C. L., & Bernstein, I. S. (1992). Intervention behavior by adult macaques: Structural and functional aspects. In A. H. Harcourt & F. B. M. de Waal (Eds.), *Coalitions and alliances in humans and other animals* (pp. 83–111). Oxford: Oxford University Press.

Ehrenkranz, J., Bliss, E., & Sheard, M. H. (1974). Plasma testosterone: Correlation with aggressive behavior and social dominance in man. *Psychosomatic Medicine, 36,* 469–475.

Ehrhardt, A. A., Epstein, R., & Money, J. (1968). Fetal androgens and female gender identity in the early treated adrenogenital syndrome. *Johns Hopkins Medical Journal, 122,* 160–167.

Eibl-Eibesfeldt, I. (1989). *Human ethology.* New York: Aldine.

Eisenberg, A. R. (1992). Conflicts between mothers and their young children. *Merrill-Palmer Quarterly, 38,* 21–41.

Eisenberg, A. R., & Garvey, C. (1981). Children's use of verbal strategies in resolving conflicts. *Discourse Processes, 4,* 149–170.

Eisenberg, N. (1986). *Altruistic emotion, cognition, and behavior.* Hillsdale, NJ: Erlbaum.

Ekman, P. (1973). *Darwin and facial expressions.* New York: Academic Press.

Ekman, P. (1984). Expression and the nature of emotion. In K. R. Scherer & P. Ekman (Eds.), *Approaches to emotion* (pp. 329–343). Hillsdale, NJ: Erlbaum.

Ekman, P. (1992). An argument for basic emotions. *Cognition and Emotion, 6,* 169–200.

Ekman, P. (1993). Facial expression and emotion. *American Psychologist, 48,* 384–392.

Ekman, P. (Ed.). (1998). *The expression of the emotions in man and animals* (2nd ed.). New York: Oxford University Press.

Ekman, P., & Davidson, R. J. (1994). *The nature of emotion: Fundamental questions.* New York: Oxford University Press.

Ekman, P., & Friesen, W. V. (1969). The repertoire of nonverbal behavior: Categories, origins, usage, and coding. *Semiotica, 1,* 49–98.

Ekman, P., & Friesen, W. V. (1975). *Unmasking the face.* Englewood Cliffs, NJ: Prentice-Hall.

Ekman, P., & Friesen, W. V. (1978). *Facial Action Coding System (FACS): A technique for the measurement of facial action.* Palo Alto: Consulting Psychologists Press.

Ekman, P., Levenson, R. W., & Friesen, W. V. (1983). Autonomic nervous system activity distinguishes among emotions. *Science, 221,* 1208–10.

Ekman, P., & Oster, H. (1979). Facial expressions of emotion. *Annual Review of Psychology, 30,* 527–554.

Elder, G. H. (1969). Appearance and education in marriage mobility. *American Sociological Review, 34,* 519–533.

Ellis, A. W. (1992). Cognitive mechanisms of face processing. *Philosophical Transactions of the Royal Society of London, B, 335,* 113–119.

Emery, N. J. (2000). The eyes have it: The neuroethology, function, and evolution of social gaze. *Neuroscience and Biobehavioral Reviews, 24,* 581–604.

Emery, N. J., Lorincz, E. N., Perrett, D. I., Oram, M. W., & Baker, C. I. (1997). Gaze following and joint attention in rhesus monkeys *(Macaca mulatta). Journal of Comparative Psychology, 111,* 286–293.

Emory, G. R., & Harris, S. J. (1978). On the directional orientation of female presents in *Macaca fascicularis. Primates, 19,* 227–229.

Epstein, R., Lanza, R. P., & Skinner, B. F. (1980). Symbolic communication between two pigeons. *Science, 207,* 543–545.

Epstein, S. (1979). The stability of behavior. I: On predicting most of the people much of the time. *Journal of Personality and Social Psychology, 37*, 1097–1126.

Epstein, S. (1983). Aggregation and beyond: Some basic issues on the prediction of behavior. *Journal of Personality, 51*, 360–392.

Erickson, K., Lindell, S., Champoux, M., Gold, P., Schulkin, J., Suomi, S. J., & Higley, J. D. (2001). Relationships between behavior and neurochemical changes in rhesus macaques during a separation paradigm. Paper presented at the annual meeting of the Society for Neuroscience.

Eron, L. D. (1987). The development of aggressive behavior from the perspective of a developing behaviorism. *American Psychologist, 42*, 435–442.

Erwin, J. (1977). Factors influencing aggressive behavior and risk of trauma in the pigtail macaque *(Macaca nemestrina). Laboratory Animal Science, 27*, 541–547.

Erwin, J., Brandt, E. M., & Mitchell, G. (1973). Attachment formation and separation in heterosexually naïve preadult rhesus monkeys *(Macaca mulatta). Developmental Psychobiology, 6*, 531–538.

Erwin, J., & Deni, R. (1979). Strangers in a strange land: Abnormal behaviors or abnormal environments? In J. Erwin, T. L. Maple, & G. Mitchell (Eds.), *Captivity and behavior: Primates in breeding colonies, laboratories, and zoos* (pp. 1–28). New York: Van Nostrand Reinhold.

Erwin, J., & Flett, M. (1974). Responses of rhesus monkeys to reunion after long-term separation. *Psychological Reports, 35*, 171–174.

Erwin, J., & Maple, T. (1976). Ambisexual behavior with male-male anal penetration in male rhesus monkeys. *Archives of Sexual Behavior, 5*, 9–14.

Erwin, J., & Mitchell, G. (1975). Initial heterosexual behavior of adolescent rhesus monkeys *(Macaca mulatta). Archives of Sexual Behavior, 4*, 97–104.

Essock, S., Gill, T. V., & Rumbaugh, D. M. (1977). Language relevant object- and color-naming tasks. In D. M. Rumbaugh (Ed.), *Language learning by a chimpanzee* (pp. 193–206). New York: Academic Press.

Essock-Vitale, S. M., & McGuire, M. T. (1985). Women's lives viewed from evolutionary perspective. II: Patterns of helping. *Ethology and Sociobiology, 6*, 155–173.

Ettlinger, G. (1988). Hand preference, ability, and hemispheric specialization: How far are these factors related in the monkey ? *Cortex, 24*, 389–398.

Evans, A., & Tomasello, M. (1986). Evidence for social referencing in young chimpanzees *(Pan troglodytes). Folia Primatologica, 47*, 49–54.

Evans, C. S. (1997). Referential signals. In D. H. Owings, M. D. Beecher, & N. S. Thompson (Eds.), *Perspectives in ethology,* Vol. 12, *Communication* (pp. 99–143). New York: Plenum Press.

Evans, S., & Hodges, J. K. (1984). Reproductive status of adult daughters in family groups of common marmosets *(Callithrix jacchus jacchus)*. *Folia Primatologica, 42,* 127–133.

Everitt, B. J., & Herbert, J. (1969). Adrenal glands and sexual receptivity in female rhesus monkeys. *Nature, 222,* 1065–66.

Everitt, B. J., & Herbert, J. (1971). The effects of dexamethasone and androgens on sexual receptivity of female rhesus monkeys. *Journal of Endocrinology, 51,* 575–588.

Everitt, B. J., Herbert, J., & Hamer, J. D. (1972). Sexual receptivity of bilaterally adrenalectomized female rhesus monkeys. *Physiology and Behavior, 8,* 409–415.

Eysenck, H. J. (1967). *The biological basis of personality.* Springfield, IL: Charles C. Thomas.

Fabre-Nys, C., Meller, R. E., & Keverne, E. B. (1982). Opiate antagonists stimulate affiliative behaviour in monkeys. *Pharmacology Biochemistry and Behavior, 16,* 653–659.

Fady, J. C. (1976). Social play: The choice of playmates observed in the young of crab-eating macaques. In J. S. Bruner, A. Jolly, & K. Silva (Eds.), *Play: Its role in development and evolution* (pp. 328–335). New York: Basic Books.

Fagot, J., & Vauclair, J. (1991). Manual laterality in nonhuman primates: A distinction between handedness and manual specialization. *Psychological Bulletin, 109,* 76–89.

Fairbanks, L. A. (1980). Relationships among adult females in captive vervet monkeys: Testing a model of rank-related attactiveness. *Animal Behaviour, 28,* 853–859.

Fairbanks, L. A. (1988a). Mother-infant behavior in vervet monkeys: Response to failure of last pregnancy. *Behavioral Ecology and Sociobiology, 23,* 157–165.

Fairbanks, L. A. (1988b). Vervet monkey grandmothers: Effects on mother-infant relationships. *Behaviour, 104,* 176–188.

Fairbanks, L. A. (1988c). Vervet monkey grandmothers: Interactions with infant grandoffspring. *International Journal of Primatology, 9,* 425–441.

Fairbanks, L. A. (1989). Early experience and cross-generational continuity of mother-infant contact in vervet monkeys. *Developmental Psychobiology, 22,* 669–681.

Fairbanks, L. A. (1990). Reciprocal benefits of allomothering for female vervet monkeys. *Animal Behaviour, 40,* 553–562.

Fairbanks, L. A. (1993). Risk-taking by juvenile vervet monkeys. *Behaviour, 124,* 57–72.

Fairbanks, L. A. (1996a). Individual differences in maternal styles: Causes and consequences for mothers and offspring. *Advances in the Study of Behavior, 25,* 579–611.

Fairbanks, L. A. (1996b). Adolescent pregnancy in vervet monkeys. Paper presented at the Sixteenth Congress of the International Primatological Society, Madison, WI.

Fairbanks, L. A. (2000a). Maternal investment throughout the life span in Old World monkeys. In P. F. Whitehead & C. J. Jolly (Eds.), *Old World monkeys* (pp. 341–367). Cambridge: Cambridge University Press.

Fairbanks, L. A. (2000b). The developmental timing of primate play: A neural selection model. In S. T. Parker, J. Langer, & M. McKinney (Eds.), *Biology, brains, and behavior* (pp. 131–158). Santa Fe, NM: SAR Press.

Fairbanks, L. A. (2000c). Dominance in vervets: Social, reproductive, and developmental correlates of rank. *American Journal of Primatology, 51,* 34.

Fairbanks, L. A., Fontenot, M. B., Phillips-Conroy, J. E., Jolly, C. J., Kaplan, J. R., & Mann, J. J. (1999). CSF monoamines, age, and impulsivity in wild grivet monkeys *(Cercopithecus aethiops aethiops). Brain, Behavior, and Evolution, 53,* 305–312.

Fairbanks, L. A., & McGuire, M. T. (1984). Determinants of fecundity and reproductive success in captive vervets. *American Journal of Primatology, 7,* 27–38.

Fairbanks, L. A., & McGuire, M. T. (1985). Relationships of vervet mothers with sons and daughters from one through three years of age. *Animal Behaviour, 33,* 40–50.

Fairbanks, L. A., & McGuire, M. T. (1986). Age, reproductive value, and dominance-related behaviour in vervet monkey females: Cross-generational influences on social relationships and reproduction. *Animal Behaviour, 34,* 1710–21.

Fairbanks, L. A., & McGuire, M. T. (1987). Mother-infant relationships in vervet monkeys: Response to new adult males. *International Journal of Primatology, 8,* 351–366.

Fairbanks, L. A., & McGuire, M. T. (1988). Long term effects of early mothering behavior on responsiveness to the environment in vervet monkeys. *Developmental Psychobiology, 21,* 711–724.

Fairbanks, L. A., & McGuire, M. T. (1993). Maternal protectiveness and response to the unfamiliar in vervet monkeys. *American Journal of Primatology, 30,* 119–129.

Fairbanks, L. A., & McGuire, M. T. (1995). Maternal condition and the quality of maternal care in vervet monkeys. *Behaviour, 132,* 733–754.

Fairbanks, L. A., McGuire, M. T., & Harris, C. L. (1982). Nohverbal interaction of patients and therapists during psychiatric interviews. *Journal of Abnormal Psychology, 91,* 109–119.

Fairbanks, L. A., Melega, W. P., Jorgensen, M. J., Kaplan, J. R., & McGuire, M. T. (2001). Social impulsivity inversely associated with CSF 5-HIAA and fluoxetine exposure in vervet monkeys. *Neuropsychopharmacology, 24,* 370–378.

Falk, D. (1978). Cerebral asymmetry in Old World monkeys. *Acta Anatomica, 101,* 334–339.

Falk, D. (1986). Endocranial casts and their significance for primate brain evolution. In D. R. Swindler and J. Erwin (Eds.), *Comparative primate biology.* Vol. 1, *Systematics, evolution, and anatomy* (pp. 477–490). New York: Alan R. Liss.

Falk, D. (1987). Hominid paleoneurology. *Annual Review of Anthropology, 16,* 13–30.

Falk, D., Cheverud, J., Vannier, M. W., & Conroy, G. C. (1986). Advanced computer graphics technology reveals cortical asymmetry in endocasts of rhesus monkeys. *Folia Primatologica, 46,* 98–103.

Falk, D., Hildebolt, C., Cheverud, J., Vannier, M., Helmkamp, C., & Konigsberg, L.

(1990). Cortical asymmetries in frontal lobes of rhesus monkeys *(Macaca mulatta)*. *Brain Research, 512,* 40–45.

Fantz, R. L. (1963). Pattern vision in newborn infants. *Science, 140,* 296–297

Farah, M. J., Tanaka, J. W., & Drain, H. M. (1995). What causes the face inversion effect? *Journal of Experimental Psychology 21,* 628–634.

Fausto-Sterling, A. (2000). *Sexing the body: Gender politics and the construction of sexuality.* New York: Basic Books.

Fawcett, K., & Muhumuza, G. (2000). Death of a wild chimpanzee community member: Possible outcome of intense sexual competition. *American Journal of Primatology, 51,* 243–247.

Fedigan, L. M. (1976). A study of roles in the Arashiyama monkeys *(Macaca fuscata).* In *Contributions to Primatology.* Vol. 9. Basel: Karger.

Feeney, J. A. (1998). Adult attachment and relationship-centered anxiety. In J. A. Simpson & W. S. Rholes (Eds.), *Attachment theory and close relationships* (pp. 189–218). New York: Guilford Press.

Fernald, A. (1991). Prosody in speech to children: Prelinguistic and linguistic functions. *Annals of Child development, 8,* 43–80.

Ferrari, P. F., Kohler, E., Fogassi, L., & Gallese, V. (2000). The ability to follow eye gaze and its emergence during development in macaque monkeys. *Proceedings of the National Academy of Sciences, USA, 97,* 13997–14002.

Fichtel, C., Hammerschmidt, K., & Jürgens, U. (2001). On the vocal expression of emotion: A multi-parametric analysis of different states of aversion in the squirrel monkey. *Behaviour, 138,* 97–116.

Field, T. (1985). Attachment as psychobiological attunement: Being on the same wavelength. In M. Reite & T. Field (Eds.), *The psychobiology of attachment and separation* (pp. 415–454). New York: Academic Press.

Field, T. (1996). Attachment and separation in young children. *Annual Review of Psychology, 47,* 541–561.

Figueredo, A. J., Cox, R. L., & Rhine, R. J. (1995). A generalizability analysis of subjective personality assessments in the stumptail macaque and the zebra finch. *Multivariate Behavioral Research, 30,* 167–197.

Filler, W., & Drezner, N. D. (1944). The results of surgical castration in women under forty. *American Journal of Obstetrics and Gynecology, 47,* 122–124.

Finkelstein, J. W., Susman, E. J., Chinchilli, V. M., D'Arcangelo, M. R., Kunselman, S. J., Schwab, J., Demers, L. M., Liben, L. S., & Kulin, H. E. (1998). Effects of estrogen or testosterone on self-reported sexual responses and behaviors in hypogonadal adolescents. *Journal of Clinical Endocrinology and Metabolism, 83,* 2281–85.

Fitch, W. T. (1997). Vocal tract length and formant frequency dispersion correlate with body size in rhesus monkeys. *Journal of the Acoustical Society of America, 102,* 1213–22.

Flack, J. C., & de Waal, F. B. M. (2000a). "Any animal whatever": Darwinian building blocks of morality in monkeys and apes. *Journal of Consciousness Studies, 7,* 1–29.

Flack, J. C., & de Waal, F. B. M. (2000b). Being nice is not a building block of morality (response to commentary discussion). In L. D. Katz (Ed.), *Evolutionary origins of morality* (pp. 67–77). Thorverton, UK: Imprint Academic.

Flavell, J. H. (1999). Cognitive development: Children's knowledge about the mind. *Annual Review of Psychology, 50,* 21–45.

Fonzi, A., Schneider, B. H., Tani, F., & Tomada, G. (1997). Predicting children's friendship status from their dyadic interaction in structured situations of potential conflict. *Child Development, 68,* 496–506.

Ford, C. V. (1984). *The somatizing disorders.* New York: Elsevier.

Foss, G. L. (1951). The influence of androgens on sexuality in women. *Lancet,* 667–669.

Foster, D. L. (1977). Luteinizing hormone and progesterone secretion during sexual maturation of the rhesus monkey: Short luteal phases during the initial menstrual cycles. *Biology of Reproduction, 17,* 584–590.

Foundas, A. L., Eure, K. F., Luevano, L. F., & Weinberger, D. R. (1998). MRI asymmetries of Broca's area: The pars triangularis and pars opercularis. *Brain and Language, 64,* 282–296.

Foundas, A. L., Faulhaber, J. R., Kulynych, J. J., Browning, C. A., & Weinberger, D. R. (1999). Hemispheric and sex-linked differences in Sylvian fissure morphology: A quantitative approach using volumetric magnetic resonance imaging. *Neuropsychiatry, Neuropsychology, and Behavioral Neurology, 12,* 1–10.

Foundas, A. L., Leonard, C. M., & Heilman, K. M. (1995). Morphologic cerebral asymmetries and handedness: The pars triangularis and planum temporale. *Archives of Neurology, 52,* 501–508.

Fouts, R. S. (1973). Acquisition and testing of gestural signs in four young chimpanzees. *Science, 180,* 978–980.

Fouts, R. S., Chowin, B., & Goodin, L. (1976). Transfer of signed responses in American Sign Language from vocal English stimuli to physical object stimuli by a chimpanzee *(Pan). Learning and Motivation, 7,* 458–475.

Fouts, R. S., & Fouts, D. H. (1989). Loulis in conversation with cross-fostered chimpanzees. In R. A. Gardner, B. T. Gardner, & T. E. van Cantfort (Eds.), *Teaching sign language to chimpanzees* (pp. 293–307). Albany: SUNY Press.

Fouts, R. S., Fouts, D. H., & van Cantfort, T. E. (1989). The infant Loulis learns signs from cross-fostered chimpanzees. In R. A. Gardner, B. T. Gardner, & T. E. van

Cantfort (Eds.), *Teaching sign language to chimpanzees* (pp. 280–292). Albany: SUNY Press.

Franco, F., & Butterworth, G. (1996). Pointing and social awareness: Declaring and requesting in the second year. *Journal of Child Language, 23,* 307–336.

Frank, M. G., Ekman, P., & Friesen, W. V. (1993). Behavioral markers and recognizability of the smile of enjoyment. *Journal of Personality and Social Psychology, 64,* 83–93.

Frank, R. H. (1988). *Passions within reason: The strategic role of the emotions.* New York: W. W. Norton.

Fredrikson, M. (1989). Psychophysiological and biochemical indices in "stress" research: Applications to psychopathology and pathophysiology. In G. Turpin (Ed.), *Handbook of clinical psychophysiology* (pp. 241–279). New York: John Wiley & Sons.

Freedman, L. Z., & Rosvold, H. E. (1962). Sexual, aggressive, and anxious behavior in the laboratory macaque. *Journal of Nervous and Mental Disease, 134,* 18–27.

French, J. A. (1981). Individual differences in play in *Macaca fuscata:* The role of maternal status and proximity. *International Journal of Primatology, 2,* 237–246.

French, J. A., & Inglett, B. J. (1991). Responses to novel social stimuli in callitrichid monkeys: a comparative perspective. In H. O. Box (Ed.), *Primate responses to environmental change* (pp. 275–294). London: Chapman and Hall.

Fridlund, A. J. (1991). Evolution and facial action in reflex, social motive, and paralanguage. *Biological Psychology, 32,* 3–100.

Fridlund, A. J. (1994). *Human facial expression: An evolutionary view.* New York: Academic Press.

Friedman, E. M., Boinski, S., & Coe, C. L. (1995). Interleukin-1 induces sleep-like behavior and alters call structure in juvenile rhesus macaques. *American Journal of Primatology, 35,* 143–153.

Friedman, M. P., Reed, S. K., & Carterette, E. C. (1971). Feature saliency and recognition memory for schematic faces. *Perceptual Psychophysics, 10,* 47–50.

Frijda, N. H. (1986). *The emotions.* New York: Cambridge University Press.

Frijda, N. H. (1993). The place of appraisal in emotion. *Cognition and Emotion, 7,* 357–387.

Frontera, J. G. (1958). Evaluation of the immediate effects of some fixatives upon the measurements of the brains of macaques. *Journal of Comparative Neurology, 109,* 417–438.

Fry, D. P. (2000). Conflict management in cross-cultural perspective. In F. Aureli & F. B. M. de Waal (Eds.), *Natural conflict resolution* (pp. 334–351). Berkeley: University of California Press.

Fry, D. P., & Bjorkqvist, K. (Eds.). (1997). *Cultural variation in conflict resolution: Alternatives to violence.* Mahwah, NJ: Erlbaum.

Fry, D. P., & Fry, C. B. (1997). Culture and conflict-resolution models: Exploring alternatives to violence. In D. P. Fry & K. Bjorkqvist (Eds.), *Cultural variation in conflict resolution: Alternatives to violence* (pp. 9–23). Mahwah, NJ: Erlbaum.

Funder, D. C. (1987). Errors and mistakes: Evaluating the accuracy of social judgment. *Psychological Bulletin, 101,* 75–90.

Funder, D. C., & Colvin, C. R. (1988). Friends and strangers: Acquaintanceship, agreement, and the accuracy of personality judgment. *Journal of Personality and Social Psychology, 55,* 149–158.

Funder, D. C., & Dobroth, K. M. (1987). Differences between traits: Properties associated with interjudge agreement. *Journal of Personality and Social Psychology, 52,* 409–418.

Funder, D. C., Kolar, D. C., & Blackman, M. C. (1995). Agreement among judges of personality: Interpersonal relations, similarity, and acquaintance. *Journal of Personality and Social Psychology, 69,* 656–672.

Funder, D. C., & Ozer, D. J. (1983). Behavior as a function of the situation. *Journal of Personality and Social Psychology, 44,* 107–112.

Furness, W. (1916). Observations on the mentality of chimpanzees and orangutans. *Proceedings of the American Philosophical Society, 45,* 281–290.

Furuhjelm, M., Karlgren, E., & Carlstrom, K. (1984). The effect of estrogen therapy on somatic and psychical symptoms in postmenopausal women. *Acta Obstetrica Gynecologica Scandinava, 63,* 655–661.

Galdikas, B. (1985). Adult male sociality and reproductive tactics among orangutans at Tanjung Puting. *Folia Primatologica, 45,* 9–24.

Galef, B. G. Jr. (1992). The question of animal culture. *Human Nature, 3,* 157–178.

Gallup, G. G. (1970). Chimpanzees: Self-recognition. *Science, 167,* 86–87.

Gallup, G. G. (1985). Do minds exist in species other than our own? *Neuroscience and Biobehavioral Reviews, 9,* 631–641.

Gallup, G. G. Jr., Nash, R. F., & Ellison, A. L. (1971). Tonic immobility as a reaction to predation: Artificial eyes as a fear stimulus for chickens. *Psychonomic Science, 23,* 79–80.

Gannon, P. J., Holloway, R. L., Broadfield, D. C., & Braun, A. R. (1998). Asymmetry of chimpanzee planum temporale: Humanlike pattern of Wernicke's brain language area homologue. *Science, 279,* 220–222.

Garcia Coll, C. T., Hoffman, J., & Oh, W. (1987). The social ecology and early parenting of Caucasian adolescent mothers. *Child Development, 58,* 955–963.

Gardner, B. T., & Gardner, R. A. (1971). Two-way communication with an infant chimpanzee. In A. M. Schrier & F. Stollnitz (Eds.), *Behavior of nonhuman primates.* Vol. 4 (pp. 117–183). New York: Academic Press.

Gardner, B. T., & Gardner, R. A. (1975). Evidence for sentence constituents in the early utterances of child and chimpanzee. *Journal of Experimental Psychology: General, 104,* 244–267.

Gardner, R. A., & Gardner, B. T. (1969). Teaching sign language to a chimpanzee. *Science, 165,* 664–672.

Gardner, R. A., & Gardner, B. T. (1978). Comparative psychology and language acquisition. *Annals of the New York Academy of Sciences, 309,* 37–76.

Gardner, R. A., & Gardner, B. T. (1984). A vocabulary test for chimpanzees *(Pan troglodytes). Journal of Comparative Psychology, 98,* 381–404.

Gardner, R. A., Gardner, B. T., & van Cantfort, T. E. (Eds.). (1989). *Teaching sign language to chimpanzees.* Albany: SUNY Press.

Garza, R. T., & Borchert, J. E. (1990). Maintaining social identity in a mixed-gender setting: Minority / majority status and cooperative / competitive feedback. *Sex Roles, 22,* 679–691.

Gauthier, I., & Logothetis, N. K. (2000). Is face recognition not so unique after all? *Cognitive Neuropsychology, 17,* 125–142.

Gauthier, I., Skudlarski, P., Gore, J. C., & Anderson, A. W. (2000). Expertise for cars and birds recruits brain areas involved in face recognition. *Nature Neuroscience, 3,* 191–197.

Geary, D. C. (1998). *Male, female: The evolution of human sex differences.* Washington, DC: American Psychological Association Press.

Geist, S. H. (1941). Androgen therapy in the human female. *Journal of Clinical Endocrinology, 1,* 154–161.

George, C., & Solomon, J. (1999). Attachment and caregiving. In J. Cassidy & P. R. Shaver (Eds.), *Handbook of attachment* (pp. 395–433). New York: Guilford Press.

George, M. S., Ketter, T. A., Gill, D. S., Haxby, J. V., Ungerleider, L. G., Herscovitch, P., & Post, R. M. (1993). Brain regions involved in recognizing facial emotion or identity: An oxygen-15 PET study. *Journal of Neuropsychiatry, 5,* 384–391.

George, M. S., Parekh, P. I., Rosinsky, N., Ketter, T. A., Kimbrell, T. A., Heilman, K. A., Herscovitch, P., & Post, R. M. (1996). Understanding emotional prosody activates right hemisphere regions. *Archives of Neurology, 53,* 665–670.

Gergely, G., Nádasdy, Z., Csibra, G., & Biró, S. (1995). Taking the intentional stance at 12 months of age. *Cognition, 56,* 165–193.

Geschwind, N., & Levitsky, W. (1968). Human brain: Left-right asymmetries in the temporal speech region. *Science, 161,* 186–187.

Ghazanfar, A. A., & Hauser, M. D. (1999). The neuroethology of primate vocal communication: Substrates for the evolution of speech. *Trends in Cognitive Science, 3,* 377–384.

Gibber, J. R., & Goy, R. W. (1985). Infant-directed behavior in young rhesus monkeys: Sex differences and effects of prenatal androgens. *American Journal of Primatology, 8,* 225–237.

Gigerenzer, G., & Hug, K. (1992). Domain-specific reasoning: Social contracts, cheating, and perspective change. *Cognition, 43,* 127–171.

Gilissen, E. (2001). Structural symmetries and asymmetries in human and chimpanzee brains. In D. Falk & K. R. Gibson (Eds.), *Evolutionary anatomy of the primate cerebral cortex* (pp. 187–215). Cambridge: Cambridge University Press.

Gill, K., & Amit, Z. (1989). Serotonin uptake blockers and voluntary alcohol consumption: A review of recent studies. *Recent Developments in Alcoholism, 7,* 225–248.

Gill, T. V. (1977). Conversations with Lana. In D. M. Rumbaugh (Ed.), *Language learning by a chimpanzee* (pp. 225–246). New York: Academic Press.

Ginsberg, S. D., Hof, P. R., McKinney, W. T., & Morrison, J. H. (1993). The noradrenergic innervation density of the monkey paraventricular nucleus is not altered by early social deprivation. *Neuroscience Letters, 158,* 130–134.

Ginther, A. J., Washabaugh, K. F., & Snowdon, C. T. (2000). Measurement of scrotum and testis size of unrestrained captive cotton-top tamarins *(Saguinus oedipus oedipus). American Journal of Primatology, 51,* 187–195.

Ginther, A. J., Ziegler, T. E., & Snowdon, C. T. (2001). Reproductive biology of captive male cotton-top tamarin monkeys as a function of social environment. *Animal Behaviour, 61,* 65–78.

Gleeson, S., Ahlers, S. T., Mansbach, R. S., Foust, J. M., & Barrett, J. E. (1989). Behavioral studies with anxiolytic drugs. VI: Effects on punished responding of drugs interacting with serotonin receptor subtypes. *Journal of Pharmacology and Experimental Therapeutics, 250,* 809–817.

Glenn, P. J. (1991/1992). Current speaker initiation of two-party shared laughter. *Research on Language and Social Interaction, 25,* 139–162.

Glick, B. B. (1979). Testicular size, testosterone level, and body weight in male *Macaca radiata:* Maturational and seasonal effects. *Folia Primatologica, 32,* 268–289.

Glick, B. B. (1980). Ontogenetic and psychobiological aspects of the mating activities of male *Macaca radiata.* In D. G. Lindburg (Ed.), *The macaques: Studies in ecology, behavior, and evolution* (pp. 345–369). New York: Van Nostrand Reinhold.

Glick, B. B., Eaton, G. G., Johnson, D. F., & Worlein, J. (1986a). Social behavior of infant and mother Japanese macaques *(Macaca fuscata):* Effects of kinship, partner sex, and infant sex. *International Journal of Primatology, 7,* 139–155.

Glick, B. B., Eaton, G. G., Johnson, D. F., & Worlein, J. M. (1986b). Development of partner preferences in Japanese macaques *(Macaca fuscata):* Effects of gender and kinship during the second year of life. *International Journal of Primatology, 7,* 467–479.

Gold, K. C., & Maple, T. L. (1994). Personality assessment in the gorilla and its utility as a management tool. *Zoo Biology, 13,* 509–522.

Goldberg, L. R., & Digman, J. M. (1994). Revealing the structure in the data: Principles of exploratory factor analysis. In S. Strack & M. Lorr (Eds.), *Differentiating normal and abnormal personality* (pp. 216–242). New York: Springer.

Golden, R. R., & Meehl, P. E. (1979). Detection of the schizoid taxon with MMPI indicators. *Journal of Abnormal Psychology, 88,* 217–233.

Goldfoot, D. A. (1977). Sociosexual behaviors of nonhuman primates during development and maturity: Social and hormonal relationships. In A. M. Schrier (Ed.), *Behavioral primatology: Advances in research & theory,* Vol. 1 (pp. 139–184). Hillsdale, NJ: Erlbaum.

Goldfoot, D. A., & Wallen, K. (1978). Development of gender role behaviors in heterosexual and isosexual groups of infant rhesus monkeys. In D. J. Chivers (Ed.), *Recent advances in primatology: Behaviour,* Vol. 1 (pp. 155–159). New York: Academic Press.

Goldfoot, D. A., Wallen, K., Neff, D. A., McBrair, M. C., & Goy, R. W. (1984). Social influences on the display of sexually dimorphic behavior in rhesus monkeys: Isosexual rearing. *Archives of Sexual Behavior, 13,* 395–412.

Goldfoot, D. A., Weigand, S. J., & Scheffler, G. (1978). Continued copulation in ovariectomized adrenal-suppressed stumptail macaques *(Macaca arctoides). Hormones and Behavior, 11,* 89–99.

Goldfoot, D. A., Westerborg–Van Loon, H., Groeneveld, W., & Slob, A. K. (1980). Behavioral and physiological evidence of sexual climax in the female stump-tailed macaque *(Macaca arctoides). Science, 208,* 1477–79.

Goldizen, A. W. (1987). Tamarins and marmosets: Communal care of offspring. In B. B. Smuts, D. L. Cheney, R. M. Seyfarth, R. W. Wrangham, & T. T. Struhsaker (Eds.), *Primate societies* (pp. 34–43). Chicago: University of Chicago Press.

Goldizen, A. W., & Terborgh, J. (1989). Demography and dispersal patterns of a tamarin population: Possible causes of delayed breeding. *American Naturalist, 134,* 208–224.

Goldsmith, H., Buss, A., Plomin, R., Rothbart, M., Thomas, A., Chess, S., Hinde, R., & McCall, R. (1987). Roundtable: What is temperament? Four approaches. *Child Development, 58,* 505–529.

Golomb, B. A., Stattin, H., & Mednick, S. (2000). Low cholesterol and violent crime. *Journal of Psychiatric Research, 34,* 301–309.

Gómez, J. C. (1990). The emergence of intentional communication as a problem-solving strategy in the gorilla. In S. T. Parker & K. R. Gibson (Eds.), *"Language" and intelligence in monkeys and apes* (pp. 333–355). Cambridge: Cambridge University Press.

Gómez, J. C. (1991). Visual behaviour as a window for reading the mind of others in primates. In A. Whiten (Ed.), *Natural theories of mind* (pp. 195–207). Oxford: Blackwell.

Gómez, J. C. (1996). Non-human primate theories of (non-human primate) minds: Some issues concerning the origins of mind-reading. In P. Carruthers & P. K. Smith (Eds.), *Theories of theories of mind* (pp. 330–343). Cambridge: Cambridge University Press.

Gómez, J. C. (1998). Assessing theory of mind with nonverbal procedures: Problems with training methods and an alternative "key" procedure. *Behavioral and Brain Sciences, 21,* 119–120.

Goodall, J. (1968). The behaviour of free-living chimpanzees in the Gombe Stream Reserve. *Animal Behaviour Monographs, 1,* 165–311.

Goodall, J. (1986). *The chimpanzees of Gombe: Patterns of behavior.* Cambridge, MA: Belknap Press of Harvard University Press.

Goosen, C., & Kortmulder, K. (1979). Relationships between faces and body motor patterns in a group of captive pigtailed macaques *(Macaca nemestrina). Primates, 20,* 221–236.

Goosen, C., & Ribbens, L. G. (1977). Allogrooming conflicts in a pair of adult stump-tailed macaques. *Proceedings of the Sixth Congress of the International Primatological Society, 318*–320.

Gopnik, A. (1993). How we know our minds: The illusion of first-person knowledge about intentionality. *Behavioral and Brain Sciences, 16,* 1–14.

Gopnik, A., & Meltzoff, A. (1997). *Words, thoughts, and theories.* Cambridge, MA: MIT Press.

Gopnik, A., & Wellman, H. M. (1992). Why the child's theory of mind really is a theory. *Mind and Language, 7,* 145–171.

Gordon, T. P. (1981). Reproductive behavior in the rhesus monkey: Social and endocrine variables. *American Zoologist, 21,* 185–195.

Gordon, T. P., Bernstein, I. S., & Rose, R. M. (1978). Social and seasonal influences on testosterone secretion in the male rhesus monkey. *Physiology and Behavior, 21,* 623–637.

Gordon, T. P., Rose, R. M., & Bernstein, I. S. (1976). Seasonal rhythm in plasma testosterone levels in the rhesus monkey *(Macaca mulatta):* A three year study. *Hormones and Behavior, 7,* 229–243.

Gordon, T. P., Rose, R. M., Grady, C. L., & Bernstein, I. S. (1979). Effects of increased testosterone secretion on the behavior of adult male rhesus living in a social group. *Folia Primatologica, 32,* 149–160.

Gosling, S. D. (2001). From mice to men: What can we learn about personality from animal research? *Psychological Bulletin, 127,* 45–86.

Gosling, S. D., & John, O. P. (1999). Personality dimensions in non-human animals: A cross-species review. *Current Directions in Psychological Science, 8,* 69–75.

Gosling, S. D., John, O. P., Craik, K. H., & Robins, R. W. (1998). Do people know how they behave? Self-reported act frequencies compared with on-line codings by observers. *Journal of Personality and Social Psychology, 74,* 1337–49.

Gottfried, A. E., Gottfried, A. W., & Bathurst, K. (1995). Maternal and dual-earner employment status and parenting. In M. H. Borstein (Ed.), *Handbook of parenting,* Vol. 2, *Biology and ecology of parenting.* (pp. 139–160). Hillsdale, NJ: Erlbaum.

Goustard, M. (1963). Introduction à l'étude de la communication vocale chez *Macaca irus. Annals des Science Naturale Zoologie, 5,* 707–748.

Gouzoules, H., Gouzoules, S., & Ashley J. (1995). Representational signaling in non-human primate vocal communication. In E. Zimmermann (Ed.), *Current topics in primate vocal communication* (pp. 235–252). New York: Plenum Press.

Gouzoules, H., Gouzoules, S., & Fedigan, L. (1982). Behavioral dominance and reproductive success in female Japanese monkeys *(Macaca fuscata). Animal Behaviour, 30,* 1138–50.

Gouzoules, H., Gouzoules, S., & Marler, P. (1986). Vocal communication: A vehicle for the study of social relationships. In R. G. Rawlins & M. J. Kessler (Eds.), *The Cayo Santiago macaques: History, behavior, and biology* (pp. 111–129). Albany: SUNY Press.

Gouzoules, S., & Gouzoules, H. (1987). Kinship. In B. B. Smuts, D. L. Cheney, R. M. Seyfarth, R. W. Wrangham, & T. T. Struhsaker (Eds.), *Primate societies* (pp. 299–305). Chicago: University of Chicago Press.

Gouzoules, S., Gouzoules, H., & Marler, P. (1984). Rhesus monkey *(Macaca mulatta)* screams: Representational signaling in the recruitment of agonistic aid. *Animal Behaviour, 32,* 182–193.

Goy, R. W. (1979). Sexual compatibility in rhesus monkeys: Predicting sexual performance of oppositely sexed pairs of adults. *Ciba Foundation Symposium, 62,* 227–255.

Goy, R. W. (1981). Differentiation of male social traits in female rhesus macaques by prenatal treatment with androgens: Variation in type of androgen, duration, and timing of treatment. In M. J. Novy & J. A. Resko (Eds.), *Fetal endocrinology* (pp. 319–339). New York: Academic Press.

Goy, R. W., Bercovitch, F. B., & McBrair, M. C. (1988). Behavioral masculinization is independent of genital masculinization in prenatally androgenized female rhesus macaques. *Hormones and Behavior, 22,* 552–571.

Goy, R. W., Kraemer, G., & Goldfoot, D. (1988). Biological influences on grooming in nonhuman primates. In D. L. Colbern & W. H. Gispen (Eds.), *Neural mechanisms and biological significance of grooming behavior,* Vol. 525 (pp. 56–68). New York: New York Academy of Science.

Goy, R. W., & Wallen, K. (1979). Experiential variables influencing play, foot-clasp mounting, and adult sexual competence in male rhesus monkeys. *Psychoneuroendocrinology, 4,* 1–12.

Graham, F. K. (1979). Distinguishing among orienting, defensive, and startle reflexes. In H. D. Kimmel, E. H. van Olst, & J. F. Orlebeke (Eds.), *The orienting reflex in humans* (pp. 137–167). Hillsdale, NJ: Erlbaum.

Grammer, K. (1990). Strangers meet: Laughter and nonverbal signs of interest in opposite-sex encounters. *Journal of Nonverbal Behavior, 14,* 209–236.

Grammer, K. (1992). Intervention in conflicts among children: Contexts and consequences. In A. H. Harcourt & F. B. M. de Waal (Eds.), *Coalitions and alliances in humans and other animals* (pp. 258–283). Oxford: Oxford University Press.

Grammer, K., & Eibl-Eibesfeldt, I. (1990). The ritualization of laughter. In W. Koch (Ed.), *Naturlichkeit der Sprache und der Kultur: Acta colloquii* (pp. 192–214). Bochum: Brockmeyer.

Gray, J. A. (1975). *Elements of a two-process theory of learning.* London: Academic Press.

Gray, J. A. (1981). A critique of Eysenck's theory of personality. In H. J. Eysenck (Ed.), *A model for personality* (pp. 246–276). New York: Springer-Verlag.

Gray, J. A. (1990). Brain systems that mediate both emotion and cognition. *Cognition and Emotion, 4,* 269–288.

Green, S. (1975). Variation of vocal pattern with social situation in the Japanese monkey *(Macaca fuscata):* A field study. In L. A. Rosenblum (Ed.), *Primate behavior: Developments in field and laboratory research* (pp. 1–101). New York: Academic Press.

Green, S. (1981). Sex differences and age gradation in vocalizations of Japanese and liontailed macaques. *American Zoologist, 21,* 165–183.

Griffin, D. R. (1981). *The question of animal awareness.* New York: Rockefeller University Press.

Griffin, D. R. (1984). *Animal thinking.* Cambridge, MA: Harvard University Press.

Griffin, D. R. (1992). *Animal minds.* Chicago: University of Chicago Press.

Grimm, R. J. (1967). Catalogue of sounds of the pigtailed macaque. *Journal of the Zoological Society of London, 152,* 361–373.

Gross, C. G. (1992). Representation of visual stimuli in inferior temporal cortex. *Philosophical Transactions of the Royal Society of London, B, 335,* 3–10.

Groves, C. P., & Humphrey, N. K. (1973). Asymmetry in gorilla skulls: Evidence of lateralized brain function? *Nature, 244,* 53–54.

Grusec, J. (1991). The socialization of altruism. In M. S. Clark (Ed.), *Prosocial behavior* (pp. 9–33). Newbury Park, CA: Sage.

Hadidian, J. M. (1979). Allo- and autogrooming in a captive Black Ape colony (*Macaca nigra Desmarest*, 1822). Ph.D. Dissertation, Pennsylvania State University,

Haig, D. (1993). Genetic conflicts in human pregnancy. *Quarterly Review of Biology, 68,* 495–532.

Hakvoort, I., & Oppenheimer, L. (1993). Children and adolescents' conceptions of peace, war, and strategies to attain peace: A Dutch case study. *Journal of Peace Research, 30,* 65–77.

Halpern, C. T., Udry, J. R., & Suchindran, C. (1997). Testosterone predicts initiation of coitus in adolescent females. *Psychosomatic Medicine, 59,* 161–171.

Halpern, C. T., Udry, J. R., & Suchindran, C. (1998). Monthly measures of salivary testosterone predict sexual activity in adolescent males. *Archives of Sexual Behavior, 27,* 445–465.

Hamilton, C. R. (1977a). An assessment of hemispheric specialization in monkeys. *Annals of the New York Academy of Sciences, 299,* 222–232.

Hamilton, C. R. (1977b). Investigations of perceptual and mnemonic lateralization in monkeys. In S. Harnad, R. W. Doty, J. Jaynes, L. Goldstein, & G. Krauthamer (Eds.), *Lateralization in the nervous system* (pp. 45–62). New York: Academic Press.

Hamilton, C. R., & Vermeire, B. A. (1983). Discrimination of monkey faces by split-brain monkeys. *Behavioural Brain Research, 9,* 263–275.

Hamilton, C. R., & Vermeire, B. A. (1988). Complementary hemispheric specialization in monkeys. *Science, 242,* 1691–94.

Hamilton, W. D. (1964). The genetic evolution of social behaviour. *Journal of Theoretical Biology, 7,* 1–16.

Hammerschmidt, K., Ansorge, V., Fischer, J., & Todt, D. (1994). Dusk calling in Barbary macaques *(Macaca sylvanus):* Demand for social shelter. *American Journal of Primatology, 32,* 277–289.

Hammerschmidt, K., & Todt, D. (1995). Individual differences in vocalisations of young Barbary macaques *(Macaca sylvanus):* A multi-parametric analysis to identify critical cues in acoustic signalling. *Behaviour, 132,* 381–399.

Hampton, R. R. (2001). Rhesus monkeys know when they remember. *Proceedings of the National Academy of Sciences, USA, 98,* 5359–62.

Hanby, J. P. (1980a). Relationships in six groups of rhesus monkeys. I: Networks. *American Journal of Physical Anthropology, 52,* 549–564.

Hanby, J. P. (1980b). Relationships in six groups of rhesus monkeys. II: Dyads. *American Journal Physical Anthropological, 52,* 565–575.

Hanby, J. P., & Brown, C. E. (1974). The development of sociosexual behaviours in Japanese macaques, *Macaca fuscata. Behaviour, 159,* 152–196.

Hansen, A. J. (1986). Fighting behavior in bald eagles: A test of game theory. *Ecology,* *67,* 787–797.

Hansen, E. W. (1976). Selective responding by recently separated juvenile rhesus monkeys to the calls of their mothers. *Developmental Psychobiology, 9,* 83–88.

Harcourt, A. H., & de Waal, F. B. M. (Eds.). (1992). *Coalitions and alliances in humans and other animals.* Oxford: Oxford University Press.

Hare, B. (2001). Can competitive paradigms increase the validity of experiments on primate social cognition? *Animal Cognition, 4,* 269–280.

Hare, B., Call, J., Agnetta, B., & Tomasello, M. (2000). Chimpanzees know what conspecifics do and do not see. *Animal Behaviour, 59,* 771–785.

Hare, B., Call, J., & Tomasello, M. (1998). Communication of food location between human and dog *(Canis familiaris). Evolution of Communication, 2,* 137–159.

Hare, B., Call, J., & Tomasello, M. (2001). Do chimpanzees know what conspecifics know and do not know? *Animal Behaviour, 61,* 139–151.

Hare, B., & Tomasello, M. (1999). Domestic dogs *(Canis familiaris)* use human and conspecific social cues to locate hidden food. *Journal of Comparative Psychology, 113,* 173–177.

Harlow, H. F. (1949). The formation of learning sets. *Psychological Review, 56,* 51–65.

Harlow, H. F. (1958). The nature of love. *American Psychologist, 13,* 673–685.

Harlow, H. F. (1959). Affectional response in the infant monkey. *Science, 130,* 421–432.

Harlow, H. F. (1965). Sexual behavior in the rhesus monkey. In F. A. Beach (Ed.), *Sex and Behavior* (pp. 234–265). New York: Krieger.

Harlow, H. F. (1969). Age-mate or peer affectional system. *Advances in the Study of Behavior, 2,* 333–383.

Harlow, H. F. (1974). *Learning to love.* New York, Jason Aronson.

Harlow, H. F., & Harlow, M. K. (1965). The affectional system. In A. M. Schrier, H. F. Harlow, & F. Stollnitz (Eds.), *Behavior of nonhuman primates,* Vol. 2 (pp. 287–334). New York: Academic Press.

Harlow, H. F., & Lauersdorf, H. E. (1974). Sex differences in passion and play. *Perspectives in Biology and Medicine, 17,* 348–360.

Harris, J. R. (1995). Where is the child's environment? A group socialization theory of development. *Psychological Review, 102,* 458–489.

Harris, L. J. (1993). Handedness in apes and monkeys: Some views from the past. In J. P. Ward & W. D. Hopkins (Eds.), *Primate laterality: Current behavioral evidence of primate asymmetries* (pp. 1–53). New York: Springer-Verlag.

Harris, P. (1991). The work of the imagination. In A. Whiten (Ed.), *Natural theories of mind* (pp. 283–304). Oxford: Blackwell.

Hart, B., & Risley, T. R. (1992). American parenting of language-learning children:

Persisting differences in family-child interactions observed in natural home environments. *Developmental Psychology, 25,* 1096–1106.

Hartman, C. G. (1931). On the relative sterility of the adolescent organism. *Science, 74,* 226–227.

Hartup, W. W. (1992). Conflict and friendship relations. In C. U. Shantz & W. W. Hartup (Eds.), *Conflict in child and adolescent development* (pp. 186–215). Cambridge: Cambridge University Press.

Hartup, W. W., French, D. C., Laursen, B., Johnston, K. T., & Ogawa, J. R. (1993). Conflict and friendship relations in middle childhood: Behavior in a closed-field situation. *Child Development, 64,* 445–454.

Hartup, W. W., Laursen, B., Stewart, M. I., & Eastenson, A. (1988). Conflict and the friendship relations of young children. *Child Development, 59,* 1590–1600.

Harvey, P. H., Martin, R. D., & Clutton-Brock, T. H. (1987). Life histories in comparative perspective. In B. B. Smuts, D. L. Cheney, R. M. Seyfarth, R. W. Wrangham, & T. T. Struhsaker (Eds.), *Primate societies* (pp. 181–196). Chicago: University of Chicago Press.

Hauser, M. D. (1991). Sources of acoustic variation in rhesus macaque *(Macaca mulatta)* vocalizations. *Ethology, 89,* 29–46.

Hauser, M. D. (1992). Costs of deception: Cheaters are punished in rhesus monkeys. *Proceedings of the National Academy of Science, USA, 89,* 12137–39.

Hauser, M. D. (1993). Right hemisphere dominance in the production of facial expression in monkeys. *Science, 261,* 475–477.

Hauser, M. D. (1996). *The evolution of communication.* Cambridge, MA: MIT Press.

Hauser, M. D. (1998a). Functional referents and acoustic similarity: Field playback experiments with rhesus monkeys. *Animal Behaviour, 55,* 1647–58.

Hauser, M. D. (1998b). A nonhuman primate's expectations about object motion and destination: The importance of self-propelled movement and animacy. *Developmental Science, 1,* 31–37.

Hauser, M. D. (1999a). Primate representations and expectations: Mental tools for navigating in a social world. In P. D. Zelazo, J. W. Astington, & D. R. Olson (Eds.), *Developing theories of intention* (pp. 169–194). Hillsdale, NJ: Erlbaum.

Hauser, M. D. (1999b). Perseveration, inhibition, and the prefrontal cortex: A new look. *Current Opinion in Neurobiology, 9,* 214–222.

Hauser, M. D., Agnetta, B., & Perez, C. (1998). Orienting asymmetries in rhesus monkeys: The effect of time-domain changes on acoustic perception. *Animal Behaviour, 56,* 41–47.

Hauser, M. D., & Andersson, K. (1994). Left hemisphere dominance for processing vocalizations in adult, but not infant, rhesus monkeys: Field experiments. *Proceeding of the National Academy of Sciences, USA, 91,* 3946–48.

Hauser, M. D., & Fairbanks, L. A. (1988). Mother-offspring conflict in vervet monkeys: Variation in response to ecological conditions. *Animal Behaviour, 36,* 802–813.

Hauser M. D., & Marler P. (1993). Food-associated calls in rhesus macaques *(Macaca mulatta).* I: Socioecological factors. *Behavioral Ecology, 4,* 194–205.

Hausfater, G. (1972). Intergroup behavior of free-ranging rhesus monkeys *(Macaca mulatta). Folia Primatologica, 18,* 78–107.

Haviland, J. M., & Lelwicka, M. (1987). The induced affect response: 10-week-old infants' responses to three emotion expressions. *Developmental Psychology, 23,* 97–104.

Haviland, J. M., & Malatesta, C. Z. (1981). The development of sex differences in non-verbal signals: Fallacies, facts, and fantasies. In C. Mayo & N. M. Henley (Eds.), *Gender and nonverbal behavior* (pp. 183–208). New York: Springer-Verlag.

Hawkes, K. (1990). Showing off: Tests of a hypothesis about men's foraging goals. *Ethology and Sociobiology, 12,* 29–54.

Hay, D. F., & Ross, H. S. (1982). The social nature of early conflict. *Child Development, 53,* 105–113.

Hayes, C. (1951). *The ape in our house.* New York: Harper.

Hayes, K. J., & Hayes, C. (1951). The intellectual development of a home-raised chimpanzee. *Proceedings of the American Philosophical Society, 95,* 105–109.

Hayes, K. J., & Hayes, C. (1952). Imitation in a home-raised chimpanzee. *Journal of Comparative and Physiological Psychology, 45,* 450–459.

Hayworth, D. (1928). The social origin and function of laughter. *Psychological Review, 35,* 367–384.

Hazan, C., & Shaver, P. R. (1994). Attachment as an organizational framework for research on close relationships. *Psychological Inquiry, 5,* 1–22.

Heath-Lange, S., Ha, J. C., & Sackett, G. P. (1999). Behavioral measurement of temperament in male nursery-reared infant macaques and baboons. *American Journal of Primatology, 47,* 43–50.

Hebb, D. O. (1946). Emotions in man and animals: An analysis of the intuitive process of recognition. *Psychological Review, 53,* 88–106.

Hebb, D. O. (1949). Temperament in chimpanzees. I: Method and analyses. *Journal of Comparative and Physiological Psychology, 42,* 192–206.

Hedricks, C., Piccinino, L. J., Udry, J. R., & Chimbira, T. H. K. (1987). Peak coital rate coincides with onset of luteinizing hormone surge. *Fertility and Sterility, 48,* 234–238.

Hedricks, C. A., Schramm, W., & Udry, J. R. (1994). Effects of creatinine correction to urinary LH levels on the timing of the LH peak and the distribution of coitus within the human menstrual cycle. *Annals of the New York Academy of Sciences, 709,* 204–206.

Heffner, H. E., & Heffner, R. S. (1984). Temporal lobe lesions and perception of species-specific vocalizations by macaques. *Science, 226,* 75–76.

Heilbronner, P. L., & Holloway, R. L. (1988). Anatomical brain asymmetries in New World and Old World monkeys. Stages of temporal lobe development in primate evolution. *American Journal of Physical Anthropology, 76,* 39–48.

Heilbronner, P. L., & Holloway, R. L. (1989). Anatomical brain asymmetry in monkeys: Frontal, temporoparietal, and limbic cortex in *Macaca. American Journal of Physical Anthropology, 80,* 203–211.

Heim, N. (1981). Sexual behavior of castrated sex offenders. *Archives of Sexual Behavior, 10,* 11–19.

Heinz, A., Higley, J. D., Gorey, J. G., Saunders, R. C., Jones, D. W., Hommer, D., Zajicek, K., Suomi, S. J., Lesch, K. P., Weinberger, D. R., & Linnoila, M. (1998). In vivo association between alcohol intoxication, aggression, and serotonin transporter availability in nonhuman primates. *American Journal of Psychiatry, 155,* 1023–28.

Hellige, J. B. (1993). *Hemispheric asymmetry: What's right and what's left.* Cambridge, MA: Harvard University Press.

Hemelrijk, C. K. (1990a). A matrix partial correlation test used in investigations of reciprocity and other social interaction patterns at the group level. *Journal of Theoretical Biology, 143,* 405–420.

Hemelrijk, C. K. (1990b). Models of, and tests for, reciprocity, unidirectionality and other interaction patterns at a group level. *Animal Behaviour, 39,* 1013–29.

Hemelrijk, C. K. (1994). Support for being groomed in long-tailed macaques, *Macaca fascicularis. Animal Behaviour, 48,* 479–481.

Hemelrijk, C. K., & Ek, A. (1991). Reciprocity and interchange of grooming and "support" in captive chimpanzees. *Animal Behaviour, 41,* 923–935.

Henzi, S. P., & Barrett, L. (1999). The value of grooming to female primates. *Primates, 40,* 47–59.

Henzi, S. P., & Barrett, L. (2002). Infants as a commodity in a baboon market. *Animal Behaviour, 63,* 915–921.

Henzi, S. P., Lycett, J. E., & Piper, S. E. (1997). Fission and troop size in a mountain baboon population. *Animal Behaviour, 53,* 525–535.

Henzi, S. P., Lycett, J. E., & Weingrill, T. (1997). Cohort size and the allocation of social effort by female mountain baboons. *Animal Behaviour, 54,* 1235–43.

Herman, L. M. (1988). The language of animal language research: Reply to Schusterman and Gisiner. *Psychological Record, 38,* 349–362.

Herman, L. M., Abichandani, S. L., Elhajj, A. N., Herman, E. Y. K., Sanchez, J. L., & Pack, A. A. (1999). Dolphins *(Tursiops truncatus)* comprehend the referential character of the human pointing gesture. *Journal of Comparative Psychology, 113,* 347–364.

Herman, L. H., Morrel-Samuels, P., & Pack, A. A. (1990). Bottlenosed dolphin and human recognition of veridical and degraded video displays of an artificial gestural language. *Journal of Experimental Psychology: General, 119,* 215–230.

Herman, L. H., Pack, A. A., & Morrel-Samuels, P. (1993). Representational and conceptual skills of dolphins. In H. L. Roitblat, L. M. Herman, & P. E. Nachtigall (Eds.), *Language and communication: Comparative perspectives* (pp. 404–442). Hillsdale, NJ: Erlbaum.

Herman, L. H., Richards, D. G., & Wolz, J. P. (1984). Comprehension of sentences by bottlenosed dolphins. *Cognition, 16,* 129–219.

Herman, R. A., Jones, B., Mann, D. R., & Wallen, K. (2000). Timing of prenatal androgen exposure: Anatomical and endocrine effects on juvenile male and female rhesus monkeys. *Hormones and Behavior, 38,* 52–66.

Herman, R. A., Measday, M. A., & Wallen, K. (2003). Sex differences in interest in infants in juvenile rhesus monkeys: Relationship to prenatal androgen. *Hormones and Behavior, 43,* 573–583.

Hetherington, E. M., & Stanley-Hagan, M. M. (1995). Parenting in divorced and re-married families. In M. H. Bornstein (Ed.), *Handbook of parenting.* Vol 3. (pp. 233–254). Hillsdale, NJ: Erlbaum.

Heyes, C. M. (1998). Theory of mind in nonhuman primates. *Behavioral and Brain Sciences, 21,* 101–148.

Heyes, C. M. (1993). Anecdotes, training, trapping, and triangulating: Do animals attribute mental states? *Animal Behaviour, 46,* 177–188.

Heyes, C. M., & Ray, E. (2000). What is the significance of imitation in animals? *Advances in the Study of Behavior, 29,* 215–245.

Hibbeln, J. R., Umhau, J. C., George, D. T., Shoaf, S. E., Linnoila, M., & Salem, N. Jr. (2000). Plasma total cholesterol concentrations do not predict cerebrospinal fluid neurotransmitter metabolites: Implications for the biophysical role of highly unsaturated fatty acids. *American Journal of Clinical Nutrition Supplement, 71,* 331–338.

Hickok, G., Bellugi, U., & Klima, E. S. (1996). The neurobiology of sign language and its implications for the neural basis of language. *Nature, 381,* 699–702

Higley, J. D., & Bennett, A. J. (1999). Central nervous system serotonin and personality as variables contributing to excessive alcohol consumption in non-human primates. *Alcohol and Alcoholism, 34,* 402–418.

Higley, J. D., Hasert, M. F., Dodson, A., Linnoila, M., & Suomi, S. J. (1992). Treatment of excessive alcohol consumption using the serotonin reuptake inhibitor Sertraline in a nonhuman primate model of alcohol abuse. Paper presented at the Research Society on Alcoholism, San Diego, June 13–18.

Higley, J. D., Hasert, M. F., Dodson, A., Suomi, S. J., & Linnoila, M. (1994). Diminished central nervous system serotonin functioning as a predictor of excessive alcohol consumption: The role of early experiences. *American Journal of Primatology, 33,* 214.

Higley, J. D., Hasert, M. F., Suomi, S. J., & Linnoila, M. (1998). The serotonin reuptake inhibitor sertraline reduces excessive alcohol consumption in nonhuman primates: Effect of stress. *Neuropsychopharmacology, 18,* 431–443.

Higley, J. D., Hopkins, W. D., Thompson, W. W., Byrne, E. A., Hirsch, R. M., & Suomi, S. J. (1992). Peers as primary attachment sources in yearling rhesus monkeys *(Macaca mulatta). Developmental Psychology, 28,* 1163–71.

Higley, J. D., King, S. T., Hasert, M. F., Champoux, M., Suomi, S. J., & Linnoila, M. (1996). Stability of interindividual differences in serotonin function and its relationship to aggressive wounding and competent social behavior in rhesus macaque females. *Neuropsychopharmacology, 14,* 67–76.

Higley, J. D., & Linnoila, M. (1997a). Low central nervous system serotonergic activity is trait-like and correlates with impulsive behavior: A nonhuman primate model investigating genetic and environmental influences on neurotransmission. *Annals of the New York Academy of Sciences, 836,* 39–56.

Higley, J. D., & Linnoila, M. (1997b). A nonhuman primate model of excessive alcohol intake: Personality and neurobiological parallels of Type I- and Type II-like alcoholism. *Recent Developments in Alcoholism, 13,* 192–219.

Higley, J. D., Linnoila, M., & Suomi, S. J. (1994). Ethological contributions: Experiential and genetic contributions to the expression and inhibition of aggression in primates. In M. Hersen, R. T. Ammerman, & L. Sisson (Eds.), *Handbook of aggressive and destructive behavior in psychiatric patients* (pp. 17–32). New York: Plenum Press.

Higley, J. D., Mehlman, P. T., Higley, S. B., Fernald, B., Vickers, J., Lindell, S. G., Taub, D. M., Suomi, S. J., & Linnoila, M. (1996). Excessive mortality in young free-ranging male nonhuman primates with low cerebrospinal fluid 5-hydroxyindoleacetic acid concentrations. *Archives of General Psychiatry, 153,* 537–543.

Higley, J. D., Mehlman, P. T., Poland, R. E., Taub, D. M., Vickers, J., Suomi, S. J., & Linnoila, M. (1996). CSF testosterone and 5-HIAA correlate with different types of aggressive behaviors. *Biological Psychiatry, 40,* 1067–82.

Higley, J. D., Mehlman, P., Taub, D., Higley, S. B., Vickers, J. H., Suomi, S. J., & Linnoila, M. (1992). Cerebrospinal fluid monoamine and adrenal correlates of aggression in free-ranging rhesus monkeys. *Archives of General Psychiatry, 49,* 436–441.

Higley, J. D., & Suomi, S. J. (1989). Temperamental reactivity in non-human primates. In. G. A. Kohnstamm, J. E. Bates, & M. K. Rothbart (Eds.), *Temperament in child-hood* (pp. 153–167). New York: Wiley.

Higley, J. D., & Suomi, S. J. (1996). Effect of reactivity and social competence on individual responses to severe stress in children: Investigations using nonhuman primates. In C. R. Pfeffer (Ed.), *Intense stress and mental disturbance in children* (pp. 1–69). Washington, DC: American Psychiatric Press.

Higley, J. D., Suomi, S. J., & Linnoila, M. (1996a). A nonhuman primate model of type II alcoholism? Part 2: Diminished social competence and excessive aggression correlates with low cerebrospinal fluid 5-hydroxyindoleacetic acid concentrations. *Alcoholism: Clinical and Experimental Research, 20,* 643–650.

Higley, J. D., Suomi, S. J., & Linnoila, M. (1996b). A nonhuman primate model of type II excessive alcohol consumption? Part 1: Low cerebrospinal fluid 5-hydroxyindoleacetic acid concentrations and diminished social competence correlate with excessive alcohol consumption. *Alcoholism: Clinical and Experimental Research, 20,* 629–642.

Hill, C. T., & Stull, D. E. (1987). Gender and self-disclosure: Strategies for exploring the issues. In V. J. Derlega & J. H. Berg (Eds.), *Self-disclosure: Theory, research, and therapy* (pp. 81–101). New York: Plenum.

Hill, K., & Hurtado, A. M. (1996). *Ache life history.* New York: Aldine De Gruyter.

Hill, R. A., & Lee, P. C. (1998). Predation risk as an influence on group size in cercopithecoid primates: Implications for social structure. *Journal of Zoology, 245,* 447–456.

Hillbrand, M., Waite, B. M., Miller, D. S., Spitz, R. T., & Lingswiler, V. M. (2000). Serum cholesterol concentrations and mood states in violent psychiatric patients: An experience sampling study. *Journal of Behavioral Medicine, 23,* 519–529.

Hinde, R. A. (1972). Concepts of emotion. Ciba Foundation Symposium. *Physiology, Emotion, and Psychosomatic Illness, 8,* 3–13.

Hinde, R. A. (1974a). Mother-infant relations in rhesus monkeys. In N. F. White (Ed.), *Ethology and psychiatry* (pp. 29–46). Toronto: University of Toronto Press.

Hinde, R. A. (1974b). *Biological bases of human social behaviour.* New York: McGraw-Hill.

Hinde, R. A. (1976). Interactions, relationships, and social structure. *Man, 11,* 1–17.

Hinde, R. A. (1979). *Towards understanding relationships.* London: Academic Press.

Hinde, R. A. (1982). The uses and limitations of studies of nonhuman primates for the understanding of human social development. In L. W. Hoffman, R. Gandelman, & H. R. Schiffman (Eds.), *Parenting: Its causes and consequences* (pp. 5–17). Hillsdale, NJ: Erlbaum.

Hinde, R. A. (1985). Expression and negotiation. In G. Zivin (Ed.), *The development of expressive behavior* (pp. 103–116). New York: Academic Press.

Hinde, R. A. (1987). *Individuals, relationships, and culture: Links between ethology and the social sciences.* Cambridge: Cambridge University Press.

Hinde, R. A. (1990). The interdependence of the behavioural sciences. *Philosophical Transactions of the Royal Society of London, B, 329,* 119–129.

Hinde, R. A. (1995). A suggested structure for a science of relationships. *Personal Relationships, 2,* 1–15.

Hinde, R. A., & Atkinson, S. (1970). Assessing the roles of social partners in maintaining mutual proximity, as exemplified by mother-infant relations in rhesus monkeys. *Animal Behaviour, 18,* 169–176.

Hinde, R. A., & Rowell, T. E. (1962). Communication by postures and facial expressions in the rhesus monkey *(Macaca mulatta). Proceedings of the Zoological Society of London, 138,* 1–21.

Hinde, R. A., & Simpson, M. J. A. (1975). Qualities of mother-infant relationships in monkeys. CIBA Foundation Symposium, *Parent-infant interaction* (pp. 39–67). Amsterdam: Elsevier.

Hinde, R. A., & Spencer-Booth, Y. (1967). The behaviour of socially living rhesus monkeys in their first two and a half years. *Animal Behaviour, 15,* 169–196.

Hinde, R. A., & Spencer-Booth, Y. (1968). The study of mother-infant interaction in captive group-living rhesus monkeys. *Proceedings of the Royal Society of London, B, 169,* 177–201.

Hinde, R. A., & Spencer-Booth, Y. (1971). Towards understanding individual differences in rhesus mother-infant interaction. *Animal Behaviour, 19,* 165–173.

Hinde, R. A., Spencer-Booth, Y., & Bruce, M. (1966). Effects of 6-day maternal deprivation on rhesus monkey infants. *Nature, 210,* 1021–33.

Hiraiwa, M. (1981). Maternal and alloparental care in a troop of free-ranging Japanese monkeys. *Primates, 22,* 309–329.

Hiraiwa-Hasegawa, M., & Hasegawa, T. (1994). Infanticide in nonhuman primates: Sexual selection and local resource competition. In S. Parmigiani & F. S. vom Saal (Eds.), *Infanticide and parental care* (pp. 137–154). Chur, Switzerland: Harwood Academic Publishers.

Hisaw, F. L., & Hisaw, F. L. Jr. (1961). Action of estrogen and progesterone on the reproductive tract of lower primates. In W. C. Young (Ed.), *Sex and Internal Secretions* (3d ed.), Vol. 1 (pp. 556–589). Baltimore: Williams and Wilkins.

Hodos, W., & Campbell, C. B. G. (1969) Scala naturae: Why there is no theory in comparative psychology. *Psychological Review, 76,* 337–350.

Hoff-Ginsberg, E. (1991). Mother-child conversation in different social classes and communicative settings. *Child Development, 62,* 782–796.

Hoff-Ginsberg, E., & Tardif, T. (1995). Socioeconomic status and parenting. In

M. H. Bornstein (Ed.), *Handbook of parenting*, Vol. 2 (pp. 161–188). Hillsdale, NJ: Erlbaum.

Hoffman, K. A., Mendoza, S. P., Hennessy, M. B., & Mason, W. A. (1995). Responses of infant titi monkeys, *Callicebus moloch*, to removal of one or both parents: Evidence for paternal attachment. *Developmental Psychobiology, 28,* 399–407.

Hogan, R., DeSoto, C. B., & Solano, C. (1977). Traits, tests, and personality research. *American Psychologist, 32,* 255–264.

Hohmann, G. M. (1989). Vocal communication of wild bonnet macaques *(Macaca radiata). Primates, 30,* 325–345.

Hohmann, G. M., & Herzog, M. O. (1985). Vocal communication in lion-tailed macaques *(Macaca silenus). Folia Primatologica, 45,* 148–178.

Holloway, R. L., & De La Coste-Lareymondie, M. C. (1982). Brain endocast asymmetry in pongids and hominids: Some preliminary findings on the paleontology of cerebral dominance. *American Journal of Physical Anthropology, 58,* 101–110.

Holman, S. D., & Goy, R. W. (1988). Sexually dimorphic transitions revealed in the relationships of yearling rhesus monkeys following the birth of siblings. *International Journal of Primatology, 9,* 113–133.

Hook-Costigan, M. A., & Rogers, L. J. (1998). Lateralized use of the mouth in production of vocalizations by marmosets. *Neuropsychologia, 36,* 1265–73.

Hopkins, W. D. (1994). Hand preference for bimanual feeding in a sample of 140 chimpanzees *(Pan troglodytes):* Ontogenetic and developmental factors. *Developmental Psychobiology, 27,* 395–408.

Hopkins, W. D. (1995a). Hand preferences for a coordinated bimanual task in 110 chimpanzees: Cross-sectional analysis. *Journal of Comparative Psychology, 109,* 291–297.

Hopkins, W. D. (1995b). Hand preferences in juvenile chimpanzees: Continuity in development. *Developmental Psychology, 31,* 619–625.

Hopkins, W. D. (1996). Chimpanzee handedness revisited: 54 years since Finch (1941). *Psychonomic Bulletin and Review, 3,* 449–457.

Hopkins, W. D. (1999). On the other hand: Statistical issues in the assessment and interpretation of hand preference data in nonhuman primates. *International Journal of Primatology, 20,* 851–866.

Hopkins, W. D., Dahl, J. F., & Pilcher, D. (2000). Birth order and left-handedness revisited: Some recent findings in chimpanzees *(Pan troglodytes)* and their implications for developmental and evolutionary models of human handedness. *Neuropsychologia, 38,* 1626–33.

Hopkins, W. D., Dahl, J. F., & Pilcher, D. (2001). Genetic influence on the expression of hand preferences in chimpanzees *(Pan troglodytes):* Evidence in support of the right shift theory and developmental instability. *Psychological Science, 12,* 299–303.

Hopkins, W. D., & de Waal, F. B. M. (1995). Behavioral laterality in captive bonobos *(Pan paniscus):* Replication and extension. *International Journal of Primatology, 16,* 261–276.

Hopkins, W. D., & Fernandez-Carriba, S. (2002). Laterality in communicative behaviors in nonhuman primates: A critical analysis. In L. Rogers & R. Andrews (Eds.), *Comparative vertebrate lateralization* (pp. 445–479). Oxford: Oxford University Press.

Hopkins, W. D., & Leavens, D. A. (1998). Hand use and gestural communication in chimpanzees *(Pan troglodytes). Journal of Comparative Psychology, 112,* 95–99.

Hopkins, W. D., & Marino, L. (2000). Asymmetries in cerebral width in nonhuman primate brains as revealed by magnetic resonance imaging (MRI). *Neuropsychologia, 38,* 493–499.

Hopkins, W. D., Marino, L., Rilling, J. K., & MacGregor, L. A. (1998). Planum temporale asymmetries in great apes as revealed by magnetic resonance imaging (MRI). *NeuroReport, 9,* 2913–18.

Hopkins, W. D., & Pearson, K. (2000). Chimpanzee *(Pan troglodytes)* handedness: Variability across multiple measures of hand use. *Journal of Comparative Psychology, 114,* 126–135.

Hopkins, W. D., & Pilcher, D. L. (2001). Neuroanatomical localization of the motor hand area with magnetic resonance imaging: The left hemisphere is larger in great apes. *Behavioral Neuroscience, 115,* 1159–64.

Hopkins, W. D., Pilcher, D. L., & MacGregor, L. (2000). Sylvian fissure asymmetries in nonhuman primates revisited: A comparative MRI study. *Brain, Behavior, and Evolution, 56,* 293–299.

Hopkins, W. D., & Rilling, J. K. (2000). A comparative MRI study of the relationship between neuroanatomical asymmetry and interhemispheric connectivity in primates: Implication for the evolution of functional asymmetries. *Behavioral Neuroscience, 114,* 739–748.

Hopkins, W. D., & Wesley, M. J. (2002). Gestural communication in chimpanzees *(Pan troglodytes):* The influence of experimenter position on gesture type and hand preference. *Laterality, 7,* 19–30.

Horr, D. A. (1977). Orang-utan maturation: Growing up in a female world. In S. Chevalier-Skolnikoff & F. E. Poirier (Eds.), *Primate bio-social development* (pp. 289–321). New York: Garland.

Horrocks, J. A., & Hunte, W. (1983a). Maternal rank and offspring rank in vervet monkeys: An appraisal of the mechanisms of rank acquisition. *Animal Behaviour, 31,* 772–782.

Horrocks, J. A., & Hunte, W. (1983b). Rank relations in vervet sisters: A critique of the role of reproductive value. *American Naturalist, 122,* 417–421.

Hrdy, S. B. (1976). Care and exploitation of nonhuman primate infants by conspecifics other than the mother. *Advances in the Study of Behavior, 6,* 101–158.

Hrdy, S. B. (1984). Assumptions and evidence regarding the sexual selection hypothesis: A reply to Boggess. In G. Hausfater & S. B. Hrdy (Eds.), *Infanticide: Comparative and evolutionary perspectives* (pp. 315–319). New York: Aldine.

Hrdy, S. B. (1999). *Mother nature: A history of mothers, infants, and natural selection.* New York: Pantheon Books.

Huesmann, L. R., Eron, L. D., Lefkowitz, M. M., & Walder, L. O. (1984). Stability of aggression over time and generations. *Developmental Psychology, 20,* 1120–34.

Iatrakis, G., Haronis, N., Sakellaropoulos, G., Kourkoubas, A., & Gallos, M. (1986). Psychosomatic symptoms of postmenopausal women with or without hormonal treatment. *Psychotherapy and Psychosomatics, 46,* 116–121.

Ifune, C. K., Vermeire, B. A., & Hamilton, C. R. (1984). Hemispheric differences in split-brain monkeys viewing and responding to videotaped recordings. *Behavioral and Neural Biology, 41,* 231–235.

Imanishi, K. (1965). The origin of the human family: A primatological approach. *Japanese Journal of Ethnology, 25,* 119–138.

Insel, T. R. (1992). Oxytocin and the neurobiology of attachment. *Behavioral and Brain Sciences, 15,* 515–516.

Itakura, S. (1993). Emotional behavior during the learning of a contingency task in a chimpanzee. *Perceptual and Motor Skills, 76,* 563–566.

Itakura, S. (1996). An exploratory study of gaze-monitoring in nonhuman primates. *Japanese Psychological Research, 38,* 174–180.

Itakura, S., Agnetta, B., Hare, B., & Tomasello, M. (1999). Chimpanzees use human and conspecific social cues to locate hidden food. *Developmental Science, 2,* 448–456.

Itakura, S., & Anderson, J. R. (1996). Learning to use experimenter-given cues during an object-choice task by a capuchin monkey. *Current Psychology of Cognition, 15,* 103–112.

Itakura, S., & Tanaka, M. (1998) Use of experimenter-given cues during object-choice tasks by chimpanzees *(Pan troglodytes),* an orangutan *(Pongo pygmaeus),* and human infants *(Homo sapiens). Journal of Comparative Psychology, 112,* 119–126.

Itani, J. (1963). Vocal communication of the wild Japanese monkey. *Primates, 4,* 11–66.

Itani, J. (1975). Twenty years with Mount Takasaki monkeys. In B. Bermant & D. G. Lindburg (Eds.), *Primate utilization and conservation* (pp. 101–125). New York: John Wiley & Sons.

Itoigawa, N. (1973). Group organization of a natural troop of Japanese monkeys and mother-infant interactions. In C. R. Carpenter (Ed.), *Behavioral regulators of behavior in primates* (pp. 229–243). Lewisburg, PA: Bucknell University Press.

Itoigawa, N. (1993). Social conflict in adult male relationships in a free-ranging group of Japanese monkeys. In W. A. Mason & S. P. Mendoza (Eds.), *Primate social conflict* (pp. 145–169). New York: SUNY Press.

Izard, C. E. (1977). *Human emotions.* New York: Plenum.

Izard, C. E. (1992). Basic emotions, relations among emotions, and emotion-cognition relations. *Psychological Review, 99,* 561–565.

Izard, C. E. (1993). Four systems for emotion activation: Cognitive and noncognitive processes. *Psychological Review, 100,* 68–90.

James, W. (1884). What is an emotion? *Mind, 9,* 188–205.

Jancke, L., & Steinmetz, H. (1994). Interhemispheric transfer time and corpus callosum size. *NeuroReport, 5,* 2385–88.

Janus, M. (1992). Interplay between various aspects in social relationships of young rhesus monkeys: Dominance, agonistic help, and affiliation. *American Journal of Primatology, 26,* 291–308.

Jay, P. (1965). The common langur of North India. In I. DeVore (Ed.), *Primate behavior: Field studies of monkeys and apes* (pp. 197–249). New York: Holt, Rinehart & Winston.

Jay, P. C. (1963). Mother-infant relations in langurs. In H. L. Rheingold (Ed.), *Maternal behavior in mammals* (pp. 282–304). New York: John Wiley & Sons.

Jensen, G. D. & Gordon, B. N. (1970). Sequences of mother-infant behavior following a facial communicative gesture of pigtail monkeys. *Biological Psychiatry, 2,* 267–272.

Jensen, G. D., & Tolman, C. W. (1962). Mother-infant relationship in the monkey *Macaca nemestrina:* The effect of brief separation on mother-infant specificity. *Journal of Comparative and Physiological Psychology, 55,* 131–136.

Jensvold, M. L. A., & Gardner, R. A. (2000). Interactive use of sign language by cross-fostered chimpanzees *(Pan troglodytes). Journal of Comparative Psychology, 114,* 335–346.

Jerison, H. J. (1973). *Evolution of the brain and intelligence.* New York: Academic Press.

Jerison, H. J. (1985). On the evolution of mind. In D. A. Oakley (Ed.), *Brain and mind* (p. 1–31). London: Metheun.

John, O. P., & Benet-Martinez, V. (2000). Measurement, scale construction, and reliability. In H. T. Reis and C. M. Judd (Eds.), *Handbook of research methods in social psychology* (pp. 339–369). Cambridge: Cambridge University Press.

John, O. P., & Gosling, S. D. (2000). Personality traits. In A. E. Kazdin (Ed.), *Encyclopedia of psychology,* Vol. 6 (pp. 140–144.) Washington, DC: American Psychological Association.

John, O. P., & Robins, R. W. (1993). Determinants of interjudge agreement on personality traits: The big five domains, observability, evaluativeness, and the unique perspective of the self. *Journal of Personality, 61,* 521–551.

John, O. P., & Srivastava, S. (1999). The Big Five trait taxonomy: History, measurement, and theoretical perspectives. In L. A. Pervin & O. P. John (Eds.), *Handbook of personality: Theory and research* (2nd ed.) (pp. 102–138). New York: Guilford Press.

Johnson, D. F., & Phoenix, C. H. (1976). Hormonal control of female sexual attractiveness, proceptivity, and receptivity in rhesus monkeys. *Journal of Comparative and Physiological Psychology, 90,* 473–482.

Johnson, P. (1976). Women and power: Toward a theory of effectiveness. *Journal of Social Issues, 32,* 99–110.

Johnson, R. L., Berman, C. M., & Malik, I. (1993). An integrative model of the lactational and environmental control of mating in female rhesus monkeys. *Animal Behaviour, 46,* 63–78.

Johnson-Laird, P. N., & Oatley, K. (1989). The language of emotions: An analysis of a semantic field. *Cognition and Emotion, 3,* 81–123.

Johnson-Laird, P. N., & Oatley, K. (1992). Basic emotions, rationality, and folk theory. *Cognition and Emotion, 6,* 201–223.

Joslin, J., Fletcher, H., & Emlen, J. (1964). A comparison of the responses to snakes of lab- and wild-reared rhesus monkeys. *Animal Behaviour, 12,* 348–352.

Judge, P. G. (1982). Redirection of aggression based on kinship in a captive group of pigtail macaques. *International Journal of Primatology, 3,* 301.

Judge, P. G. (1991). Dyadic and triadic reconciliation in pigtail macaques *(Macaca nemestrina). American Journal of Primatology, 23,* 225–237.

Judge, P. G. (2000). Coping with crowded conditions. In F. Aureli & F. B. M. de Waal (Eds.), *Natural conflict resolution* (pp. 129–154). Berkeley: University of California Press.

Judge, P. G., Bernstein, I. S., & Ruehlmann, T. E. (1997). Reconciliation and other postconflict behavior in juvenile rhesus macaques *(Macaca mulatta). American Journal of Primatology, 42,* 120.

Judge, P. G., & de Waal, F. B. M. (1993). Conflict avoidance among rhesus monkeys: Coping with short-term crowding. *Animal Behaviour, 46,* 221–232.

Judge, P. G., & de Waal, F. B. M. (1997). Rhesus monkey behaviour under diverse population densities: Coping with long-term crowding. *Animal Behaviour, 54,* 643–662.

Jürgens, U. (1979). Vocalization as an emotional indicator: A neuroethological study in the squirrel monkey. *Behaviour, 69,* 88–117.

Jürgens, U. (1995). Neuronal control of vocal production in non-human and human primates. In E. Zimmerman, J. D. Newmann, & U. Jürgens (Eds.), *Current topics in primate vocal communication* (pp. 199–206). New York: Plenum Press.

Jürgens, U. (1998). Common features in the vocal expression of emotion in human and non-human primates. In S. Santi, J. Guaitella, C. Cavé, & G. Konopczynski (Eds.), *Oralité et gestualité* (pp. 153–158). Paris: L'Harmattan.

Kagan, J. (1992). The meanings of attachment. *Behavioral and Brain Sciences, 15*, 517–518.

Kagan, J. (1998). *Galen's prophecy.* Boulder, CO: Westview Press.

Kagan, J., Reznick, J. S., & Gibbons, J. (1989). Inhibited and uninhibited types of children. *Child Development, 60*, 838–845.

Kalin, N. H., Davidson, R. J., Irwin, W., Warner, G., Orendi, J. L., Sutton, S. K., Mock, B. J., Sorenson, J. A., Lowe, M., & Turski, P. A. (1997). Functional magnetic resonance imaging studies of emotional processing in normal and depressed patients: Effects of venlafaxine. *Journal of Clinical Psychiatry, 58 (Supplement 16)*, 32–39.

Kalin, N. H., Larson, C., Shelton, S. E., & Davidson, R. J. (1998). Asymmetric frontal brain activity, cortisol, and behavior associated with fearful temperament in rhesus monkeys. *Behavioral Neuroscience, 112*, 286–292.

Kalin, N. H., Shelton, S. E., & Snowdon, C. T. (1992). Affiliative vocalizations in infant rhesus monkeys *(Macaca mulatta). Journal of Comparative Psychology, 106*, 254–261.

Kalin, N. H., Shelton, S. E., & Takahashi, L. K. (1991). Defensive behaviors in infant rhesus monkeys: Ontogeny and context-dependent selective expression. *Child Development, 62*, 1175–83.

Kanazawa, S. (1996). Recognition of facial expressions in a Japanese monkey. *Primates, 37*, 25–38.

Kaplan, J. (1977a). Perceptual properties of attachment in surrogate-reared and mother-reared squirrel monkeys. In S. Chevalier-Skolnikoff & F. E. Poirier (Eds.), *Primate bio-social development* (pp. 225–234). New York: Garland.

Kaplan, J. R. (1977b). Patterns of fight interference in free-ranging rhesus monkeys. *American Journal of Physical Anthropology, 47*, 279–287.

Kaplan, J. R., Manuck, S. B., & Shively, C. (1991). The effects of fat and cholesterol on social behavior in monkeys. *Psychosomatic Medicine, 53*, 634–642.

Kaplan, J. R., Phillips-Conroy, J., Fontenot, M. B., Jolly, C. J., Fairbanks, L. A., & Mann, J. J. (1999). Cerebrospinal fluid monoaminergic metabolites differ in wild anubis and hybrid *(Anubis hamadryas)* baboons: Possible relationships to life history and behavior. *Neuropsychopharmacology, 20*, 517–524.

Kaplan, J. R., Shively, C. A., Fontenot, M. B., Morgan, T. M., Howell, S. M., Manuck, S. B., Muldoon, M. F., & Mann, J. J. (1994). Demonstration of an association among dietary cholesterol, central serotonergic activity, and social behavior in monkeys. *Psychosomatic Medicine, 56*, 479–484.

Karin-D'Arcy, R., & Povinelli, D. J. (under review). Do chimpanzees know what each other see? A closer look.

Kaufman, I. C. (1974). Mother-infant relations in monkeys and humans: A reply to Professor Hinde. In N. F. White (Ed.), *Ethology and psychiatry* (pp. 47–68). Toronto: University of Toronto Press.

Kaufman, I. C., & Rosenblum, L. A. (1966). A behavioral taxonomy for *Macaca nemest-rina* and *Macaca radiata:* Based on longitudinal observation of family groups in the laboratory. *Primates, 7,* 205–258.

Kaufman, I. C., & Rosenblum, L. A. (1967). The reaction to separation in monkeys: Anaclitic depression and conservation-withdrawal. *Psychosomatic Medicine, 29,* 649–675.

Kaufmann, J. H. (1967). Social relations of adult males in a free-ranging band of rhesus monkeys. In S. A. Altmann (Ed.), *Social communication among primates* (pp. 73–98). Chicago: University of Chicago Press.

Kawai, M. (1958). On the rank system in a natural group of Japanese monkeys. I: The basic and dependent rank. *Primates, 1,* 111–130.

Kawai, M. (1965). Newly acquired precultural behavior of the natural troop of Japanese macaques. *Primates, 6,* 1–10.

Keating, C. F., & Keating, E. G. (1993). Monkeys and mug shots: Cues used by rhesus monkeys *(Macaca mulatta)* to recognize a human face. *Journal of Comparative Psychology, 107,* 131–139.

Keenan, K., & Shaw, D. S. (1994). The development of aggression in toddlers: A study of low-income families. *Journal of Abnormal Child Psychology, 22,* 53–77.

Keil, F. C., & Wilson, R. A. (2000). *Explanation and cognition.* Cambridge, MA: MIT Press.

Kellogg, W. N., & Kellogg, L. A. (1933). *The ape and the child: A study of environmental influences upon early behavior.* New York: McGraw-Hill.

Kennedy, G. E., & Keeney, V. Y. (1988). The extended family revisited: Grandparents rearing grandchildren. *Child Psychiatry and Human Development, 19,* 26–35.

Kennedy, J. S. (1992). *The new anthropomorphism.* Cambridge: Cambridge University Press.

Kenny, D. A. (1991). A general model of consensus and accuracy in interpersonal perception. *Psychological Review, 98,* 155–163.

Kenny, D. A. (1994). *Interpersonal perception: A social relations analysis.* New York: Guilford Press.

Kenny, D. A., & Albright, L. (1987). Accuracy in interpersonal perception: A social relations analysis. *Psychological Bulletin, 102,* 390–402.

Kenny, D. A., Albright, L., Malloy, T. E., & Kashy, D. A. (1994). Consensus in interpersonal perception: Acquaintance and the big five. *Psychological Bulletin, 116,* 245–258.

Kenrick, D. T., & Funder, D. C. (1988). Profiting from controversy: Lessons from the person-situation debate. *American Psychologist, 43,* 23–34.

Kessler, R. C., McGonagle, K. A., Zhao, S., Nelson, C. B., Hughes, M., Eshleman, S., Wittchen, H. U., & Kendler, K. S. (1994). Lifetime and 12-month prevalence of

DSM-III-R psychiatric disorders in the United States. *Archives of General Psychiatry, 51*, 8–19.

Keverne, E. B., Eberhart, J. A., & Meller, R. E. (1983). Plasma testosterone, sexual, and aggressive behavior in social groups of talapoin monkeys. In H. D. Steklis & A. S. King (Eds.), *Hormones, drugs, and social behavior in primates* (pp. 33–55). New York: Spectrum.

Keverne, E. B., Martensz, N. D., & Tuite, B. (1989). Beta-endorphin concentrations in cerebrospinal fluid of monkeys are influenced by grooming relationships. *Psychoneuroendocrinology, 14*, 155–161.

Killen, M., & de Waal, F. B. M. (2000). The evolution and development of morality. In F. Aureli & F. B. M. de Waal (Eds.), *Natural conflict resolution* (pp. 352–372). Berkeley: University of California Press.

Killen, M., & Turiel, E. (1991). Conflict resolution in preschool social interactions. *Early Education and Development, 2*, 240–255.

King, J. E., & Figueredo, A. J. (1997). The five-factor model plus dominance in chimpanzee personality. *Journal of Research in Personality, 31*, 257–271.

Kingstone, A., Friesen, C. K., & Gazzaniga, M. S. (2000). Reflexive joint attention depends on lateralized cortical connections. *Psychological Science, 11*, 159–166.

Kirkevold, B. C., Lockard, J. S., & Heestand, J. E. (1982). Developmental comparisons of grimace and play mouth in infant pigtail macaques *(Macaca nemestrina)*. *American Journal of Primatology, 3*, 277–283.

Kitko, R., Gesser, D., & Owren, M. J. (1999). Noisy screams of macaques may function to annoy conspecifics. *Journal of the Acoustical Society of America, 106*, 2221.

Klaus, M. H., & Kennell, J. H. (1976). *Maternal-infant bonding.* St. Louis: Mosby.

Koeppel, L. B., Montagne-Miller, Y., O'Hair, D., & Cody, M. J. (1993). Friendly? Flirting? Wrong? In P. J. Kalbfleisch (Ed.), *Interpersonal communication: Evolving interpersonal relationships* (pp. 13–32). Hillsdale, NJ: Erlbaum.

Koford, C. B. (1963). Rank of mothers and sons in bands of rhesus monkeys. *Science, 141*, 356–357.

Koford, C. B. (1965). Population dynamics of rhesus monkeys on Cayo Santiago. In I. DeVore (Ed.), *Primate behavior: Field studies of monkeys and apes* (pp. 53–110). New York: Holt, Rinehart & Winston.

Kohlberg, L. (1981). *The philosophy of moral development: Moral stages and the idea of justice.* San Francisco: Harper & Row.

Köhler, W. (1925). *The mentality of the apes.* London: Routledge and Kegan Paul.

Kolb, B., & Taylor, L. (1990). Neocortical substrates of emotional behavior. In N. L. Stein, B. Leventhal, & T. Trabasso (Eds.), *Psychological and biological approaches to emotion* (pp. 115–144). Hillsdale, NJ: Erlbaum.

Kondo-Ikemura, K., & Waters, E. (1995). Maternal behavior and infant security in

Old World monkeys: Conceptual issues and a methodological bridge between human and nonhuman primate research. In E. Waters, B. Vaughn, G. Posada, & K. Kondo-Ikemura (Eds.), *Caregiving, cultural, and cognitive perspectives on secure-base behavior and working models* (pp. 97–110). Chicago: University of Chicago Press.

Kostan, K. M., & Snowdon, C. T. (2002). Attachment and social preferences in cooperatively-reared cotton-top tamarins. *American Journal of Primatology, 57*, 131–139.

Koyama, N. F., & Dunbar, R. I. M. (1996). Anticipation of conflict by chimpanzees. *Primates, 37*, 79–86.

Kraemer, G. W. (1992). A psychobiological theory of attachment. *Behavioral and Brain Sciences, 15*, 493–541.

Kraemer, G. W. (1997). Psychobiology of early social attachment in rhesus monkeys. Clinical implications. *Annals of the New York Academy of Science, 807*, 401–418.

Kraemer, G. W., & Clarke, A. S. (1990). The behavioral neurobiology of self-injurious behavior in rhesus monkeys. *Progress in Neuro-Psycho-Pharmacology and Biological Psychiatry, 14*, S141–S168.

Krause, M. A., & Fouts, R. S. (1997). Chimpanzee *(Pan troglodytes)* pointing: Hand shapes, accuracy, and the role of eye gaze. *Journal of Comparative Psychology, 111*, 330–336.

Krebs, J. R., & Dawkins, R. (1984) Animal signals: Mind-reading and manipulation. In J. R. Krebs & N. B. Davies (Eds.), *Behavioural ecology: An evolutionary approach* (2nd ed.) (pp. 380–402). Oxford: Blackwell Scientific.

Kruesi, M. J., Rapoport, J. L., Hamburger, S., Hibbs, E., Potter, W. Z., Lenane, M., & Brown, G. L. (1990). Cerebrospinal fluid monoamine metabolites, aggression, and impulsivity in disruptive behavior disorders of children and adolescents. *Archives of General Psychiatry, 47*, 419–426.

Kruglanski, A. W. (1989). The psychology of being "right": The problem of accuracy in social perception and cognition. *Psychological Bulletin, 106*, 395–409.

Küderling, I., Evans, C. S., Abbott, D. H., Pryce, C. R., & Epple, G. (1995). Differential excretion of urinary estrogen by breeding females and daughters in the red-bellied tamarin *(Saguinus labiatus)*. *Folia Primatologica, 64*, 140–145.

Kuester, J., & Paul, A. (1992). Influence of male competition and female mate choice on male mating success in Barbary macaques *(Macaca sylvanus)*. *Behaviour, 120*, 192–217.

Kummer, H. (1995). *In quest of the sacred baboon: A scientist's journey.* Princeton: Princeton University Press.

Kummer, H., Anzenberger, G., & Hemelrijk, C. K. (1996). Hiding and perspective taking in long-tailed macaques *(Macaca fascicularis)*. *Journal of Comparative Psychology, 110*, 97–102.

Kurland, J. A. (1977). Kin selection in the Japanese monkey. *Contributions to Primatology,* Vol. 12. Basel: Karger.

Kurzban, R., & Leary, M. R. (2001). Evolutionary origins of stigmatization: The functions of social exclusion. *Psychological Bulletin, 127,* 187–208.

Kuyk, K., Dazey, J., & Erwin, J. (1977). Play patterns of pigtail monkey infants: Effects of age and peer presence. *Journal of Biological Psychology, 18,* 20–23.

Kwan, M., Greenleaf, W. J., Mann, J., Crapo, L., & Davidson, J. M. (1983). The nature of androgen action on male sexuality: A combined laboratory-self-report study on hypogonadal men. *Journal of Clinical Endocrinology and Metabolism, 57,* 557–562.

Kyes, R. C., Botchin, M. B., Kaplan, J. R., Manuck, S. B., & Mann, J. J. (1995). Aggression and brain serotonergic responsivity: Response to slides in male macaques. *Physiology and Behavior, 57,* 205–208.

Kyes, R. C., & Candland, D. K. (1987). Baboon *(Papio hamadryas)* visual preferences for regions of the face. *Journal of Comparative Psychology, 101,* 345–348.

Lacy, R. C., & Sherman, P. W. (1983). Kin recognition by phenotypic matching. *American Naturalist, 121,* 489–512.

Lamb, M. E. (2000). The history of research on father involvement: An overview. *Marriage and Family Review, 29,* 23–42.

Lancaster, J. B. (1975). *Primate behavior and the emergence of human culture.* New York: Holt, Rinehart, & Winston.

Lancaster, J. B., Kaplan, H. S., Hill, K., & Hurtado, A. M. (2000). The evolution of life history, intelligence, and diet among chimpanzees and human foragers. *Perspectives in Ethology, 13,* 47–72.

Langer, J. (1998). Phylogenetic and ontogenetic origins of cognition: Classification. In J. Langer & M. Killen (Eds.), *Piaget, evolution, and development* (pp. 33–54). Mahwah, NJ: Erlbaum.

Langer, J. (2000). Comparative mental development. *Journal of Adult Development, 7,* 23–30.

Langton, S. R. H., & Bruce, V. (1999). Reflexive visual orienting in response to the social attention of others. *Visual Cognition, 6,* 541–567.

Lanzetta, J. T., & McHugo, G. J. (1989). Facial expressive and psychophysiological correlates of emotion. In G. Gainotti & C. Caltagirone (Eds.), *Emotions and the dual brain* (pp. 91–118). Berlin: Springer-Verlag.

Laudenslager, M. L., & Boccia, M. L. 1996. Some observations on psychosocial stressors, immunity, and individual differences in nonhuman primates. *American Journal of Primatology, 39,* 205–221.

Laursen, B. (1993). Conflict management among close peers. In B. Laursen (Ed.), *Close friendships in adolescence,* Vol. 60 (pp. 39–54). San Francisco: Jossey-Bass Publishers.

Lazarus, R. S. (1991). *Emotion and adaptation.* New York: Oxford University Press.

Leavens, D. A., Aureli, F., Hopkins, W. D., & Hyatt, C. W. (2001). Effects of cognitive challenge on self-directed behaviors by chimpanzees *(Pan troglodytes). American Journal of Primatology, 55*, 1–14.

Leavens, D. A., & Hopkins, W. D. (1998). Intentional communication by chimpanzees: A cross-sectional study of the use of referential gestures. *Developmental Psychology, 34*, 813–822.

Leavens, D. A., & Hopkins, W. D. (1999). The whole-hand point: The structure and function of pointing from a comparative perspective. *Journal of Comparative Psychology, 113*, 417–425.

Leavens, D. A., Hopkins, W. D., & Bard, K. A. (1996). Indexical and referential pointing in chimpanzees *(Pan troglodytes). Journal of Comparative Psychology, 110*, 346–353.

LeDoux, J. E. (1994). Emotion-specific physiological activity: Don't forget about CNS physiology. In P. Ekman & R. J. Davidson (Eds.), *The nature of emotion: Fundamental questions* (pp. 248–251). New York: Oxford University Press.

LeDoux, J. E. (1995). In search of an emotional system in the brain: Leaping from fear to emotion and consciousness. In M. S. Gazzaniga (Ed.), *The cognitive neurosciences* (pp. 1049–61). Cambridge, MA: MIT Press.

LeDoux, J. E. (1996). *The emotional brain.* New York: Simon and Schuster.

LeDoux, J. E. (2000). Emotion circuits in the brain. *Annual Review of Neuroscience, 23*, 155–184.

Lee, P. C. (1984). Ecological constraints on the social development of vervet monkeys. *Behaviour, 91*, 245–262.

Lehrman, D. S. (1974). Can psychiatrists use ethology? In N. F. White (Ed.), *Ethology and psychiatry* (pp. 187–196). Toronto: University of Toronto Press.

Leiblum, S., Bachmann, G., Kemmann, E., Colburn, D., & Swartzman, L. (1983). Vaginal atrophy in the postmenopausal woman: The importance of sexual activity and hormones. *Journal of the American Medical Association, 249*, 2195–98.

Leighton, D. R. (1987). Gibbons: Territoriality and monogamy. In B. B. Smuts, D. L. Cheney, R. M. Seyfarth, R. W. Wrangham, & T. T. Struhsaker (Eds.), *Primate societies* (pp. 135–145). Chicago: University of Chicago Press.

Leinonen, L., Linnankoski, I., Laakso, M. L., & Aulanko, R. (1991). Vocal communication between species: Man and macaque. *Language and Communication, 11*, 241–262.

LeMay, M. (1976). Morphological cerebral asymmetries of modern man, fossil man and nonhuman primate. *Annals of the New York Academy of Sciences, 280*, 349–366.

LeMay, M., Billig, M. S., & Geschwind, N. (1982). Asymmetries of the brains and skulls of nonhuman primates. In E. Armstrong & D. Falk (Eds.), *Primate brain evolution: Methods and concepts* (pp. 263–277). New York: Plenum Press.

LeMay, M., & Geschwind, N. (1975). Hemispheric differences in the brains of great apes. *Brain, Behavior, and Evolution, 11*, 48–52.

Lenzenweger, M. F. (1999a). Stability and change in personality disorder features. *Archives of General Psychiatry, 56*, 1009–15.

Lenzenweger, M. F. (1999b). Deeper into the schizotypy taxon: On the robust nature of maximum covariance analysis. *Journal of Abnormal Psychology, 108*, 182–187.

Le Prell, C. G., & Moody, D. B. (2000). Factors influencing the salience of temporal cues in the discrimination of Japanese monkey coo calls. *Journal of Experimental Psychology: Animal Behavior Processes, 26*, 261–273.

Lesch, K. P., Meyer, J., Glatz, K., Flügge, G., Hinney, A., Hebebrand, J., Klauch, S. M., Poustka, A., Poustka, F., Bengel, D., Mössner, R., Riederer, P., & Heils, A. (1997). The 5-HT transporter gene-linked polymorphic region (5-HTTLPR) in evolutionary perspective: Alternative biallelic variation in rhesus monkeys. *Journal of Neural Transmission, 104*, 1259–66.

Leslie, A. M. (1984). Infant perception of a manual pick-up event. *British Journal of Developmental Psychology, 2*, 19–32.

Leslie, A. M. (1994). ToMM, ToBy, and Agency: Core architecture and domain specificity. In L. Hirschfeld & S. Gelman (Eds.), *Mapping the mind: Domain specificity in cognitive and culture* (pp. 119–148). Cambridge: Cambridge University Press.

Levenson, R. W. (1992). Autonomic nervous system differences among emotions. *Psychological Science, 3*, 23–27.

Leveroni, C. L., & Berenbaum, S. A. (1998). Early androgen effects on interest in infants: Evidence from children with congenital adrenal hyperplasia. *Developmental Neuropsychology, 14*, 321–340.

Levine, S., & Wiener, S. G. (1988). Psychoendocrine aspects of mother-infant relationships in nonhuman primates. *Psychoneuroendocrinology, 13*, 143–154.

Levine, S., Wiener, S. G., Coe, C. L., Bayart, F. E., & Hayashi, K. T. (1987). Primate vocalization: A psychobiological approach. *Child Development, 58*, 1408–19.

Levy, J., Heller, W., Banich, M. T., & Burton, L. A. (1983). Asymmetry of perception in free viewing of chimeric faces. *Brain and Cognition, 2*, 404–419.

Lewis, M., & Rosenblum, L. A. (Eds.) (1974). *The effect of the infant on its caregiver.* New York: John Wiley & Sons.

Liang, B., Zhang, S., & Wang, L. (2000). Development of sexual morphology, physiology, and behaviour in Sichuan golden monkeys, *Rhinopithecus roxellana. Folia Primatologica, 71*, 413–416.

Lidberg, L., Tuck, J. R., Åsberg, M., Scalia-Tomba, G. P., & Bertilsson, L. (1985). Homicide, suicide, and CSF 5-HIAA. *Acta Psychiatrica Scandinavica, 71*, 230–236.

Lieberman, P. (1984). *The biology and evolution of language.* Cambridge, MA: Harvard University Press.

Lilienfeld, S. O., Gershon, J., Duke, M., Marino, L., & de Waal, F. B. M. (1999). A preliminary investigation of the construct of psychopathic personality (psychopathy) in chimpanzees *(Pan troglodytes). Journal of Comparative Psychology, 113*, 365–375.

Lillard, A. S. (1998). Ethnopsychologies: Cultural variations in theories of mind. *Psychological Bulletin, 123*, 3–32.

Lillehei, R. A., & Snowdon, C. T. (1978). Individual and situational differences in the vocalizations of young stumptail macaques *(Macaca arctoides). Behaviour, 65*, 270–281.

Limongelli, L., Boysen, S. T., & Visalberghi, E. (1995). Comprehension of cause-effect relations in a tool-using task by chimpanzees *(Pan troglodytes). Journal of Comparative Psychology, 109*, 18–26.

Limson, R., Goldman, D., Roy, A., Lamparski, D., Ravitz, B., Adinoff, B., & Linnoila, M. (1991). Personality and cerebrospinal fluid monoamine metabolites in alcoholics and controls. *Archives of General Psychiatry, 48*, 437–441.

Lindburg, D. G. (1971). The rhesus monkey in North India: An ecological and behavioral study. In L. A. Rosenblum (Ed.), *Primate behavior: Developments in field and laboratory research*, Vol. 2 (pp. 1–106). New York: Academic Press.

Lindburg, D. G. (1973). Grooming behavior as a regulator of social interaction in rhesus monkeys. In C. R. Carpenter (Ed.), *Behavioral regulators of behavior in primates* (pp. 124–148). Lewisburg, PA: Bucknell University Press.

Lindburg, D. G. (1991). Ecological requirements of macaques. *Laboratory Animal Science, 41*, 315–322.

Lindman, R., Järvinen, P., & Vidjeskog, J. (1987). Verbal interactions of aggressively and nonaggressively predisposed males in a drinking situation. *Aggressive Behavior, 13*, 187–196.

Linn, G. S., Mase, D., LaFrancois, D., O'Keeffe, R. T., & Lifshitz, K. (1995). Social and menstrual cycle phase influences on the behavior of group-housed *Cebus apella. American Journal of Primatology, 35*, 41–57.

Linnoila, M., Virkkunen, M., Scheinin, M., Nuutila, A., Rimon, R., & Goodwin, F. K. (1983). Low cerebrospinal fluid 5-hydroxyindoleacetic acid concentration differentiates impulsive from nonimpulsive violent behavior. *Life Sciences, 33*, 2609–14.

Little, B. R. (1996). Free traits, personal projects, and idio-tapes: Three tiers for personality psychology. *Psychological Inquiry, 7*, 340–344.

Littlefield, C. H., & Rushton, J. P. (1986). When a child dies: The sociobiology of bereavement. *Journal of Personality and Social Psychology, 51*, 797–802.

Ljungberg, T., Westlund, K., & Forsberg, A. J. L. (1999). Conflict resolution in 5-year-old boys: Does post-conflict affiliative behavior have a reconciliatory role? *Animal Behaviour, 58*, 1007–16.

Locke, K. D., Locke, E. A., Morgan, G. A. Jr., & Zimmerman, R. R. (1964a). Dimensions of social interactions among infant rhesus monkeys. *Psychological Reports, 15, 339–349.*

Locke, K. D., Morgan, G. A., Jr., & Zimmermann, R. R. (1964b). Method for observing social interactions in groups of infant rhesus monkeys. *Psychological Reports, 14, 83–91.*

Locke-Haydon, J., & Chalmers, N. R. (1983). The development of infant care-giver relationships in captive common marmosets *(Callithrix jaccus). International Journal of Primatology, 4, 63–81.*

Loeser, A. A. (1940). Subcutaneous implantation of female and male hormone tablet form in women. *British Medical Journal, 1, 479–482.*

Loevinger, J. (1957). Objective tests as instruments of psychological theory. *Psychological Reports, 3, 635–694.*

Loevinger, J. (1994). Has psychology lost its conscience? *Journal of Personality Assessment, 62, 2–8.*

Logothetis, N. K., Guggenberger, H., Peled, S., & Pauls, J. (1999). Functional imaging of the monkey brain. *Nature Neuroscience, 2, 555–562.*

Lopez, J. F., Vazquez, D. M., Zimmer, C. A., Little, K. Y., & Watson, S. J. (2001). Chronic unpredictable stress and antidepressant modulation of mineralocorticoid, and glucocorticoid receptors. *Society for Neuroscience Abstracts, 27.*

Lorenz, K. (1966). *On aggression.* London: Methuen.

Lorincz, E. N., Baker, C. I., & Perrett, D. I. (1999). Visual cues for attention following in rhesus monkeys. *Cahiers de Psychologie Cognitive, 18, 973–1003.*

Lott, D. F. (1991). *Intraspecific variation in the social system of wild vertebrates.* Cambridge: Cambridge University Press.

Lovejoy, J., & Wallen, K. (1988). Sexually dimorphic behavior in group-housed rhesus monkeys *(Macaca mulatta)* at 1 year of age. *Psychobiology, 16, 348–356.*

Lovejoy, J., & Wallen, K. (1990). Adrenal suppression and sexual initiation in group-living female rhesus monkeys. *Hormones and Behavior, 24, 256–269.*

Low, B. S. (1989). Cross-cultural patterns in the training of children: An evolutionary perspective. *Journal of Comparative Psychology, 103, 311–319.*

Lowe Vandell, D. (2000). Parents, peer groups, and other socializing influences. *Developmental Psychology, 36, 699–710.*

Lunn, S. F., Recio, R., Morris, K., & Fraser, H. M. (1994). Blockade of the neonatal rise in testosterone by a gonadotropin-releasing-hormone antagonist: Effects on timing of puberty and sexual behavior in the male marmoset monkey. *Journal of Endocrinology, 141, 439–447.*

Lycett, J. S., Henzi, S. P., & Barrett, L. (1998). Maternal investment in mountain baboons and the hypothesis of reduced care. *Behavioral Ecology and Sociobiology, 42, 49–56.*

Lykken, D. T. (1995). *The antisocial personalities.* Hillsdale, NJ: Erlbaum.

Lyons, D. M., Kim, S., Schatzberg, A. F., & Levine, S. (1998). Postnatal foraging demands alter adrenocortical activity and psychosocial development. *Developmental Psychobiology, 32,* 285–291.

Lytton, H., & Romney, D. M. (1991). Parents' different socialization of boys and girls: A meta-analysis. *Psychological Bulletin, 109,* 267–296.

Maccoby, E. E. (1992). The role of parents in the socialization of children: An historical overview. *Developmental Psychology, 6,* 1006–17.

Maccoby, E. E. (1998). *The two sexes: Growing up apart, coming together.* Cambridge, MA: Belknap Press of Harvard University Press.

Maccoby, E. E. (2000a). Parenting and its effects on children: On reading and misreading behavior genetics. *Annual Review of Psychology, 51,* 1–27.

Maccoby, E. E. (2000b). Perspectives on gender development. *International Journal of Behavioral Development, 24,* 398–406.

Maccoby, E. E., & Jacklin, C. N. (1974). *The psychology of sex differences.* Stanford: Stanford University Press.

Maccoby, E. E., & Jacklin, C. N. (1987). Gender segregation in childhood. In E. H. Reese (Ed.), *Advances in child development and behavior,* Vol. 20 (pp. 239–287). New York: Academic Press.

Maccoby, E. E., & Martin, J. A. (1983). Socialization in the context of the family: Parent-child interaction. In E. M. Hetherington (Ed.), *Handbook of child psychology,* Vol. 4 (pp. 1–101). New York: John Wiley.

Macedonia J. M., & Evans C. S. (1993) Variation among mammalian alarm call systems and the problem of meaning in animal signals. *Ethology, 93,* 177–197.

MacLean, P. D. (1952). Some psychiatric implications of physiological studies on frontotemporal portion of limbic system (visceral brain). *Electroencephalography and Clinical Neurophysiology, 4,* 407–418.

MacLean, P. D. (1990). *The triune brain in evolution: Role in paleocerebral functions.* New York: Plenum Press.

MacNeilage, P. F. (1998). The frame/content theory of evolution of speech production. *Behavioral and Brain Sciences, 21,* 499–511.

MacNeilage, P. F., Studdert-Kennedy, M. G., & Lindblom, B. (1987). Primate handedness reconsidered. *Behavioral and Brain Sciences, 10,* 247–303.

Maes, M., & Meltzer, H. Y. (1995). The serotonin hypothesis of major depression. In F. E. Bloom & D. J. Kupfer (Eds.), *Psychopharmacology: The fourth generation of progress* (pp. 933–944). New York: Raven Press.

Maestripieri, D. (1993a). Maternal anxiety in rhesus macaques *(Macaca mulatta).* I: Measurement of anxiety and identification of anxiety-eliciting situations. *Ethology, 95,* 19–31.

Maestripieri, D. (1993b). Maternal anxiety in rhesus macaques *(Macaca mulatta)*. II: Emotional bases of individual differences in mothering style. *Ethology, 95,* 32–42.

Maestripieri, D. (1993c). Vigilance costs of allogrooming in macaque mothers. *American Naturalist, 141,* 744–753.

Maestripieri, D. (1994a). Mother-infant relationships in three species of macaques *(Macaca mulatta, M. nemestrina, M. arctoides).* I: Development of the mother-infant relationship in the first three months. *Behaviour, 131,* 75–96.

Maestripieri, D. (1994b). Influence of infants on female social relationships in monkeys. *Folia Primatologica, 63,* 192–202.

Maestripieri, D. (1995a). First steps in the macaque world: Do rhesus mothers encourage their infants' independent locomotion? *Animal Behaviour, 49,* 1541–1549.

Maestripieri, D. (1995b). Maternal encouragement in nonhuman primates and the question of animal teaching. *Human Nature, 6,* 361–378.

Maestripieri, D. (1996a). Gestural communication and its cognitive implications in pigtail macaques *(Macaca nemestrina). Behaviour, 133,* 997–1022.

Maestripieri, D. (1996b). Social communication among captive stumptail macaques *(Macaca arctoides). International Journal of Primatology, 17,* 785–802.

Maestripieri, D. (1996c). Primate cognition and the bared-teeth display: A reevaluation of the concept of formal dominance. *Journal of Comparative Psychology, 110,* 402–405.

Maestripieri, D. (1996d). Maternal encouragement of infant locomotion in pigtail macaques *(Macaca nemestrina). Animal Behaviour, 51,* 603–610.

Maestripieri, D. (1997). Gestural communication in macaques: Usage and meaning of nonvocal signals. *Evolution of Communication, 1,* 193–222.

Maestripieri, D. (1998a). Social and demographic influences on mothering styles in pigtail macaques. *Ethology, 104,* 379–385.

Maestripieri, D. (1998b). Parenting styles of abusive mothers in group-living rhesus macaques. *Animal Behaviour, 55,* 1–11.

Maestripieri, D. (1998c). The evolution of male-infant interactions in the Tribe Papionini (Primates: Cercopithecidae) *Folia Primatologica, 69,* 247–251.

Maestripieri, D. (1999a). The biology of human parenting: Insights from nonhuman primates. *Neuroscience and Biobehavioral Reviews, 23,* 411–422.

Maestripieri, D. (1999b). Fatal attraction: Interest in infants and infant abuse in rhesus macaques. *American Journal of Physical Anthropology, 110,* 17–25.

Maestripieri, D. (2000a). Determinants of affiliative interactions between adult males and lactating females in pigtail macaques. *Ethology, 106,* 425–439.

Maestripieri, D. (2000b). Measuring temperament in rhesus macaques: Consistency and change in emotionality over time. *Behavioural Processes, 49,* 167–171.

Maestripieri, D. (2001a). Is there mother-infant bonding in primates? *Developmental Review, 21,* 93–120.

Maestripieri, D. (2001b). Biological bases of maternal attachment. *Current Directions in Psychological Science, 10,* 80–83.

Maestripieri, D. (2001c). Intraspecific variability in parenting styles of rhesus macaques: The role of the social environment. *Ethology, 107,* 237–248.

Maestripieri, D. (2002). Parent-offspring conflict in primates. *International Journal of Primatology, 23,* 923–951.

Maestripieri, D., & Call, J. (1996). Mother-infant communication in primates. *Advances in the Study of Behavior, 25,* 613–642.

Maestripieri, D., & Carroll, K. A. (1998). Child abuse and neglect: Usefulness of the animal data. *Psychological Bulletin, 123,* 211–223.

Maestripieri, D., Jovanovic, T., & Gouzoules, H. (2000). Crying and infant abuse in rhesus monkeys. *Child Development, 71,* 301–309.

Maestripieri, D., Martel, F. L., Nevison, C. M., Simpson, M. J. A., & Keverne, E. B. (1992). Anxiety in rhesus monkey infants in relation to interactions with their mother and other social companions. *Developmental Psychobiology, 24,* 571–581.

Maestripieri, D., & Megna, N. L. (2000). Hormones and behavior in rhesus macaque abusive and nonabusive mothers: Mother-infant interactions. *Physiology & Behavior, 71,* 43–49.

Maestripieri, D., & Pelka, S. (2002). Sex differences in interest in infants across the life-span: A biological adaptation for parenting? *Human Nature, 13,* 327–344.

Maestripieri, D., Ross, S. R. & Megna, N. L. (2002). Mother-infant interactions in western lowland gorillas *(Gorilla gorilla gorilla):* Spatial relationships, communication, and opportunities for social learning. *Journal of Comparative Psychology, 116,* 219–227.

Maestripieri, D., Schino, G., Aureli, F., & Troisi, A. (1992). A modest proposal: Displacement activities as an indicator of emotions in primates. *Animal Behaviour, 44,* 967–979.

Maestripieri, D., & Wallen, K. (1997). Affiliative and submissive communication in rhesus macaques. *Primates, 38,* 127–138.

Maestripieri, D., Wallen, K., & Carroll, K. A. (1997). Infant abuse runs in families of group-living pigtail macaques. *Child Abuse & Neglect, 21,* 465–471.

Maggioncalda, A. N., Czekala, N. M., & Sapolsky, R. M. (2000). Growth hormone and thyroid stimulating hormone concentrations in captive male orangutans: Implications for understanding developmental arrest. *American Journal of Primatology, 50,* 67–76.

Maggioncalda, A. N., Sapolsky, R. M., & Czekala, N. M. (1999). Reproductive hormone profiles in captive male orangutans: Implications for understanding developmental arrest. *American Journal of Physical Anthropology, 109,* 19–32.

Main, M., Kaplan, N., & Cassidy, J. (1985). Security in infancy, childhood, and adulthood: A move to the level of representation. In I. Bretherton & E. Waters (Eds.), Growing points of attachment theory and research. *Monographs of the Society for Research in Child Development, 50,* 66–104.

Main, M., & Solomon, J. (1990). Procedures for identifying infants as Disorganized/Disoriented during the Ainsworth Strange Situation. In M. T. Greenberg, D. Cicchetti, & E. M. Cummings (Eds.), *Attachment in the preschool years* (pp. 121–160). Chicago: University of Chicago Press.

Mandler, G. (1975). *Mind and emotion.* New York: John Wiley.

Mann, D. R., Akinbami, M. A., Gould, K. G., Paul, K., & Wallen, K. (1998). Sexual maturation in male rhesus monkeys: Importance of neonatal testosterone exposure and social rank. *Journal of Endocrinology, 156,* 493–501.

Mann, D. R., Davis-DaSilva, M., Wallen, K., Coan, P., Evans, D. E., & Collins, D. C. (1984). Blockade of neonatal activation of the pituitary-testicular axis with continuous administration of a gonadotropin-releasing hormone agonist in male rhesus monkeys. *Journal of Clinical Endocrinology and Metabolism, 59,* 207–211.

Mann, D. R., Gould, K. G., Collins, D. C., & Wallen, K. (1989). Blockade of neonatal activation of the pituitary-testicular axis: Effect on peripubertal luteinizing-hormone and testosterone secretion and on testicular development in male monkeys. *Journal of Clinical Endocrinology and Metabolism, 68,* 600–607.

Mann, J. (1992). Nurturance or negligence: Maternal psychology and behavioral preference among preterm twins. In J. Barkow, L. Cosmides, & J. Tooby (Eds.), *The adapted mind* (pp. 367–390). New York: Oxford University Press.

Mann, J. J., & Kapur, S. (1995). A dopaminergic hypothesis of major depression. *Clinical Neuropharmacology, 18 (Supplement 1),* S57–S65.

Mansdotter, S., Epple, G., & Küderling, I. (1992). Age-related changes in ovarian morphology of the South American tamarin *Saguinus fuscicollis (Callitrichidae). Journal of Zoology, 227,* 239–255.

Manson, J. H. (1993). Sons of low-ranking female rhesus macaques can attain high dominance rank in their natal groups. *Primates, 34,* 285–288.

Manson, J. H., & Wrangham, R. W. (1991). Intergroup aggression in chimpanzees and humans. *Current Anthropology, 32,* 369–390.

Manuck, S. B., Kaplan, J. R., & Clarkson, T. B. (1986). Atherosclerosis, social dominance, and cardiovascular reactivity. In T. H. Schmidt, T. M. Dembrosky, & G. Blumchen (Eds.), *Biological and psychological factors in cardiovascular disease* (pp. 461–475). Berlin: Springer-Verlag.

Marks, I. M., & Nesse, R. M. (1994). Fear and fitness: An evolutionary analysis of anxiety disorders. *Ethology and Sociobiology, 15,* 247–261.

Marler, P. (1978). Primate vocalization: Affective or symbolic? In G. Bourne (Ed.), *Progress in Ape Research* (pp. 85–96). New York: Academic Press.

Marler, P., & Evans, C. S. (1997). Communication signals of animals: Contributions of emotion and reference. In U. Segerstrale & P. Molnar (Eds.), *Nonverbal communication: Where nature meets culture* (pp. 151–170). New York: Erlbaum.

Marriott, B. M. (1988). Time budgets of rhesus monkeys *(Macaca mulatta)* in a forest habitat in Nepal and on Cayo Santiago. In J. E. Fa & C. H. Southwick (Eds.), *Ecology and behavior of food-enhanced primate groups* (pp. 199–228). New York: Alan R. Liss.

Martau, P. A., Caine, N. G., & Candland, D. K. (1985). Reliability of the Emotions Profile Index, primate form, with *Papio hamadryas, Macaca fuscata,* and two *Saimiri* species. *Primates, 26,* 501–505.

Martel, F. L., Nevison, C. M., Simpson, M. J. A., & Keverne, E. B. (1995). Effects of opioid receptor blockade on the social behavior of rhesus monkeys living in large family groups. *Developmental Psychobiology, 28,* 71–84.

Martensz, N. D., Vellucci, S. V., Fuller, L. M., Everitt, B. J., Keverne, E. B., & Herbert, J. (1987). Relation between aggressive behaviour and circadian rhythms in cortisol and testosterone in social groups of talapoin monkeys. *Journal of Endocrinology, 115,* 107–120.

Martin, P., & Bateson, P. (1993). *Measuring behaviour: An introductory guide* (2nd ed.). Cambridge: Cambridge University Press.

Martins, E. P. (1996). *Phylogenies and the comparative method in animal behavior.* New York: Oxford University Press.

Masataka, N. (1985). Development of vocal recognition of mothers in infant Japanese macaques. *Developmental Psychobiology, 18,* 107–114.

Masataka, N. (1992). Attempts by animal caretakers to condition Japanese macaque vocalizations result inadvertently in individual-specific calls. In T. Nishida, W. C. McGrew, P. Marler, M. Pickford, & F. B. M. de Waal (Eds.), *Topics in primatology.* Vol. 1. *Human origins* (pp. 271–278). Tokyo: University of Tokyo Press.

Masataka, N., & Fujita, K. (1989). Vocal learning of Japanese and rhesus monkeys. *Behaviour, 109,* 191–199.

Masataka, N., & Thierry, B. (1993). Vocal communication of Tonkean macaques in confined environments. *Primates, 34,* 169–180.

Mascie-Taylor, C. G. N. (1990). The biology of social class. In C. G. N. Mascie-Taylor (Ed.), *Biosocial aspects of social class* (pp. 117–142). Oxford: Oxford University Press.

Mason, W. A. (1985). Experiential influences on the development of expressive behaviors in rhesus monkeys. In G. Zivin (Ed.), *The development of expressive behavior* (pp. 117–152). New York: Academic Press.

Mason, W. A., & Capitanio, J. P. (1988). Formation and expression of filial attachment in rhesus monkeys raised with living and inanimate mother substitutes. *Developmental Psychobiology, 21,* 401–430.

Mason, W. A., Hill, S. D., & Thomsen, C. E. (1974). Perceptual aspects of filial attachment in monkeys. In N. F. White (Ed.), *Ethology and psychiatry* (pp. 84–93). Ontario: Toronto University Press.

Mason, W. A., & Mendoza, S. P. (1998). Generic aspects of primate attachments: Parents, offspring, and mates. *Psychoneuroendocrinology, 23,* 765–778.

Massey, A. (1977). Agonistic aids and kinship in a group of pigtail macaques. *Behavioral Ecology and Sociobiology, 2,* 31–40.

Matsumura, S. (1996). Postconflict affiliative contacts between former opponents among wild moor macaques *(Macaca maurus). American Journal of Primatology, 38,* 211–219.

Matsuzawa, T. (1985). Color naming and classification in a chimpanzee *(Pan troglodytes). Journal of Human Evolution, 14,* 283–291.

Matsuzawa, T. (1989). Spontaneous pattern construction in a chimpanzee. In P. Heltne & L. A. Marquardt (Eds.), *Understanding chimpanzees.* (pp. 252–265). Cambridge, MA: Harvard University Press.

Matsuzawa, T. (Ed.). (2001). *Primate origins of human behavior and cognition.* New York: Springer-Verlag.

Matthews, K. A., Batson, C. D., Horn, J., & Rosenman, R. H. (1981). "Principles in his nature which interest him in the fortunes of others . . .": The heritability of empathic concern for others. *Journal of Personality, 49,* 237–247.

Mattsson, A., Schalling, D., Olweus, D., Löw, H., & Svensson, J. (1980). Plasma testosterone, aggressive behavior, and personality dimensions in young male delinquents. *Journal of the American Academy of Child Psychiatry, 19,* 476–490.

Mayagoitia, L., Santillan-Doherty, A. M., Lopez-Vergara, L., & Mondragon-Ceballos, R. (1993). Affiliation tactics prior to a period of competition in captive groups of stumptail macaques. *Ethology, Ecology, and Evolution, 5,* 435–446.

Mayr, E. (1963). *Animal species and evolution.* Cambridge, MA: Belknap Press of Harvard University Press.

Mazur, A. (1983). Hormones, aggression, and dominance in humans. In B. B. Svare (Ed.), *Hormones and aggressive behavior* (pp. 563–576). New York: Plenum Press.

McAdams, D. P. (1995). What do we know when we know a person? *Journal of Personality, 63,* 365–396.

McAdams, D. P. (1996). Personality, modernity, and the storied self: A contemporary framework for studying persons. *Psychological Inquiry, 7,* 295–321.

McBride, W. J., Murphy, J. M., Lumeng, L., & Li, T. K. (1989). Serotonin and ethanol preference. *Recent Developments in Alcoholism, 7,* 187–209.

McCarthy, G., Puce, A., Belger, A., & Allison, T. (1999). Electrophysiological studies of human face perception II: Response properties of face-specific potentials generated in occipitotemporal cortex. *Cerebral Cortex, 9,* 431–444.

McComas, H. C. (1923). The origin of laughter. *Psychological Review, 30,* 45–56.

McCoy, N. L., & Davidson, J. M. (1985). A longitudinal study of the effects of menopause on sexuality. *Maturitas, 7,* 203–210.

McCrae, R. R. (1982). Consensual validation of personality traits: Evidence from self-reports and ratings. *Journal of Personality and Social Psychology, 43,* 293–303.

McCrae, R. R., & Costa, P. T. Jr., (1999). A five-factor theory of personality. In L. A. Pervin & O. P. John (Eds.), *Handbook of personality theory and research* (pp. 139–153). New York: Guilford Press.

McCrae, R. R., Costa, P. T. Jr., Ostendorf, F., Angleitner, A., Hrebíckova, M., Avia, M. D., Sanz, J., Sánchez-Bernardos, M. L., Kusdil, M. E., Woodfield, R., Saunders, P. R., & Smith, P. B. (2000). Nature over nurture: Temperament, personality, and life span development. *Journal of Personality and Social Psychology, 78,* 173–186.

McFarland, D. (1966). On the causal and functional significance of displacement activities. *Zeitschrift für Tierpsychologie, 23,* 217–235.

McFarland Symington, M. (1990). Fission-fusion social organization in *Ateles* and *Pan. International Journal of Primatology, 11,* 47–61.

McGrew, W. C. (1992). *Chimpanzee material culture: Implications for human evolution.* Cambridge: Cambridge University Press.

McGrew, W. C. (1998). Culture in non-human primates? *Annual Review of Anthropology, 27,* 301–328.

McGrew, W. C., & Marchant, L. F. (1997). On the other hand: Current issues in and meta-analysis of the behavioral laterality of hand function in nonhuman primates. *Yearbook of Physical Anthropology, 40,* 201–232.

McGrew, W. C., & Tutin, C. E. G. (1978). Evidence for a social custom in wild chimpanzees? *Man, 13,* 234–251.

McGuire, M. T., Brammer, G. L., & Raleigh, M. J. (1983). Animal models: Are they useful in the study of psychiatric disorders? In K. A. Miczek (Ed.), *Ethopharmacology: Primates models of neuropsychiatric disorders* (pp. 313–328). New York: Alan Liss.

McGuire, M. T., Marks, I., Nesse, R. M., & Troisi, A. (1992). Evolutionary biology: a basic science for psychiatry? *Acta Psychiatrica Scandinavica, 86,* 89–96.

McGuire, M. T., Raleigh, M. J., & Pollack, D. B. (1994). Personality features in vervet monkeys: The effects of sex, age, social status, and group composition. *American Journal of Primatology, 33,* 1–13.

McGuire, M. T., & Troisi, A. (1998). *Darwinian psychiatry.* New York: Oxford University Press.

McGuire, M. T., Troisi, A., & Raleigh, M. J. (1997). Depression in evolutionary context. In S. Baron-Cohen (Ed.), *The Maladapted mind: Classic readings in evolutionary psychopathology* (pp. 255–282). Hove, UK: Psychology Press.

McKenna, J. J. (1979). Aspects of infant socialization, attachment, and maternal caregiving patterns among primates: A cross-disciplinary review. *Yearbook of Physical Anthropology, 22,* 250–286.

McKenna, J. J. (1990). Evolution and the sudden infant death syndrome (SIDS). Part III: Infant arousal and parent-infant co-sleeping. *Human Nature, 1,* 291–330.

McKinley, J., & Sambrook, T. D. (2000). Use of human-given cues by domestic dogs *(Canis familiaris)* and horses *(Equus caballus). Animal Cognition, 3,* 13–22

McKinney, D. H., & Donaghy, W. C. (1993). Dyad gender structure, uncertainty reduction, and self-disclosure during initial interaction. In P. J. Kalbfleisch (Ed.), *Interpersonal communication: Evolving interpersonal relationships* (pp. 33–50). Hillsdale, NJ: Erlbaum.

McKinney, M. L. (1998). Cognitive evolution by extending brain development: On recapitulation, progress, and other heresies. In J. Langer & M. Killen (Eds.), *Piaget, evolution, and development.* (pp. 9–31). Mahwah, NJ: Erlbaum.

McKinney, W. T. (2000). Animal research and its relevance to psychiatry. In B. J. Sadock & V. A. Sadock (Eds.), *Kaplan and Sadock's Comprehensive textbook of psychiatry* (7th ed.), Vol. 1 (pp. 545–562). Philadelphia: Lippincott, Williams, and Wilkins.

McKinsey Crittenden, P., Lang, C., Hartl Claussen, A., & Partridge, M. F. (2000). Relations among mothers' dispositional representations of parenting. In P. McKinsey Crittenden & A. Hartl Claussen (Eds.), *The organization of attachment relationships: Maturation, culture, and context* (pp. 214–233). Cambridge: Cambridge University Press.

Meaney, M. J., Lozos, E., & Stewart, J. (1990). Infant carrying by nulliparous female vervet monkeys *(Cercopithecus aethiops). Journal of Comparative Psychology, 104,* 377–381.

Mehlman, P. T., & Chapais, B. (1988). Differential effects of kinship, dominance, and the mating season on female allogrooming in a captive group of *Macaca fuscata. Primates, 29,* 195–217.

Mehlman, P. T., Higley, J. D., Faucher, I., Lilly, A. A., Taub, D. M., Vickers, J., Suomi, S. J., & Linnoila, M. (1994). Low CSF 5-HIAA concentrations and severe aggression and impaired impulse control in nonhuman primates. *Archives of General Psychiatry, 151,* 1485–91.

Mehlman, P. T., Higley, J. D., Faucher, I., Lilly, A. A., Taub, D. M., Vickers, J., Suomi, S. J., & Linnoila, M. (1995). Correlation of CSF 5-HIAA concentration with sociality and the timing of emigration in free-ranging primates. *Archives of General Psychiatry, 152,* 907–913.

Mehlman, P. T., Higley, J. D., Fernald, B. J., Sallee, F. R., Suomi, S. J., & Linnoila, M. (1997). CSF 5-HIAA, testosterone, and sociosexual behaviors in free-ranging male rhesus macaques in the mating season. *Psychiatry Research, 72,* 89–102.

Mehlman, P. T., Westergaard, G. C., Hoos, B. J., Sallee, F. R., Marsh, S., Suomi, S. J., Linnoila, M., & Higley, J. D. (2000). CSF 5-HIAA and nighttime activity in free-ranging primates. *Neuropsychopharmacology, 22,* 210–218.

Meikle, B. B., & Vessey, S. H. (1988). Maternal dominance rank and lifetime survivorship of male and female rhesus monkeys. *Behavioral Ecology and Sociobiology, 23,* 379–383.

Melnick, D. J., & Pearl, M. C. (1987). Cercopithecines in multimale groups: Genetic diversity and population structure. In B. B. Smuts, D. L. Cheney, R. M. Seyfarth, R. W. Wrangham, & T. T. Struhsaker (Eds.), *Primate societies* (pp. 121–134). Chicago: University of Chicago Press.

Meltzoff, A. N. (1995). Understanding the intentions of others: Reenactment of intended acts by 18-month-old children. *Developmental Psychology, 31,* 838–850.

Mende, W., Herzel, H., & Wermke, K. (1990). Bifurcations and chaos in newborn infant cries. *Physics Letters, A, 145,* 418–424.

Mendelson, M. J., Haith, M. M., & Goldman-Rakic, P. S. (1982). Face scanning and responsiveness to social cues in infant rhesus monkeys. *Developmental Psychology, 18,* 222–228.

Mendoza, S. P., & Mason, W. A. (1986). Contrasting responses to intruders and to involuntary separation by monogamous and polygynous New World monkeys. *Physiology and Behavior, 38,* 795–801.

Mendoza, S. P., Smotherman, W. P., Miner, M. T., Kaplan, J., & Levine, S. (1978). Pituitary-adrenal response to separation in mother and infant squirrel monkeys. *Developmental Psychobiology, 11,* 169–175.

Mendoza-Granados, D., & Sommer, V. (1995). Play in chimpanzees of the Arnhem Zoo: Self-serving compromises. *Primates, 36,* 57–68.

Menzel, E. W., Jr., Davenport, R. K., & Rogers, C. M. (1970). The development of tool using in wild born and restriction reared chimpanzees. *Folia Primatologica, 12,* 273–283.

Meyer-Bahlburg, H. F., Feldman, J. F., Cohen, P., & Ehrhardt, A. A. (1988). Perinatal factors in the development of gender-related play behavior: Sex hormones versus pregnancy complications. *Psychiatry, 51*, 260–271.

Michael, R. P., & Bonsall, R. W. (1979). Hormones and the sexual behavior of rhesus monkeys. In C. Beyer (Ed.), *Endocrine control of sexual behavior* (pp. 279–302). New York: Raven Press.

Michael, R. P., & Wilson, M. (1973). Changes in the sexual behaviour of male rhesus monkeys *(M. mulatta)* at puberty: Comparisons with the behaviour of adults. *Folia Primatologica, 19*, 384–403.

Michael, R. P., & Zumpe, D. (1970). Rhythmic changes in the copulatory frequency of rhesus monkeys *(Macaca mulatta)* in relation to the menstrual cycle and a comparison with the human cycle. *Journal of Reproduction and Fertility, 21*, 199–201.

Miczek, K. A. (Ed.) (1983). *Ethopharmacology: Primates models of neuropsychiatric disorders.* New York: Alan Liss.

Miczek, K. A., & Donat, P. (1990). Brain 5-HT system and inhibition of aggressive behavior. In T. Archer, P. Bevan, & A. Cools (Eds.), *Behavioral pharmacology of 5-HT* (pp. 117–144). Hillsdale, NJ: Erlbaum.

Miczek, K. A., Mos, J., & Olivier, B. (1989). Brain 5-HT and inhibition of aggressive behavior in animals: 5-HIAA and receptor subtypes. *Psychopharmacology Bulletin, 25*, 399–403.

Miklósi, A., Polgárdi, R., Topál, J., & Csányi, V. (1998). Use of experimenter-given cues in dogs. *Animal Cognition, 1*, 113–121.

Miles, H. L. (1983). Apes and language: The search for communicative competence. In J. de Luce & H. T. Wilder (Eds.), *Language in primates: Perspectives and implications* (pp. 43–61). New York: Springer.

Miles, H. L. (1990). The cognitive foundations for reference in a signing orangutan. In S. T. Parker & K. R. Gibson (Eds.), *"Language" and intelligence in monkeys and apes* (pp. 511–539). Cambridge: Cambridge University Press.

Miles, H. L. (1994). Me Chantek: The development of self-awareness in signing orangutan. In S. T. Parker, R. W. Mitchell, & M. L. Boccia (Eds.), *Self-awareness in animals and humans* (pp. 254–272). Cambridge: Cambridge University Press.

Milford, P. A. (1980). Perception of laughter and its acoustical properties (Ph.D. dissertation, Pennsylvania State University). *Dissertation Abstracts International, 41*, 3779.

Miller, G. E., Cohen, S., Rabin, B. S., Skoner, D. P., & Doyle, W. J. (1999). Personality and tonic cardiovascular, neuroendocrine, and immune parameters. *Brain, Behavior, and Immunity, 13*, 109–123.

Miller, L. C., Bard, K. A., Juno, C. J., & Nadler, R. D. (1986). Behavioral responsiveness of young chimpanzees *(Pan troglodytes)* to a novel environment. *Folia Primatologica, 47*, 128–142.

Miller, L. C., Bard, K. A., Juno, C. J., & Nadler, R. D. (1990). Behavioral responsiveness to strangers in young chimpanzees *(Pan troglodytes)*. *Folia Primatologica, 55,* 142–155.

Miller, N. E. (1959). Liberalisation of basic S-R concepts: Extension to conflict, motivation, and social learning. In S. Koch (Ed.), *Psychology: A study of a science.* Study 1, Vol. 2 (pp. 196–292). New York: McGraw-Hill.

Miller, R. E. (1971). Experimental studies of communication in the monkey. In L. A. Rosenblum (Ed.), *Primate behavior: Developments in field and laboratory research* (pp. 139–175). New York: Academic Press.

Miller, R. E., Murphy, J. V., & Mirsky, I. A. (1959). Relevance of facial expression and posture as cues in communication of affect between monkeys. *Archives of General Psychiatry, 1,* 480–488.

Mills, J., & Clark, M. S. (1982). Communal and exchange relationships. In L. Wheeler (Ed.), *Review of personality and social psychology,* Vol. 3 (pp. 121–144). Beverly Hills, CA: Sage Publications.

Mineka, S., & Suomi, S. J. (1978). Social separation in monkeys. *Psychological Bulletin, 85,* 1376–1400.

Mischel, W. (1968). *Personality and assessment.* New York: Wiley.

Mischel, W., & Peake, P. K. (1982). Beyond déjà vu in the search for cross-situational consistency. *Psychological Review, 89,* 730–755.

Missakian, E. A. (1974). Mother-offspring grooming relations in rhesus monkeys. *Archives of Sexual Behavior, 3,* 135–141.

Mitani, J., Gros-Louis, J., & Macedonia, J. (1996). Selection for acoustic variability within the vocal repertoire of wild chimpanzees. *International Journal of Primatology, 17,* 569–583.

Mitani, M. (1986). Voiceprint identification and its application to sociological studies of wild Japanese monkeys *(Macaca fuscata yakui). Primates, 27,* 397–412.

Mitchell, G. D. (1968). Attachment differences in male and female infant monkeys. *Child Development, 39,* 611–620.

Mitchell, P. (1997). *Introduction to theory of mind: Children, autism, and apes.* London: Arnold.

Moberg, G. P. (Ed.). (1985). *Animal stress.* Bethesda, MD: Animal Physiological Society.

Moely, B., Skarin, E., & Weil, K. (1979). Sex differences in competition-cooperation behavior of children at two age levels. *Sex Roles, 5,* 329–342.

Moffat, S. D., Hampson, E., & Lee, D. H. (1998). Morphology of the planum temporale and corpus callosum in left handers with evidence of left and right hemisphere speech representation. *Brain, 121,* 2369–79.

Mohnot, S. M. (1971). Some aspects of social changes and infant-killing in the hanuman langur, *Presbytis entellus,* in Western India. *Mammalia, 35,* 175–198.

Money, J., & Ehrhardt, A. A. (1972). *Man and woman, boy and girl.* Baltimore: John Hopkins University Press.

Montemayor, R., & Hanson, E. (1985). A naturalistic view of conflict between adolescents and their parents and siblings. *Journal of Early Adolescence, 5,* 23–30.

Morris, D. (1962). *The biology of art.* London: Methuen.

Morris, R. D., & Hopkins, W. D. (1993). Perception of human chimeric faces by chimpanzees *(Pan troglodytes):* Evidence for a right hemisphere asymmetry. *Brain and Cognition, 21,* 111–122.

Moskowitz, D. S. (1990). Convergence of self-reports and independent observers: Dominance and friendliness. *Journal of Personality and Social Psychology, 58,* 1096–1106.

Moskowitz, D. S., Schwartzman, A. E., & Ledingham, J. E. (1985). Stability and change in aggression and withdrawal in middle childhood and early adolescence. *Journal of Abnormal Psychology, 94,* 30–41.

Moskowitz, D. S., & Schwarz, J. C. (1982). Validity comparison of behavior counts and ratings by knowledgeable informants. *Journal of Personality and Social Psychology, 42,* 518–528.

Muehlenhard, C. L., Korlewski, M. A., Andrews, S. L., & Burdick, C. A. (1986). Verbal and nonverbal cues that convey interest in dating: Two studies. *Behavior Therapy, 17,* 404–419.

Muldoon, M. F., Rossouw, J. E., Manuck, S. B., Glueck, C. J., Kaplan, J. R., & Kaufman, P. G. (1993). Low or lowered cholesterol and risk of death from suicide and trauma. *Metabolism, 42 (Supplement 1),* 45–56.

Muroyama, Y. (1991a). Mutual reciprocity of grooming in female Japanese macaques *(Macaca fuscata). Behaviour, 119,* 161–170.

Muroyama, Y. (1991b). Role choice in the sequence of grooming interaction of Japanese monkeys. In A. Ehara, T. Kimura, O. Takenaka, & M. Iwamoto (Eds.), *Primatology Today* (pp. 159–161). Amsterdam: Elsevier.

Muroyama, Y. (1995). Developmental changes in mother-offspring grooming in Japanese macaques. *American Journal of Primatology, 37,* 57–64.

Muroyama, Y. (1996). Decision making in grooming by Japanese macaques *(Macaca fuscata). International Journal of Primatology, 17,* 817–830.

Murray, L. E. (1998). The effects of group structure and rearing strategy on personality in chimpanzees *(Pan troglodytes)* at Chester, London ZSL, and Twycross Zoos. *International Zoo Yearbook, 36,* 97–108.

Myowa, M., & Matsuzawa, T. (2000). Imitation of intentional manipulatory actions in chimpanzees *(Pan troglodytes). Journal of Comparative Psychology, 114,* 381–391.

Nadler, R. D. (1982). Laboratory research on sexual behavior and reproduction of gorillas and orang-utans. *American Journal of Primatology (Supplement 1),* 57–66.

Nadler, R. D., Wallis, J., Roth-Meyer, C., Cooper, R. W., & Baulieu, E. E. (1987). Hormones and behavior of prepubertal and peripubertal chimpanzees. *Hormones and Behavior, 21*, 118–131.

Nagell, K., Olguin, R. S., & Tomasello, M. (1993). Processes of social learning in the tool use of chimpanzees *(Pan troglodytes)* and human children *(Homo sapiens). Journal of Comparative Psychology, 107*, 174–186.

Nahm, F. K. D., Perret, A., Amaral, D. G., & Albright, T. D. (1997). How do monkeys look at faces? *Journal of Cognitive Neuroscience, 9*, 611–623.

Nakamichi, M. (1989). Sex differences in social development during the first 4 years in a free-ranging group of Japanese monkeys, *Macaca fuscata. Animal Behaviour, 38*, 737–748.

Nakamichi, M., & Yoshida, A. (1986). Discrimination of mother by infant among Japanese macaques *(Macaca fuscata). International Journal of Primatology, 7*, 481–489.

Namboodiri, M. A., Sugden, D., Klein, D. C., Tamarkin, L., & Mefford, I. N. (1985). Serum melatonin and pineal indoleamine metabolism in a species with a small day/night N-acetyltransferase rhythm. *Comparative Biochemistry and Physiology (B), 80*, 731–736.

Nash, V. J., & Chamove, A. S. (1981). Personality and dominance behavior in stumptailed macaques. In A. B. Chiarelli & R. S. Corruccini (Eds.), *Primate behavior and sociobiology* (pp. 88–92). New York: Springer-Verlag.

Nelson, D. A. (1984). Communication of intentions in agonistic contexts by the pigeon guillemot, *Ceppus columba. Behaviour, 88*, 145–189.

Nelson, J., & Aboud, F. E. (1985). The resolution of social conflict between friends. *Child Development, 56*, 1009–17.

Nesse, R. M. (1990). Evolutionary explanations of emotions. *Human Nature, 1*, 261–289.

Nesse, R. M. (2000). Is depression an adaptation? *Archives of General Psychiatry, 57*, 14–20.

Nevison, C. M., Brown, G. R., & Dixson, A. F. (1997). Effects of altering testosterone in early infancy on social behavior in captive yearling rhesus monkeys. *Physiology and Behavior, 62*, 1397–1403.

Newman, J. D. (1991). Vocal manifestations of anxiety and their pharmacological control. In S. E. File (Ed.), *Psychopharmacology of anxiolytics and antidepressants* (pp. 251–260). New York: Pergamon Press.

Newman, J. D. (1995). Vocal ontogeny in macaques and marmosets: Convergent and divergent lines of development. In E. Zimmermann, J. D. Newman, & U. Jürgens (Eds.), *Current topics in primate vocal communication* (pp. 73–97). New York: Plenum Press.

Newman, J. D., & Farley, M. J. (1995). An ethologically based, stimulus and gendersensitive nonhuman primate model for anxiety. *Progress in Neuro-Psychopharmacology and Biological Psychiatry, 19*, 677–685.

Newman, J. D., & Symmes, D. (1974). Vocal pathology in socially deprived monkeys. *Developmental Psychobiology, 7,* 351–358.

Newman, M. E., Shapira, B., & Lerer, B. (1998). Evaluation of central serotonergic function in affective and related disorders by the fenfluramine challenge test: A critical review. *International Journal of Neuropsychopharmacology, 1,* 49–69.

NICHD (1997). The effects of infant care on infant mother attachment security: Results of the NICHD study of early child care. *Child Development, 68,* 860–879.

Nicholls, J. G., Licht, B. G., & Pearl, R. A. (1982). Some dangers of using personality questionnaires to measure personality. *Psychological Bulletin, 92,* 572–580.

Nieuwenhuijsen, K., Bonke-Jansen, M., Broekhuijzen, E., de Neef, K. J., van Hooff, J. A., van der Werff ten Bosch, J. J., & Slob, A. K. (1988). Behavioral aspects of puberty in group-living stumptail monkeys *(Macaca arctoides). Physiology and Behavior, 42,* 255–264.

Nieuwenhuijsen, K., Bonke-Jansen, M., de Neef, K. J., van der Werff ten Bosch, J. J., & Slob, A. K. (1987). Physiological-aspects of puberty in group-living stumptail monkeys *(Macaca arctoides). Physiology and Behavior, 41,* 37–45.

Nieuwenhuijsen, K., Lammers, A. J. J. C., de Neef, K. J., & Slob, A. K. (1985). Reproduction and social rank in female stumptail macaques *(Macaca arctoides). International Journal of Primatology, 6,* 77–99.

Nieuwenhuijsen, K., Slob, A., & van der Werff ten Bosch, J. (1988). Gender-related behaviors in group-living stumptail macaques. *Psychobiology, 16,* 357–371.

Nikulina, E. M., Augustinovich, D. F., & Popova, N. K. (1992). Role of 5HT1A receptors in a variety of kinds of aggressive behavior in wild rats and counterparts selected for low defensiveness to man. *Aggressive Behavior, 18,* 357–364.

Ninan, P. T., Insel, T. M., Cohen, R. M., Cook, J. M., Skolnik, P., & Paul, S. M. (1982). Benzodiazepine receptor-mediated experimental "anxiety" in primates. *Science, 218,* 1332–34.

Nishida, T. (1980). The leaf-clipping display: A newly discovered expressive gesture in wild chimpanzees. *Journal of Human Evolution, 9,* 117–128.

Nissen, H. W. (1931). A field study of the chimpanzee: Observations of chimpanzee behavior and environment in western French Guinea. *Comparative Psychology Monographs, 8,* 1–122.

Nöe, R., de Waal, F. B. M., & van Hooff, J. A. R. A. M. (1980). Types of dominance in a chimpanzee colony. *Folia Primatologica, 34,* 90–110.

Nöe, R., van Schaik, C. P., & van Hooff, J. A. R. A. M. (1991). The market effect: An explanation for pay-off asymmetries among collaborating animals. *Ethology, 87,* 97–118.

Norman, W. T., & Goldberg, L. R. (1966). Raters, ratees, and randomness in personality structure. *Journal of Personality and Social Psychology, 4,* 681–691.

Novak, M. A., O'Neill, P., & Suomi, S. J. (1992). Adjustments and adaptations to indoor and outdoor environments: Continuity and change in young adult rhesus monkeys. *American Journal of Primatology, 28,* 125–138.

Nowak, M. A., & Sigmund, K. (1998). Evolution of indirect reciprocity by image scoring. *Nature, 393,* 573–577.

Nunn, C. L., & Barton, R. A. (2001). Comparative methods for studying primate adaptation and allometry. *Evolutionary Anthropology, 10,* 81–98.

Nwokah, E. E., Davies, P., Islam, A., Hsu, H. C., & Fogel, A. (1993). Vocal affect in three-year-olds: A quantitative acoustic analysis of child laughter. *Journal of the Acoustical Society of America, 94,* 3076–90.

O'Brien, T. G. (1993). Allogrooming behaviour among adult female wedge-capped capuchin monkeys. *Animal Behaviour, 46,* 499–510.

O'Connor, L. E., Berry, J. W., King, J. E., Landau, V., & Pederson, A. (2001). Psychopathology and subjective well-being in chimpanzees. *Poster presented at the annual meeting of the American Psychological Association,* San Francisco.

Ohbuchi, K., Kameda, M., & Agarie, N. (1989). Apology as aggression control: Its role in mediating appraisal and response to harm. *Journal of Personality and Social Psychology, 56,* 219–227.

Öhman, A. (1993). Fear and anxiety as emotional phenomena: Clinical phenomenology, evolutionary perspectives, and information-processing mechanisms. In M. Lewis & J. M. Haviland (Eds.), *Handbook of emotions* (pp. 511–536). New York: Guilford Press.

Oki, J., & Maeda, Y. (1973). Grooming as a regulator of behavior in Japanese macaques. In C. R. Carpenter (Ed.), *Behavioral regulators of behavior in primates* (pp. 149–163). Lewisburg, PA: Bucknell University Press.

Olivares, R., Michalland, S., & Aboitiz, F. (2000). Cross-species and intraspecies morphometric analysis of the corpus callosum. *Brain, Behavior, and Evolution, 55,* 37–43.

Olivier, B., & Mos, J. (1990). Serenics, serotonin, and aggression. *Progress in Clinical and Biological Research, 361,* 203–230.

Olivier, B., Mos, J., Tulp, M., Schipper, J., & Bevan, P. (1990). Modulatory action of serotonin in aggressive behavior. In T. Archer, P. Bevan, & A. Cools (Eds.), *Behavioral pharmacology of 5-HT* (pp. 89–116). Hillsdale, NJ: Erlbaum.

Olweus, D. (1979). Stability of aggressive reaction patterns in males: A review. *Psychological Bulletin, 86,* 852–875.

Olweus, D. (1980). The consistency issue in personality psychology revisited—with special reference to aggression. *British Journal of Social and Clinical Psychology, 19,* 377–390.

Olweus, D. (1984). Development of stable aggressive reaction patterns in males. *Advances in the Study of Aggression, 1,* 103–137.

Olweus, D. (1986). Aggression and hormones: Behavioral relationship with testosterone and adrenaline. In D. Olweus, J. Block, & M. Radke-Yarrow (Eds.), *Development of antisocial and prosocial behavior* (pp. 51–72). New York: Academic Press.

Olweus, D., Mattsson, A., Schalling, D., & Löw, H. (1980). Testosterone, aggression, physical, and personality dimensions in normal adolescent males. *Psychosomatic Medicine, 42,* 253–269.

Olweus, D., Mattsson, A., Schalling, D., & Löw, H. (1988). Circulating testosterone levels and aggression in adolescent males: A causal analysis. *Psychosomatic Medicine, 50,* 261–272.

Ono, T., & Nishijo, H. (2000). Neurophysiological basis of emotion in primates: Neuronal responses in the monkey amygdala and anterior cingulate cortex. In M. S. Gazzaniga (Ed.), *The new cognitive neurosciences* (2nd ed.) (pp. 1099–14). Cambridge, MA: MIT Press.

Ortony, A., & Turner, T. J. (1990). What's basic about basic emotions? *Psychological Review, 97,* 315–331.

O'Scalaidhe, S. P., Wilson, F. A. W., & Goldman-Rakic, P. S. (1997). Areal segregation of face-processing neurons in prefrontal cortex. *Science, 278,* 1135–38.

Osterman, K., Bjorkqvist, K., Lagerspetz, K. M. J., Landau, S. F., Fraczek, A., & Pastorelli, C. (1997). Sex differences in styles of conflict resolution: A developmental and cross-cultural study with data from Finland, Israel, Italy, and Poland. In D. P. Fry & K. Bjorkqvist (Eds.), *Cultural variation in conflict resolution: Alternatives to violence* (pp. 185–197). Mahwah, NJ: Erlbaum.

Ostwald, P. F. (1972). The sounds of infancy. *Developmental Medicine and Child Neurology, 14,* 350–361.

Overman, W. H., & Doty, R. W. (1982). Hemispheric specialization displayed by man but not macaques in the analysis of faces. *Neuropsychologia, 20,* 113–128.

Owings, D. H., & Morton, E. S. (1997). The role of information in communication: An assessment/management approach. In D. H. Owings, M. D. Beecher, & N. S. Thompson (Eds.), *Perspectives in ethology.* Vol. 12. *Communication* (pp. 359–390). New York: Plenum.

Owren, M. J., & Bachorowski, J.-A. (2001). Smiling, laughter, and cooperative relationships: An attempt to account for human expressions of positive emotions based on "selfish gene" evolution. In T. Mayne & G. A. Bonanno (Eds.), *Emotion: Current issues and future development* (pp. 152–191). New York: Guilford.

Owren, M. J., & Bernacki, R. H. (1988). The acoustic features of vervet monkey alarm calls. *Journal of the Acoustical Society of America, 83,* 1927–35.

Owren, M. J., & Casale, T. M. (1994). Variations in fundamental frequency peak position in Japanese macaque *(Macaca fuscata)* coo calls. *Journal of Comparative Psychology, 108,* 291–297.

Owren, M. J., Dieter, J. A., Seyfarth, R. M., & Cheney, D. L. (1992). "Food" calls produced by adult female rhesus *(Macaca mulatta)* and Japanese *(M. fuscata)*
macaques, their normally raised offspring, and offspring cross-fostered between
species. *Behaviour, 120,* 218–231.

Owren, M. J., Dieter, J. A., Seyfarth, R. M., & Cheney, D. L. (1993). Vocalizations
of rhesus *(Macaca mulatta)* and Japanese *(M. fuscata)* macaques cross-fostered
between species show evidence of only limited modification. *Developmental Psychobiology, 26,* 389–406.

Owren, M. J., & Nederhouser, M. (2000). Functional implications of rhesus macaque
scream acoustics. Paper presented at the annual meeting of the Animal Behavior
Society, Atlanta.

Owren, M. J., & Rendall, D. (1997). An affect-conditioning model of nonhuman primate signaling. In D. H. Owings, M. D. Beecher, & N. S. Thompson (Eds.), *Perspectives in ethology.* Vol. 12. *Communication* (pp. 299–346). New York: Plenum.

Owren, M. J., & Rendall, D. (2001). Sound on the rebound: Bringing form and function back to the forefront in understanding nonhuman primate vocal signaling.
Evolutionary Anthropology, 10, 58–71.

Owren, M. J., Seyfarth, R. M., & Cheney, D. L. (1997). The acoustic features of vowel-
like grunt calls in chacma baboons *(Papio cynocephalus ursinus):* Implications for
production processes and functions. *Journal of the Acoustical Society of America,
101,* 2951–63.

Packer, C., & Pusey, A. E. (1979). Female aggression and male membership in troops
of Japanese macaques and olive baboons. *Folia Primatologica, 31,* 212–218.

Palombit, R. A. (1992). A preliminary study of vocal communication in wild long-
tailed macaques *(Macaca fascicularis).* I: Vocal repertoire and call emission. *International Journal of Primatology, 13,* 143–182.

Palombit, R. A., Cheney, D. L., & Seyfarth, R. M. (1999). Male grunts as mediators of
social interaction with females in wild chacma baboons *(Papio cynocephalus ursinus). Behaviour, 136,* 221–242.

Palombit, R. A., Seyfarth, R. M., & Cheney, D. L. (1997). The adaptive value of
"friendships" to female baboons: Experimental and observational evidence. *Animal Behaviour, 54,* 599–614.

Panksepp, J. (1989). The psychobiology of emotions: The animal side of human
feelings. In G. Gainotti & C. Caltagirone (Eds.), *Emotions and the dual brain*
(pp. 31–55). Berlin: Springer-Verlag.

Panksepp, J. (1994). The clearest physiological distinctions between emotions will be
found among the circuits of the brain. In P. Ekman & R. J. Davidson (Eds.), *The
nature of emotion: Fundamental questions* (pp. 258–260). New York: Oxford University Press.

Panksepp, J. (1998). *Affective neuroscience: The foundation of human and animal emotions.* New York: Oxford University Press.

Panksepp, J., Nelson, E., & Bekkedal, M. (1997). Brain systems for the mediation of social separation-distress and social-reward. *Annals of the New York Academy of Sciences, 807,* 78–100.

Parkel, D. A., White, R. A., & Warner, H. (1977). Implications of the Yerkes technology for mentally retarded human subjects. In D. M. Rumbaugh (Ed.), *Language learning by a chimpanzee: The Lana Project* (pp. 273–286). New York: Academic Press.

Parker, S. T., & Gibson, K. R. (1979). A developmental model for the evolution of language and intelligence in early hominids. *Behavioral and Brain Sciences, 2,* 367–408.

Parker, S. T., & McKinney, M. L. (1999). *Origins of intelligence: The evolution of cognitive development in monkeys, apes, and humans.* Baltimore: Johns Hopkins University Press.

Parr, L. A. (2001). Cognitive and physiological markers of emotional awareness in chimpanzees. *Animal Cognition, 4,* 223–229.

Parr, L. A., & de Waal, F. B. M. (1999). Visual kin recognition in chimpanzees. *Nature, 399,* 647–648.

Parr, L. A., Dove, T. A., & Hopkins, W. D. (1998). Why faces may be special: Evidence of the inversion effect in chimpanzees *(Pan troglodytes). Journal of Cognitive Neuroscience, 10,* 615–622.

Parr, L. A., & Hopkins, W. D. (2000). Brain temperature asymmetries and emotional perception in chimpanzees, *Pan troglodytes. Physiology and Behavior, 71,* 363–371.

Parr, L. A., Hopkins, W. D., & de Waal, F. B. M. (1998). The perception of facial expressions in chimpanzees *(Pan troglodytes). Evolution of Communication, 2,* 1–23.

Parr, L. A., Winslow, J. T., & Hopkins, W. D. (1999). Is the inversion effect in rhesus monkeys face specific? *Animal Cognition, 2,* 123–129.

Parr, L. A., Winslow, J. T., Hopkins, W. D., & de Waal, F. B. M. (2000). Recognizing facial cues: Individual recognition in chimpanzees *(Pan troglodytes)* and rhesus monkeys *(Macaca mulatta). Journal of Comparative Psychology, 114,* 47–60.

Passini, F. T., & Norman, W. T. (1966). A universal conception of personality structures? *Journal of Personality and Social Psychology, 56,* 493–505.

Paton, D., & Caryl, P. G. (1986). Communication by agonistic displays. I: Variation in information content between samples. *Behaviour, 18,* 23–239.

Patterson, F. G., Bonvillian, J. D., Reynolds, P. C., & Maccoby, E. E. (1975). Mother and peer attachment under conditions of fear in rhesus monkeys *(Macaca mulatta). Primates, 16,* 75–81.

Patterson, F. L. (1978). The gestures of a gorilla: Language acquisition in another pongid. *Brain and Language, 5,* 72–97.

Patterson, F. L., & Linden, E. (1981). *The education of Koko.* New York: Holt, Rinehart, & Winston.

Paul, A. (1989). Determinants of male mating success in a large group of Barbary macaques *(Macaca sylvanus)* at Affenberg Salem. *Primates, 30,* 344–349.

Paul, A., & Kuester, J. (1985). Intergroup transfer and incest avoidance in semifree-ranging Barbary macaques *(Macaca sylvanus)* at Salem (FRG). *American Journal of Primatology, 8,* 317–322.

Paul, A., & Kuester, J. (1988). Life-history patterns of Barbary macaques *(Macaca sylvanus)* at Affenberg Salem. In J. E. Fa & C. H. Southwick (Eds.), *Ecology and behavior of food-enhanced primate groups* (pp. 199–228). New York: Alan R. Liss.

Paul, A., & Thommen, D. (1984). Timing of birth, female reproductive success, and infant sex-ratio in semifree-ranging Barbary macaques *(Macaca sylvanus). Folia Primatologica, 42,* 2–16.

Pavani, S., Maestripieri, D., Schino, G., Turillazzi, P. G., & Scucchi, S. (1991). Factors influencing scratching behavior in long-tailed macaques. *Folia Primatologica, 57,* 34–38.

Pavlov, I. P. (1955). *Selected works.* Moscow: Foreign Languages Publishing House.

Pearl, M. C., & Schulman, S. R. (1983). Techniques for the analysis of social structure in animal societies. *Advances in the Study of Behavior, 13,* 107–146.

Peignot, P., & Anderson, J. R. (1999). Use of experimenter-given manual and facial cues by gorillas *(Gorilla gorilla)* in an object-choice task. *Journal of Comparative Psychology, 113,* 253–260.

Pellis, S. & Pellis, V. (1996). On knowing it's only play: The role of play signals in play fighting. *Aggression and Violent Behavior, 1,* 249–268.

Pepperberg, I. M. (1990a). Cognition in an African Grey parrot *(Psittacus erithacus):* Further evidence for comprehension of categories and labels. *Journal of Comparative Psychology, 104,* 41–52.

Pepperberg, I. M. (1990b). Conceptual abilities of some nonprimate species, with an emphasis on the African Grey parrot. In S. T. Parker & K. R. Gibson (Eds.), *"Language" and intelligence in monkeys and apes: Comparative developmental perspectives* (pp. 469–507). Cambridge: Cambridge University Press.

Pepperberg, I. M. (1999). *The Alex studies: Cognitive and communicative abilities of Grey parrots.* Cambridge, MA: Harvard University Press.

Pereira, M. E. (1986). Maternal recognition of juvenile offspring coo vocalizations in Japanese macaques. *Animal Behaviour, 34,* 935–937.

Pereira, M. E. (1988). Effects of age and sex on intra-group spacing behaviour in juvenile savannah baboons, *Papio cynocephalus cynocephalus. Animal Behaviour, 36,* 184–204.

Pereira, M. E. (1989). Agonistic interactions of juvenile savanna baboons. II: Agonistic support and rank acquisition. *Ethology, 80,* 152–171.

Pereira, M. E. (1995). Development and social dominance among group-living primates. *American Journal of Primatology, 37,* 143–175.

Pereira, M. E., & Altmann, J. (1985). Development of social behavior in free-living nonhuman primates. In E. S. Watts (Ed.), *Nonhuman primate models for human growth and development* (pp. 217–309). New York: Alan Liss.

Pereira, M. E., & Kappeler, P. M. (2000). Divergent social patterns in two primitive primates. In F. Aureli & F. B. M. de Waal (Eds.), *Natural conflict resolution* (pp. 318–320). Berkeley: University of California Press.

Perlman, M., & Ross, H. S. (1997). The benefits of parent intervention in children's disputes: An examination of concurrent changes in children's fighting styles. *Child Development, 64,* 690–700.

Perner, J. (1991). *Understanding the representational mind.* Oxford: Oxford University Press.

Perrett, D. I., Hietanen, J. K., Oram, M. W., & Benson, P. J. (1992). Organization and functions of cells responsive to faces in the temporal cortex. *Philosophical Transactions of the Royal Society of London, B, 335,* 23–30.

Perry, S., & Manson, J. H. (1995). A comparison of the mating behavior of adolescent and adult female rhesus macaques *(Macaca mulatta). Primates, 36,* 27–39.

Persky, H., Charney, N., Lief, H. I., O'Brien, C. P., Miller, W. R., & Strauss, D. (1978). The relationship of plasma estradiol level to sexual behavior in young women. *Psychosomatic Medicine, 40,* 523–535.

Persky, H., Lief, H. I., Strauss, D., Miller, W. R., & O'Brien, C. P. (1978). Plasma testosterone level and sexual behavior of couples. *Archives of Sexual Behavior, 7,* 157–173.

Pervin, L. A. (1999). Epilogue: Constancy and change in personality theory and research. In L. A. Pervin & O. P. John (Eds.), *Handbook of personality theory and research* (pp. 689–704). New York: Guilford Press.

Petersen, M. R., Beecher, M. D., Zoloth, S. R., Moody, D. B., & Stebbins, W. C. (1978). Neural lateralization of species-specific vocalizations in Japanese macaques *(Macaca fuscata). Science, 202,* 324–327.

Petit, O., Abegg, C., & Thierry, B. (1997). A comparative study of aggression and conciliation in three cercopithecine monkeys *(Macaca fuscata, Macaca nigra, Papio papio). Behaviour, 134,* 415–432.

Petit, O., & Thierry, B. (1992). Affiliative function of the silent bared-teeth display in moor macaques *(Macaca maurus):* further evidence for the particular status of Sulawesi macaques. *International Journal of Primatology, 13,* 97–105.

Petit, O., & Thierry, B. (1994). Reconciliation in a group of Guinea baboons. In J. J. Roeder, B. Thierry, J. R. Anderson, & N. Herrenschmidt (Eds.), *Current primatology*. Vol. 2. *Social development, learning, and behaviour* (pp. 137–145). Strasbourg: Presses de l' Université Louis Pasteur.

Petit, O., & Thierry, B. (2000). Do impartial interventions in conflicts occur in monkeys and apes? In F. Aureli & F. B. M. de Waal (Eds.), *Natural conflict resolution* (pp. 267–269). Berkeley: University of California Press.

Petitto, L. A., & Seidenberg, M. S. (1979). On the evidence for linguistic abilities in signing apes. *Brain and Language, 8,* 162–183.

Pfaff, D. W. (1999). *Drive: Neurobiological and molecular mechanisms of sexual motivation.* Cambridge, MA: MIT Press.

Phillips, A. T., Wellman, H. M., & Spelke, E. S. (2002). Infants' ability to connect gaze and emotional expression to intentional action. *Cognition, 85,* 53–78.

Phoenix, C. H. (1973). The role of testosterone in the sexual behavior of laboratory male rhesus. *Primate Reproductive Behavior, 2,* 99–122.

Phoenix, C. H. (1974a). Effects of dihyrotestosterone on sexual behavior on castrated male rhesus monkeys. *Physiology and Behavior, 12,* 1045–55.

Phoenix, C. H. (1974b). Prenatal testosterone in the nonhuman primate and its consequences for behavior. In R. Friedman, R. Richart, & R. Vande Wiele (Eds.), *Sex differences in behavior* (pp. 19–32). New York: John Wiley & Sons.

Phoenix, C. H. (1977). Factors influencing sexual performance in male rhesus monkeys. *Journal of Comparative and Physiological Psychology, 91,* 697–710.

Phoenix, C. H., Copenhaver, K. H., & Brenner, R. M. (1976). Scanning electron microscopy of penile papillae in intact and castrated rats. *Hormones and Behavior, 7,* 217–227.

Phoenix, C. H., Slob, A. K., & Goy, R. W. (1973). Effects of castration and replacement therapy on sexual behavior of adult male rhesuses. *Journal of Comparative and Physiological Psychology, 84,* 472–481.

Piaget, J. (1954). *The construction of reality in the child.* New York: Norton.

Pilcher, D. L., Hammock, E. D., & Hopkins, W. D. (2001). Cerebral volumetric asymmetries in nonhuman primates: A magnetic resonance imaging study. *Laterality, 6,* 165–179.

Pizzella, V., Tecchio, F., Romani, G. L., & Rossini, P. M. (1999). Functional localization of the sensory hand area with respect to the motor central gyrus knob. *NeuroReport, 10,* 3809–14.

Plant, T. M. (1986). A striking sex difference in the gonadotropin response to gonadectomy during infantile development in the rhesus monkey *(Macaca mulatta)*. *Endocrinology, 119,* 539–545.

Plant, T. M. (1994). Puberty in primates. In E. Knobil & J. D. Neill (Eds.), *The physiology of reproduction* (2nd ed.) (pp. 453–485). New York: Raven Press.

Plomin, R. (1994). *Genetics and experience. The interplay between nature and nurture.* London: Sage Publications.

Plomin, R., & Rutter, M. (1998). Child development, molecular genetics, and what to do with genes once they are found. *Child Development, 69,* 1223–42.

Plooij, F. X. (1978). Some basic traits of language in wild chimpanzees? In A. Lock (Ed.), *Action, gesture, and symbol: The emergence of language* (pp. 111–131). London: Academic Press.

Plooij, F. X. (1979). How wild chimpanzee babies trigger the onset of mother-infant play and what the mother makes of it. In M. Bullowa (Ed.), *Before speech: The beginning of interpersonal communication* (pp. 223–243). Cambridge: Cambridge University Press.

Plooij, F. X. (1984). *The behavioral development of free-living chimpanzee babies and infants.* Norwood, NJ: Ablex.

Plutchik, R. (1980). *Emotion: A psychoevolutionary synthesis.* New York: Harper & Row.

Plutchik, R., & Kellerman, H. (1974). *Emotions Profile Index.* Los Angeles: Western Psychological Services.

Pohl, C. R., deRidder, C. M., & Plant, T. M. (1995). Gonadal and nongonadal mechanisms contribute to the prepubertal hiatus in gonadotropin secretion in the female rhesus monkey *(Macaca mulatta). Journal of Clinical Endocrinology and Metabolism, 80,* 2094–2101.

Poirier, F. E. (1972). Introduction. In F. E. Poirier (Ed.), *Primate socialization* (pp. 3–28). New York: Random House.

Poirier, F. E. (1974). Colobine aggression: A review. In R. L. Holloway (Ed.), *Primate aggression, territoriality, and xenophobia: A comparative perspective* (pp. 123–158). New York: Academic Press.

Polan, H. J., & Hofer, M. (1999). Psychobiological origins of infant attachment and separation responses. In J. Cassidy & P. R. Shaver (Eds.), *Handbook of attachment* (pp. 162–180). New York: Guilford Press.

Pope, N. S., Wilson, M. E., & Gordon, T. P. (1987). The effect of season on the induction of sexual behavior by estradiol in female rhesus monkeys. *Biology of Reproduction, 36,* 1047–54.

Pope, T. R. (1998). Effects of demographic change on group kin structure and gene dynamics of populations of red howling monkeys. *Journal of Mammalogy, 79,* 692–712.

Popova, N. K., Kulikov, A. V., Augustinovich, D. F., Voitenko, N. N., & Trut, L. N. (1997). Effect of domestication of the silver fox on the main enzymes of serotonin metabolism and serotonin receptors. *Genetika, 33,* 370–374.

Popova, N. K., Kulikov, A. V., Nikulina, E. M., Kozlachkova, E. Y., & Maslova, G. B. (1991). Serotonin metabolism and serotonergic receptors in Norway rats selected for low aggressiveness towards man. *Aggressive Behavior, 17,* 207–213.

Popova, N. K., Voitenko, N. N., Kulikov, A. V., & Avgustinovich, D. F. (1991). Evidence for the involvement of central serotonin in mechanism of domestication of silver foxes. *Pharmacology Biochemistry and Behavior, 40,* 751–756.

Porac, C., & Coren, S. (1981). *Lateral preferences and human behavior.* New York: Springer.

Porges, S. W. (1991). Vagal tone: An autonomic mediator of affect. In J. A. Garber & K. A. Dodge (Eds.) *The development of affect regulation and dysregulation* (pp. 111–128). New York: Cambridge University Press.

Porges, S. W. (1997). Emotion: An evolutionary by-product of the neural regulation of the autonomic nervous sysyem. In C. S. Carter, B. Kirkpatrick, & I. Lenderhendler (Eds.), *The integrative neurobiology of affiliation* (pp. 62–77). New York: New York Academy of Sciences.

Potegal, M., & Davidson, R. J. (1997). Young children's post-tantrum affiliation with their parents. *Aggressive Behavior, 23,* 329–342.

Povinelli, D. J. (1993). Reconstructing the evolution of mind. *American Psychologist, 48,* 493–509.

Povinelli, D. J. (1994). Comparative studies of animal mental state attribution: A reply to Heyes. *Animal Behaviour, 48,* 239–241.

Povinelli, D. J. (1999). Social understanding in chimpanzees: New evidence from a longitudinal approach. In P. D. Zelazo, J. W. Astington, & D. R. Olson (Eds.), *Developing theories of intention* (pp. 195–225). Hillsdale, NJ: Erlbaum.

Povinelli, D. J. (2000). *Folk physics for apes: The chimpanzee's theory of how the world works.* Oxford: Oxford University Press.

Povinelli, D. J. (2001). On the possibilities of detecting intentions prior to understanding them. In B. Malle, D. Baldwin, & L. Moses (Eds.), *Intentionality: A key to human understanding* (pp. 225–248). Cambridge, MA: MIT Press.

Povinelli, D. J., Bering, J. M., & Giambrone, S. (2000). Toward a science of other minds: Escaping the argument by analogy. *Cognitive Science, 24,* 509–541.

Povinelli, D. J., Bering, J. M., & Giambrone, S. (in press). Chimpanzee "pointing": Another error of the argument by analogy? In S. Kita (Ed.), *Pointing: Where language, culture, and cognition meet.* Mahwah, NJ: Erlbaum.

Povinelli, D. J., Bierschwale, D. T., & Cech, C. G. (1999). Comprehension of seeing as a referential act in young children but not juvenile chimpanzees. *British Journal of Developmental Psychology, 17,* 37–60.

Povinelli, D. J., & Davis, D. R. (1994). Differences between chimpanzees *(Pan troglodytes)* and humans *(Homo sapiens)* in the resting state of the index finger: Implications for pointing. *Journal of Comparative Psychology, 108,* 134–139.

Povinelli, D. J., & deBlois, S. (1992). Young children's *(Homo sapiens)* understanding of knowledge formation in themselves and others. *Journal of Comparative Psychology, 106,* 228–238.

Povinelli, D J., & Dunphy-Lelii, S. (2001). Do chimpanzees seek explanations? Preliminary comparative investigations. *Canadian Journal of Experimental Psychology, 55,* 93–101.

Povinelli, D. J., Dunphy-Lelii, S., Reaux, J. E., & Mazza, M. P. (2002). Psychological diversity in chimpanzees and humans: New longitudinal assessments of chimpanzees' understanding of attention. *Brain, Behavior, and Evolution, 59,* 33–53.

Povinelli, D. J., & Eddy, T. J. (1996a). What young chimpanzees know about seeing. *Monographs of the Society for Research in Child Development, 61,* 1–152.

Povinelli, D. J., & Eddy, T. J. (1996b). Chimpanzees: Joint visual attention. *Psychological Science, 7,* 129–135.

Povinelli, D. J., & Eddy, T. J. (1996c). Factors influencing young chimpanzees' recognition of attention. *Journal of Comparative Psychology, 110,* 336–345.

Povinelli, D. J., & Eddy, T. J. (1997). Specificity of gaze-following in young chimpanzees. *British Journal of Developmental Psychology, 15,* 213–222.

Povinelli, D. J., & Giambrone, S. (1999). Inferring other minds: Failure of the argument by analogy. *Philosophical Topics, 27,* 167–201.

Povinelli, D. J., & Giambrone, S. (2001). Reasoning about beliefs: A human specialization? *Child Development, 72,* 691–695.

Povinelli, D. J., & Godfrey, L. R. (1993). The chimpanzee's mind: How noble in reason? How absent of ethics? In M. Nitecki & D. Nitecki (Eds.), *Evolutionary ethics* (pp. 277–324). Albany: SUNY Press.

Povinelli, D. J., Nelson, K. E., & Boysen, S. T. (1990). Inferences about guessing and knowing by chimpanzees *(Pan troglodytes). Journal of Comparative Psychology, 104,* 203–210.

Povinelli, D. J., & O'Neill, D. K. (2000). Do chimpanzees use their gestures to instruct each other in cooperative situations? In S. Baron-Cohen, H. Tager-Flusberg, & D. J. Cohen (Eds.), *Understanding other minds: Perspectives from developmental cognitive neuroscience* (pp. 459–487). Oxford: Oxford University Press.

Povinelli, D. J., Parks, K. A., & Novak, M. A. (1991). Do rhesus monkeys *(Macaca mulatta)* attribute knowledge and ignorance to others? *Journal of Comparative Psychology, 105,* 318–325.

Povinelli, D. J., Perilloux, H. K., Reaux, J. E., & Bierschwale, D. T. (1998). Young and juvenile chimpanzees' *(Pan troglodytes)* reactions to intentional versus accidental and inadvertent actions. *Behavioural Processes, 42,* 205–218.

Povinelli, D. J., & Prince, C. G. (1998). When self met other. In M. Ferrari & R. J. Sternberg (Eds.), *Self-awareness: Its nature and development* (pp. 37–107). New York: Guilford Press.

Povinelli, D. J., Reaux, J. E., Bierschwale, D. T., Allain, A. D., & Simon, B. B. (1997). Exploitation of pointing as a referential gesture in young children but not adolescent chimpanzees. *Cognitive Development, 12,* 423–461.

Povinelli, D. J., Rulf, A. B., & Bierschwale, D. T. (1994). Absence of knowledge attribution and self-recognition in young chimpanzees *(Pan troglodytes). Journal of Comparative Psychology, 108,* 74–80.

Povinelli, D. J., Theall, L. A., Reaux, J. E., & Dunphy-Lelii, S. (in press). Chimpanzees spontaneously modify the direction of their gestural signals to match the attentional orientation of others. *Animal Behaviour.*

Premack, D. (1971). On the assessment of language competence in the chimpanzee. In A. M. Schrier & F. Stollnitz (Eds.), *Behavior of nonhuman primates,* Vol. 4 (pp. 185–228). New York: Academic Press.

Premack, D. (1976). *Intelligence in ape and man.* Hillsdale, NJ: Erlbaum.

Premack, D. (1986). *Gavagai.* Cambridge, MA: MIT Press.

Premack, D. (1988). "Does the chimpanzee have a theory of mind?" revisited. In R. W. Byrne & A. Whiten (Eds.), *Machiavellian intelligence: Social expertise and the evolution of intellect in monkeys, apes, and humans* (pp. 160–179). New York: Oxford University Press.

Premack, D. (1990). The infant's theory of self-propelled objects. *Cognition, 36,* 1–16.

Premack, D., & Premack, A. J. (1983). *The mind of an ape.* New York: Norton.

Premack, D., & Woodruff, G. (1978). Does the chimpanzee have a theory of mind? *Behavioral and Brain Sciences, 4,* 515–526.

Preston, S. D., & de Waal, F. B. M. (2002). Empathy: Its ultimate and proximate bases. *Behavioral and Brain Sciences.*

Preuschoft, S. (1992). "Laughter" and "smile" in Barbary macaques *(Macaca sylvanus). Ethology, 91,* 220–236.

Preuschoft, S., & van Hooff, J. A. R. A. M. (1995). Homologizing primate facial displays: A critical review of methods. *Folia Primatologica, 65,* 121–137.

Preuschoft, S., & van Hooff, J. A. R. A. M. (1997). The social function of "smile" and "laughter": Variation across primate species and societies. In U. Segerstrale & P. Molnar (Eds.), *Nonverbal communication: Where nature meets culture* (pp. 171–189). New York: Erlbaum.

Preuss, T. M. (1995). The argument from animals to humans in cognitive neuroscience. In M. S. Gazzaniga (Ed.), *The cognitive neurosciences* (pp. 1227–41). Cambridge, MA: MIT Press.

Price, J., Sloman, L., Gardner, R. Jr., Gilbert, P., & Rhode, P. (1994). The social competition hypothesis of depression. *British Journal of Psychiatry, 164,* 309–315.

Provine, R. R., & Fischer, K. R. (1989). Laughing, smiling, and talking: Relation to sleeping and social context in humans. *Ethology, 83,* 295–305.

Provine, R. R., & Yong, Y. L. (1991). Laughter: A stereotyped human vocalization. *Ethology, 89,* 115–124.

Prud'homme, J., & Chapais, B. (1993). Aggressive interventions and matrilineal dominance relations in semifree-ranging Barbary macaques. *Primates, 34,* 271–283.

Pryce, C. R. (1996). Socialization, hormones, and the regulation of maternal behavior in nonhuman simian primates. In J. S. Rosenblatt & C. T. Snowdon (Eds.), *Parental care: Evolution, mechanisms, and adaptive significance* (pp. 643–683). San Diego: Academic Press.

Puce, A., Allison, T., & McCarthy, G. (1999). Electrophysiological studies of human face perception. III: Effects of top-down processing on face-specific potentials. *Cerebral Cortex, 9,* 445–458.

Putallaz, M., & Sheppard, B. H. (1990). Social status and children's orientations to limited resources. *Child Development, 61,* 2022–27.

Raleigh, M. J., Brammer, G. L., & McGuire, M. T. (1983). Male dominance, serotonergic systems, and the behavioral and physiological effects of drugs in vervet monkeys *(Cercopithecus aethiops sabaeus).* In K. A. Miczek (Ed.), *Ethopharmacology: Primate models of neuropsychiatric disorders* (pp. 185–197). New York: Alan R. Liss.

Raleigh, M. J., Brammer, G. L., McGuire, M. T., & Yuwiler, A. (1985). Dominant social status facilitates behavioral effects of serotonergic agonists. *Brain Research, 348,* 274–282.

Raleigh, M. J., Brammer, G. L., Ritvo, E. R., Geller, E., McGuire, M. T., & Yuwiler, A. (1986). Effects of chronic fenfluramine on blood serotonin, cerebrospinal fluid metabolites, and behavior in monkeys. *Psychopharmacology, 90,* 503–508.

Raleigh, M. J., Brammer, G. L., Yuwiler, A., Flannery, J. W., & McGuire, M. T. (1980). Serotonergic influences on the social behavior of vervet monkeys *(Cercopithecus aethiops sabaeus). Experimental Neurology, 68,* 322–334.

Raleigh, M. J., & McGuire, M. T. (1986). Animal analogues of ostracism: Biological mechanisms and social consequences. *Ethology and Sociobiology, 7,* 53–66.

Raleigh, M. J., & McGuire, M. T. (1990). Social influences on endocrine function in male vervet monkeys. In T. E. Ziegler & F. B. Bercovitch (Eds.), *Socioendocrinology of primate reproduction* (pp. 95–111). New York: Wiley-Liss.

Raleigh, M. J., & McGuire, M. T. (1994). Serotonin, aggression, and violence in vervet monkeys. In R. D. Masters & M. T. McGuire (Eds.), *The neurotransmitter revolution* (pp. 129–145). Carbondale: Southern Illinois University Press.

Raleigh, M. J., McGuire, M. T., & Brammer, G. L. (1989). Subjective assessment of behavioral style: Links to overt behavior and physiology in vervet monkeys. *American Journal of Primatology, 18*, 161–162.

Raleigh, M. J., McGuire, M. T., Brammer, G. L., Pollack, D. B., & Yuwiler, A. (1991). Serotonergic mechanisms promote dominance acquisition in adult male vervet monkeys. *Brain Research, 559*, 181–190.

Raleigh, M. J., & Steklis, H. D. (1981). Effect of orbitofrontal and temporal neocortical lesion on the affiliative behavior of vervet monkeys *(Cercopithecus aethiops sabaeus)*. *Experimental Neurology, 73*, 378–379.

Raleigh, M. J., Steklis, H. D., Ervin, F. R., Kling, A. S., & McGuire, M. T. (1979). The effects of orbitofrontal lesions on the aggressive behavior of vervet monkeys *(Cercopithecus aethiops sabaeus)*. *Experimental Neurology, 66*, 158–168.

Rapisarda, J. J., Bergman, K. S., Steiner, R. A., & Foster, D. L. (1983). Response to estradiol inhibition of tonic luteinizing-hormone secretion decreases during the final stage of puberty in the rhesus monkey. *Endocrinology, 112*, 1172–79.

Rasmussen, K. L. R. (1983). Age-related variation in the interactions of adult females with adult males in yellow baboons. In R. A. Hinde (Ed.), *Primate social relationships* (pp. 47–53). Boston: Blackwell Scientific Publications.

Rawlins, R. G., & Kessler, M. J. (1986). The history of the Cayo Santiago colony. In R. G. Rawlins & M. J. Kessler (eds), *The Cayo Santiago macaques* (pp. 13–45). Albany: SUNY Press.

Reaux, J. E., Theall, L. A., & Povinelli, D. J. (1999). A longitudinal investigation of chimpanzees' understanding of visual perception. *Child Development, 70*, 275–290.

Redican, W. K. (1982). An evolutionary perspective on human facial displays. In P. Ekman (Ed.), *Emotion and the human face* (pp. 212–281). Cambridge: Cambridge University Press.

Reiman, E. M., Lane, R. D., Ahern, G. L., Schwartz, G. E., & Davidson, R. J. (2000). Positron Emission Tomography in the study of emotion, anxiety, and anxiety disorders. In R. D. Lane & L. Nadel (Eds.), *Cognitive neuroscience of emotion* (pp. 389–406. New York: Oxford University Press.

Reiner, W. G. (1996). Case study: Sex reassignment in a teenage girl. *Journal of American Academy of Child and Adolescent Psychiatry, 35*, 799–803.

Reisner, I. R., Mann, J. J., Stanley, M., Huang, Y. Y., & Houpt, K. A. (1996). Comparison of cerebrospinal fluid monoamine metabolite levels in dominant-aggressive and non-aggressive dogs. *Brain Research, 714*, 57–64.

Reite, M. (1985). Implantable biotelemetry and social separation in monkeys. In G. P. Moberg (Ed.), *Animal stress* (pp. 141–160). Bethesda, MD: American Physiological Society.

Reite, M., & Capitanio, J. P. (1985). On the nature of social separation and social attachment. In M. Reite & T. Field (Eds.), *The psychobiology of attachment and separation* (pp. 223–255). New York: Academic Press.

Reite, M., Kaemingk, K., & Boccia, M. L. (1989). Maternal separation in bonnet monkey infants: Altered attachment and social support. *Child Development, 60,* 473–480.

Reite, M., & Short, R. (1980) A biobehavioral developmental profile (BDP) for the pig-tailed monkey. *Developmental Psychobiology, 13,* 243–284.

Rejeski, W. J., Brubaker, P. H., Herb, R. A., Kaplan, J. R., & Koritnik, D. (1988). Anabolic steroids and aggressive behavior in cynomolgus monkeys. *Journal of Behavioral Medicine, 11,* 95–105.

Rejeski, W. J., Gregg, E., Kaplan, J. R., & Manuck, S. B. (1990). Anabolic-androgenic steroids: Effects on social behavior and baseline heart rate. *Health Psychology, 9,* 774–791.

Ren, R., Yan, K., Su, Y., Qi, H., Liang, B., Bao, W., & de Waal, F. B. M. (1991). The reconciliation behavior of golden monkeys *(Rhinopithecus roxellanae roxellanae)* in small breeding groups. *Primates, 32,* 321–327.

Rendall, D. (1996). *Social communication and vocal recognition in free-ranging rhesus monkeys (Macaca mulatta).* Ph.D. dissertation, University of California, Davis.

Rendall, D. (under review). The function of agonistic scream vocalizations in primates: Recruiting aid or repelling attack?

Rendall, D., & Di Fiore, A. (1995). The road less traveled: Phylogenetic perspectives in primatology. *Evolutionary Anthropology, 4,* 43–52.

Rendall, D., & Owren, M. J. (2002). Animal vocal communication: Say what? In M. Bekoff, C. Allen, & G. Burghardt (Eds.), *The cognitive animal* (pp. 307–314). Cambridge, MA: MIT Press.

Rendall, D., Owren, M. J., & Rodman, P. S. (1998). The role of vocal tract filtering in identity cueing in rhesus monkey *(Macaca mulatta)* vocalizations. *Journal of the Acoustical Society of America, 103,* 602–614.

Rendall, D., Rodman, P. S., & Emond, R. E. (1996). Vocal recognition of individuals and kin in free-ranging rhesus monkeys. *Animal Behaviour, 51,* 1007–15.

Repacholi, B. M., & Gopnik, A. (1997). Early reasoning about desires: Evidence from 14- and 18-month-olds. *Developmental Psychology, 33,* 12–21.

Rescorla, R. A., & Wagner, A. R. (1972). A theory of Pavlovian conditioning: Variations in the effectiveness of reinforcement and nonreinforcement. In A. H. Black & W. F. Prokasy (Eds.), *Classical conditioning.* Vol. 2. *Current research and theory* (pp. 64–99). New York: Appleton-Century-Crofts.

Resko, J. A., Ellinwood, W. E., Pasztor, L. M., & Buhl, A. E. (1980). Sex steroids in the umbilical circulation of fetal rhesus monkeys from the time of gonadal differentiation. *Journal of Clinical Endocrinology and Metabolism, 50,* 900–905.

Resko, J. A., Goy, R. W., Robinson, J. A., & Norman, R. L. (1982). The pubescent rhesus monkey: Some characteristics of the menstrual cycle. *Biology of Reproduction, 27,* 354–361.

Reynolds, V. (1970). Roles and role change in monkey society: The consort relationship of rhesus monkeys. *Man, 5,* 450–465.

Rhine, R. (1973). Variation and consistency in the social behavior of two groups of stumptail macaques *(Macaca arctoides). Primates, 14,* 21–35.

Richards, J. C., Hof, A., & Alvarenga, M. (2000). Serum lipids and their relationships with hostility and angry affect and behaviors in men. *Health Psychology, 19,* 393–398.

Riede, T., Wilden, I., & Tembrock, G. (1997). Subharmonics, biphonations, and frequency jumps: Common components of mammalian vocalization or indicators for disorders? *Zeitschrift für Saugetierkunde, 62,* 198–203.

Rilling, J. K., Gutman, D. A., Zeh, T. R., Pagnoni, G., Bezus, G. S., & Kilts, C. D. (2002). A neural basis for social cooperation. *Neuron, 35,* 395–405.

Rilling, J. K., & Insel, T. R. (1999). Differential expansions of neural projection systems in primate brain evolution. *NeuroReport, 10,* 1453–59.

Rilling, J. K., Winslow, J. T., O'Brien, D., Gutman, D. A., Hoffman, J. M., & Kilts, C. D. (2001). Neural correlates of maternal separation in rhesus monkeys. *Biological Psychiatry, 49,* 146–157.

Rimm-Kaufman, S. E., & Kagan, J. (1996). The psychological significance of changes in skin temperature. *Motivation and Emotion, 20,* 63–78.

Ringo, J. L., Doty, R. W., Demeter, S., & Simard, P. Y. (1994). Time is of essence: A conjecture that hemispheric specialization arises from interhemispheric conduction delay. *Cerebral Cortex, 4,* 331–343.

Rinn, W. E. (1984). The neuropsychology of facial expression: A review of the neurological and psychological mechanisms for producing facial expressions. *Psychological Bulletin, 95,* 52–77.

Rissman, E. F. (1991). Frank A. Beach Award: Behavioral endocrinology of the female musk shrew. *Hormones and Behavior, 25,* 125–127.

Ristau, C. A. (1991a). *Cognitive ethology: The minds of other animals.* Hillsdale, NJ: Erlbaum.

Ristau, C. A. (1991b). Aspects of the cognitive ethology of an injury-feigning bird, the piping plover. In C. A. Ristau (Ed.), *Cognitive ethology: The minds of other animals* (pp. 91–126). Hillsdale, NJ: Erlbaum.

Robertson, J. (1953). Some responses of young children to loss of maternal care. *Nursing Times, 49,* 382–386.

Robinson, R. G., & Downhill, J. E. (1995). Lateralization of psychopatology in response to focal brain injury. In R. J. Davidson & K. Hugdahl (Eds.), *Brain asymmetry* (pp. 693–711). Cambridge, MA: MIT Press.

Rodseth L., Wrangham, R. W., Harrigan, A. M., & Smuts, B. B. (1991). The human community as a primate society. *Current Anthropology, 32,* 221–254.

Rohde, P. (2001). The relevance of hierarchies, territories, defeat for depression in humans: Hypotheses and clinical predictions. *Journal of Affective Disorders, 65,* 221–230.

Rolls, E. T. (1990). A theory of emotion, and its application to understanding the neural basis of emotion. *Cognition and Emotion, 4,* 161–190.

Rolls, E. T. (1995). A theory of emotion and consciousness, and its application to understanding the neural basis of emotion. In M. S. Gazzaniga (Ed.), *The cognitive neurosciences* (pp. 1091–1106). Cambridge, MA: MIT Press.

Rolls, E. T. (1999). *The brain and emotion.* Oxford: Oxford University Press.

Rolls, E. T. (2000). Précis of The brain and emotion. *Behavioral and Brain Sciences, 23,* 177–234.

Romer, D., & Revelle, W. (1984). Personality traits: Facts or fiction? A critique of the Shweder and D'Andrade systematic distortion hypothesis. *Journal of Personality and Social Psychology, 47,* 1028–42.

Romski, M. A., Sevcik, R. A., & Joyner, S. E. (1984). Nonspeech communication systems: Implications for mentally retarded children. *Topics in Language Disorders, 5,* 66–81.

Romski, M. A., Sevcik, M. A., & Rumbaugh, D. M. (1985). Retention of symbolic communication skills in five severely retarded persons. *American Journal of Mental Deficiency, 89,* 441–444.

Rose, R. M., Bernstein, I. S., & Gordon, T. P. (1975). Consequences of social conflict on plasma testosterone levels in rhesus monkeys. *Psychosomatic Medicine, 37,* 50–61.

Rose, R. M., Bernstein, I. S., Gordon, T. P., & Lindsley, J. G. (1978). Changes in testosterone and behavior during adolescence in the male rhesus monkey. *Psychosomatic Medicine, 40,* 60–70.

Rose, R. M., Gordon, T. P., & Bernstein, I. S. (1972). Plasma testosterone levels in the male rhesus: Influences of sexual and social stimuli. *Science, 178,* 643–645.

Rose, R. M., Gordon, T. P., & Bernstein, I. S. (1978). Diurnal variation in plasma testosterone and cortisol in rhesus monkeys living in social groups. *Journal of Endocrinology, 76,* 67–74.

Rose, R. M., Holaday, J. W., & Bernstein, I. S. (1971). Plasma testosterone, dominance rank, and aggressive behaviour in male rhesus monkeys. *Nature, 231,* 366–368.

Rosenblum, L. A., & Alpert, S. (1974). Fear of strangers and specificity of attachment in monkeys. In M. Lewis & L. A. Rosenblum (Eds.), *The origins of fear* (pp. 165–193). New York: John Wiley.

Rosenblum, L. A., Coplan, J. D., Friedman, S., Bassoff, T., Gorman, J. M., & Andrews, M. W. (1994). Adverse early experiences affect noradrenergic and serotonergic functioning in adult primates. *Biological Psychiatry, 35,* 221–227.

Rosenblum, L. A., & Kaufman, I. C. (1968). Variations in infant development and response to maternal loss in monkeys. *American Journal of Orthopsychiatry, 38,* 418–426.

Rosenblum, L. A., & Paully, G. S. (1984). The effects of varying environmental demands on maternal and infant behavior. *Child Development, 55,* 305–314.

Rosenfeld, S. A., & Van Hoesen, G. W. (1979). Face recognition in the rhesus monkey. *Neuropsychologia, 17,* 503–509.

Rosenthal, R., & Rubin, D. B. (1982). A simple, general purpose display of magnitude of experimental effect. *Journal of Educational Psychology, 74,* 166–169.

Ross, H., Telsa, C., Kenyon, B., & Lollis, S. (1990). Maternal intervention in toddler peer conflict: The socialization of principles of justice. *Developmental Psychology, 26,* 994–1003.

Rostal, D. C., & Eaton, G. G. (1983). Puberty in male Japanese macaques *(Macaca fuscata):* Social and sexual behavior in a confined troop. *American Journal of Primatology, 4,* 135–141.

Rowell, T. E., & Chism, J. (1986). The ontogeny of sex differences in the behavior of patas monkeys. *International Journal of Primatology, 7,* 83–107.

Rowell, T. E., & Hinde, R. A. (1962). Vocal communication by the rhesus monkey *(Macaca mulatta). Proceeding of the Zoological Society of London, 138,* 279–294.

Rowell, T. E., Hinde, R. A., & Spencer-Booth, Y. (1964). "Aunt"-infant interactions in captive rhesus monkeys. *Animal Behaviour, 12,* 219–226.

Rowell, T. E., Wilson, C., & Cords, M. (1991). Reciprocity and partner preference in grooming of female blue monkeys. *International Journal of Primatology, 12,* 319–336.

Roy, A., Pickar, D., Linnoila, M., Doran, A. R., Ninan, P., & Paul, S. M. (1985). Cerebrospinal fluid monoamine and monoamine metabolite concentrations in melancholia. *Psychiatry Research, 15,* 281–292.

Rubin, J. Z., Pruitt, D. G., & Kim, S. H. (1994). *Social conflict: Escalation, stalemate and settlement* (2nd ed.). New York: McGraw-Hill.

Rumbaugh, D. M. (Ed.). (1977). *Language learning by a chimpanzee: The Lana Project.* New York: Academic Press.

Rumbaugh, D. M., & Gill, T. V. (1976). The mastery of language-type skills by the chimpanzee *(Pan). Annals of the New York Academy of Sciences, 280,* 562–578.

Rumbaugh, D. M., Gill, T. V., & von Glasersfeld, E. C. (1973). Reading and sentence completion by a chimpanzee *(Pan)*. *Science, 182,* 731–733.

Rumbaugh, D. M., Hopkins, W. D., Washburn, D. A., & Savage-Rumbaugh, E. S. (1989). Lana chimpanzee learns to count by "Numath": A summary of a video-taped experimental report. *Psychological Record, 39,* 459–470.

Rumbaugh, D. M., & Pate, J. L. (1984). The evolution of primate cognition: A comparative perspective. In H. L. Roitblat, T. G. Bever, & H. S. Terrace (Eds.), *Animal cognition* (pp. 569–587). Hillsdale, NJ: Erlbaum.

Rumbaugh, D. M., & Savage-Rumbaugh, E. S. (1994). Language in comparative perspective. In N. J. Mackintosh (Ed.), *Animal learning and cognition* (pp. 307–333). San Diego: Academic Press.

Rumbaugh, D. M., & Savage-Rumbaugh, E. S. (1999). Language, nonhuman. In G. Adelman & B. H. Smith (Eds.), *Encyclopedia of Neuroscience* (pp. 1013–15). Amsterdam: Elsevier.

Rumbaugh, D. M., Savage-Rumbaugh, E. S., & Washburn, D. A. (1996). Toward a new outlook on primate learning and behavior: Complex learning and emergent processes in comparative perspective. *Japanese Psychological Research, 38,* 113–125.

Rumbaugh, D. M., & Washburn, D. A. (1993). Counting by chimpanzees and ordinality judgments by macaques in video-formatted tasks. In S. T. Boysen & E. J. Capaldi (Eds.), *The development of numerical competence: Animal and human models* (pp. 87–106). Hillsdale, NJ: Erlbaum.

Rumbaugh, D. M., Washburn, D. A., & Hillix, W. A. (1996). Respondents, operants, and emergents: Toward an integrated perspective on behavior. In K. Pribram & J. King (Eds.), *Learning as a self-organizing process* (pp. 57–73). Hillsdale, NJ: Erlbaum.

Ruppenthal, G., Arling, G., Harlow, H., Sackett, G., & Suomi, S. (1976). A 10-year perspective of motherless-mother monkey behavior. *Journal of Abnormal Psychology, 85,* 341–349.

Ruscio, A. M., Borkevec, T. D., & Ruscio, J. (2001). A taxometric investigation of the latent structure of worry. *Journal of Abnormal Psychology, 110,* 413–422.

Rushton, J. P. (1980). *Altruism, socialization, and society.* Englewood Cliffs, NJ: Prentice-Hall.

Rushton, J. P. (1989). Genetic similarity, human altruism, and group selection. *Behavioral and Brain Sciences, 12,* 503–559.

Russell, A., Aloa, V., Feder, T., Glover, A., Miller, H., & Palmer, G. (1998). Sex-based parenting styles in a sample with preschool children. *Australian Journal of Psychology, 50,* 89–99.

Russell, A., & Saebel, J. (1997). Mother-son, mother-daughter, father-son, and father-daughter: Are they distinct relationships? *Developmental Review, 17,* 111–147.

Russell, C. L., Bard, K. A. & Adamson, L. B. (1997). Social referencing by young chimpanzees *(Pan troglodytes). Journal of Comparative Psychology, 111,* 185–193.

Russell, J. A. (1994). Is there universal recognition of emotion from facial expression? A review of the cross-cultural studies. *Psychological Bulletin 115,* 102–141.

Russell, J. A., & Fernandez-Dols, J. M. (Eds.). (1997). *The psychology of facial expressions.* Cambridge: Cambridge University Press.

Russon, A. E., & Galdikas, B. M. F. (1993). Imitation in ex-captive orangutans. *Journal of Comparative Psychology, 107,* 147–161.

Saal, F. E., Johnson, C. B., & Weber, N. (1989). Friendly or sexy? It may depend on whom you ask. *Psychology of Women Quarterly, 13,* 263–276.

Sackett, G. P. (1966). Monkeys reared in isolation with pictures as visual input: Evidence for an innate releasing mechanism. *Science, 154,* 1468–73.

Sackett, G. P. (1968). Abnormal behavior in laboratory-reared rhesus monkeys. In M. W. Fox (Ed.), *Abnormal behavior in animals* (pp. 293–331). St. Louis: W. B. Saunders.

Sackett, G. P. (1973). Innate mechanisms in behavior. In C. R. Carpenter (Ed.), *Behavioral regulators of behavior in primates* (pp. 56–67). Lewisburg, PA: Bucknell University Press.

Sackett, G. P. (1975). Unlearned responses, differential rearing experiences, and the development of attachments by rhesus monkeys. In L. A. Rosenblum (Ed.), *Primate behavior: Developments in field and laboratory research* (pp. 112–140). New York: Academic Press.

Sackett, G. P., Griffin, G. A., Pratt, C., Joslyn, W. D., & Ruppenthal, G. (1967). Mother-infant and adult female choice behavior in rhesus monkeys after various rearing experiences. *Journal of Comparative and Physiological Psychology, 63,* 376–381.

Sackin, S., & Thelen, E. (1984). An ethological study of peaceful associative outcomes to conflict in preschool children. *Child Development, 55,* 1098–1102.

Sade, D. S. (1965). Some aspects of parent-offspring and sibling relations in a group of rhesus monkeys, with a discussion of grooming. *American Journal of Physical Anthropology, 23,* 1–18.

Sade, D. S. (1967). Determinants of dominance in a group of free-ranging rhesus monkeys. In S. A. Altmann (Ed.), *Social communication among primates* (pp. 99–114). Chicago: University of Chicago Press.

Sade, D. S. (1972). Sociometrics of *Macaca mulatta.* I: Linkages and cliques in grooming matrices. *Folia Primatologica, 18,* 196–223.

Sade, D. S., Altmann, M., Loy, J., Hausfater, G., & Breuggeman, J. A. (1988). Sociometrics of *Macaca mulatta.* II: Decoupling centrality and dominance in rhesus monkey social networks. *American Journal of Physical Anthropology, 77,* 409–425.

Sade, D. S., Cushing, K., Cushing, P., Dunaif, J., Figueroa, A., Kaplan, J. R., Lauer, C.,

Rhodes, D., & Schneider, J. (1976). Population dynamics in relation to social structure on Cayo Santiago. *Yearbook of Physical Anthropology, 20,* 253–262.

Salmon, U., & Geist, S. (1943). Effect of androgens upon libido in women. *Journal of Neuroendocrinology, 3,* 441–448.

Saltzman, W., Schultz-Darken, N. J., & Abbott, D. H. (1997). Familial influences on ovulatory function in common marmosets *(Callithrix jacchus). American Journal of Primatology, 41,* 159–177.

Sambrook, T. D., Whiten, A., & Strum, S. C. (1995). Priority of access and grooming patterns of females in a large and small group of olive baboons. *Animal Behaviour, 50,* 1667–82.

Sapolsky, R. M. (1983). Endocrine aspects of social instability in the olive baboon *(Papio anubis). American Journal of Primatology, 5,* 365–379.

Sapolsky, R. M. (1992). *Stress, the aging brain, and the mechanisms of neuron death.* Cambridge, MA: MIT Press.

Sapolsky, R. M., Alberts, S. C., & Altmann, J. (1997). Hypercortisolism associated with social subordinance or social isolation among wild baboons. *Archives of General Psychiatry, 54,* 1137–43.

Sapolsky, R. M., & Ray, J. (1989). Styles of dominance and their physiological correlates among wild baboons. *American Journal of Primatology, 18,* 1–9.

Sarrel, P., Dobay, B., & Wiita, B. (1998). Estrogen and estrogen-androgen replacement in postmenopausal women dissatisfied with estrogen-only therapy. *Journal of Reproductive Medicine, 43,* 847–856.

Savage, A., Ziegler, T. E., & Snowdon, C. T. (1988). Sociosexual development, pair bond formation, and mechanisms of fertility suppression in female cotton-top tamarins *(Saguinus oedipus oedipus). American Journal of Primatology, 14,* 345–359.

Savage, E. S., Wilkerson, B. J., & Bakeman, R. (1977). Spontaneous gestural communication among conspecifics in the pygmy chimpanzee *(Pan paniscus).* In G. H. Bourne (Ed.), *Progress in ape research* (pp. 97–116). New York: Academic Press.

Savage-Rumbaugh, E. S. (1986). *Ape language: From conditioned response to symbol.* New York: Columbia University Press.

Savage-Rumbaugh, E. S. (1987). A new look at ape language: Comprehension of vocal speech and syntax. *Nebraska Symposium on Motivation, 35,* 201–255.

Savage-Rumbaugh, E. S. (1991). Language learning in the bonobo: How and why they learn. In N. Krasnegor, D. M. Rumbaugh, R. L. Schiefelbusch, & M. Studdert-Kennedy (Eds), *Biological and behavioral determinants of language development* (pp. 209–233). Hillsdale, NJ: Erlbaum.

Savage-Rumbaugh, E. S., & Brakke, K. E. (1996). Animal language: Methodological and interpretive issues. In M. Beckoff & D. Jamison (Eds.), *Readings in animal cognition* (269–288). Cambridge, MA: MIT Press.

Savage-Rumbaugh, E. S., & Lewin, R. (1994). *Kanzi: The ape at the brink of the human mind*. New York: John Wiley & Sons.

Savage-Rumbaugh, E. S., McDonald, K., Sevcik, R. A., Hopkins, W. D., & Rubert, E. (1986). Spontaneous symbol acquisition and communicative use by pygmy chimpanzees *(Pan paniscus)*. *Journal of Experimental Psychology: General, 115,* 211–235.

Savage-Rumbaugh, E. S., Murphy, J., Sevcik, R. A., Brakke, K. E., Williams, S. L., & Rumbaugh, D. M. (1993). Language comprehension in ape and child. *Monographs for the Society for Research in Child Development, 58,* 1–221.

Savage-Rumbaugh, E. S., Pate, J. L., Lawson, J., Smith, S. T., & Rosenbaum, S. (1983). Can a chimpanzee make a statement? *Journal of Experimental Psychology: General, 112,* 457–492.

Savage-Rumbaugh, E. S., Romski, M. A., Hopkins, W. D., & Sevcik, R. A. (1989). Symbol acquisition and use by *Pan troglodytes, Pan paniscus,* and *Homo sapiens.* In L. A. Marquardt & P. G. Heltne (Eds.), *Understanding chimpanzees,* (pp. 266–295). Cambridge, MA: Harvard University Press.

Savage-Rumbaugh, E. S., & Rumbaugh, D. M. (1979). Chimpanzee problem comprehension: Insufficient evidence. *Science, 206,* 1201–2.

Savage-Rumbaugh, E. S., Rumbaugh, D. M., & Boysen, S. (1978a). Linguistically mediated tool use and exchange by chimpanzees *(Pan troglodytes)*. *Behavioral and Brain Sciences, 4,* 539–554.

Savage-Rumbaugh, E. S., Rumbaugh, D. M., & Boysen, S. (1978b). Symbolic communication between two chimpanzees *(Pan troglodytes)*. *Science, 201,* 641–644.

Savage-Rumbaugh, E. S., Rumbaugh, D. M., & Boysen, S. T. (1978c). Sarah's problems in comprehension. *Behavioral and Brain Sciences, 1,* 555–557.

Savage-Rumbaugh, E. S., Rumbaugh, D. M., & Boysen, S. (1980). Do apes use language? *American Scientist, 68,* 49–61.

Savage-Rumbaugh, E. S., Rumbaugh, D. M., Smith, S. T., & Lawson, J. (1980). Reference: The linguistic essential. *Science, 210,* 922–924.

Savage-Rumbaugh, E. S., Sevcik, R. A., Brakke, K. E., Rumbaugh, D. M., & Greenfield, P. (1990). Symbols: Their communicative use, comprehension, and combination by bonobos *(Pan paniscus)*. In C. Rovee-Collier & L. P. Lipsett (Eds.), *Advances in infancy research,* Vol 6 (pp. 221–278). Norwood, NJ: Ablex.

Savage-Rumbaugh, E. S., Sevcik, R. A., & Hopkins, W. D. (1988). Symbolic cross-modal transfer in two species of chimpanzees. *Child Development, 59,* 617–625.

Savage-Rumbaugh, E. S., & Wilkerson, B. J. (1978). Socio-sexual behavior in *Pan paniscus* and *Pan troglodytes:* A comparative study. *Journal of Human Evolution, 7,* 327–344.

Savage-Rumbaugh, E. S., Williams, S. L., Furuichi, T., & Kano, T. (1996). Language perceived: *Paniscus* branches out. In W. C. McGrew, L. F. Marchant, & T. Nishida (Eds.), *Great ape societies* (173–184). New York: Cambridge University Press.

Scaramella, T. J., & Brown, W. A. (1978). Serum testosterone and aggressiveness in hockey players. *Psychosomatic Medicine, 40,* 262–265.

Schachter, S., & Singer, J. E. (1962). Cognitive, social, and physiological determinants of emotional state. *Psychological Review, 69,* 279–399.

Schaffner, C. M. (1991). Aggression and post-conflict behavior in red-bellied tamarins *(Saguinus labiatus).* Master's thesis, Bucknell University.

Schaffner, C. M., & Caine, N. G. (2000). The peacefulness of cooperatively breeding primates. In F. Aureli & F. B. M. de Waal (Eds.), *Natural conflict resolution* (pp. 155–169). Berkeley: University of California Press.

Schellenberg, J. A. (1996). *Conflict resolution: Theory, research, and practice.* Albany: SUNY Press.

Scherer, K. R. (1994). Affect bursts. In S. H. M van Goozen, N. E. van de Poll, & J. A. Sergeant (Eds.), *Emotions: Essays on emotion theory* (pp. 161–193). Hillsdale, NJ: Erlbaum.

Scherer, K. R., & Kappas, A. (1988). Primate vocal expression of affective state. In D. Todt, P. Goedeking, & D. Symmens (Eds.), *Primate vocal communication* (pp. 171–194). Berlin: Springer-Verlag.

Schino, G. (2001). Grooming, competition, and social rank among female primates: A meta-analysis. *Animal Behaviour, 62,* 265–271.

Schino, G., D'Amato, F., & Troisi, A. (1995). Mother-infant relationships in Japanese macaques: Sources of inter-individual variation. *Animal Behaviour, 49,* 151–158.

Schino, G., Maestripieri, D., Scucchi, S., & Turilazzi, P. G. (1990). Social tension in familiar and unfamiliar pairs of long-tail macaques. *Behaviour, 113,* 264–272.

Schino, G., Perretta, G., Taglioni, A., Monaco, V., & Troisi, A. (1996). Primate displacement activities as an ethopharmacological model of anxiety. *Anxiety, 2,* 186–191.

Schino, G., Rosati, L., & Aureli, F. (1998). Intragroup variation in conciliatory tendency in captive Japanese macaques. *Behaviour, 135,* 897–912.

Schino, G., Scucchi, S., Maestripieri, D., & Turillazzi, P. G. (1988). Allogrooming as a tension-reduction mechanism: A behavioral approach. *American Journal of Primatology, 16,* 43–50.

Schino, G., Speranza, L., & Troisi, A. (2001). Early maternal rejection and later social anxiety in juvenile and adult Japanese macaques. *Developmental Psychobiology, 38,* 186–190.

Schino, G., & Troisi, A. (1992). Opiate receptor blockade in juvenile macaques: Effect on affiliative interactions with their mothers and group companions. *Brain Research, 576,* 125–130.

Schino, G., Troisi, A., Perretta, G., & Monaco, V. (1991). Measuring anxiety in nonhuman primates: Effects of lorazepam on macaque scratching. *Pharmacology Biochemistry and Behavior, 38,* 889–891.

Schneider, M. L., Moore, C. F., Suomi, S. J., & Champoux, M. (1991). Laboratory assessment of temperament and environmental enrichment in rhesus monkey infants *(Macaca mulatta)*. *American Journal of Primatology, 25,* 137–155.

Schulman, S. R. (1983). Analysis of social structure. In R. A. Hinde (Ed.), *Primate social relationships: An integrated approach* (pp. 221–225). Oxford: Blackwell Scientific.

Schusterman R. J., & Gisiner R. (1988). Artificial language comprehension in dolphins and sea lions: The essential cognitive skills. *Psychological Record, 38,* 311–348.

Schusterman, R. J., & Krieger, K. (1984). California sea lions are capable of semantic comprehension. *Psychological Record, 34,* 3–23.

Schusterman, R. J., & Krieger, K. (1986). Artificial language comprehension and size transposition by a California sea lion *(Zalophus californianus)*. *Journal of Comparative Psychology, 100,* 348–355.

Seidenberg, M. S., & Petitto, L. A. (1979). Signing behavior in apes: A critical review. *Cognition, 7,* 177–215.

Sergent, J., & Signoret, J. (1992). Functional and anatomical decomposition of face processing: Evidence from prosopagnosia and PET study of normal subjects. *Philosophical Transactions of the Royal Society of London, B, 335,* 55–62.

Setchell, J. M., & Dixson, A. F. (2001a). Changes in the secondary sexual adornments of male mandrills *(Mandrillus sphinx)* are associated with gain and loss of alpha status. *Hormones and Behavior, 39,* 177–184.

Setchell, J. M., & Dixson, A. F. (2001b). Arrested development of secondary sexual adornments in subordinate adult male mandrills *(Mandrillus sphinx)*. *American Journal of Physical Anthropology, 115,* 245–252.

Sevcik, R. A., Romski, M. A., & Wilkinson, K. M. (1991). Roles of graphic symbols in the language acquisition process for persons with severe cognitive disabilities. *Augmentative and Alternative Communication, 7,* 161–170.

Seyfarth, R. M. (1977). A model of social grooming among adult female monkeys. *Journal of Theoretical Biology, 65,* 671–698.

Seyfarth, R. M. (1983). Grooming and social competition in primates. In R. A. Hinde (Ed.), *Primate social relationships: An integrated approach* (pp. 182–190). Oxford: Blackwell Scientific.

Seyfarth, R. M. (1987). Vocal communication and its relation to language. In B. B. Smuts, D. L. Cheney, R. M. Seyfarth, R. W. Wrangham, & T. T. Struhsaker (Eds.), *Primate societies* (pp. 440–451). Chicago: University of Chicago Press.

Seyfarth, R. M., & Cheney, D. L. (1984). Grooming, alliances, and reciprocal altruism in vervet monkeys. *Nature, 308,* 541–543.

Seyfarth, R. M., & Cheney, D. L. (1988). Empirical tests of reciprocity theory: Problems in assessment. *Ethology and Sociobiology, 9,* 181–187.

Seyfarth, R. M., Cheney, D. L., & Marler P. (1980a). Monkey responses to three different alarm calls: Evidence of predator classification and semantic communication. *Science, 210,* 801–803.

Seyfarth, R. M., Cheney, D. L., & Marler P. (1980b). Vervet monkey alarm calls: Semantic communication in a free-ranging primate. *Animal Behaviour, 28,* 1070–94.

Shafer, D. D. (1993). Patterns of hand preference in gorillas and children. In J. P. Ward & W. D. Hopkins (Eds.), *Primate laterality: Current behavioral evidence of primate asymmetries* (pp. 267–283). New York: Springer-Verlag.

Shafer, D. D. (1997). Hand preference behaviors shared by two groups of captive bonobos. *Primates, 38,* 303–313.

Shantz, C. U. (1987). Conflicts between children. *Child Development, 58,* 283–305.

Sherwin, B. B. (1985). Changes in sexual behavior as a function of plasma sex steroid levels in post-menopausal women. *Maturitas, 7,* 225–233.

Sherwin, B. B. (1991). The impact of different doses of estrogen and progestin on mood and sexual behavior in postmenopausal women. *Journal of Clinical Endocrinology and Metabolism, 72,* 336–343.

Sherwin, B. B., & Gelfand, M. M. (1987). The role of androgen in the maintenance of sexual functioning in oophorectomized women. *Psychosomatic Medicine, 49,* 397–409.

Sherwin, B. B., Gelfand, M. M., & Brender, W. (1985). Androgen enhances sexual motivation in females: A prospective, crossover study of sex steriod administration in the surgical menopause. *Psychosomatic Medicine, 47,* 339–351.

Shettleworth, S. J. (1998). *Cognition, evolution, and behaviour.* Oxford: Oxford University Press.

Shifren, J. L., Braunstein, G. D., Simon, J. A., Casson, P. R., Buster, J. E., Redmond, G. P., Burki, R. E., Ginsburg, E. S., Rosen, R. C., Leiblum, S. R., Caramelli, K. E., & Mazer, N. A. (2000). Transdermal testosterone treatment in women with impaired sexual function after oophorectomy. *New England Journal of Medicine, 343,* 682–688.

Shirek-Ellefson, J. (1972). Social communication in some Old World monkeys and gibbons. In P. Dolhinow (Ed.), *Primate patterns* (pp. 297–311). New York: Holt, Rinehart & Winston.

Shively, C. A. (1998). Social subordination stress, behavior, and central monoaminergic function in female cynomolgus monkeys. *Biological Psychiatry, 44,* 882–891.

Shively, C. A., Fontenot, M. B., & Kaplan, J. R. (1995). Social status, behavior, and central serotonergic responsivity in female cynomolgus monkeys. *American Journal of Primatology, 37,* 333–340.

Shively, C. A., Grant, K. A., Ehrenkaufer, R. L., Mach, R. H., & Nader, M. A. (1997). Social stress, depression, and brain dopamine in female cynomolgus monkeys. *Annals of the New York Academy of Sciences, 807,* 574–577.

Shotland, R. L., & Craig, J. M. (1988). Can men and women differentiate between friendly and sexually interested behavior? *Social Psychology Quarterly, 51*, 66–73.

Shweder, R. A. (1982a). Beyond self-constructed knowledge: The study of culture and morality. *Merrill-Palmer Quarterly, 28*, 41–69.

Shweder, R. A. (1982b). Fact and artifact in trait perception: The systematic distortion hypothesis. In B. A. Maher & W. B. Maher (Eds.), *Progress in experimental personality research*, Vol. 11 (pp. 65–99). New York: Academic Press.

Shweder, R. A., & D'Andrade, R. G. (1980). The systematic distortion hypothesis. In R. A. Shweder (Ed.), *Fallible judgment in behavioral research: New directions for the methodology of social and behavioral science*, Vol. 4 (pp. 37–58). San Francisco: Jossey-Bass.

Sibley, C. G., & Ahlquist, J. E. (1987). DNA hybridization evidence of hominid phylogeny: Results from an expanded data set. *Journal of Molecular Evolution, 26*, 99–121.

Sicotte, P. (1995). Interpositions in conflicts between males in bimale groups of mountain gorillas. *Folia Primatologica, 65*, 14–24.

Siddiqui, A. A., & Ross, H. S. (1999). How do sibling conflicts end? *Early Education and Development, 10*, 315–332.

Silk, J. B. (1982). Altruism among female *Macaca radiata:* Explanations and analysis of patterns of grooming and coalition formation. *Behaviour, 79*, 162–188.

Silk, J. B. (1983). Local resource competition and facultative adjustment of sex ratios in relation to competitive abilities. *American Naturalist, 121*, 56–66.

Silk, J. B. (1992). The patterning of intervention among male bonnet macaques: Reciprocity, revenge, and loyalty. *Current Anthropology, 33*, 318–325.

Silk, J. B. (1997). The function of peaceful post-conflict contacts among primates. *Primates, 38*, 265–279.

Silk, J. B. (1998). Making amends: Adaptive perspectives on conflict remediation in monkeys, apes, and humans. *Human Nature, 9*, 341–368.

Silk, J. B. (2002). Using the "F" word in primatology. *Behaviour, 139*, 421–446.

Silk, J. B., Cheney, D. L., & Seyfarth, R. M. (1996). The form and function of post-conflict interactions between female baboons. *Animal Behaviour, 52*, 259–268.

Silk, J. B., Seyfarth, R. M., & Cheney, D. L. (1999). The structure of social relationships among female savanna baboons in Moremi Reserve, Botswana. *Behaviour, 136*, 679–703.

Simonds, P. E. (1965). The bonnet macaque in South India. In I. DeVore (Ed.), *Primate behavior: Field studies of monkeys and apes* (pp. 175–196). New York: Holt, Rinehart & Winston.

Simons, R. C., & Bielert, C. F. (1973). An experimental study of vocal communication between mother and infant monkeys *(Macaca nemestrina). American Journal of Physical Anthropology, 38*, 455–462.

Simpson, J. A. (1999). Attachment theory in modern evolutionary perspective. In J. Cassidy & P. R. Shaver (Eds.), *Handbook of attachment* (pp. 115–140). New York: Guilford Press.

Simpson, J. A., & Rholes, W. S. (2000). Caregiving, attachment theory, and the connection theoretical orientation. *Psychological Inquiry, 11*, 114–117.

Simpson, M. J. A., & Datta, S. B. (1990). Predicting infant enterprise from early relationships in rhesus macaques. *Behaviour, 116*, 42–63.

Simpson, M. J. A., & Simpson, A. E. (1986). The emergence and maintenance of interdyad differences in the mother-infant relationships of rhesus macaques: a correlational study. *International Journal of Primatology, 7*, 379–399.

Singer, J. E. (1964). The use of manipulative strategies: Machiavellianism and attractiveness. *Sociometry, 27*, 128–150.

Slob, A. K., & Schenck, P. E. (1986). Heterosexual experience and isosexual behavior in laboratory-housed male stumptailed macaques *(Macaca arctoides). Archives of Sexual Behavior, 15*, 261–268.

Smith, C. A., & Pope, L. K. (1992). Appraisal and emotion: The interactional contributions of dispositional and situational factors. In M. S. Clark (Ed.), *Emotion and social behavior* (pp. 32–62). Newbury Park, CA: Sage Publications.

Smith, E. A., Borgerhoff Mulder, M., & Hill, K. (2000). Evolutionary analyses of human behaviour: A commentary on Daly & Wilson. *Animal Behaviour, 60*, F21–F26.

Smith, E. O. (1973). A further description of the control role in pigtail macaques, *Macaca nemestrina. Primates, 14*, 413–419.

Smith, M. S., Kish, B. J., & Crawford, C. B. (1987). Inheritance of wealth as human kin investment. *Ethology and Sociobiology, 8*, 171–182.

Smith, O. A., Astley, C. A., Chesney, M. A., Taylor, D. J., & Spelman, F. A. (1986). Personality, stress and cardiovascular disease: Human and nonhuman primates. In B. Lown, A. Malliani, & M. Prosdomici (Eds.), *Neural mechanisms and cardiovascular disease* (pp. 471–484). Padua: Liviana Press.

Smith, O. A., Astley, C. A., Spelman, F. A., Golanov, E. V., Bowden, D. M., Chesney, M. A., & Chalyan, V. G. (2000). Cardiovascular responses in anticipation of changes in posture and locomotion. *Brain Research Bulletin, 53*, 69–76.

Smith, O. A., Astley, C. A., Spelman, F. A., Golanov, E. V., Chalyan, V. G., Bowden, D. M., & Taylor, D. J. (1993). Integrating behavior and cardiovascular responses: posture and locomotion. I: Static analysis. *American Journal of Physiology, 265*, R1458–68.

Smith, P. K. (1995). Grandparenthood. In M. H. Bornstein (ed.), *Handbook of parenting*, Vol. 3 (pp. 89–112). Hillsdale, NJ: Erlbaum.

Smith, R. M., & Rubenstein, B. B. (1940). Adolescence of macaques. *Endocrinology, 26*, 667–679.

Smith W. J. (1977). *The behavior of communicating: An ethological approach.* Cambridge, MA: Harvard University Press.

Smith, W. J. (1997). The behavior of communicating, after twenty years. In D. H. Owings, M. D. Beecher, & N. S. Thompson (Eds.), *Perspectives in ethology.* Vol. 12. *Communication* (pp. 7–53). New York: Plenum Press.

Smucny, D. A., Price, C. S., & Byrne, E. A. (1997). Post-conflict affiliation and stress reduction in captive rhesus macaques. *Advances in Ethology, 32,* 157.

Smuts, B. B. (1985). *Sex and friendship in baboons.* New York: Aldine.

Smuts, B. B. (1987). Gender, aggression, and influence. In B. B. Smuts, D. L. Cheney, R. M. Seyfarth, R. W. Wrangham, & Struhsaker, T. T. (Eds.). *Primate societies* (pp. 400–412). Chicago: University of Chicago Press.

Smuts, B. B., Cheney, D. L., Seyfarth, R. M., Wrangham, R. W., & T. T. Struhsaker (Eds.), (1987). *Primate societies.* Chicago: University of Chicago Press.

Smuts, B. B., & Watanabe, J. M. (1990). Social relationships and ritualized greetings in adult male baboons *(Papio cynocephalus anubis). International Journal of Primatology, 11,* 147–172.

Snowdon, C. T. (1996). Infant care in cooperatively breeding species. In J. S. Rosenblatt & C. T. Snowdon (Eds.), *Parental care: Evolution, mechanisms, and adaptive significance* (pp. 643–683). San Diego: Academic Press.

Snowdon, C. T., Elowson, A. M., & Roush, R. S. (1997). Social influences on vocal development in New World primates. In C. T. Snowdon & M. Hausberger (Eds.), *Social influences on vocal development* (pp. 234–248). New York: Cambridge University Press.

Snowdon, C. T., Ziegler, T. E., & Widowski, T. M. (1993). Further hormonal suppression of eldest daughter cotton-top tamarins following birth of infants. *American Journal of Primatology, 31,* 11–21.

Sommer, V. (1994). Infanticide among the langurs of Jodhpur: Testing the sexual selection hypothesis with long-term record. In S. Parmigiani & F. S. vom Saal (Eds.), *Infanticide and parental care* (pp. 155–198). Chur, Switzerland: Harwood Academic Publishers.

Soubrié, P. (1986). Reconciling the role of central serotonin neurons in human and animal behavior. *Behavioral and Brain Sciences, 9,* 319–364.

Southwick, C. H., Siddiqi, M. F., Farooqui, M. Y., & Pal, B. C. (1974). Xenophobia among free-ranging rhesus groups in India. In R. L. Holloway (Ed.), *Primate aggression, territoriality, and xenophobia: A comparative perspective* (pp. 185–210). New York: Academic Press.

Southwick, C. H., Beg, M. A., & Siddiqi, M. R. (1965). Rhesus monkeys in North India. In I. DeVore (Ed.), *Primate behavior: Field studies of monkeys and apes* (pp. 111–159). New York: Holt, Rinehart & Winston.

Spencer, H. (1887). *The factors of organic evolution.* London: Williams & Norgate.

Spencer-Booth, Y., & Hinde, R. A. (1969). Tests of behavioural characteristics for rhesus monkeys. *Behaviour, 33,* 180–211.

Spinozzi, G., Castorina, M. G., & Truppa, V. (1998). Hand preferences in unimanual and coordinated-bimanual tasks by tufted capuchin monkeys *(Cebus apella). Journal of Comparative Psychology, 112,* 183–191.

Spitz, R. A. (1946). Anaclitic depression: An inquiry into the genesis of psychiatric conditions in early childhood. *Psychoanalytic Study of the Child, 2,* 313–342.

Sroufe, L. A. (1996). *Emotional development.* Cambridge: Cambridge University Press.

Sroufe, L. A., & Waters, E. (1976). The ontogenesis of smiling and laughter: A perspective on the organization of development in infancy. *Psychological Review, 83,* 173–189.

Sroufe, L. A., & Waters, E. (1977). Attachment as an organizational construct. *Child Development, 48,* 1184–99.

Stammbach, E. (1988). Group responses to specially skilled individuals in a *Macaca fascicularis* group. *Behaviour, 107,* 241–266.

Stanford, C. B. (1995). Chimpanzee hunting behavior and human evolution. *American Scientist, 83,* 256–261.

Staub, E. (1978). *Positive social behavior and morality.* Vol. 1. *Social and personal influences.* New York: Academic Press.

Stebbins, W. C., & Moody, D. B. (1994). How monkeys hear the world: Auditory perception in nonhuman primates. In R. R. Fay & A. N. Popper (Eds.), *Comparative hearing: Mammals* (pp. 97–133). New York: Springer-Verlag.

Steegmans, P. H., Fekkes, D., Hoes, A. W., Bak, A. A., van der Does, E., & Grobbee, D. E. (1996). Low serum cholesterol concentration and serotonin metabolism in men. *British Medical Journal, 312,* 221.

Steenbeek, R., Sterck, E. H. M., De Vries, H., & Van Hooff, J. A. R. A. M. (2000). Cost and benefits of the one-male, age-graded and all-male phases in wild Thomas's langur groups. In P. M. Kappeler (Ed.), *Primate males: Causes and consequences of variation in group composition* (pp. 130–145). New York: Cambridge University Press.

Stefanacci, L., Reber, P., Costanza, J ., Wong, E., Buxton, R., Zola, S., Squire, L., & Albright, T. (1998). fMRI of monkey visual cortex. *Neuron, 20,* 1051–57.

Stein, N. L., & Oatley, K. (1992). Basic emotions: Theory and measurement. *Cognition and Emotion, 6,* 161–168.

Steklis, H. D., Brammer, G. L., Raleigh, M. J., & McGuire, M. T. (1985). Serum testosterone, male dominance, and aggression in captive groups of vervet monkeys *(Cercopithecus aethiops sabaeus). Hormones and Behavior, 19,* 154–165.

Stephan, H., Frahm, H., & Baron, G. (1981). New and revised data on volumes of brain structures in insectivores and primates. *Folia Primatologica, 35,* 1–29.

Sterck, E. H. M., Watts, D. P., & van Schaik, C. P. (1997). The evolution of female social relationships in nonhuman primates. *Behavioral Ecology and Sociobiology, 41,* 291–309.

Stevenson-Hinde, J. (1983). Individual characteristics and the social situation. In R. A. Hinde (Ed.), *Primate social relationships: An integrated approach* (pp. 28–35). Sunderland, MA: Sinauer Associates.

Stevenson-Hinde, J., Stillwell-Barnes, R., & Zunz, M. (1980a). Subjective assessment of rhesus monkeys over four successive years. *Primates, 21,* 66–82.

Stevenson-Hinde, J., Stillwell-Barnes, R., & Zunz, M. (1980b). Individual differences in young rhesus monkeys: Consistency and change. *Primates, 21,* 498–509.

Stevenson-Hinde, J., & Zunz, M. (1978). Subjective assessment of individual rhesus monkeys. *Primates, 19,* 473–482.

Stevenson-Hinde, J., Zunz, M., & Stillwell-Barnes, R. (1980). Behaviour of one-year-old rhesus monkeys in a strange situation. *Animal Behaviour, 28,* 266–277.

Stewart, K. J., & Harcourt, A. H. (1987). Gorillas: Variation in female relationships. In B. B. Smuts, D. L.Cheney, R. M. Seyfarth, R. W. Wrangham, & T. T. Struhsaker, (Eds.), *Primate societies* (pp. 155–164). Chicago: University of Chicago Press.

Strayer, F. F. (1992). The development of agonistic and affiliative structures in preschool play groups. In J. Silverberg & J. P. Gray (Eds.), *Aggression and peacefulness in humans and other primates* (pp. 150–171). Oxford: Oxford University Press.

Strayer, F. F., & Noel, J. M. (1986). The prosocial and antisocial functions of preschool aggression: An ethological study of triadic conflict among young children. In C. Zahn-Waxler, M. Cummings, & R. Iannotti (Eds.), *Altruism and aggression* (pp. 107–134). Cambridge: Cambridge University Press.

Strayer, F. F., & Strayer, J. (1976). An ethological analysis of social agonism and dominance relations among preschool children. *Child Development, 47,* 980–989.

Strier, K. B. (1992). Causes and consequences of nonaggression in the woolly spider monkey, or muriqui *(Brachyteles arachnoides).* In J. Silverberg & J. P. Gray (Eds.), *Aggression and peacefulness in humans and other primates* (pp. 100–116). Oxford: Oxford University Press.

Strier, K. B. (1994). Myth of the typical primate. *Yearbook of Physical Anthropology, 37,* 233–271.

Struhsaker T. T. (1967). Auditory communication in vervet monkeys *(Cercopithecus aethiops).* In S. A. Altmann (Ed.), *Social communication among primates* (pp. 281–324). Chicago: University of Chicago Press.

Struhsaker, T. T. (1971). Social behavior of mother and infant vervet monkeys *(Cercopithecus aethiops). Animal Behaviour, 19,* 233–250.

Strum, S. S. (1987). *Almost human: A journey into the world of baboons.* New York: Norton.

Suddendorf, T. (1999). The rise of the metamind. In M. C. Corballis & S. E. G. Lea

(Eds.), *The descent of mind: Psychological perspectives on hominid evolution* (pp. 218–260). New York: Oxford University Press.

Suddendorf, T., & Whiten, A. (2001). Mental evolution and development: Evidence for secondary representation in children, great apes, and other animals. *Psychological Bulletin, 127,* 629–650.

Sugiura, H. (1998). Matching of acoustic features during the vocal exchange of coo calls by Japanese macaques. *Animal Behaviour, 55,* 673–687.

Sugiyama, Y. (1965). On the social change of hanuman langurs, *Presbytis entellus,* in their natural conditions. *Primates, 6,* 381–418.

Sugiyama, Y. (1968). The ecology of the liontailed macaque *(Macaca silenus Linnaeus):* A pilot study. *Journal of the Bombay Natural History Society, 65,* 283–293.

Sugiyama, Y. (1976). Life history of Japanese macaques. *Advances in the Study of Behavior, 7,* 255–284.

Sugiyama, Y. (1997). Social tradition and the use of tool-composites by wild chimpanzees. *Evolutionary Anthropology, 6,* 23–27.

Suomi, S. J. (1982a). In memoriam: Harry Harlow (1905–1981). *American Journal of Primatology, 2,* 319–342.

Suomi, S. J. (1982b). Abnormal behavior and primate models of psychopathology. In J. L. Fobes & J. E. King (Eds.), *Primate behavior* (pp. 171–215). New York: Academic Press.

Suomi, S. J. (1987). Genetic and maternal contributions to individual differences in rhesus monkey biobehavioral development. In N. Krasnegor, E. Blass, M. Hofer, & W. Smotherman (Eds.), *Perinatal development: A psychobiological perspective* (pp. 397–420). New York: Academic Press.

Suomi, S. J. (1991). Up-tight and laid-back monkeys: Individual differences in the response to social challenger. In S. Brauth, W. Hall, & R. Dooling (Eds.), *Plasticity of development* (pp. 27–56). Cambridge, MA: MIT Press.

Suomi, S. J. (1995). Influence of attachment theory on ethological studies of biobehavioral development in nonhuman primates. In S. Goldberg, R. Muir, & J. Kerr (Eds.), *Attachment theory: Social, developmental, and clinical perspectives* (pp. 185–201). Hillsdale, NJ: Analytic Press.

Suomi, S. J. (1997). Early determinants of behaviour: Evidence from primate studies. *British Medical Bulletin, 53,* 170–184.

Suomi, S. J. (1999). Attachment in rhesus monkeys. In J. Cassidy & P. R. Shaver (Eds.), *Handbook of attachment* (pp. 181–197). New York: Guilford Press.

Suomi, S. J., Kraemer, G. W., Baysinger, C. M., & DeLizio, R. D. (1981). Inherited and experiential factors associated with individual differences in anxious behavior displayed by rhesus monkeys. In D. F. Klein & J. Rabkin (Eds.), *Anxiety: New research and changing concepts* (pp. 179–200). New York: Raven Press.

Suomi, S. J., Novak, M. A., & Well, A. (1996). Aging in rhesus monkeys: Different windows on behavioral continuity and change. *Developmental Psychology, 32,* 1116–28.

Suomi, S. J., & Ripp, C. (1983). A history of motherless mother monkey mothering at the University of Wisconsin Primate Laboratory. In M. Reite & N. Caine (Eds.), *Child abuse: The nonhuman primate data* (pp. 49–77). New York: Alan Liss.

Surakka, V., & Hietanen, J. K. (1998). Facial and emotional reactions to Duchenne and non-Duchenne smiles. *International Journal of Psychophysiology, 29,* 23–33.

Suter, K. J., Pohl, C. R., & Plant, T. M. (1998). The pattern and tempo of the pubertal reaugmentation of open-loop pulsatile gonadotropin-releasing hormone release assessed indirectly in the male rhesus monkey *(Macaca mulatta). Endocrinology, 139,* 2774–83.

Suter, K. J., Pohl, C. R., & Plant, T. M. (1999). Indirect assessment of pulsatile gonadotropin-releasing hormone release in agonadal prepubertal rhesus monkeys *(Macaca mulatta). Journal of Endocrinology, 160,* 35–41.

Swartz, K. B. (1982). Issues in the measurement of attachment in nonhuman primates. *Journal of Human Evolution, 11,* 237–245.

Symons, D. (1978). *Play and aggression: A study of rhesus monkeys.* New York: Columbia University Press.

Tanaka, I. (1992). Three phases of lactation in free-ranging Japanese macaques. *Animal Behaviour, 44,* 129–139.

Tanner, J. E., & Byrne, R. W. (1993). Concealing facial evidence of mood: Perspective-taking in a captive gorilla? *Primates, 34,* 451–457.

Tanner, J. E., & Byrne, R. W. (1996). Representation of action through iconic gesture in a captive lowland gorilla. *Current Anthropology, 37,* 162–173.

Tanner, J. E., & Byrne, R. W. (1999). The development of spontaneous gestural communication in a group of zoo-living lowland gorillas. In S. T. Parker, R. W. Mitchell, & H. L. Miles (Eds.), *The mentalities of gorillas and orangutans* (pp. 211–239). Cambridge: Cambridge University Press.

Tarr, M. J., & Gauthier, I. (2000). FFA: a flexible fusiform area for subordinate-level visual processing automatized by expertise. *Nature Neuroscience, 3,* 764–769.

Tellegen, A. (1991). Personality traits: Issues of definition, evidence, and assessment. In D. Cicchetti & W. M. Grove (Eds.), *Thinking clearly about psychology: Essays in honor of Paul Everett Meehl,* Vol. 2 (pp. 10–35). Minneapolis: University of Minnesota Press.

Temerlin, M. K. (1975). *Lucy: Growing up human.* London: Souvenir Press.

Terasawa, E., & Fernandez, D. L. (2001). Neurobiological mechanisms of the onset of puberty in primates. *Endocrine Reviews, 22,* 111–151.

Terrace, H. (1979). *Nim: A chimpanzee who learned sign language.* New York: Simon & Schuster.

Terrace, H. S., Petitto, L. A., Sanders, R. J., & Bever, T. G. (1979). Can an ape create a sentence? *Science, 206*, 891–900.

Terrace, H. S., Straub, R. O., Bever, T. G., & Seidenberg, M. S. (1977). Representation of a sequence by a pigeon. *Bulletin of the Psychonomic Society, 10*, 269.

Terry, R. (1970). Primate grooming as a tension reduction mechanism. *Journal of Psychology, 76*, 129–136.

Thase, M. E. (2000). Mood disorders: Neurobiology. In B. J. Sadock & V. A. Sadock (Eds.), *Kaplan and Sadock's Comprehensive textbook of psychiatry* (7th ed.), Vol. 1 (pp. 1319–28). Philadelphia: Lippincott Williams and Wilkins.

Theall, L. A., & Povinelli, D. J. (1999). Do chimpanzees tailor their attention-getting behaviors to fit the attentional states of others? *Animal Cognition, 2*, 207–214.

Thierry, B. (1984). Clasping behavior in *Macaca tonkeana. Behaviour, 89*, 1–28.

Thierry, B. (1985). Pattern of agonistic interactions in three species of macaque *(Macaca mulatta, M. fascicularis, M. tonkeana). Aggressive Behavior, 11*, 223–233.

Thierry, B., Demaria, C., Preuschoft, S., & Desportes, C. (1989). Structural convergence between silent bared-teeth display and relaxed open-mouth display in the tonkean macaque *(Macaca tonkeana). Folia Primatologica, 52*, 178–184.

Thompson, C. R., & Church, R. M. (1980). An explanation of the language of a chimpanzee. *Science, 208*, 313–314.

Thompson, N. S. (1997). Communication and natural design. In D. H. Owings, M. D. Beecher, & N. S. Thompson (Eds.), *Perspectives in ethology.* Vol. 12. *Communication* (pp. 391–415). New York: Plenum Press.

Tinbergen, N. (1952). "Derived activities": Their causation, biological significance, origin, and emancipation during evolution. *Quarterly Review of Biology, 27*, 1–32.

Tinbergen, N. (1963). On the aims and methods of ethology. *Zeitschrift für Tierpsychologie, 20*, 410–433.

Todt, D. (1988). Serial calling as a mediator of interaction processes: Crying in primates. In D. Todt, P. Goedeking, & D. Symmes (Eds.), *Primate vocal communication* (pp. 88–107). Berlin: Springer-Verlag.

Tokuda, I., Riede, T., Neubauer, J., Owren, M. J., & Herzel, H. (2002). Nonlinear prediction of irregular animal vocalizations. *Journal of the Acoustical Society of America, 111*, 2908–19.

Tomasello, M. (1990). Cultural transmission in the tool use and communicatory signaling of chimpanzees? In S. T. Parker & K. R. Gibson (Eds.), *"Language" and intelligence in monkeys and apes* (pp. 274–311). Cambridge: Cambridge University Press.

Tomasello, M. (1996). Do apes ape? In B. G. Galef Jr. & C. M. Heyes (Eds.), *Social learning in animals: The roots of culture* (pp. 319–346). New York: Academic Press.

Tomasello, M. (1999). *The cultural origins of human cognition.* Cambridge, MA: Harvard University Press.

Tomasello, M., & Call, J. (1997). *Primate cognition.* Oxford: Oxford University Press.

Tomasello, M., Call, J., & Gluckman, A. (1997). The comprehension of novel communicative signs by apes and human children. *Child Development, 68,* 1067–81.

Tomasello, M., Call, J., & Hare, B. (1998). Five primate species follow the visual gaze of conspecifics. *Animal Behaviour, 55,* 1063–69.

Tomasello, M., Call, J., Nagell, K., Olguin, R., & Carpenter, M. (1994). The learning and use of gestural signals by young chimpanzees: A trans-generational study. *Primates, 35,* 137–154.

Tomasello, M., Call, J., Warren, J., Frost, G. T., Carpenter, M., & Nagell, K. (1997). The ontogeny of chimpanzee gestural signals: A comparison across groups and generations. *Evolution of Communication, 1,* 223–259.

Tomasello, M., & Camaioni, L. (1997). A comparison of the gestural communication of apes and human infants. *Human Development, 40,* 7–24.

Tomasello, M., George, B. L., Kruger, A. C., Farrar, M. J., & Evans, A. (1985). The development of gestural communication in young chimpanzees. *Journal of Human Evolution, 14,* 175–186.

Tomasello, M., Gust, D., & Frost, G. T. (1989). A longitudinal investigation of gestural communication in young chimpanzees. *Primates, 30,* 35–50.

Tomasello, M. Hare, B., & Agnetta, B. (1999). Chimpanzees follow gaze direction geometrically. *Animal Behaviour, 58,* 769–777.

Tomasello, M., Kruger, A. C., & Ratner, H. H. (1993). Cultural learning. *Behavioral and Brain Sciences, 16,* 495–552.

Tomasello, M., Savage-Rumbaugh, E. S., & Kruger, A. C. (1993). Imitative learning of actions on objects by children, chimpanzees, and enculturated chimpanzees. *Child Development, 64,* 1688–1705.

Tomkins, S. S. (1962). *Affect, imagery, consciousness.* Vol. 1. *The positive affects.* New York: Springer.

Tomkins, S. S. (1963). *Affect, imagery, consciousness.* Vol. 2. *The negative affects.* New York: Springer.

Tomonaga, M., Itakura, S., & Matsuzawa, T. (1993). Superiority of conspecific faces and reduced inversion effect in face perception by a chimpanzee. *Folia Primatologica, 61,* 110–114.

Tooby, J., & Cosmides, L. (1989). Adaptation versus phylogeny: The role of animal psychology in the study of human behavior. *International Journal of Comparative Psychology, 2,* 175–188.

Tooby, J., & Cosmides, L. (1992). The psychological foundations of culture. In

J. H. Barkow, L. Cosmides, & J. Tooby (Eds.), *The adapted mind* (pp. 19–136). New York: Oxford University Press.

Tooby, J., & Cosmides, L. (1996). Friendship and the banker's paradox: Other pathways to the evolution of adaptations for altruism. *Proceedings of the British Academy, 88,* 119–143.

Topal, J., Miklosi, A., Csany, V., & Doka, A. (1998). Attachment behavior in dogs *(Canis familiaris):* A new application of Ainsworth's (1969) Strange Situation Test. *Journal of Comparative Psychology, 112,* 219–229.

Tovee, M. J., Rolls, E. T., & Azzopardi, P. (1994). Translation invariance in the response to faces of single neurons in the temporal visual cortical areas of the alert macaque. *Journal of Neurophysiology, 72,* 1049–60.

Trevarthen, C., & Aitken, K. J. (2001). Infant intersubjectivity: Research, theory, and clinical applications. *Journal of Child Psychology and Psychiatry, 42,* 3–48.

Triggs, W. J., Calvanio, R., & Levine, M. (1997). Transcranial magnetic stimulation reveals a hemispheric asymmetry correlate of intermanual differences in motor performance. *Neuropsychologia, 35,* 1355–63.

Trivers, R. L. (1971). The evolution of reciprocal altruism. *Quarterly Review of Biology 46, 35–57.*

Trivers, R. L. (1972). Parental investment and sexual selection. In B. Campbell (Ed.), *Sexual selection and the descent of man, 1871–1971* (pp. 136–179). Chicago: Aldine.

Trivers, R. L. (1974). Parent-offspring conflict. *American Zoologist, 14,* 249–264.

Trivers, R. L., & Willard, D. E. (1973). Natural selection of parental ability to vary the sex ratio of offspring. *Science 179,* 90–92.

Troisi, A. (1994). The relevance of ethology for animal models of psychiatric disorders: A clinical perspective. In S. J. Cooper & C. A. Hendrie (Eds.), *Ethology and psychopharmacology* (pp. 329–340). Chichester, UK: John Wiley and Sons.

Troisi, A. (1999). Ethological research in clinical psychiatry: The study of nonverbal behavior during interviews. *Neuroscience and Biobehavioral Reviews, 23,* 905–913.

Troisi, A. (2002). Displacement activities as a behavioral measure of stress in nonhuman primates and human subjects. *Stress, 5,* 47–54

Troisi, A., Aureli, F., Piovesan, P., & D'Amato, F. R. (1989). Severity of early separation and later abusive mothering in monkeys: What is the pathogenic threshold? *Journal of Child Psychology and Psychiatry, 30,* 277–284.

Troisi, A., & D'Amato, F. R. (1984). Ambivalence in monkey mothering: Infant abuse combined with maternal possessiveness. *Journal of Nervous and Mental Disease, 172,* 105–108.

Troisi, A., & D'Amato, F. R. (1991). Anxiety in the pathogenesis of primate infant abuse: A pharmacological study. *Psychopharmacology, 103,* 571–572.

Troisi, A., & D'Amato, F. R. (1994). Mechanisms of primate infant abuse: The maternal anxiety hypothesis. In S. Parmigiani & F. S. vom Saal (Eds.), *Infanticide and parental care* (pp. 199–210). Chur, Switzerland: Harwood Academic Publishers.

Troisi, A., & McGuire, M. T. (1991). Deception and somatizing disorders. In C. N. Stefanis, A. D. Rabavilas, & C. R. Soldatos (Eds.), *Psychiatry: A world perspective*, Vol. 3 (pp. 973–978). Amsterdam: Excerpta Medica.

Troisi, A., & McGuire, M. T. (1998). Evolution and mental health. In H. S. Friedman (Ed.), *Encyclopedia of mental health*, Vol. 2 (pp. 173–181). San Diego: Academic Press.

Troisi, A., & Schino, G. (1987). Environmental and social influences on autogrooming behaviour in a captive group of Java monkeys. *Behaviour, 100*, 292–302.

Troisi, A., Schino, G., D'Antoni, M., Pandolfi, N., Aureli, F., & D'Amato, F. R. (1991). Scratching as a behavioral index of anxiety in macaque mothers. *Behavioral and Neural Biology, 56*, 307–313.

Troisi, A., Spalletta, G., & Pasini, A. (1998). Non-verbal behavior deficits in schizophrenia: An ethological study of drug-free patients. *Acta Psychiatrica Scandinava, 97*, 109–115.

Tschudin, A., Call, J., Dunbar, R. I. M., Harris, G., & van der Elst, C. (2001). Comprehension of signs by dolphins *(Tursiops truncatus)*. *Journal of Comparative Psychology, 115*, 100–105.

Turkhan, J. S. (1989). Pavlovian conditioning: The new hegemony. *Behavioral and Brain Sciences, 12*, 121–179.

Udry, J. R., Billy, J. O., Morris, N. M., Groff, T. R., & Raj, M. H. (1985). Serum androgenic hormones motivate sexual behavior in adolescent boys. *Fertility and Sterility, 43*, 90–94.

Utian, W. H. (1972). The true clinical features of postmenopause and oophorectomy and their responses to oestrogen therapy. *South African Medical Journal, 46*, 732–737.

van de Rijt-Plooij, H. H. C., & Plooij, F. X. (1987). Growing independence, conflict, and learning in mother-infant relations in free-ranging chimpanzees. *Behaviour, 101*, 1–86.

van Goozen, S. H. M., Wiegant, V. M., Endert, E., Helmond, F. A., & Van de Poll, N. E. (1997). Psychoendocrinological assessment of the menstrual cycle: The relationship between hormones, sexuality, and mood. *Archives of Sexual Behavior, 26*, 359–382.

van Hooff, J. A. R. A. M. (1962). Facial expressions in higher primates. *Symposia of the Zoological Society of London, 8*, 97–125.

van Hooff, J. A. R. A. M. (1967). The facial displays of the Catarrhine monkeys and apes. In D. Morris (Ed.), *Primate ethology* (pp. 7–68). London: Weidenfield.

van Hooff, J. A. R. A. M., (1970). A component analysis of the structure of the social behavior of a semi-captive chimpanzee group. *Experientia, 26,* 549–550.

van Hooff, J. A. R. A. M. (1972). A comparative approach to the phylogeny of laughter and smile. In R. A. Hinde (Ed.), *Nonverbal communication* (pp. 209–241). Cambridge: Cambridge University Press.

van Hooff, J. A. R. A. M. (1973). A structural analysis of the social behavior of a semi-captive group of chimpanzees. In M. von Cranach & I. Vine (Eds.), *Social communication and movement* (pp. 75–162). London: Academic Press.

van Hooff, J. A. R. A. M. (1976). The comparison of facial expression in man and higher primates. In M. von Cranach (Ed.), *Methods of inference from animal to human behaviour* (pp. 165–196). Chicago: Aldine.

van Hooff, J. A. R. A. M., & Aureli, F. (1994). Social homeostasis and the regulation of emotion. In S. H. M. van Goozen, N. E. van de Poll, & J. A. Sergeant (Eds.), *Emotions: Essays on emotion theory* (pp. 197–217). Hillsdale, NJ: Erlbaum.

van Ijzendoorn, M. H., & Sagi, A. (1999). Cross-cultural patterns of attachment: Universal and contextual dimensions. In J. Cassidy, & P. R. Shaver (Eds.), *Handbook of attachment* (pp. 713–734). New York: Guilford Press.

van Noordwijk, M. A. (1985). Sexual behavior of Sumatran long-tailed macaques *(Macaca fascicularis)*. *Zeitschrift für Tierpsychologie, 70,* 277–296.

van Noordwijk, M. A., & van Schaik, C. P. (1985). Male migration and rank acquisition in wild long-tailed macaques *(Macaca fascicularis)*. *Animal Behaviour, 33,* 849–861.

van Noordwijk, M. A., & van Schaik, C. P. (2001). Career moves: Transfer and rank challenge decisions by male long-tailed macaques. *Behaviour, 138,* 359–395.

van Schaik, C. P. (1989). The ecology of social relationships amongst female primates. In V. Standen & R. Foley (Eds.), *Comparative socioecology: The behavioural ecology of humans and other mammals* (pp. 195–218). Oxford: Blackwell.

van Schaik, C. P., & Aureli, F. (2000). The natural history of valuable relationships in primates. In F. Aureli & F. B. M. de Waal (Eds.), *Natural conflict resolution* (pp. 307–333). Berkeley: University of California Press.

van Schaik, C. P., & Janson, C. H. (2000). *Infanticide by males and its implications.* Cambridge: Cambridge University Press.

van Schaik, C. P., & van Hooff, J. A. R. A. M. (1983). On the ultimate causes of primate social systems. *Behaviour, 85,* 91–117.

Varley, M. A., & Symmes, D. (1966). The hierarchy of dominance in a group of macaques. *Behaviour, 27,* 54–75.

Vea, J. J., & Sabater-Pi, J. (1998). Spontaneous pointing behaviour in the wild pygmy chimpanzee *(Pan paniscus)*. *Folia Primatologica, 69,* 289–290.

Veney, S. L., & Rissman, E. F. (2000). Immunolocalization of androgen receptors and aromatase enzyme in the adult musk shrew brain. *Neuroendocrinology, 72,* 29–36.

Verbeek, P., & de Waal, F. B. M. (1997). Postconflict behavior of captive brown capuchins in the presence and absence of attractive food. *International Journal of Primatology, 18,* 703–725.

Verbeek, P., & de Waal, F. B. M. (2001). Peacemaking among preschool children. *Peace and Conflict, 7,* 5–28.

Verbeek, P., Hartup, W. W., & Collins, W. A. (2000). Conflict management in children and adolescents. In F. Aureli & F. B. M. de Waal (Eds.), *Natural conflict resolution* (pp. 34–53). Berkeley: University of California Press.

Vermeire, B. A. & Hamilton, C. R. (1998a). Effects of facial identity, facial expression, and subject's sex on laterality in monkeys. *Laterality, 3,* 1–19.

Vermeire, B. A., & Hamilton, C. R. (1998b). Inversion effect for faces in split-brain monkeys. *Neuropsychologia, 36,* 1003–14.

Vermeire, B. A., Hamilton, C. R., & Erdmann, A. L. (1998). Right hemispheric superiority in split-brain monkeys for learning and remembering facial discriminations. *Behavioral Neuroscience, 112,* 1048–61.

Vespo, J. E., & Caplan, M. (1993). Preschoolers' differential conflict behavior with friends and acquaintances. *Early Education and Development, 4,* 45–53.

Vick, S. J., & Anderson, J. R. (2000). Learning and limits of use of eye gaze by capuchin monkeys *(Cebus apella)* in an object-choice task. *Journal of Comparative Psychology, 114,* 200–207.

Vinden, P. G., & Astington, J. W. (2000). Culture and understanding other minds. In S. Baron-Cohen, H. Tager-Flusberg, & D. Cohen (Eds.), *Understanding other minds: Perspectives from autism and cognitive neuroscience* (pp. 503–520). Oxford: Oxford University Press.

Virgin, C. E., & Sapolsky, R. M. (1997). Styles of male social behavior and their endocrine correlates among low-ranking baboons. *American Journal of Primatology, 42,* 25–39.

Virkkunen, M., De Jong, J., Bartko, J., Goodwin, F. K., & Linnoila, M. (1989). Relationship of psychobiological variables to recidivism in violent offenders and impulsive fire setters. A follow-up study. *Archives of General Psychiatry, 46,* 600–603.

Virkkunen, M., Kallio, E., Rawlings, R., Tokola, R., Poland, R. E., Guidotti, A., Nemeroff, C., Bissette, G., Kalogeras, K., Karonen, S. L., & Linnoila, M. (1994). Personality profiles and state aggressiveness in Finnish alcoholic, violent offenders, fire setters, and healthy volunteers. *Archives of General Psychiatry, 51,* 28–33.

Virkkunen, M., Nuutila, A., Goodwin, F. K., & Linnoila, M. (1987). Cerebrospinal fluid monoamine metabolite levels in male arsonists. *Archives of General Psychiatry, 44,* 241–217.

Virkkunen, M., Rawlings, R., Tokola, R., Poland, R. E., Guidotti, A., Nemeroff, C., Bissette, G., Kalogeras, K., Karonen, S. L., & Linnoila, M. (1994). CSF biochemistries, glucose metabolism, and diurnal activity rhythms in alcoholic, violent offenders, fire setters, and healthy volunteers. *Archives of General Psychiatry, 51,* 20–27.

Visalberghi, E., & Fragaszy, D. (1990). Do monkeys ape? In S. T. Parker, & K. R. Gibson (Eds.), *"Language" and intelligence in monkeys and apes* (pp. 247–273). Cambridge: Cambridge University Press.

Visalberghi, E., Fragaszy, D. M., & Savage-Rumbaugh, S. (1995). Performance in tool-using task by common chimpanzees *(Pan troglodytes),* bonobos *(Pan paniscus),* an orangutan *(Pongo pygmaeus),* and capuchin monkeys *(Cebus apella). Journal of Comparative Psychology, 109,* 52–60.

Visalberghi, E., & Limongelli, L. (1994). Lack of comprehension of cause-effect relations in tool-using capuchin monkeys *(Cebus apella). Journal of Comparative Psychology, 108,* 15–22.

Visalberghi, E., & Trinca, L. (1989). Tool use in capuchin monkeys: Distinguishing between performing and understanding. *Primates, 30,* 511–521.

Vitale, A. F., Visalberghi, E., & De Lillo, C. (1991). Responses to a snake model in captive crab-eating macaques *(Macaca fascicularis)* and captive tufted capuchins *(Cebus apella). International Journal of Primatology, 12,* 277–286.

Voland, E. (1989). Differential parental investment: Some ideas on the contact area of European history and evolutionary biology. In V. Standen & R. A. Foley (Eds.), *Comparative socioecology: The behavioural ecology of humans and other mammals* (pp. 391–403). Oxford: Blackwell Scientific Publications.

Voydanoff, P., & Donnelly, B. W. (1990). *Adolescent sexuality and pregnancy.* Newbury Park, CA: Sage Publications.

Wallen, K. (1982). Influence of female hormonal state on rhesus sexual behavior varies with space for social interaction. *Science, 217,* 375–377.

Wallen, K. (1990). Desire and ability: Hormones and the regulation of female sexual behavior. *Hormones and Behavior, 14,* 233–241.

Wallen, K. (1996). Nature needs nurture: The interaction of hormonal and social influences on the development of behavioral sex differences in rhesus monkeys. *Hormones and Behavior, 30,* 364–378.

Wallen, K. (2000a). The development of hypothalamic control of sexual behavior. In J. P. Bourguignon & T. M. Plant (Eds.), *Onset of puberty in perspective* (pp. 377–388). Amsterdam: Elsevier Science.

Wallen, K. (2000b). Risky business: Social context and hormonal modulation of primate sexual desire. In K. Wallen & J. E. Schneider (Eds.), *Reproduction in context* (pp. 289–324). Cambridge, MA: MIT Press.

Wallen, K. (2001). Sex and context: Hormones and primate sexual motivation. *Hormones and Behavior, 40,* 339–357.

Wallen, K., Bielert, C., & Slimp, J. (1977). Foot clasp mounting in the prepubertal rhesus monkey: Social and hormonal influences, *Primate bio-social development: Biological, social, and ecological determinants* (pp. 439–461). New York: Garland.

Wallen, K., Goldfoot, D. A., & Goy, R. W. (1981). Peer and maternal influences on the expression of foot-clasp mounting by juvenile male rhesus monkeys. *Developmental Psychobiology, 14*, 299–309.

Wallen, K., & Goy, R. W. (1977). Effects of estradiol benzoate, estrone, and propionates of testosterone or dihydrotestosterone on sexual and related behaviors of ovariectomized rhesus monkeys. *Hormones and Behavior, 9*, 228–248.

Wallen, K., Herman, R. A., & Zehr, J. L. (2001). Late gestation androgen hypermasculinizes juvenile sexual behavior in male rhesus monkeys. Paper presented at the International Academy of Sex Research, Montreal.

Wallen, K., Maestripieri, D., & Mann, D. R. (1995). Effects of neonatal testicular suppression with a GnRH antagonist on social behavior in group-living juvenile rhesus monkeys. *Hormones and Behavior, 29*, 322–337.

Wallen, K., Mann, D. R., Davis-DaSilva, M., Gaventa, S., Lovejoy, J. C., & Collins, D. C. (1986). Chronic gonadotropin-releasing hormone agonist treatment suppresses ovulation and sexual behavior in group-living female rhesus monkeys. *Physiology and Behavior, 36*, 369–375.

Wallen, K., & Parsons, W. (1998a). Androgen may increase sexual motivation in estrogen-treated ovariectomized rhesus monkeys by increasing estrogen availability. Paper presented at the Serono International Symposium on Biology of Menopause, Newport Beach, CA.

Wallen, K., & Parsons, W. A. (1998b). Sexual behavior in same-sexed nonhuman primates: Is it relevant to understanding human homosexuality? *Annual Review of Sex Research, 8*, 195–223.

Wallen, K., & Tannenbaum, P. L. (1997). Hormonal modulation of sexual behavior and affiliation in rhesus monkeys. *Annals of the New York Academy of Sciences, 807*, 185–202.

Wallen, K., & Winston, L. A. (1984). Social complexity and hormonal influences on sexual behavior in rhesus monkeys *(Macaca mulatta)*. *Physiology and Behavior, 32*, 629–637.

Wallen, K., Winston, L. A., Gaventa, S., Davis-DaSilva, M., & Collins, D. C. (1984). Periovulatory changes in female sexual behavior and patterns of ovarian steroid secretion in group-living rhesus monkeys. *Hormones and Behavior, 18*, 431–450.

Walters, J. (1981). Inferring kinship from behaviour: Maternity determinations in yellow baboons. *Animal Behaviour, 29*, 126–136.

Walters, J. R., & Seyfarth, R. M. (1987). Conflict and cooperation. In B. B. Smuts, D. L. Cheney, R. M. Seyfarth, R. W. Wrangham, & T. T. Struhsaker (Eds.), *Primate societies* (pp. 306–317). Chicago: University of Chicago Press.

Ward, J. P., & Hopkins, W. D. (Eds.) (1993). *Primate laterality: Current behavioral evidence of primate asymmetries.* New York: Springer-Verlag.

Warren, J. M. (1980). Handedness and laterality in humans and other animals. *Physiological Psychology, 8,* 351–359.

Watanabe, G., & Terasawa, E. (1989). *In vivo* release of luteinizing-hormone releasing hormone increases with puberty in the female rhesus monkey. *Endocrinology, 125,* 92–99.

Watson, D., Clark, L. A., & Harkness, A. R. (1994). Structures of personality and their relevance to psychopathology. *Journal of Abnormal Psychology, 103,* 18–31.

Watson, D., & Tellegen, A. (1985). Toward a consensual structure of mood. *Psychological Bulletin, 98,* 219–235.

Watson, S. L., & Ward, J. P. (1996). Temperament and problem solving in the small-eared bushbaby *(Otolemur garnettii). Journal of Comparative Psychology, 110,* 377–385.

Watts, D. P. (1994). Agonistic relationships between female mountain gorillas *(Gorilla gorilla beringei). Behavioral Ecology and Sociobiology, 34,* 347–358.

Watts, D. P. (1995). Post-conflict social events in wild mountain gorillas *(Mammalia, Hominoidea).* I: Social interactions between opponents. *Ethology, 100,* 139–157.

Watts, D. P. (1997). Agonistic interventions in wild mountain gorilla groups. *Behaviour, 134,* 23–57.

Watts, D. P. (2002). Reciprocity and interchange in the social relationships of wild male chimpanzees. *Behaviour, 139,* 343–370.

Watts, D. P., Colmenares, F., & Arnold, K. (2000). Redirection, consolation, and male policing: How targets of aggression interact with bystanders. In F. Aureli & F. B. M. de Waal (Eds.), *Natural conflict resolution* (pp. 281–301). Berkeley: University of California Press.

Watts, D. P., & Mitani, J. C. (2001). Boundary patrols and intergroup encounters in wild chimpanzees. *Behaviour, 138,* 299–327.

Watts, D., & Pusey, A. (1993). Behavior of juvenile and adolescent great apes. In M. E. Pereira & L. A. Fairbanks (Eds.), *Juvenile primates* (pp. 148–167). New York: Oxford University Press.

Waxenberg, S. E., Drellich, M. G., & Sutherland, A. M. (1959). The role of hormones in human behavior. I: Changes in female sexuality after adrenalectomy. *Journal of Clinical Endocrinology, 19,* 193–202.

Waxer, P. H. (1977). Nonverbal cues for anxiety: An examination of emotional leakage. *Journal of Abnormal Psychology, 86,* 306–314.

Weaver, A. C., & de Waal, F. B. M. (2000). The development of reconciliation in brown capuchins. In F. Aureli & F. B. M. de Waal (Eds.), *Natural conflict resolution* (pp. 216–218). Berkeley: University of California Press.

Weber, M. (1921/1968). *Economy and society.* Translated and edited by G. Roth & C. Wittich. New York: Bedminster Press.

Weisfeld, G. E. (1993). The adaptive value of humor and laughter. *Ethology and Sociobiology, 14,* 141–169.

Weiss, A., King, J. E., & Figueredo, A. J. (2000). The heritability of personality in chimpanzees *(Pan troglodytes). Behavior Genetics, 30,* 213–221.

Weiss, D. S., & Mendelsohn, G. A. (1986). An empirical demonstration of the implausibility of the semantic similarity explanation of how trait ratings are made and what they mean. *Journal of Personality and Social Psychology, 50,* 595–601.

Weiss, L., & Lowenthal, M. F. (1975). Life-course perspective on friendship. In M. Thurnher & D. Chiraboga (Eds.), *Four stages of life* (pp. 48–61). San Francisco: Jossey-Bass.

Weiss, R. S. (1982). Attachment in adult life. In C. Murray Parkes & J. Stevenson-Hinde (Eds.), *The place of attachment in human behavior* (pp. 171–184). New York: Basic Books.

Wellman, H. M., Cross, D., & Watson, J. (2001). Meta-analysis of theory-of-mind development: The truth about false belief. *Child Development, 72,* 655–684.

Wellman, H. M., & Phillips, A. T. (2001). Developing intentional understandings. In B. Malle, D. Baldwin, & L. Moses (eds.), *Intentionality: A key to human understanding* (pp. 125–148). Cambridge, MA: MIT Press.

Westergaard, G. C., Champoux, M., & Suomi, S. J. (1997). Hand preference in infant rhesus macaques *(Macaca mulatta). Child Development, 68,* 387–393.

Westergaard, G. C., Kuhn, H. E., & Suomi, S. J. (1998). Bipedal posture and hand preference in humans and other primates. *Journal of Comparative Psychology, 112,* 56–63.

Westergaard, G. C., Mehlman, P. T., Hoos, B., Suomi, S. J., & Higley, J. D. (under review). Serotonergic influences on life history outcomes in free-ranging male primates.

Westergaard, G. C., Mehlman, P. T., Shoaf, S. E., Suomi, S. J., & Higley, J. D. (1999). CSF 5-HIAA and aggression in female primates: Species and interindividual differences. *Psychopharmacology, 146,* 440–446.

Westergaard, G. C., & Suomi, S. J. (1996). Hand preference for a bimanual task in tufted capuchins *(Cebus apella)* and rhesus macaques *(Macaca mulatta). Journal of Comparative Psychology, 110,* 406–411.

Westergaard, G. G., & Suomi, S. J. (1997). Lateral bias in capuchin monkeys *(Cebus apella):* Concordance between parents and offspring. *Developmental Psychobiology, 31,* 143–147.

Westergaard, G. C., Suomi, S. J., Higley, J. D., & Mehlman, P. T. (1999). CSF 5-HIAA and aggression in female macaque monkeys: Species and interindividual differences. *Psychopharmacology, 146,* 440–446.

White, P. C., New, M. I., & Dupont, B. (1987). Congenital adrenal hyperplasia (pt. 1). *New England Journal of Medicine, 316,* 1519–24.

Whiten, A. (1993a). The evolution and development of emotional states, emotional expressions, and emotion-reading in human and non-human primates. Paper presented at the Mellow Symposium, "Emotion: Culture, Psychology, and Biology," Emory University, Atlanta.

Whiten, A. (1993b). Evolving a theory of mind: The nature of non-verbal mentalism in other primates. In S. Baron-Cohen, H. Tager-Flusberg, & D. Cohen (Eds.) *Understanding other minds* (pp. 367–396). Oxford: Oxford University Press.

Whiten, A. (1994) Grades of mindreading. In C. Lewis & P. Mitchell (Eds.), *Children's early understanding of mind* (pp. 47–70). Hillsdale, NJ: Erlbaum.

Whiten, A. (1996). When does smart behaviour-reading become mind-reading? In P. Carruthers, & P. K. Smith (Eds.), *Theories of theories of mind* (pp. 277–292). Cambridge: Cambridge University Press.

Whiten, A. (1997). The Machiavellian mindreader. In A. Whiten & R. W. Byrne (Eds.), *Machiavellian intelligence,* Vol. 2 (pp. 144–173). Cambridge: Cambridge University Press.

Whiten, A. (1999a). The evolution of deep social mind in humans. In M. C. Corballis & S. E. G. Lea (Eds.), *The descent of mind* (pp. 155–175). Oxford: Oxford University Press.

Whiten, A. (1999b). Parental encouragement in *Gorilla* in comparative perspective: Implications for social cognition and the evolution of teaching. In S. T. Parker, R. W. Mitchell, & H. L. Miles (Eds.), *The mentalities of gorillas and orangutans: Comparative perspectives* (pp. 342–366). Cambridge: Cambridge University Press.

Whiten, A. (2000a). Primate culture and social learning. *Cognitive Science, 24,* 477–508.

Whiten, A. (2000b). Chimpanzee cognition and the question of mental re-representation. In D. Sperber (Ed.), *Metarepresentation: A multidisciplinary perspective* (pp. 139–167). Oxford: Oxford University Press.

Whiten, A. (2001). Tool tests challenge chimpanzees. *Nature, 409,* 133.

Whiten, A., & Byrne, R. W. (1988). Tactical deception in primates. *Behavioral and Brain Sciences, 11,* 233–244.

Whiten, A., Custance, D. M., Gomez, J. C., Teixidor, P., & Bard, K. A. (1996). Imitative learning of artificial fruit processing in children *(Homo sapiens)* and chimpanzees *(Pan troglodytes). Journal of Comparative Psychology, 110,* 3–14.

Whiten, A., Goodall, J., McGrew, W. C., Nishida, T., Reynolds, V., Sugiyama, Y., Tutin, C. E. G., Wrangham, R. W., & Boesch, C. (1999). Cultures in chimpanzees. *Nature, 399,* 682–685.

Whiten, A., & Ham, R. (1992). On the nature and evolution of imitation in the animal kingdom: Reappraisal of a century of research. *Advances in the Study of Behavior, 21,* 239–283.

Whiting, B. B. (1980). Culture and social behavior. A model for the development of social behavior. *Ethos, 8,* 95–116.

Whitten, P. L. (1987). Infants and adult males. In Smuts, B. B., Cheney, D. L., Seyfarth, R. M., Wrangham, R. W., & Struhsaker, T. T. (Eds.), *Primate societies* (pp. 343–357). Chicago: University of Chicago Press.

Wickings, E. J., & Dixson, A. F. (1992). Testicular function, secondary sexual development, and social status in male mandrills *(Mandrillus sphinx). Physiology and Behavior, 52,* 909–916.

Widdig, A., Nurnberg, P., Krawczak, M., Streich, W. J., & Bercovitch, F. B. (2001). Paternal relatedness and age proximity regulate social relationships among adult female rhesus macaques. *Proceedings of the National Academy of Science, USA, 98,* 13769–73.

Widdig, A., Streich, W. J., & Tembrock, G. (2000). Coalition formation among male Barbary macaques *(Macaca sylvanus). American Journal of Primatology, 50,* 37–51.

Wiggins, J. S. (1973). *Personality and prediction: Principles of personality assessment.* Reading, MA: Addison-Wesley.

Wiggins, J. S. (1979). A psychological taxonomy of trait-descriptive terms: The interpersonal domain. *Journal of Personality and Social Psychology, 37,* 395–412.

Wilden, I., Herzel, H., Peters, G., & Tembrock, G. (1998). Subharmonics, biphonation, and deterministic chaos in mammal vocalizations. *Bioacoustics, 9,* 171–196.

Wilke, H., & Lanzetta, J. T. (1970). The obligation to help: The effects of amount of prior help on subsequent helping behavior. *Journal of Experimental Social Psychology, 6,* 483–493.

Wilke, H., & Lanzetta, J. T. (1982). The obligation to help: Factors affecting response to help received. *European Journal of Social Psychology, 12,* 315–319.

Williams, G. C. (1966). *Adaptation and natural selection.* Princeton: Princeton University Press.

Wilson, D. S., & Sober, E. (1994). Reintroducing group selection to the human behavioral sciences. *Behavioral and Brain Sciences, 17,* 585–654.

Wilson, E. O. (1978). *On human nature.* Cambridge, MA: Harvard University Press.

Wilson, M. E. (1992). Factors determining the onset of puberty. In A. A. Gerall, H. Moltz, & I. L. Ward (Eds.), *Sexual Differentiation,* Vol. 11 (pp. 275–312). New York: Plenum Press.

Wilson, M. E., & Gordon, T. P. (1980). Age differences in the duration of mating period of female rhesus monkeys. *Developmental Psychobiology, 13,* 637–642.

Wilson, M. E., Gordon, T. P., Blank, M. S., & Collins, D. C. (1984). Timing of sexual maturity in female rhesus monkeys *(Macaca mulatta)* housed outdoors. *Journal of Reproduction and Fertility, 70,* 625–633.

Wilson, M. E., Gordon, T. P., & Collins, D. C. (1982a). Age differences in copulatory behavior and serum 17 beta-estradiol in female rhesus monkeys. *Physiology and Behavior, 28,* 733–737.

Wilson, M. E., Gordon, T. P., & Collins, D. C. (1982b). Variation in ovarian steroids associated with the annual mating period in female rhesus monkeys *(Macaca mulatta). Biology of Reproduction, 27,* 530–539.

Wilson, M. E., Gordon, T. P., & Collins, D. C. (1986). Ontogeny of luteinizing-hormone secretion and first ovulation in seasonal breeding rhesus monkeys. *Endocrinology, 118,* 293–301.

Wilson, M. E., Walker, M. L., & Gordon, T. P. (1984). Effects of age, lactation, and repeated cycles on rhesus monkey copulatory intervals. *American Journal of Primatology, 7,* 21–26.

Wimmer, H., & Perner, J. (1983). Beliefs about beliefs: Representation and constraining function of wrong beliefs in young children's understanding of deception. *Cognition, 13,* 103–128.

Witelson, S. F. (1985). The brain connection: The corpus callosum is larger in left-handers. *Science, 299,* 665–668.

Witelson, S. F. (1989). Hand and sex differences in the isthmus and genu of the human corpus callosum: A postmortem morphological study. *Brain, 112,* 799–835.

Wolfe, L. (1978). Age and sexual behavior of Japanese macaques *(Macaca fuscata). Archives of Sexual Behavior, 7,* 55–68.

Woodward, A. L. (1998). Infants selectively encode the goal of an actor's reach. *Cognition 69,* 1–34.

Woodward, A. L., Sommerville, J. A., & Guajardo, J. J. (2001). How infants make sense of intentional action. In B. Malle, D. Baldwin, & L. Moses (Eds.), *Intentionality: A key to human understanding* (pp. 149–170). Cambridge, MA: MIT Press.

Worlein, J. M., Eaton, G. G., Johnson, D. F., & Glick, B. B. (1988). Mating season effects on mother-infant conflict in Japanese macaques, *Macaca fuscata. Animal Behaviour, 36,* 1472–81.

Wrangham, R. W. (1980). An ecological model of female-bonded primate groups. *Behaviour, 75,* 262–300.

Wrangham, R. W. (1986). Ecology and social relationships in two species of chimpanzee. In D. I. Rubenstein & R. W. Wrangham (Eds.), *Ecological aspects of social evolution: Birds and mammals* (pp. 352–378). Princeton: Princeton University Press.

Wrangham, R. W., Chapman, C. A., Clark-Arcadi, A. P., & Isabirye-Basuta, G. (1996). Social ecology of Kanyawara chimpanzees: Implications for understanding the costs of great ape groups. In W. C. McGrew, L. F. Marchant, & T. Nishida (Eds.)., *Great ape societies* (pp. 45–57). Cambridge: Cambridge University Press.

Wrangham, R. W., McGrew, W. C., de Waal, F. B. M., & Heltne, P. G. (Eds.). (1994). *Chimpanzee cultures*. Cambridge, MA: Harvard University Press.

Wright, A. A. & Roberts, W. A. (1996). Monkey and human face perception: Inversion effects for human faces but not for monkey faces or scenes. *Journal of Cognitive Neuroscience, 8,* 278–290.

Wright, P. H. (1982). Men's friendships, women's friendships, and the alleged inferiority of the latter. *Sex Roles, 8,* 1–20.

Wu, F. C. W., Brown, D. C., Butler, G. E., Stirling, H. F., & Kelnar, C. J. H. (1993). Early morning plasma testosterone is an accurate predictor of imminent pubertal development in prepubertal boys. *Journal of Clinical Endocrinology and Metabolism, 76,* 26–31.

Yarn, D. H. (2000). Law, love, and reconciliation: Searching for natural conflict resolution in *Homo sapiens*. In F. Aureli & F. B. M. de Waal (Eds.), *Natural conflict resolution* (pp. 54–70). Berkeley: University of California Press.

Yeni-Komshian, G. H., & Benson, D. A. (1976). Anatomical study of cerebral asymmetry in the temporal lobe of humans, chimpanzees, and rhesus monkeys. *Science, 192,* 387–389.

Yerkes, R. M. (1925). *Almost human*. New York: Century Company.

Yerkes, R. M. (1933). Genetic aspects of grooming, a socially important primate behavior pattern. *Journal of Social Psychology, 4,* 3–25.

Yerkes, R. M. (1934). Modes of behavioral adaptation in chimpanzee to multiple-choice problems. *Comparative Psychology Monographs, 10,* 1–108.

Yerkes, R. M. (1939). The life history and personality of the chimpanzee. *American Naturalist, 73,* 97–112.

Yerkes, R. M., & Tomilin, M. I. (1935). Mother-infant relations in chimpanzees. *Journal of Comparative Psychology, 20,* 321–348.

Yerkes, R. M., & Yerkes A. W. (1936). Nature and conditions of avoidance (fear) responses in chimpanzees. *Journal of Comparative Psychology, 21,* 53–66.

Yin, R. K. (1969). Looking at upside-down faces. *Journal of Experimental Psychology, 81,* 141–145.

Yodyingyuad, U., Eberhart, J. A., & Keverne, E. B. (1982). Effects of rank and novel females on behaviour and hormones in male talapoin monkeys. *Physiology and Behavior, 28,* 995–1005.

Young, W. C. (1937). The vaginal smear picture, sexual receptivity, and the time of ovulation in the guinea pig. *The Anatomical Record, 67,* 305–325.

Yousry, T. A., Schmid, U. D., Alkadhi, H., Schmidt, D., Peraud, A., Buettner, A., &

Winkler, P. (1997). Localization of the motor hand area to a knob on the precentral gyrus: A new landmark. *Brain, 120,* 141–157

Zahavi, A. (2000). Altruism: The unrecognized selfish traits. *Journal of Consciousness Studies, 7,* 253–256.

Zahavi, A. & Zahavi, A. (1997). *The handicap principle.* Oxford: Oxford University Press.

Zahn-Waxler, C., Radke-Yarrow, M., Wagner, E., & Chapman, M. (1992). Development of concern for others. *Developmental Psychology, 28,* 126–136.

Zajicek, K. B., Higley, J. D., Suomi, S. J., & Linnoila, M. (1997). Rhesus macaques with high CSF 5-HIAA concentrations exhibit early sleep onset. *Psychiatry Research, 73,* 15–25.

Zajonc, R. B. (1980). Feeling and thinking: Preferences need no inferences. *American Psychologist, 35,* 151–175.

Zehr, J. L. (1998). Sex differences in play and affiliation during social development in pigtail macaques (Macaca nemestrina). Master's thesis, Emory University.

Zehr, J. L., Maestripieri, D., & Wallen, K. (1998). Estradiol increases female sexual initiation independent of male responsiveness in rhesus monkeys. *Hormones and Behavior, 33,* 95–103.

Zehr, J. L., Tannenbaum, P. L., Jones, B., & Wallen, K. (2000). Peak occurrence of female sexual initiation predicts day of conception in rhesus monkeys *(Macaca mulatta). Reproduction Fertility and Development, 12,* 397–404.

Zeller, A. (1996). The inter-play of kinship organisation and facial communication in the macaques. In J. E. Fa & D. G. Lindburg (Eds.), *Evolution and ecology of macaque societies* (pp. 527–550). Cambridge: Cambridge University Press.

Zeller, A. C. (1980). Primate facial gestures: A study of communication. *International Journal of Human Communication, 13,* 565–606.

Zentall, T. R. (1996). An analysis of imitative learning in animals. In B. G. Galef Jr. & C. M. Heyes (Eds.), *Social learning in animals: The roots of culture* (pp. 221–243). New York: Academic Press.

Zeskind, P. S., & Lester, B. (1978). Acoustic features and auditory perception of the cries of newborns with prenatal and perinatal complications. *Child Development, 49,* 580–589.

Ziegler, T. E., Snowdon, C. T., & Uno, H. (1990). Social interactions and determinants of ovulation in tamarins *(Saguinus).* In T. E. Ziegler & F. B. Bercovitch (Eds.), *Socioendocrinology of primate reproduction: Monographs in primatology,* Vol. 13. (pp. 113–133). New York: Wiley-Liss.

Zilles, K., Armstrong, E., Moser, K. H., Schleicer, A., & Stephan, H. (1989). Gyrification in the cerebral cortex of primates. *Brain, Behavior, and Evolution, 34,* 143–150.

Zilles, K., Dabringhaus, A., Geyer, S., Amunts, K., Qu, M., Schleicher, A., Gilissen, E.,

Schlaug, G., & Steinmetz, H. (1996). Structural asymmetries in the human fore-brain and the forebrain of non-human primates and rats. *Neuroscience and Biobehavioral Reviews, 20,* 593–605.

Zuberbühler, K., Cheney, D. L., & Seyfarth, R. M. (1999). Conceptual semantics in a nonhuman primate. *Journal of Comparative Psychology, 113,* 33–42.

Zuckerman, S. (1932). *The social life of monkeys and apes.* London: Kegan Paul Trench Trubner.

Zumpe, D., & Michael, R. P. (1990). Effects of the presence of a second male on pair tests of captive cynomolgus monkeys *(Macaca fascicularis):* Role of dominance. *American Journal of Primatology, 22,* 145–158.

Zupanc, G. K. H., & Lamprecht, J. (2000). Towards a cellular understanding of motivation: Structural reorganization and biochemical switching as key mechanisms of behavioral plasticity. *Ethology, 106,* 467–477.

Zvoch, K. (1999). Family type and investment in education: A comparison of genetic and stepparent families. *Evolution and Human Behavior, 20,* 453–464.

Contributors

Filippo Aureli
School of Biological and Earth Sciences,
Liverpool John Moores University, Liverpool, U.K.

Jo-Anne Bachorowski
Department of Psychology, Vanderbilt University, Nashville, Tenn.

Michael J. Beran
Language Research Center, Georgia State University, Atlanta, Ga.

Jesse M. Bering
Department of Psychology, Florida Atlantic University, Boca Raton, Fla.

Josep Call
Department of Comparative and Developmental Psychology,
Max Planck Institute for Evolutionary Anthropology, Leipzig, Germany

Claudio Cantalupo

Division of Psychobiology, Yerkes Regional Primate Research Center,
Emory University, Atlanta, Ga.

Lynn A. Fairbanks

Neuropsychiatric Institute, University of California Los Angeles,
Los Angeles, Calif.

Samuel D. Gosling

Department of Psychology, University of Texas, Austin, Tex.

Franklynn C. Graves

Department of Psychology, Emory University, Atlanta, Ga.

Rebecca A. Herman

Department of Psychology, Emory University, Atlanta, Ga.

J. Dee Higley

Laboratory of Clinical Studies, DICBR, National Institute on Alcohol Abuse
and Alcoholism, NIH Animal Center, Poolesville, Md.

William D. Hopkins

Department of Psychology, Berry College, and Division
of Psychobiology, Yerkes Regional Primate Research Center,
Emory University, Atlanta, Ga.

Peter G. Judge

Department of Psychology, Bucknell University, Lewisburg, Pa.

Scott O. Lilienfeld

Department of Psychology, Emory University, Atlanta, Ga.

Dario Maestripieri

Institute for Mind and Biology, University of Chicago, Chicago, Ill.

Lori Marino

Center for Behavioral Neuroscience, Emory University, Atlanta, Ga.

Michael J. Owren

Department of Psychology, Cornell University, Ithaca, N.Y.

Lisa A. Parr

Center for Behavioral Neuroscience, Emory University, Atlanta, Ga.

Dawn L. Pilcher

Division of Psychobiology, Yerkes Regional Primate, Research Center, Emory University, Atlanta, Ga.

Daniel J. Povinelli

Cognitive Evolution Group, University of Southwestern Louisiana, Lafayette, La.

Drew Rendall

Department of Psychology, University of Lethbridge, Lethbridge, Canada

James R. Roney

Committee on Human Development, University of Chicago, Chicago, Ill.

Duane M. Rumbaugh

Language Research Center and Departments of Psychology and Biology, Georgia State University, Atlanta, Ga.

E. Sue Savage-Rumbaugh

Language Research Center and Department of Biology, Georgia State University, Atlanta, Ga.

Michael Tomasello

Department of Comparative and Developmental Psychology, Max Planck Institute for Evolutionary Anthropology, Leipzig, Germany

Alfonso Troisi

Cattedra di Psichiatria, Università di Roma Tor Vergata, Rome, Italy

Kim Wallen

Department of Psychology, Emory University, Atlanta, Ga.

Andrew Whiten

Scottish Primate Research Group, School of Psychology,
University of St. Andrews, St. Andrews, U.K.

Julia L. Zehr

Department of Psychology, Emory University, Atlanta, Ga.

Index